THEORETICAL FOUNDATIONS
of
DIGITAL IMAGING
Using MATLAB®

CHAPMAN & HALL/CRC MATHEMATICAL AND COMPUTATIONAL IMAGING SCIENCES

Series Editors

Chandrajit Bajaj
Center for Computational Visualization
The University of Texas at Austin

Guillermo Sapiro
Department of Electrical
and Computer Engineering
University of Minnesota

Aims and Scope

This series aims to capture new developments and summarize what is known over the whole spectrum of mathematical and computational imaging sciences. It seeks to encourage the integration of mathematical, statistical and computational methods in image acquisition and processing by publishing a broad range of textbooks, reference works and handbooks. The titles included in the series are meant to appeal to students, researchers and professionals in the mathematical, statistical and computational sciences, application areas, as well as interdisciplinary researchers involved in the field. The inclusion of concrete examples and applications, and programming code and examples, is highly encouraged.

Published Titles

Theoretical Foundations of Digital Imaging Using MATLAB®
by Leonid P. Yaroslavsky

Rough Fuzzy Image Analysis: Foundations and Methodologies
by Sankar K. Pal and James F. Peters

Proposals for the series should be submitted to the series editors above or directly to:
CRC Press, Taylor & Francis Group
3 Park Square, Milton Park, Abingdon, OX14 4RN, UK

THEORETICAL FOUNDATIONS of DIGITAL IMAGING Using MATLAB®

Leonid P. Yaroslavsky

CRC Press
Taylor & Francis Group
Boca Raton London New York

CRC Press is an imprint of the
Taylor & Francis Group, an **informa** business

A CHAPMAN & HALL BOOK

CRC Press
Taylor & Francis Group
6000 Broken Sound Parkway NW, Suite 300
Boca Raton, FL 33487-2742

© 2013 by Taylor & Francis Group, LLC
CRC Press is an imprint of Taylor & Francis Group, an Informa business

Library of Congress Cataloging-in-Publication Data

IAroslavskii, L. P. (Leonid Pinkhusovich)
      Theoretical foundations of digital imaging using MATLAB / Leonid P.
      Yaroslavsky.
      pages cm. -- (Chapman & Hall/CRC mathematical and computational imaging
sciences series)
      Includes bibliographical references and index.
      ISBN 978-1-4398-6140-0 (hardback)
      1. Image processing--Digital techniques. 2. Three-dimensional imaging. 3.
MATLAB. I. Title.

TA1637.I224 2013
006.6--dc23   2012028835

Visit the Taylor & Francis Web site at
http://www.taylorandfrancis.com

and the CRC Press Web site at
http://www.crcpress.com

To Dima and Yaro on their 25th and 5th birthdays

Contents

Preface

With the advent and ubiquitous spreading of digital imaging, a new profession has emerged: imaging engineering. This book is intended as a textbook for studying the theoretical foundations of digital imaging to master this profession. It is based on the experience accumulated by the present author for 50 years of working in the field and teaching various courses in digital image processing and digital holography in the Russian Academy of Sciences, the National Institutes of Health (Bethesda, Maryland, USA), the University of Erlangen-Nuremberg (Germany), the Tampere University of Technology (Tampere, Finland), Agilent Labs (Palo Alto, California, USA), the Autonomous University of Barcelona (Spain), the Institute of Henri Poincare (Paris, France), the Kiryu University (Kiryu, Japan), and the University of Tel Aviv (Israel).

The book is addressed to young students who opt to pursue a scientific and research career in imaging science and engineering. The most outstanding minds of mankind, such as Galileo Galilei, René Descartes, Isaac Newton, James Clerk Maxwell, and many other scientists and engineers contributed to this branch of modern science and technology. At least 12 Nobel Prizes have been awarded for contributions directly associated with image science and imaging devices, and a majority of others would not be possible without one or the other imaging methods. You will be in good company, dear reader. Let this book help you to become a master of digital imaging.

A number of MATLAB programs are available at the Download section of this book's web page on the CRC Press website (http://www.crcpress.com/product/isbn/9781439861400).

MATLAB® and Simulink® are registered trademarks of The MathWorks, Inc. For product information, please contact:

The MathWorks, Inc.
3 Apple Hill Drive
Natick, MA 01760-2098 USA
Tel: 508 647 7000
Fax: 508-647-7001
E-mail: info@mathworks.com
Web: www.mathworks.com

Author

Leonid P. Yaroslavsky is a fellow of the Optical Society of America, MS (1961, Faculty of Radio Engineering, Kharkov Polytechnic Institute, Kharkov, Ukraine), PhD (1968, Institute for Information Transmission Problems, Moscow, Russia), and Dr. Sc. Habilitatus in Physics Mathematics (1982, State Optical Institute, Saint Petersburg, Russia). From 1965 to 1983, he headed a group for digital image processing and digital holography at the Institute for Information Transmission Problems, Russian Academy of Sciences, which in particular carried out digital processing of images transmitted from the spacecraft Mars-4, Mars-5, Venera-9, and Venera-10 and obtained the first color images of the surface of Mars and the first panoramas of the surface of Venus. From 1983 to 1995, he headed the Laboratory of Digital Optics at the institute. From 1995 to 2008, he was a professor at the Faculty of Engineering, Tel Aviv University, where, at present, he is a professor emeritus. He was also a visiting professor at the University of Erlangen, Germany; National Institutes of Health, Bethesda, Maryland, USA; Institute of Optics, Orsay, France; Institute Henri Poincare, Paris, France; International Center for Signal Processing, Tampere University of Technology, Tampere, Finland; Agilent Laboratories, Palo Alto, California, USA; Gunma University, Kiryu, Japan; and the Autonomous University of Barcelona, Spain. He has supervised 20 PhD candidates and is an author and editor of several books and more than 100 papers on digital image processing and digital holography.

1

Introduction

Imaging Goes Digital

The history of science is, to a considerable extent, the history of invention, development, and perfecting imaging methods and devices. Evolution of imaging systems can be traced back to rock engravings and to ancient mosaics thousands of years ago (Figure 1.1).

Apparently, the very first "imaging devices" were humans, painters. Great artists, such as Leonardo da Vinci, Michelangelo, Albrecht Dürer, and many others, not only created outstanding masterpieces of art, but also actually pioneered imaging science and engineering (Figure 1.2).

The first artificial imaging device was, seemingly, camera-obscura ("pinhole camera," Latin for "dark room") that dates back to Aristotle (384–322 BCE) and Euclid (365–265 BCE). Then, in the thirteenth century, methods for polishing lenses were invented and eye glasses were widely used in Europe by the mid-fifteenth century. In the first few years of the seventeenth century, a decisive breakthrough took place, when Galileo Galilei (1564–1642) in October 1609, apparently using his knowledge of laws of light refraction, greatly improved the magnification of three-powered "spyglasses" unveiled not long before in the Netherlands and built a 20-powered instrument. He directed it to the sky and immediately discovered mountains on the Moon, four satellites of Jupiter, rings of Saturn, phases of Venus, and nebular patches in stars. It was the beginning of the scientific revolution of the seventeenth century. Since then, the pace of evolution of imaging science and imaging devices has become numbered in decades rather than in centuries.

In the 1630s, René Descartes published the Dioptrique (the *Optics*) that summarized contemporary knowledge on topics such as the law of refraction, vision, and the rainbow.

In the late 1660s, Isaac Newton discovered that white light is composed of a spectrum of colors and built his first reflecting telescope that allowed avoiding color aberrations and building telescopes with much greater magnification than was possible with refractive Galilean telescopes.

In the 1670s, Robert Hook and Anthony Leeuwenhoek introduced a microscope, invented new methods for grinding and polishing microscope lenses, and discovered cells and bacteria.

FIGURE 1.1
Details from the mosaic floor of the Petra Church.

(a)   (b)

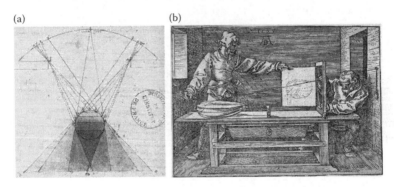

FIGURE 1.2
Drawing By Leonardo Da Vinci illustrating treatment of light and shade (a) and a woodcut by Albrecht Dürer showing an artist using Dürer's drawing machine to paint a lute (b).

The next decisive breakthrough happened in the 1830s through the 1840s, when photography was invented, which for the first time solved the problem of converting images into pictures that can be stored, copied, and sent by mail. This invention had a very profound influence on the development of our civilization, from people's everyday life to science and to art. Photography has become a branch of industry and a profession that served people's need to memorize the images. After the invention of photography, the art of painting became a pure art, which stimulated the birth of new art trends such as impressionism. Photographic plates became a major means in experimental science, which eventually led to almost all outstanding discoveries made in the nineteenth and the twentieth centuries.

In the 1890s, photographic plates enabled discoveries of x-rays by Wilhelm Conrad Roentgen (the Nobel Prize laureate, 1901) and radioactivity by Antoine Henri Becquerel (the Nobel Prize laureate, 1903) at the end of the nineteenth century. These discoveries, in their turn, almost immediately gave birth to new imaging techniques, x-ray imaging and radiography.

The 1890s were remarkable years in the history of science. Apart from discoveries of x-rays and radioactivity, these were also the years of the discovery of radio, the major breakthrough in communication and information transmission, and of the invention of motion pictures, which solved the problem of imaging moving objects. With the invention of motion pictures, the new, for imaging, principal concept of time-sampled images was introduced and realized.

X-rays were discovered by W.C. Roentgen in experiments with cathode rays. Cathode rays were discovered in 1859 by the German mathematician and physicist Julius Plücker, who used vacuum tubes invented in 1855 by the German inventor, Heinrich Geissler. These tubes eventually led to the invention of the cathode ray tube (CRT) by Karl Braun in 1897 and finally brought about the development of electronic television and electron microscopy.

In 1907, the Russian professor Boris Rosing used a CRT in the receiver of a television system that, at the camera end, made use of mirror-drum scanning. Rosing transmitted crude geometrical patterns onto the television screen and was the first inventor to do so using a CRT. Then, in the 1920s through the 1940s, Vladimir Zworykin, a former student of Rosing, working first for Westinghouse and later for RCA in the United States, and Philo Farnsworth, working independently in San Francisco, brought about the birth of purely electronic television. Electronic television has finally won in the 30 years of competition with electromechanical television based on using, for image transmission, the image scanning disk, invented in 1883 by a German student, Paul Nipkow. This engineering solution turned to be a dead end in the evolution of television, although the idea itself of converting images to time signals by means of image row-wise scanning had a principal value and was eventually implemented in a much more efficient way as image scanning by easily controlled electron beam. There is no need to tell the present generation the role that television plays in our civilization.

The victory of electronic television heralded the birth of electronic imaging. One more outstanding representative of electronic imaging devices was electronic microscope. The first electron microscope was designed in 1931 by by the German engineer E. Ruska. In 1986, E. Ruska was awarded the Nobel Prize in physics "for his fundamental work in electron optics, and for the design of the first electron microscope." Electron microscopes depend on electrons rather than light to view objects and because of this electron microscopes make it possible to view objects with a resolution far beyond the resolution of optical microscopes.

Since the 1940s, the pace of evolution of imaging devices has become numbered in years. In 1947, the British (native of Hungary) scientist Dennis

Gabor, while working to improve the resolution of electron microscopes, invented an optical method for recording and reconstructing amplitudes and phases of coherent light radiation. He coined the term holography for this method, meaning that it is the method for recording and reproducing whole information carried by optical signals. For this invention, he was awarded the Nobel Prize in physics in 1971.

In the 1950s, synthetic aperture and side-looking radars were invented, which opened ways for creating images of objects in radio frequency band of electromagnetic radiation. The principle of synthetic aperture radars is actually the same principle of recording of amplitude and phase of radiation wavefronts as that of holography, and in 1962, Emmett Leith and Juris Upatnieks of the University of Michigan, USA, recognized from their work in side-looking radar that holography could be used as a three-dimensional (3D) visual medium. They read Gabor's paper and "simply out of curiosity" decided to duplicate Gabor's technique using the recently invented laser and the "off-axis" technique borrowed from their work in the development of side-looking radar. The result was the first laser transmission hologram of 3D objects. These transmission holograms produced images with clarity and realistic depth but required laser light to view the holographic image. Also in 1962, on another side of the globe, Yuri N. Denisyuk from the Russian Academy of Sciences combined holography with the 1908 Nobel laureate Gabriel Lippmann's work in natural color photography. Denisyuk's method produced reflection holograms which, for the first time, did not need the use of coherent laser light for image reconstruction and could be viewed in light from an ordinary incandescent light bulb. Holography was the first example of what we can call transform domain imaging.

In 1969, at Bell Labs, USA, George Smith and Willard Boyle invented the first CCDs or charge-coupled devices. A CCD is a semiconductor electronic memory that can be charged by light. CCDs can hold a charge corresponding to variable shades of light, which makes them useful as imaging devices for cameras, scanners, and fax machines. Because of its superior sensitivity, the CCD has revolutionized the field of electronic imaging. In 2009, George Smith and Willard Boyle were awarded the Nobel Prize for physics for their work on the CCD.

In 1971–1972, the British engineer Godfrey Hounsfield of EMI Laboratories, England, and the South Africa-born physicist Allan Cormack of Tufts University, Massachusetts, USA, working independently, invented a new imaging method that allowed building images of slices of bodies from sets of their x-ray projections taken from different angles. Reconstruction of images from their projections required special computer processing of projections, and the method obtained the name "computer-assisted tomography." Computer tomography (CT) revolutionized medical imaging and in few years tens and later hundreds and thousands of CT scanner were installed in medical hospitals all over the world. In 1979, Hounsfield and Cormack were awarded the Nobel Prize in medicine.

In the 1979s, Gerd Binnig and Heinrich Rohrer, IBM Research Division, Zurich Research Laboratory, Switzerland, invented the scanning tunneling microscope that gives 3D images of objects down to the atomic level. Binnig and Rohrer were awarded the Nobel Prize in physics in 1986 for this invention. The powerful scanning tunneling microscope is the strongest microscope to date. With this invention, imaging techniques, which before this were based on electromagnetic radiation, conquered the quantum world of atomic forces.

But this is not the end of the "evolution of imaging" story. The years after the 1970s and the 1980s were the years, when imaging began changing rapidly from completely analog to digital. The first swallow in this process was CT, in which images of body slices are computed from projection data, though its germs one can notice (or find, or trace) in methods of crystallography emerged from the discovery of diffraction of x-rays by Max von Laue (the Nobel Prize laureate, 1914) in the beginning of the twentieth century. Although Von Laue's motivation was not creating a new imaging technique, but rather proving the wave nature of x-rays, photographic records of x-ray diffraction on crystals, or lauegrams, had very soon become the main imaging tool in crystallography because using lauegrams, one can numerically reconstruct the spatial structure of atoms in crystals. This eventually led to one of the greatest discoveries of twentieth century, the discovery, by J. Watson and Fr. Crick, of the Molecular Structure of DNA—the Double Helix (the Nobel Prize in Physiology and Medicine, 1962).

Digital imaging was born at the junction of imaging and communication. The first reports on digital images date back to the 1920s, when images were sent for newspapers by telegraph over submarine cable between London and New York. Image transmission was always a big challenge to communication because it required communication channels of very large capacity. In the 1950s, the needs to transmit television over long distances demanded to compress TV signals as much as possible. By that time, the first digital computers were available, at least for large companies and research institutions, and researchers started using them for investigations in image compression. For these purposes, the first image input–output devices for computers were developed in the late 1950s through the early 1960s. Satellite communication and space research that have been sky rocketing since the first "Sputnik" in 1957 greatly stimulated works in digital image processing.

In 1964–1971, computers were used at the Jet Propulsion Laboratory (Pasadena, California, USA) for improving the quality of the first images of the Moon surface transmitted, in digital form, from the US spacecraft Ranger-7 and images of Mars transmitted from the US spacecrafts Mariner-4 (1964), Mariner 7 (1969), and Mariner 9 (1971). In 1973–1976, the first color images of the surface of Mars and the first panoramas from the surface of Venus were published in USSR Academy of Sciences, which were obtained using digital image processing of data transmitted from spacecrafts Mars-4, Mars-5, Venus 9, and Venus 10 launched in the former USSR. By the late 1970s, digital image processing had become the main tool in the processing of satellite images for space exploration and remote sensing.

The availability of digital computers by the 1960s and the new opportunities this offered for information processing could not pass yet another fast-growing child of the 1960s, holography. In 1966–1968, a German professor Adolf Lohmann, while working in San Diego University, California, USA, invented computer-generated holograms and in the late 1960s through the early 1970s, the first experiments with numerical reconstruction of optical holograms were reported. This was the birth of digital holography.

By the mid-1970s, the first image processing "mini-computer"-based workstations and high-quality gray-scale and color computer displays were created. It would be no exaggeration to say that needs for processing, storage, and displaying images were one of the main, if not the major, impetuses in the development of personal computers (PC), which emerged in the early 1980s and became the main stream in the computer industry. In the late 1980s through the early 1990s, the industry of dedicated image processing boards for minicomputers and PC emerged, and no PC has been sold since that time without a "video board."

The digital revolution affected the industry of image-capturing devices as well. In 1972, Texas Instruments patented a filmless electronic camera, the first to do so. In August 1981, Sony released the first commercial electronic camera. Since the mid-1970s, Kodak has invented several solid-state image sensors that converted light to digital pictures for professional and home consumer use. In 1991, Kodak released the first professional digital camera system (DCS) with a 1.3 megapixel sensor. This has become possible not only because megapixel CCDs were developed by this time, but also because image compression methods, foundations of which were laid in the 1950s through the 1960s, have reached a stage, when their implementation has become possible, thanks to the emergence of the appropriate computational hardware. In 1992, the international Joint Photographic Experts Group issued the first JPEG standard for compression of still images, and in 1993 the Motion Picture Expert Group issued the first MPEG standard for compression of video. Nowadays, digital photographic and video cameras and JPEG and MPEG standards are overwhelmingly used in all ranges of image-related activities from space telescopes to mobile phones.

At the beginning of the twenty-first century, the era of photography and analog electronic imaging gave way to the era of digital imaging. Digital imaging has overpowered the evolution of imaging devices because

- It is much more cheap, productive, and versatile than the analog imaging.
- Acquiring and processing of image data is most natural when the data are represented, handled, and stored in a digital form. In the same way as money is the general equivalent in economy, digital signals are the general equivalent in information handling. Thanks to its universal nature, digital signals are the ideal means for integrating various informational systems.

- It enables using digital computers for image processing and analysis. No hardware modifications are necessary anymore for solving different tasks. With the same hardware, one can build an arbitrary problem solver by simply selecting or designing an appropriate code for the computer.

Images are now overwhelmingly created, stored, transmitted, and processed in a digital form. The history has completed its spiral cycle from the art of making ancient mosaics, as shown in Figure 1.1, which were in fact the first digital pictures, to modern digital imaging technology.

Briefly about the Book Structure

The father of information theory and, generally, of modern communication theory Claude Shannon and his coauthor H.W. Bode wrote in their paper "A simplified derivation of linear least square smoothing and prediction theory" [1]:

> In a classical report written for the National Defence Research Council [2], Wiener has developed a mathematical theory of smoothing and prediction of considerable importance in communication theory. A similar theory was independently developed by Kolmogoroff [3] at about the same time. Unfortunately the work of Kolmogoroff and Wiener involves some rather formidable mathematics—Wiener's yellow-bound report soon came to be known among bewildered engineers as "The Yellow Peril"—and this has prevented the wide circulation and use that theory deserves. In this paper the chief results of the smoothing theory will be developed by a new method which, while not as rigorous or general as the methods of Wiener and Kolmogoroff, has the advantage of greater simplicity, particularly for readers with background of electrical circuit theory. The mathematical steps in the present derivation have, for the most part, a direct physical interpretation, which enables one to see intuitively what mathematics is doing.

This approach was this author's guideline in writing this book. The book concept is to expose the theory of digital imaging in its integrity, from image formation to image perfecting, as a specific branch of engineering sciences and to present theoretical foundations of digital imaging as comprehensively as possible avoiding unnecessarily formidable mathematics as much as possible and providing motivations and physical interpretation to all mathematical entities and derivations. The book is intended to be, for imaging engineers, similar to texts on theoretical mechanics are for mechanical engineers or texts on communication theory are for communication engineers.

Including this introductory chapter, the book consists of eight chapters. The book assumes some familiarity with relevant mathematics. The necessary mathematical preliminaries are provided in the second chapter. Chapter 3 presents the basic contents of the book. It addresses the very first problem of digital imaging, the problem of converting images into digital signals that can be stored, transmitted, and processed in digital computers. Chapter 4 is devoted to the problem of adequate representation of image transformations in computers. The main emphasis is made on keeping a correspondence between the original analog nature of image transformations and their computer implementations. In Chapter 5, the concept of computational imaging is illustrated by several specific and instructive examples: image reconstruction from sparse or nonuniform samples data, digital image formation by means of numerical reconstruction of holograms, virtual image formation by means of computer-generated display holograms, and computational image formation from sensor data obtained without imaging optics. Chapter 6 introduces methods for image perfect resampling and building, in computers, continuous image models. Chapter 7 treats the problem of statistically optimal estimation of image numerical parameters. As a typical and representative example, the task of optimal localization of targets in images is discussed. In Chapter 8, methods of optimal linear and nonlinear filtering for image perfecting and enhancement are discussed. All chapters are supplied with short introductory synopses.

In order to illustrate the major results and facilitate their deeper understanding, the book offers a number of exercises supported by demo programs in MATLAB®. Those results and algorithms that are not supported by a dedicated demo program can be easily implemented in MATLAB as well.

We tried to keep the book self-containing and to not overload readers by the needs to refer to other sources. As a rule, only references to classical milestone publications are provided to give credit to their authors. All other references, when necessary, can be easily found by Google search using keywords provided in the text. In order to make reading and understanding of the book easier, all formula derivations that require more than two to three lines of equations are moved to appendixes and contain all needed details.

References

1. H.W. Bode, C.E. Shannon, A simplified derivation of linear least square smoothing and prediction theory, *Proceedings of I.R.E.*, 38, 4, 417–425, 1950.
2. N. Wiener, *The Interpolation, Extrapolation, and Smoothing of Stationary Time Series*, National Defence Research Committee; reprinted as a book, J. Wiley and Sons, Inc., New York, NY, 1949.
3. A Kolmogoroff, Interpolation und Extrapolation von Stationären Zufällige Folgen, *Bull. Acad. Sci. (USSR) Sér. Math.* 5, 3–14, 1941.

2

Mathematical Preliminaries

In this chapter, we outline the basic notions of signal theory, which will be used later throughout the book. In the section "Mathematical Models in Imaging," we introduce notions of signals and signal space. In the section "Signal Transformations," two basic classes of signal transformations are introduced: linear and element-wise nonlinear ones. In the section "Imaging Systems and Integral Transforms," major imaging transforms are detailed and in the section "Statistical Models of Signals and Transformations," the basics of the statistical approach to treatment of signals and their transformations are reviewed.

Mathematical Models in Imaging

Primary Definitions

In the context of this book, images are signals, or information carriers. The main peculiarity of images as signals that distinguish them from other signals is that images are accessible for visual perceptions.

Images are generated by dedicated technical devices, such as microscopes, telescopes, photographic and video cameras, tomographic machines, and alike, which we call imaging devices. The process of generating images is called imaging.

Signals and signal-processing methods are treated through mathematical models of mathematical functions, which specify relationships between physical parameters of imaged objects, such as their capability to reflect or absorb radiation used to create images, and parameters of the object physical space, such as spatial coordinates and/or of time.

Figure 2.1 represents a classification diagram of signals as mathematical functions and associated terminology. In terms of the function values, functions can be scalar, that is characterized by scalar values, or vectorial, that is, characterized by multiple values treated as value vector components. Examples of scalar image signals are monochrome gray-scale images. Examples of vectorial image signals are two-component signals that represent orthogonal components of polarized electromagnetic wave fields, pairs of left/right stereoscopic images, color images represented by their red, green, and blue components, and multispectral images that contain more than three components.

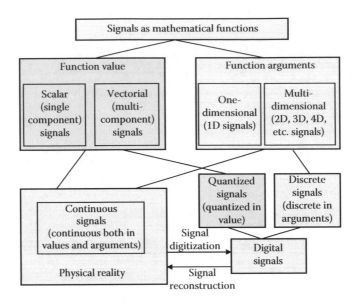

FIGURE 2.1
Classification of signals as mathematical functions.

In terms of function arguments, signals can be one-dimensional (1D) or multidimensional functions (2D, 3D, etc.). Examples of 1D signals are audio signals, physical measurements as functions of time, rows of image scans, and alike. Images are usually modeled as 2D functions of coordinates in image plane; video can be regarded as 3D functions of image plane coordinates and time. Volumetric data that represent 3D scenes are functions of three spatial coordinates.

In describing physical reality, we conventionally disregard quantum effects and use functions that are continuous in both values and arguments; hence, the notion of continuous signals, or analog signals, that can take arbitrary values and are defined in continuous coordinates. They are regarded as a complete analog of physical objects.

The numbers we store and operate within digital computers can be classified as digital signals. They are produced from continuous signals by a procedure, which we call digitization, implemented in special devices, analog-to-digital converters. Signal digitization always, explicitly or implicitly, assumes that inverse conversion of the digital signal into the corresponding analog signal is possible using special devices called digital-to-analog converters.

There also exist intermediate classes of signals: discrete signals, which are ordered sets of numbers that can take arbitrary values, and quantized signals, which are continuous functions of their arguments but can assume only specific fixed, or quantized, values.

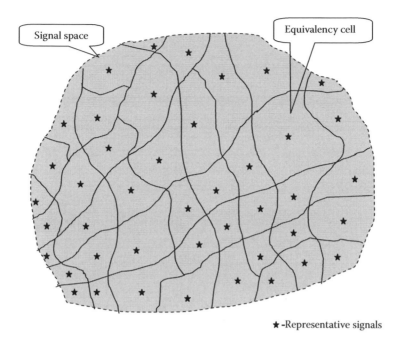

★ -Representative signals

FIGURE 2.2
Signal space, equivalency cells, and representative signals (stars).

Generally, each digital signal can be regarded as a single number. In practice, digital signals are represented as ordered sets of quantized numbers that may be considered as digits of this single number. It is in this sense that digital signals can be regarded as being both discrete and quantized.

In order to make the signal theory more intuitional, it is very convenient to give signals a geometric interpretation by regarding them as points in some space, called the signal space (Figure 2.2). In such treatment, each point of the signal space corresponds to a certain signal and vice versa. Signal space of continuous signals is a continuum, that is, its signals are not countable.

Using the notion of the signal space, one can easily portray the relationship between continuous and digital signals. Let us split signal space into a set of nonoverlapping subspaces, or cells, which together cover the entire space, and assign to each cell a numerical index. Each index will correspond to all signals within the corresponding cell and, for each signal, one can find an index of the cell to which the signal belongs.

Suppose now that all cells are formed in such a way that all signals within one cell can be regarded by the signal end user as identical, or indistinguishable, for a given specific application. We will call such cells equivalence cells. In each equivalence cell, one can select a signal that will serve, in a given particular application, as a representative signal that substitutes all signals within this cell. In such a treatment, conversion of a continuous signal into a corresponding digital signal (analog-to-digital conversion) is assigning to the continuous signal the index of the equivalence cell to which it belongs. We will call this process general digitization. The inverse process of converting digital signal to analog signal (signal reconstruction) is then generating, for each number (digital signal), a representative signal of the corresponding equivalence cell. In practice, partition of the signal space into a set of equivalence cells is "hard-wired" in analog-to-digital converters and assigning a representative signal to a number is "hard-wired" in digital-to-analog converters.

Linear Signal Space, Basis Functions, and Signal Representation as Expansion over a Set of Basis Functions

An important feature of physical signals such as image signals is that they can be added, subtracted, and amplified or attenuated. Signal space, in which operations of signal addition and multiplication by scalars are defined, is called linear signal space.

In linear signal space, one can specify a certain number N of fixed signals and use them to generate an arbitrary large set of other signals as a weighted sum, with certain weight coefficients $\{\alpha_k\}$, of these signals called basis signals (basis functions):

$$a(x) = \sum_{k=0}^{N-1} \alpha_k \phi_k(x). \tag{2.1}$$

It is only natural to select the set of basis functions from linearly independent functions, that is, functions that cannot be represented as a linear combination of the rest of the functions from the set; otherwise, the set of basis functions will be redundant. The signal space formed of signals generated as a linear combination according to Equation 2.1 of N linearly independent basis functions is referred to as an N-dimensional linear space.

Each signal $a(x)$ in N-dimensional signal space corresponds to a unique linear combination of basis functions $\{\phi_k(x)\}$, that is, fully defined by a unique set of scalar coefficients $\{\alpha_k\}$. The ordered set of scalar coefficients of decomposition of a given signal over the given basis function is the signal discrete representation with respect to this basis function.

Signal representation as an expansion over a set of basis functions (Equation 2.1) is meaningful only as far as, for each signal $a(x)$, its representation coefficients $\{\alpha_k\}$ for the given set of basis function $\{\phi_k(x)\}$ can be found. To this goal, the concepts of scalar product and orthogonality of functions are introduced. The scalar product (inner product and dot product) $a_1(x) \circ a_2(x)$ of two functions $a_1(x)$ and $a_2(x)$ is a scalar calculated as an integral of their point-wise product with the second multiplicand taken complex conjugate when multiplicands are complex valued signals:

$$a_1(x) \circ a_2(x) = \int_X a_1(x) a_2^*(x) \, dx, \qquad (2.2)$$

where $*$ is the complex conjugacy symbol. This computation method is a mathematical model of natural signal transformations that take place in signal sensors. The scalar product of a signal with itself

$$a(x) \circ a(x) = \int_X a(x) a^*(x) dx = \int_X |a(x)|^2 \, dx \qquad (2.3)$$

is called signal energy, the term adopted from the definition of energy in mechanics and electrical engineering. In functional analysis, this integral is also called function norm. Functions whose scalar product is zero are called mutually orthogonal functions.

Having defined a scalar product, one can use it to establish a simple correspondence between signals and their representation coefficients for the given basis function. Let $\{\phi_k(x)\}$ be basis functions and $\{\varphi_l(x)\}$ be a set of reciprocal functions, which are mutually orthogonal to $\{\phi_k(x)\}$ and normalized such that

$$\phi_k(x) \circ \varphi_l(x) = \delta_{k,l} = \begin{cases} 0, & k \neq l \\ 1, & k = l, \end{cases} \qquad (2.4)$$

where symbol $\delta_{k,l}$ is called the Kronecker delta. Then the signal k-th representation coefficient α_k over basis function $\{\phi_k(x)\}$ can be found as a scalar product of the signal and the corresponding k-th reciprocal basis function $\{\varphi_k(x)\}$:

$$a(x) \circ \varphi_k(x) = \left(\sum_{l=0}^{N-1} \alpha_l \phi_l(x) \right) \circ (\varphi_k(x)) = \sum_{l=0}^{N-1} \alpha_l (\phi_l(x) \circ \varphi_k(x)) = \sum_{l=0}^{N-1} \alpha_l \delta_{k,l} = \alpha_k.$$

$$(2.5)$$

One can compose basis function $\{\phi_k(x)\}$ of mutually orthogonal functions of a unit norm:

$$\phi_k(x) \circ \phi_l(s) = \delta_{k,l}. \qquad (2.6)$$

Such basis reciprocal to itself is called orthonormal basis.

Given representations $\{\alpha_k\}$ and $\{\beta_k\}$ of signals $a(x)$ and $b(x)$ over an orthonormal basis function $\{\phi_k(x)\}$, one can calculate signal scalar product and signal energy (norm) through these coefficients as

$$a(x) \circ b(x) = \left(\sum_{k=0}^{N-1} \alpha_k \phi_k(x) \right) \circ \left(\sum_{l=0}^{N-1} \beta_l \phi_l(x) \right) = \sum_{k=0}^{N-1} \sum_{l=0}^{N-1} (\alpha_k \phi_k(x)) \circ (\beta_l \phi_l(x))$$

$$= \sum_{k=0}^{N-1} \alpha_k \beta_k^* (\phi_k(x) \circ \phi_l(x)) = \sum_{k=0}^{N-1} \sum_{l=0}^{N-1} \alpha_k \beta_k^* \delta(k,l) = \sum_{k=0}^{N-1} \alpha_k \beta_k^* \qquad (2.7)$$

$$\int_X |a(x)|^2 \, dx = \sum_{k=0}^{N-1} |\alpha_k|^2 ; \quad \int_X |b(x)|^2 \, dx = \sum_{k=0}^{N-1} |\beta_k|^2 . \qquad (2.8)$$

From Equations 2.7 and 2.8, it follows that representations $\left\{ \alpha_k^{(1)} \right\}, \left\{ \beta_k^{(1)} \right\}$, and $\left\{ \alpha_k^{(2)} \right\}, \left\{ \beta_k^{(2)} \right\}$ of signals $a(x)$ and $b(x)$ over two orthonormal bases $\left\{ \phi_k^{(1)}(x) \right\}$ and $\left\{ \phi_k^{(2)}(x) \right\}$ are linked by the relationship

$$\sum_{k=0}^{N-1} \alpha_k^{(1)} \left(\beta_k^{(1)} \right)^* = \sum_{k=0}^{N-1} \alpha_k^{(2)} \left(\beta_k^{(2)} \right)^* . \qquad (2.9)$$

Equations 2.7 through 2.9 are versions of the so-called *Parseval's relationship.* Up to now, it was assumed that signals are continuous. For discrete signals defined by their elements (samples), a convenient mathematical model is that of vectors. Let $\{\alpha_k\}$ be a set of samples of a discrete signal, $k = 0,1,\ldots,N-1$. They define a vector column $\mathbf{a} = \{a_k\}$. Also let $\left\{ \vec{\phi}_r = \{\phi_{r,k}\} \right\}$ be a set of linearly independent vectors such that

$$\sum_{r=0}^{N-1} \lambda_r \vec{\phi}_r \neq 0 \qquad (2.10)$$

for arbitrary nonzero coefficients $\{\lambda_r \neq 0\}$. One can select this set as a basis for representing any signal vector

$$\mathbf{a}_N = \sum_{r=0}^{N-1} \alpha_r \vec{\phi}_r . \qquad (2.11)$$

An ordered set of basis vectors $\left\{ \vec{\phi}_r \right\}$ can be regarded as a basis matrix

$$\mathbf{T}_N = \{\phi_{k,r}\} \qquad (2.12)$$

and signal expansion of Equation 2.11 can be regarded as a matrix product of the vector column of coefficients $\boldsymbol{\alpha} = \{\alpha_r\}$ by the basis matrix

$$\mathbf{a}_N = \mathbf{T}_N \boldsymbol{\alpha}_N. \tag{2.13}$$

In the signal-processing jargon, matrix $\mathbf{T}_N$ is called transform matrix and signal representation defined by Equation 2.12 is called discrete signal transform. If the transform matrix $\mathbf{T}_N$ has its inverse matrix $\mathbf{T}_N^{-1}$ such that

$$\mathbf{T}_N \mathbf{T}_N^{-1} = \mathbf{I}_N, \tag{2.14}$$

where I is the identity matrix:

$$\mathbf{I}_N = \{\delta_{k-r}\}, \tag{2.15}$$

the vector of signal representation coefficients can be found as

$$\boldsymbol{\alpha}_N = \mathbf{T}_N^{-1} \mathbf{a}_N. \tag{2.16}$$

Such matrix treatment of discrete signals assumes that signal scalar product $a \circ b$ is defined as a matrix product

$$\mathbf{a} \circ \mathbf{b} = \mathbf{a}^{tr} \cdot \mathbf{b}^* = \sum_{k=0}^{N-1} a_k b_k^*, \tag{2.17}$$

where the symbol "*tr*" denotes transposed matrix.

Transforms $\mathbf{T}_N$ and $\mathbf{T}_N^{-1}$ are mutually orthogonal. The transform matrix that is orthogonal to its transpose

$$\mathbf{T}_N \cdot \mathbf{T}_N^{tr} = \mathbf{I}_N \tag{2.18}$$

is called the orthogonal matrix. The transform matrix $\mathbf{T}_N$ orthogonal to its complex conjugate and transposed copy $\left(\mathbf{T}_N^*\right)^{tr}$

$$\mathbf{T}_N \cdot \left(\mathbf{T}_N^*\right)^{tr} = \mathbf{I}_N \tag{2.19}$$

is called the unitary transform matrix. For unitary transforms, the Parseval's relationships hold:

$$\mathbf{a} \circ \mathbf{b} = \boldsymbol{\alpha} \circ \boldsymbol{\beta} = \sum_{k=0}^{N-1} a_k b_k^* = \sum_{r=0}^{N-1} \alpha_r \beta_r^*; \quad \mathbf{a} \circ \mathbf{a} = \boldsymbol{\alpha} \circ \boldsymbol{\alpha} = \sum_{k=0}^{N-1} |a_k|^2 = \sum_{r=0}^{N-1} |\alpha_r|^2. \tag{2.20}$$

The discrete representation of continuous signals (Equation 2.1) can be extended to their integral representation if one substitutes in Equations 2.1 basis function index k by a continuous variable, say f, and replaces the sum by the integral

$$a(x) = \int_F \alpha(f)\phi(x,f)\,df.\tag{2.21}$$

One can also extend this approach to the definition of $\alpha(f)$ by introducing reciprocal functions $\phi(f,x)$ to obtain an analog of Equation 2.5:

$$\alpha(f) = \int a(x)\phi(f,x)\,dx.\tag{2.22}$$

Function $\alpha(f)$ is called the integral transform of signal $a(x)$, or its spectrum over continuous basis function $\{\phi(f,x)\}$ called the transform kernel. The reciprocity condition for functions $\phi(x,f)$ and $\phi(f,x)$ or the condition of transform invertibility can be obtained by substituting Equation 2.22 into Equation 2.21:

$$a(x) = \int_F \left[\int_X a(\xi)\phi(f,\xi)\,d\xi\right]\phi(x,f)\,df = \int_X a(\xi)\,d\xi\int_F \phi(f,\xi)\phi(x,f)\,df$$

$$= \int_X a(\xi)\delta(x,\xi)\,d\xi,\tag{2.23}$$

where

$$\delta(x,\xi) = \int_F \phi(x,f)\phi(f,\xi)\,df.\tag{2.24}$$

Function $\delta(x,\xi)$, defined by Equation 2.24, is called the delta-function (Dirac delta). Dirac delta is a continuous analog of the Kronecker delta (Equation 2.4). It symbolizes orthogonality of kernels $\phi(x,f)$ and $\phi(f,x)$, or, in other words, invertibility of the integral transforms of Equations 2.21 and 2.22. Equation 2.23 can also be regarded as a special case of the signal integral representation with the delta-function as the integral transform kernel.

Considering Equation 2.24 as an analog to Equation 2.6, one can extend the concept of the orthogonal basis to continuous functions as well: basis kernels that satisfy the relationship

$$\int_F \phi(x,f)\phi^*(x,f)\,df = \delta(x,\xi)\tag{2.25}$$

are called self-conjugate. For spectra $\alpha_1(f)$ and $\alpha_2(f)$ of signals $a_1(x)$ and $a_2(x)$ over self-conjugate bases, the following relationship holds:

$$\int_X a_1(x)a_2^*(x)\,dx = \int_F \alpha_1(f)\alpha_2^*(f)\,df, \qquad (2.26)$$

In particular, this means that signal integral transform with self-conjugated kernel preserves signal energy:

$$\int_X |a(x)|^2\,dx = \int_F |\alpha(x)|^2\,df. \qquad (2.27)$$

Equations 2.26 and 2.27 are a continuous analog to the Parseval's relation for signal discrete representation (Equation 2.9).

Signal Transformations

In sensory and imaging systems, image signals undergo various transformations, which convert one signal to another. In general, one can mathematically treat signal transformations as a mapping in the signal space. For digital signals, this mapping, in principle, can be specified in the form of a look-up table. However, even for digital signals, this form of the specification is not practically feasible owing to the enormous volume of the digital signal space (a rough evaluation of it can be seen in Chapter 3). For discrete and continuous signals, such a specification is all the more not constructive.

This is why in practice signal transformations are mathematically described and technically implemented as a combination of certain "elementary" transformations, each of which is specified with the help of a relatively small subset of all feasible input–output pairs. The most important of these are point-wise transformations and linear transformations.

Point-wise signal transformation is transformation that results in a signal whose value in each point of its argument (coordinate) depends only on the value of the transformed signal in the same point:

$$b(x) = TF[a(x)], \qquad (2.28)$$

that is, it is specified as a single argument mathematical function of the signal value. This function $TF(.)$ is referred to as the system transfer function. If the transfer function is the same in all points in signal coordinates, the transformation is called homogeneous transformation. Generally, the transformations are inhomogeneous.

A typical example of a point-wise transformation is conversion of energy of light radiation into electrical charge or current in cells of photo-electrical light sensors. Units performing element-wise signal conversion are widely used in models of imaging systems.

Transformation $L(\cdot)$ is a linear transformation if it satisfies the super-position principle, which states that the result of transformation of a sum of several signals can be found by summation of results of the transformation of individual signals:

$$L\left(\sum_{k=0}^{N-1} \alpha_k a_k(x)\right) = \sum_{k=0}^{N-1} \alpha_k \, L(a_k(x)). \qquad (2.29)$$

For linear transforms, one can introduce the notion of the sum of linear transforms, that is, $\sum_n L_n$, $n = 1, 2, \ldots$. The physical equivalent of the sum of linear transforms is a parallel connection of units realizing the transform items as it is shown in the flow diagram in Figure 2.3.

Assuming the possibility of applying linear transformation to signals, which are themselves results of a linear transformation, one can also introduce the notion of the product of linear transforms, $\prod_n L_n$. The physical equivalent of the transformation product is a series (cascade) connection of units realizing the transform factors (Figure 2.3b). Thanks to the linearity of the transforms, their multiplication is distributive with respect to addition:

$$L_1(L_2 + L_3) = L_1 L_2 + L_1 L_3; \quad (L_1 + L_2)L_3 = L_1 L_3 + L_2 L_3. \qquad (2.30)$$

If a linear transformation L performs a one-to-one mapping of a signal, then there exists an inverse transformation L^{-1} such that

$$L^{-1}(L(\mathbf{a})) = \mathbf{a}. \qquad (2.31)$$

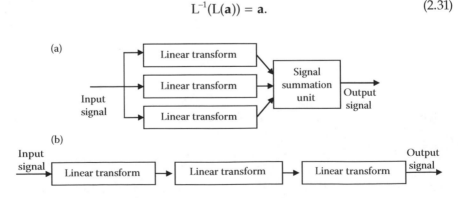

FIGURE 2.3
Flow diagrams of sum (a) and product (b) of linear transforms.

A linear transformation $\mathbf{b} = \mathbf{L}(\mathbf{a})$ of discrete signals is fully characterized by a matrix of coefficients $\Lambda = \{\lambda_{k,n}\}$ that links the input $\mathbf{a} = \{\alpha_k\}$ and output $\mathbf{b} = \{\beta_k\}$ pairs of the linear transformation coefficients, respectively (see Appendix):

$$\beta_n = \mathbf{L}(\mathbf{a}) = \Lambda a = \left\{ \beta_n = \sum_k \lambda_{k,n}\alpha_k \right\}. \tag{2.32}$$

Thus, in order to describe a linear transformation of a discrete signal of N samples, it suffices to specify a matrix of N^2 numbers. Equation 2.32 implies that, in the general case, linear transforms are not point-wise, and that they become so only if the transformation matrix is a diagonal one, that is $\{\lambda_{k,n} = \lambda_k \delta_{k-n}\}$.

For continuous signals, integral transforms are mathematical models of linear transformations. Let a continuous signal $a(x)$ be specified by means of its spectrum $\alpha(f)$ on basis function $\{\phi^{(a)}(x,f)\}$ as it is described by Equations 2.21 and 2.22 and, in a special case of delta-function basis, by Equation 2.23. Then the result $b(\xi) = \mathbf{L}a(x)$ of applying to this signal a linear transformation $\mathbf{L}$ is fully specified by the response $h(\xi, f) = \mathbf{L}(\phi^{(a)}(x, f))$ of the linear transformation to the transform kernel

$$b(\xi) = \mathbf{L}\left[\int_F \alpha(f)\phi^{(a)}(x,f)\,df \right] = \int_F \alpha(f)\mathbf{L}\left[\phi^{(a)}(x,f)\right]df = \int_F \alpha(f)h(\xi,f)\,df. \tag{2.33}$$

For the representation of the linear transformation output signal $b(\xi)$ also via its spectrum $\beta(f)$ over some basis functions $\phi^{(b)}(f,\xi)$

$$\beta(f) = \int_\Xi b(\xi)\phi^{(b)}(f,\xi)\,d\xi \tag{2.34}$$

spectra $\alpha(f)$ and $\beta(f)$ of a signal $a(x)$ and of its linear transformation $b(\xi)$ are also related with an integral transform

$$\beta(f) = \int_F \alpha(p)H(f,p)\,dp. \tag{2.35}$$

Its kernel $H(f,p)$ is, in its turn, also an integral transform, with kernel $\phi^{(b)}(f,\xi)$, of the linear transformation response $h(\xi,f)$ to basis functions $\phi^{(a)}(f,\xi)$, in which the spectrum of the signal $a(x)$ is specified

$$H(f,p) = \int h(\xi,p)\phi^{(b)}(f,\xi)\,d\xi. \tag{2.36}$$

Signal integral representation (Equation 2.23) via delta-function is of special importance for the characterization of the linear transformations. For such a representation, the relationship between output signal $b(x)$ and input signal $a(x)$ of a linear transformation (Equation 2.34) is reduced to

$$b(\xi) = \int_X a(x)h(x,\xi)\,dx, \qquad (2.37)$$

where $h(x,\xi) = \mathbf{L}(\delta(x,\xi))$ is a linear transformation response to a delta function. It is called the *impulse response* or the *point spread function* (*PSF*) of the linear system that implements this transformation. The process of linear transformation is called *linear filtering*, and a unit that performs it is called *linear filter*.

Imaging Systems and Integral Transforms

Direct Imaging and Convolution Integral

In describing imaging systems, it is always assumed that there exist a physical or imaginary object plane and an image plane and that the primary goal of imaging systems is reproducing, in image plane, distribution of light reflectance, transmittance, or whatever other physical property of the object in the object plane.

Many imaging devices, such as optical microscopes and telescopes, x-ray, gamma cameras, scanning electron and acoustic microscopes, and scanned proximity probe (atomic force, tunnel) microscopes, work on the principle of direct imaging of each point in object plane into a corresponding point in image plane. Consider the schematic diagram of photographic and TV imaging cameras shown in Figure 2.4.

Imaging optics is a linear system that performs linear signal transformation. It is usually characterized by its PSF $h(x,y;\xi,\eta)$, which links an image $b(x,y)$ in the image plane and light intensity distribution $a(\xi,\eta)$ in the object plane:

$$b(x,y) = \int_X \int_Y a(\xi,\eta)h(x,y;\xi,\eta)\,d\xi\,d\eta, \qquad (2.38)$$

where integration is carried out over the object plane (X, Y). In general, the image of a point source depends on the source location in the object plane, that is, the PSF is space variant.

An important special case is that of space invariant imaging systems, in which the image of a point source does not depend on the point source

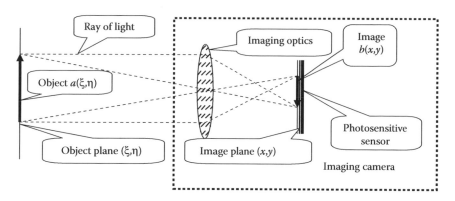

FIGURE 2.4
Schematic diagram of photographic and TV imaging cameras.

location. For space invariant systems, PSF is a function of the difference of coordinates in object and image planes, and imaging Equation 2.38 for space invariant systems takes the form

$$b(x,y) = \int\limits_{-\infty}^{\infty} \int\limits_{-\infty}^{\infty} a(\xi,\eta) h(x-\xi, y-\eta)\, d\xi\, d\eta \qquad (2.39)$$

called 2D convolution integral. Correspondingly, 1D convolution integral is defined as

$$b(x) = \int\limits_{-\infty}^{\infty} a(\xi) h(x-\xi)\, d\xi. \qquad (2.40)$$

PSF is one of the basic specification characteristics of imaging systems. It is used for certification of imaging systems and characterizes the system capability to resolve objects, that is, imaging system resolving power. Given the noise level in the signal sensor, it determines the accuracy of locating objects in images. We will discuss this issue in detail in Chapter 7.

The assumption of imaging system space invariance is used in the description and analysis of imaging systems and image restoration most frequently. It enables simplifications both in system analysis and in image processing. If the system's PSF is not shift invariant, one can, for simplifying image analysis and processing, convert images from the original coordinate system into a transformed one, in which system PSF becomes space invariant. After space invariant processing in the transformed coordinate system is completed, image should be converted back to the original coordinates. This principle is illustrated by a flow diagram in Figure 2.5.

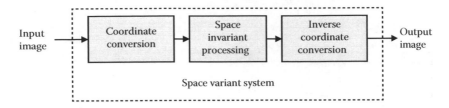

FIGURE 2.5
The principle of space invariant representation of space variant systems.

Integral transforms described by Equations 2.38 and 2.39 are, as a rule, not invertible: in general, one cannot perfectly restore system input image $a(\xi,\eta)$ from the output image $b(x,y)$. The restoration task is one of the fundamental tasks of image processing. It belongs to the so-called inverse problems. Methods of image restoration will be discussed in Chapter 8.

Multiresolution Imaging: Wavelet Transforms

It is frequently convenient to combine applying convolutional signal transforms to signals $a(x)$ taken at an arbitrary scale and to perform multiresolution analysis in this way. This leads to integral transforms that was given the name of wavelet transforms.

Consider 1D convolution integral for signal $a(x)$ taken in scaled coordinates σx:

$$\bar{a}(\sigma; x) = \int_{-\infty}^{\infty} a(\sigma\xi) h_{WL}(x - \xi)\, d\xi, \qquad (2.41)$$

where σ is a scale parameter. This convolution with PSF $h_{WL}(.)$ depends on both scale parameter σ and coordinate x. In Equation 2.41, signal is taken at different scales and convolution kernel is kept the same. One can now translate signal scaling into convolution kernel scaling by changing variables $\sigma\xi \to \xi$ and obtain

$$\bar{a}(\sigma; x) = \frac{1}{\sigma} \int_{-\infty}^{\infty} a(\xi) h_{WL}\left(\frac{\sigma x - \xi}{\sigma} \right) d\xi \qquad (2.42)$$

or, in scaled coordinates $\sigma x \to x$:

$$\bar{a}(\sigma; x) = \frac{1}{\sigma} \int_{-\infty}^{\infty} a(\xi) h_{WL}\left(\frac{x - \xi}{\sigma} \right) d\xi. \qquad (2.43)$$

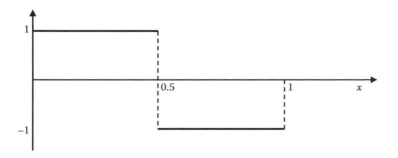

FIGURE 2.6
"Mother wavelet" of Haar transform.

Integral transform defined by Equation 2.43 is called the wavelet transform.

In 1D wavelet transforms, transform domain argument of the kernel function has two components: coordinate shift ξ and scale σ. There exist many wavelet transforms that differ from one another in the selection of the transform kernel $h_{WL}(\cdot)$ called the "mother wavelet." This selection is the main issue in the design of wavelet transforms and is governed by the requirement of the invertibility of the transform and by implementation issues. One of the simplest examples of the wavelet transforms is Haar transform. The kernel of Haar transform (Haar transform mother wavelet) is a bipolar function shown in Figure 2.6 and is defined as

$$h_{WL}(\sigma;x) = \mathrm{rect}\left(\frac{2x}{\sigma}\right) - \mathrm{rect}\left(\frac{2x-1}{\sigma}\right) \tag{2.44}$$

where rect(.) is a *rect-function*:

$$\mathrm{rect}(x) = \begin{cases} 1, & 0 \le x < 1 \\ 0, & \text{otherwise} \end{cases}. \tag{2.45}$$

Two-dimensional and multidimensional wavelets are usually defined as separable functions of coordinates.

Imaging in Transform Domain and Diffraction Integrals

The invention of holography by D. Gabor [1] heralded a new stage in the evolution of imaging methods: imaging in the transform domain. Instead of recording an image of an object, D. Gabor suggested recording the amplitude and the phase of the wavefront of coherent radiation propagated from the object to the recording device. He named the recorded wavefront as *hologram*.

In order to describe a mathematical model of wave propagation and recording holograms, consider the 1D model of free space wave propagation shown in Figure 2.7.

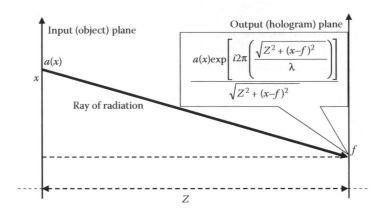

FIGURE 2.7
Free space wave propagation point spread function.

Let the complex amplitude of the wavefront of a monochromatic radiation in coordinate x at the object (input) plane be $a(x) = |a(x)| \exp(i\Phi(x))$, where $|a(x)|$ is the wavefront amplitude equal to the square root of its intensity and $\Phi(x)$ is called the wavefront "phase." Let the "output" (hologram) plane, in which the hologram is recorded, also be situated at a distance Z from the object plane. The wavefront, which traveled from a point x in the object plane to a point with coordinate f in the hologram plane a distance $\sqrt{Z^2 + (x - f)^2}$, will be attenuated proportionally to this distance, and a phase delay $2\pi\sqrt{Z^2 + (x - f)^2}/\lambda$ is obtained, where λ is the wavelength of the coherent radiation. Then, one can conclude that the PSF of free space wave propagation is

$$h(x, f) = \frac{\exp(i2\pi\sqrt{Z^2 + (x - f)^2}/\lambda)}{\sqrt{Z^2 + (x - f)^2}}. \qquad (2.46)$$

Accordingly, 2D PSF of free space wave propagation from input plane (x, y) to output plane (f_x, f_y) is

$$h(x, y; f_x, f_y) = \frac{\exp(i2\pi\sqrt{Z^2 + (x - f_x)^2 + (y - f_y)^2}/\lambda)}{\sqrt{Z^2 + (x - f_x)^2 + (y - f_y)^2}}. \qquad (2.47)$$

The relationship between complex amplitudes $a(x,y)$ and $\alpha(f_x,f_y)$ of the wavefront in the input and output planes, respectively, is then the integral transform

$$\alpha(f_x, f_y) = \int\limits_{-\infty}^{\infty} \int\limits_{-\infty}^{\infty} a(x, y) \frac{\exp(i2\pi\sqrt{Z^2 + (x - f_x)^2 + (y - f_y)^2}/\lambda)}{\sqrt{Z^2 + (x - f_x)^2 + (y - f_y)^2}} dx\, dy. \qquad (2.48)$$

This integral transform is called *Kirchhoff's integral*.

Assume now that the size of the object and dimensions of the area, over which the hologram is recorded, are substantially smaller than the distance Z between the input object plane (x, y) and the hologram plane (f_x, f_y), that is, approximated relationships:

$$\sqrt{Z^2 + (x - f_x)^2 + (y - f_y)^2} \cong Z$$

$$\exp\left(i2\pi \frac{\sqrt{Z^2 + (x - f_x)^2 + (y - f_y)^2}}{\lambda}\right) = \exp\left(i2\pi \frac{Z}{\lambda}\sqrt{1 + \frac{(x - f_x)^2 + (y - f_y)^2}{Z^2}}\right)$$

$$\cong \exp\left[i\pi \frac{(x - f_x)^2 + (y - f_y)^2}{\lambda Z}\right] \qquad (2.49)$$

hold. In this approximation, Kirchhoff's integral is reduced, after omitting an unessential scalar multiplier, to its *Fresnel approximation*:

$$\alpha(f_x, f_y) = \int_{-\infty}^{\infty} \int_{-\infty}^{\infty} a(x, y)\exp\left[i\pi \frac{(x - f_x)^2 + (y - f_y)^2}{\lambda Z}\right]dx\, dy. \qquad (2.50)$$

that describes the so-called near zone diffraction. Equation 2.50 can alternatively be written as

$$\alpha(f_x, f_y)\exp\left(-i\pi \frac{f_x^2 + f_y^2}{\lambda Z}\right)$$

$$= \int_{-\infty}^{\infty} \int_{-\infty}^{\infty} a(x, y)\exp\left(i\pi \frac{x^2 + y^2}{\lambda Z}\right)\exp\left(-i2\pi \frac{xf_x + yf_y}{\lambda Z}\right)dx\, dy. \qquad (2.51)$$

If, furthermore, distance Z is so large that one can neglect phase variations $[\pi(f_x^2 + f_y^2)/\lambda Z]$ and $[\pi(x^2 + y^2)/\lambda Z]$, we arrive at the *far zone diffraction*, or *Fraunhofer approximation* of Kirchhoff's integral:

$$\alpha(f_x, f_y) = \int_{-\infty}^{\infty} \int_{-\infty}^{\infty} a(x, y)\exp\left(-i2\pi \frac{xf_x + yf_y}{\lambda Z}\right)dx\, dy. \qquad (2.52)$$

Equation 2.52 also describes the relationship between wavefront complex amplitudes in front and rear focal planes of lenses. Consider a 1D model of a lens shown in Figure 2.8.

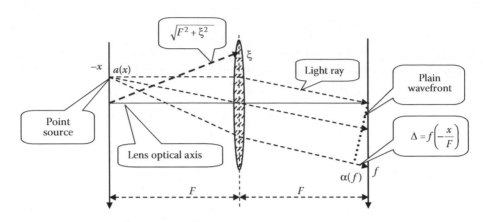

FIGURE 2.8
Schematic diagram of wave propagation between focal planes of a lens.

Lens is fully defined by its property to convert light from a point source in the lens focal plane into a parallel beam with plain wavefront tilted according to the position of the source with respect to lens optical axis. One can use this property to determine the PSF of a lens. One can see from the diagram in Figure 2.8 that, at point f in the lens rear focal plane, the plane wavefront originated from a point source of monochromatic light in coordinate ($-x$) is delayed with respect to the point at the lens optical axis by distance $\Delta = f \sin(-x/F)$, where F is the lens focal length. In the so-called paraxial approximation, when $x \ll F$, $\Delta = -xf/F$. This implies that the PSF of optical systems, whose input and output planes are front and rear focal planes of a lens, is

$$h(x, f) = \exp(i2\pi\Delta/\lambda) = \exp\left(-i2\pi\frac{fx}{\lambda F}\right) \qquad (2.53)$$

and complex amplitudes $a(x)$ and $\alpha(f)$ of monochromatic light wavefront in frontal and rear focal planes of a lens are, therefore, linked by the relationship

$$a(f) = \int_X a(x)\exp\left(-i2\pi\frac{fx}{\lambda F}\right)dx, \qquad (2.54)$$

or, in 2D denotation

$$a(f_x, f_y) = \int\int_{X\,Y} a(x, y)\exp\left(-i2\pi\frac{f_x x + f_y y}{\lambda F}\right)dx\,dy. \qquad (2.55)$$

From the schematic diagram in Figure 2.8, it follows that lenses can also be considered as spatial light phase modulators that add to the phase of the incident light a component, which is a quadratic function of the distance from the lens optical axis. Indeed, a spherical wave at point ξ in the lens plane emanated from a point source at the optical axis in the input plane has a phase defined by the distance $\sqrt{F^2 + \xi^2}$ it propagated. Hence, its complex amplitude is

$$A \propto \exp\left(i2\pi \frac{\sqrt{F^2 + \xi^2}}{\lambda} \right). \tag{2.56}$$

In the paraxial approximation, when the size of the lens is much smaller than its focal distance, wave amplitude is approximately

$$A \propto A_0 \exp\left(i\pi \frac{\xi^2}{\lambda F} \right). \tag{2.57}$$

Because this spherical wave is converted by the lens into a plain wavefront, complex transparency of the lens is then inverse of the function defined by Equation 2.57:

$$T = \exp\left(-i\pi \frac{\xi^2}{\lambda F} \right). \tag{2.58}$$

One can rewrite Equations 2.50, 2.52, 2.54, 2.56, and 2.57 in dimensionless coordinates normalized by $\sqrt{\lambda Z}$ or, correspondingly, by $\sqrt{\lambda F}$. In such normalized coordinates, obtain, after neglecting inessential constant factors

$$\alpha(f_x, f_y) = \int_{-\infty}^{\infty} \int_{-\infty}^{\infty} a(x,y) \frac{\exp\left\{ -i\pi\sqrt{Z_\lambda^2 + z_\lambda[(x - f_x)^2 + (y - f_y)]^2} \right\}}{\sqrt{Z_\lambda^2 + Z_\lambda[(x - f_x)^2 + (y - f_y)]^2}} dx\, dy, \tag{2.59}$$

where $Z_\lambda = Z/\lambda$ is a dimensionless distance parameter

$$\alpha(f_x, f_y) = \int_{-\infty}^{\infty} \int_{-\infty}^{\infty} a(x,y) \exp\{ -i\pi[(x - f_x)^2 + (y - f_y)^2] \}\, dx\, dy \tag{2.60}$$

and

$$\alpha(f) = \int_{-\infty}^{\infty} a(x) \exp(i2\pi f x)\, dx \tag{2.61}$$

$$a(f_x, f_y) = \int\limits_{-\infty}^{\infty} \int\limits_{-\infty}^{\infty} a(x, y) \exp[i2\pi(f_x x + f_y y)] \, \mathrm{d}x \, \mathrm{d}y. \tag{2.62}$$

We will refer to Equation 2.59 as *integral Kirchhoff transform*. Integral transform defined by Equation 2.60 is called *integral Fresnel transform*. Integral transforms defined by Equations 2.61 and 2.62 are 1D and 2D *integral Fourier transforms*.

Fresnel transform is very closely connected with Fourier transform and can be reduced to it by means of phase modulation of signal and its spectrum with chirp-functions $\mathrm{chirp}(x) = \exp(-i\pi x^2)$ and $\mathrm{chirp}(f) = \exp(-i\pi f^2)$:

$$\alpha_{\mathrm{Fr}}(f) = \exp(-i\pi f^2) \int\limits_{-\infty}^{\infty} a(x) \exp(-i\pi x^2) \exp(i2\pi f x) \, \mathrm{d}x \tag{2.63}$$

Fresnel transform can also be regarded as a convolutional transform and, as such, it can also be linked with Fourier transform through the convolution theorem (see the next section): Fresnel transform $\alpha_{\mathrm{Fr}}(f)$ of a signal $a(x)$ can be found as inverse Fourier transform

$$\alpha_{\mathrm{Fr}}(f) = \int\limits_{-\infty}^{\infty} \alpha_{\mathrm{F}}(p) \exp(i\pi p^2) \exp(-i2\pi f p) \, \mathrm{d}p \tag{2.64}$$

of a product of the signal Fourier spectrum $\alpha_{\mathrm{F}}(p)$ and the Fourier transform of the chirp-function which, to the accuracy of an irrelevant constant, is (see Appendix) function $\mathrm{chirp}^*(p) = \exp(i\pi p^2)$:

$$\int\limits_{-\infty}^{\infty} \exp(-i\pi\omega^2 x^2) \exp(i2\pi p x) \, \mathrm{d}x \propto \exp\left(i\pi \frac{p^2}{\omega^2}\right). \tag{2.65}$$

The transform defined by Equation 2.64 is called *angular spectrum propagation transform*.

Integral Kirchhoff transform (Equation 2.59) is also a convolution and can, therefore, also be represented, using the convolution theorem, which we present in the section "Properties of the Integral Fourier Transform" that follows, via Fourier transform:

$$\alpha_K(f_x, f_y) = \int\limits_{-\infty}^{\infty} \int\limits_{-\infty}^{\infty} \alpha_{\mathrm{F}}(p_x, p_y) K(p_x, p_y) \exp[-i2\pi(f_x p_x + f_y p_y)] \, \mathrm{d}p_x \, \mathrm{d}p_y, \tag{2.66}$$

where $\alpha_F(p_x,p_y)$ is the signal Fourier spectrum and kernel $K(p_x,p_y)$ is the Fourier transform of the kernel of the Kirchhoff integral transform

$$K(p_x,p_y)$$

$$= \int\limits_{-\infty}^{\infty}\int\limits_{-\infty}^{\infty} \frac{\exp\left\{-i\pi\sqrt{d_\lambda^2 + d_\lambda[(x-f_x)^2 + (y-f_y)]^2}\right\}}{\sqrt{d_\lambda^2 + d_\lambda[(x-f_x)^2 + (y-f_y)]^2}} \exp[-i2\pi(f_x p_x + f_y p_y)]\,dx\,dy.$$

(2.67)

Properties of the Integral Fourier Transform

Integral Fourier transform plays an exceptional role in image processing. Apart from its fundamental role in mathematical representation of signal transformations in wave propagation and image formation described in the previous section, it has, in its discrete representation as discrete Fourier transform (DFT, see Chapter 4), a remarkable property of efficient computation via *fast Fourier transform*. Integral Fourier transform also represents a very convenient tool for linking all linear transforms. In this section, we overview the main properties of the integral Fourier transform.

In order to make derivations more compact, we will, when it will be convenient, treat 2D coordinates (x, y) and (f_x, f_y) as two-component vectors $\mathbf{x} = [x \quad y]$ and $\mathbf{f} = [f_x \quad f_y]$. In such denotation, the 2D integral Fourier transform is written as

$$\alpha(\mathbf{f}) = \int\limits_{-\infty}^{\infty} a(\mathbf{x})\exp(i2\pi\mathbf{x}^{tr}\mathbf{f})\,d\mathbf{x}, \tag{2.68}$$

where $\mathbf{x}^{tr}$ is transposed vector x. Equation 2.68 can also be generally regarded as a multidimensional integral transform assuming that x and f can be, in general, more than two component vectors. We call this transform the *direct integral Fourier transform*.

The most important properties of the integral Fourier transform are the following.

Invertibility

Integral Fourier transform is invertible. It is proved in the theory of Fourier integral that signal $a(\mathbf{x})$ can be found from its Fourier spectrum $\alpha(\mathbf{f})$ as

$$a(\mathbf{x}) = \lim_{F\to\infty}\int\limits_F \alpha(\mathbf{f})\exp(-i2\pi\mathbf{f}^{tr}\cdot\mathbf{x})\,d\mathbf{f} = \int\limits_{-\infty}^{\infty} \alpha(\mathbf{f})\exp(-i2\pi\mathbf{f}^{tr}\cdot\mathbf{x})\,d\mathbf{f}. \tag{2.69}$$

Integral

$$a(\mathbf{x}) = \int_{-\infty}^{\infty} \alpha(\mathbf{f}) \exp[-i2\pi \mathbf{f}^{tr} \cdot \mathbf{x}] \, d\mathbf{f} \qquad (2.70)$$

is called the inverse integral Fourier transform. Equations 2.69 and 2.70 imply that the kernel of the integral Fourier transform is self-conjugate and delta function, which symbolizes the transform invertibility, is defined as

$$\delta(\mathbf{x}, \xi) = \lim_{F \to \infty} \int_{F} \exp[i2\pi \mathbf{f}^{tr} \cdot (\mathbf{x} - \xi)] \, d\mathbf{f}. \qquad (2.71)$$

In a special case of 1D integral Fourier transform:

$$\delta(x, \xi) = \lim_{F \to \infty} \int_{-F}^{F} \exp[i2\pi f(x - \xi)] \, df = \lim_{F \to \infty} \{2F \operatorname{sinc}[2\pi F(x - \xi)]\}, \qquad (2.72)$$

where

$$\operatorname{sinc}(x) = \frac{\sin x}{x} \qquad (2.73)$$

is the so-called *sinc-function*. According to this definition, sinc-function is the Fourier transform

$$\operatorname{sinc}(2\pi Fx) = \int_{-\infty}^{\infty} \operatorname{rect}\left(\frac{f + F}{2F}\right) \exp(i2\pi fx) \, dx \qquad (2.74)$$

of the rect-function (Equation 2.45).

The invertibility of integral Fourier transform is a mathematical idealization. In practice, integration in Fourier transform is always implemented in finite limits defined by aperture of optical systems, dimensions of lenses, and similar factors. These limitations are referred to as bandwidth limitations. Owing to the bandwidth limitations, artifacts known as Gibb's effects may appear in the form of oscillations at signal sharp edges. Figure 2.9 illustrates the appearance of oscillations caused by signal Fourier spectrum band-limitation for a signal in the form of a rectangular wave. As one can see, the oscillations do not disappear on widening the signal bandwidth; they only become tighter.

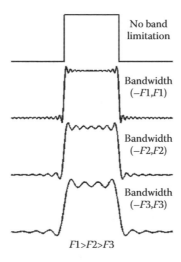

$F1>F2>F3$

FIGURE 2.9
Illustration of signal oscillation caused by signal band limitation. From top to bottom: signal with no band limitation and signals with more and more severe band limitations.

Separability

It follows from the definition of 2D integral Fourier transform (Equation 2.62) that it is separable to two 1D transforms:

$$a(f_x, f_y) = \int_{-\infty}^{\infty} \int_{-\infty}^{\infty} a(x, y) \exp[i2\pi(f_x x + f_y y)] \, dx \, dy$$

$$= \int_{-\infty}^{\infty} \exp(i2\pi f_y y) \, dy \int_{-\infty}^{\infty} a(x, y) \exp(i2\pi f_x x) \, dx. \quad (2.75)$$

Shift Theorem: Absolute value of signal Fourier spectrum is invariant to signal shift. Signal shift causes only linear modulation of the phase of the spectrum.

Proof. From Equation 2.61, it follows that

$$a(\mathbf{x} - \mathbf{x}_0) = \int_{-\infty}^{\infty} \alpha(\mathbf{f}) \exp[-i2\pi \mathbf{f}^{tr} \cdot (\mathbf{x} - \mathbf{x}_0)] \, d\mathbf{f}$$

$$= \int_{-\infty}^{\infty} [\alpha(\mathbf{f}) \exp(i2\pi \mathbf{f}^{tr} \cdot \mathbf{x}_0)] \exp[-i2\pi \mathbf{f}^{tr} \cdot \mathbf{x}] \, d\mathbf{f} \quad (2.76)$$

which implies that if Fourier spectrum of $a(\mathbf{x})$ is $\alpha(\mathbf{f})$, spectrum of $a(\mathbf{x} - \mathbf{x}_0)$ is $\alpha(\mathbf{f}) \exp(i2\pi \mathbf{f}^{tr} \cdot \mathbf{x}_0)$, and that the absolute value of signal Fourier spectrum is invariant to the signal shift.

Rotation Theorem: Signal rotation results in the same rotation of its spectrum.

Proof. In vector-matrix denotation of signal and spectrum coordinates, signal rotation by an angle ϑ can be represented as multiplication of coordinate vector $\mathbf{x}$ by a rotation matrix $\mathbf{Rot}_\vartheta$

$$\mathbf{Rot}_\vartheta = \begin{bmatrix} \cos\vartheta & -\sin\vartheta \\ \sin\vartheta & \cos\vartheta \end{bmatrix}. \tag{2.77}$$

Then, from Equations 2.68 and 2.77, we obtain, using matrix identity $\mathbf{f}^{tr} \cdot \mathbf{Rot}_\vartheta = ((\mathbf{Rot}_\vartheta)^{tr}\mathbf{f})^{tr} = (\mathbf{Rot}_{-\vartheta}\mathbf{f})^{tr}$ that

$$a(\mathbf{Rot}_\vartheta \cdot \mathbf{x}) = \int_{-\infty}^{\infty} \alpha(\mathbf{f})\exp(-i2\pi\mathbf{f}^{tr} \cdot (\mathbf{Rot}_\vartheta \cdot \mathbf{x}))\,d\mathbf{f}$$

$$= \int_{-\infty}^{\infty} \alpha(\mathbf{f})\exp(-i2\pi(\mathbf{f}^{tr} \cdot \mathbf{Rot}_\vartheta) \cdot \mathbf{x})\,d\mathbf{f} = \int_{-\infty}^{\infty} \alpha(\mathbf{f})\exp(-i2\pi((\mathbf{Rot}_\vartheta \cdot \mathbf{f})^{tr} \cdot \mathbf{x}))\,d\mathbf{f}$$

$$= \int_{-\infty}^{\infty} \alpha(\mathbf{f})\exp(-i2\pi((\mathbf{Rot}_{-\vartheta} \cdot \mathbf{f})^{tr} \cdot \mathbf{x}))\,d\mathbf{f}. \tag{2.78}$$

Changing variables $\mathbf{Rot}_{-\vartheta} \cdot \mathbf{f} \to \tilde{\mathbf{f}}$ and using the fact that the Jacobean of this coordinate conversion is unity

$$J(\mathbf{Rot}_{-\vartheta}) = \cos^2\vartheta + \sin^2\vartheta = 1 \tag{2.79}$$

finally obtain

$$a(\mathbf{Rot}_\vartheta \cdot \mathbf{x}) = \int_{-\infty}^{\infty} \alpha(\mathbf{Rot}_\vartheta \cdot \tilde{\mathbf{f}})\exp(-i2\pi\tilde{\mathbf{f}} \cdot \mathbf{x})\,d\tilde{\mathbf{f}}. \tag{2.80}$$

Therefore, if the Fourier spectrum of 2D signal $a(\mathbf{x})$ is $\alpha(\mathbf{f})$, the spectrum of rotated signal $a(\mathbf{Rot}_\vartheta \cdot \mathbf{x})$ is $\alpha(\mathbf{Rot}_\vartheta \cdot \mathbf{f})$.

Scaling Theorem: Scaling signal coordinates σ times causes $1/\sigma$ scaling of its spectrum magnitude and spectrum coordinates.

Proof. From Equation 2.68, obtain

$$a(\sigma\mathbf{x}) = \int_{-\infty}^{\infty} \alpha(\mathbf{f})\exp[-i2\pi\mathbf{f} \cdot (\sigma\mathbf{x})]\,d\mathbf{f} = \int_{-\infty}^{\infty} \alpha(\mathbf{f})\exp[-i2\pi(\sigma\mathbf{f}) \cdot \mathbf{x}]\,d\mathbf{f}$$

$$= \frac{1}{\sigma}\int_{-\infty}^{\infty} \alpha\left(\frac{1}{\sigma}\mathbf{f}\right)\exp(-i2\pi\tilde{\mathbf{f}} \cdot \mathbf{x})\,d\tilde{\mathbf{f}}. \tag{2.81}$$

Therefore, if the Fourier spectrum of signal $a(x)$ is $\alpha(f)$, the Fourier spectrum of scaled signal $a(\sigma x)$ is $(1/\sigma)\alpha((a/\sigma)f)$.

Convolution Theorem: The inverse Fourier transform of product $\alpha(f)\beta(f)$ of spectra of two signals $a(x)$ and $b(x)$ is convolution $\int_{-\infty}^{\infty} a(\xi)b(x-\xi)\,d\xi$ of these signals.

Proof.

$$\int_{-\infty}^{\infty} \alpha(f)\beta(f)\exp(-i2\pi f^{tr}x)\,df = \int_{-\infty}^{\infty}\left[\int_{-\infty}^{\infty} a(\xi)\exp(i2\pi f^{tr}\xi)\,d\xi\right]\beta(f)\exp(-i2\pi f^{tr}x)\,df$$

$$= \int_{-\infty}^{\infty} a(\xi)\left[\int_{-\infty}^{\infty} \beta(f)\exp(-i2\pi f^{tr}(x-\xi))\,df\right]d\xi$$

$$= \int_{-\infty}^{\infty} a(\xi)b(x-\xi)\,d\xi. \qquad (2.82)$$

Convolution theorem links integral Fourier transform and convolution integral. It shows that image convolution can also be treated in terms of modulation of image Fourier spectra by Fourier transform $FR(f)$ of the convolution integral kernel, PSF $h(x)$ of the space invariant imaging system:

$$FR(f) = \int_{-\infty}^{\infty} h(x)\exp(i2\pi fx)\,dx. \qquad (2.83)$$

Function $FR(f)$ is called the system *frequency response* or *modulation transfer function*.

Symmetry Properties

Signal and their spectra symmetry properties straightforwardly follow from the definition of the Fourier transform and are summarized in the following equations, in which symbol * means complex conjugation.

For even (odd) signals:

$$a(x) = \pm a(-x) \xleftrightarrow{\text{Fourier transform}} \alpha(f) = \pm\alpha(-f). \qquad (2.84)$$

For purely real (imaginary) signals:

$$a(x) = \pm a^*(x) \xleftrightarrow{\text{Fourier transform}} \alpha(f) = \pm\alpha^*(-f). \qquad (2.85)$$

For even purely real (imaginary) signals:

$$a(\mathbf{x}) = a(-\mathbf{x}) = \pm a^*(\mathbf{x}) \xleftrightarrow{\text{Fourier transform}} \alpha(\mathbf{f}) = \alpha(-\mathbf{f}) = \pm\alpha^*(-\mathbf{f}). \quad (2.86)$$

For odd purely real (imaginary) signals:

$$a(\mathbf{x}) = -a(-\mathbf{x}) = \pm a^*(\mathbf{x}) \xleftrightarrow{\text{Fourier transform}} \alpha(\mathbf{f}) = -\alpha(-\mathbf{f}) = \pm\alpha^*(-\mathbf{f}). \quad (2.87)$$

Proofs of these relationships are provided in Appendix.

Parseval's Relationship: The Fourier spectra of signals preserve signal energy

Proof.

$$\int_{-\infty}^{\infty} |a(\mathbf{x})|^2 \, d\mathbf{x} = \int_{-\infty}^{\infty} a(\mathbf{x})a^*(\mathbf{x}) \, d\mathbf{x} = \int_{-\infty}^{\infty} a(\mathbf{x}) \left[\int_{-\infty}^{-\infty} \alpha(\mathbf{f}) \exp(-i2\pi \mathbf{f}\mathbf{x}) \, d\mathbf{f} \right]^* d\mathbf{x}$$

$$= \int_{-\infty}^{\infty} \alpha^*(\mathbf{f}) \, d\mathbf{f} \int_{-\infty}^{-\infty} a(\mathbf{x}) \exp(i2\pi \mathbf{f}\mathbf{x}) \, d\mathbf{x} = \int_{-\infty}^{\infty} \alpha(\mathbf{f})\alpha^*(\mathbf{f}) \, d\mathbf{f} = \int_{-\infty}^{\infty} |\alpha(\mathbf{f})|^2 \, d\mathbf{f}.$$

$$(2.88)$$

Transforms in Sliding Window (Windowed Transforms) and Signal Sub-Band Decomposition

Up to now, Fourier integral transform as well as other integral transforms were defined as "global transforms" of signals over infinite intervals. In reality, however, only finite fractions of the signals selected by the signal sensor area are to be or should be analyzed. This can be mathematically modeled by signal windowing. In this way, we arrive at windowed transforms:

$$\alpha(x, f) = \int_{-\infty}^{\infty} w(\xi)a(x - \xi)\phi(f, \xi) \, d\xi, \quad (2.89)$$

where $\phi(f, \xi)$ is a transform kernel and $w(\xi)$ is a window function such that it is close to 1 within a certain vicinity to point $\xi = 0$ and then more or less rapidly decays to zero:

$$w(0) = 1; \quad \lim_{\xi \to \infty} w(\xi) = 0. \quad (2.90)$$

Windowed transforms of signal $a(x)$ are defined for every point x in signal coordinate system and are usually computed in a process of regular signal scanning along its coordinate. It is in this sense that they will be referred to as sliding window transforms.

One of the most important sliding window transforms is sliding window Fourier transform (SWFT). It is defined as

$$\alpha(x, f) = \int\limits_{-\infty}^{\infty} w(\xi)a(x - \xi)\exp(i2\pi f\xi)\, d\xi. \tag{2.91}$$

SWFT of 1D signals is a function of two variables: coordinate x in signal domain and frequency f in Fourier domain. At each window position x, it represents what one can call signal local spectrum, that is, spectrum of a (windowed) signal segment centered at x. This spectrum, being regarded as a function of coordinate x that defines window running position and of frequency f, is called signal space–frequency representation (time–frequency representation, in the communication engineering jargon).

From the inverse Fourier transform of the windowed Fourier transform spectrum $\alpha(x, f)$, one can obtain

$$w(\xi)a(x - \xi) = \int\limits_{-\infty}^{\infty} \alpha(x, f)\exp(-i2\pi f\xi)\, df. \tag{2.92}$$

By virtue of Equation 2.90, it follows from Equation 2.92 that

$$a(x) = \int\limits_{-\infty}^{\infty} \alpha(x, f)\, df. \tag{2.93}$$

Equation 2.93 can, therefore, be regarded as inverse SWFT.

The window function of SWFT can be an arbitrary function that satisfies Equation 2.90. A special case of SWFT with the window function

$$w(\xi) = \exp\left(-\frac{\xi^2}{2\omega^2}\right) \tag{2.94}$$

with ω as a window width parameter is known as Gabor transform.

One can give one more and very instructive interpretation to windowed Fourier transform. Equation 2.91 represents a convolution integral with convolution kernel $w(\xi)\exp(i2\pi f\xi)$. Therefore, according to the convolution theorem, Fourier transform of the signal local spectrum $\alpha(x, f)$ at frequency f over variable x

$$A(p, f) = \int\limits_{-\infty}^{\infty} \alpha(x, f)\exp(i2\pi px)\, dx \tag{2.95}$$

is a product

$$A(p, f) = \alpha(p)W_f(p) \tag{2.96}$$

of the signal "global" Fourier spectrum

$$\alpha(p) = \int_{-\infty}^{\infty} a(x)\exp(i2\pi px)\,dx \tag{2.97}$$

and Fourier transform $W_f(p)$:

$$W_f(p) = \int_{-\infty}^{\infty} [w(\xi)\exp(i2\pi f\xi)]\exp(i2\pi p\xi)\,d\xi$$

$$= \int_{-\infty}^{\infty} w(\xi)\exp[i2\pi(p + f)]\,d\xi = W(p + f) \tag{2.98}$$

of the convolution kernel $w(\xi)\exp(i2\pi f\xi)$, that is, the Fourier spectrum of the window function $w(\xi)$ shifted by f. Therefore, signal space–frequency representation $\alpha(x, f)$ as a function of x for frequency parameter f can be regarded as the signal's "sub-band" formed by extracting signal frequency components around frequency f by means of a spectral window function $W(p + f)$ as it is described by Equation 2.96.

This is also true for a general sliding window transformation (Equation 2.89): sliding window signal transformation results in signal sub-bands formed by means of windowing, in the frequency domain, of signal spectrum by a function equal to Fourier transform of the corresponding windowed basis functions:

$$W_f(p) = \int_{-\infty}^{\infty} w(\xi)\phi(f, \xi)\exp(i2\pi p\xi)\,d\xi. \tag{2.99}$$

It is in this sense that we call sliding window signal transformations signal *sub-band decomposition*.

As one can see from the definition of wavelet transform given by Equation 2.43, wavelet transforms are, for each particular scale, signal convolutions as well. Therefore, they can also be treated as signal sub-band decompositions formed, in the frequency domain, by Fourier transform of the wavelet mother function on the corresponding scale:

$$W_s(p) = \int_{-\infty}^{\infty} \phi\left(\frac{\xi}{\sigma}\right)\exp(i2\pi p\xi)\,d\xi. \tag{2.100}$$

This property establishes a link between wavelet and sliding window transforms.

Imaging from Projections and Radon Transform

One of the most important relatively recent findings in image science and technology is the discovery that image can be formed from its projections. This discovery has resulted in various methods of computed tomography and manifested the first signs of computational imaging. Image projection and reconstruction from projections are described by direct and inverse Radon transforms.

Consider an illustrative diagram of object parallel beam projection tomography shown in Figure 2.10. According to this figure, the projection of a 2D signal $a(x, y)$, specified in a Cartesian coordinate system (x, y), onto an axis ξ rotated with respect to the axis x by the angle ϑ is defined as

$$Pr(\vartheta, \xi) = \iint_{X\,Y} a(x, y)\delta(\xi - x\cos\vartheta - y\sin\vartheta)\,dx\,dy. \qquad (2.101)$$

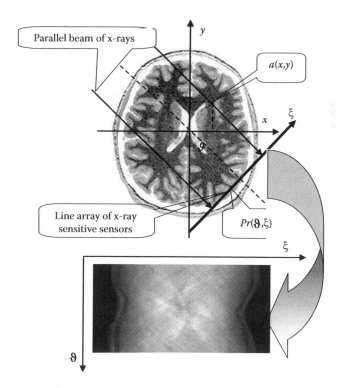

FIGURE 2.10
Parallel beam projection tomography.

This integral transform that links object signal and its projection is called the direct Radon transform. Projection signal $Pr(\vartheta,\xi)$ is called the *sinogram*.

Radon transform and Fourier transform of 2D objects are linked with each other. To show this, compute 1D Fourier transform, over coordinate ξ, of projection $Pr(\vartheta,\xi)$ defined by Equation 2.101:

$$
\begin{aligned}
\alpha_\vartheta(f) &= \int_{-\infty}^{\infty} Pr(\vartheta,\xi)\exp(i2\pi f\xi)\, d\xi \\[2mm]
&= \int_{-\infty}^{\infty} \left\{ \iint_{X\,Y} a(x,y)\delta(\xi - x\cos\vartheta - y\sin\vartheta)\, dx\, dy \right\} \exp(i2\pi f\xi)\, d\xi \\[2mm]
&= \iint_{X\,Y} a(x,y)\exp[i2\pi(fx\cos\vartheta + fy\sin\vartheta)]\, dx\, dy \\[2mm]
&= \alpha(f_x = f\cos\vartheta, f_y = f\sin\vartheta),
\end{aligned}
\tag{2.102}
$$

where $\alpha(f_x, f_y)$ is 2D Fourier spectrum of $a(x, y)$. Equation 2.102 is a mathematical formulation of the *projection theorem* that links Radon and Fourier transforms:

1D Fourier transforms $\alpha_\vartheta(f)$ of object projections $Pr(\vartheta,\xi)$ is a cross section $\alpha(f\cos\vartheta, f\sin\vartheta)$ of 2D object Fourier spectrum $\alpha(f_x, f_y)$ taken at angle ϑ.

From the projection theorem, it follows that inverse Radon transform can be computed as 2D inverse Fourier transform of projections' spectra $\alpha_\vartheta(f)$ in polar coordinate system (f, ϑ):

$$
\begin{aligned}
a(x,y) &= \int_{-\pi}^{\pi}\int_{-\infty}^{\infty} \alpha_\vartheta(f)\exp[-i2\pi(f\cos\vartheta + f\sin\vartheta)]\, d\vartheta\, df \\[2mm]
&= \int_{-\pi}^{\pi}\int_{-\infty}^{\infty} \alpha(f\cos\vartheta, f\sin\vartheta)\exp[-i2\pi(f\cos\vartheta + f\sin\vartheta)]\, d\vartheta\, df.
\end{aligned}
\tag{2.103}
$$

This method of computing inverse Radon transform is called the Fourier reconstruction method. Yet another method of inverting Radon transform, the filtered back projection method, is also derived from the interrelation between Fourier and Radon transforms.

2D function $a(x,y)$ can be found from its 2D Fourier spectrum $\alpha(f_x, f_y)$ as

$$
a(x,y) = \int_{-\infty}^{\infty}\int_{-\infty}^{\infty} \alpha(f_x, f_y)\exp[-i2\pi(f_x x + f_y x)]\, df_x\, df_y.
\tag{2.104}
$$

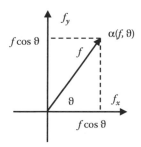

FIGURE 2.11
Cartesian and polar coordinate systems in Fourier transform domain.

In polar coordinate system (f, θ) in the frequency domain (see Figure 2.11)

$$f_x = f\cos\theta; \quad f_y = f\sin\theta; \quad df_x\, df_y = |f|\, df\, d\theta; \quad \theta \in [0,\pi] \qquad (2.105)$$

obtain

$$u(x,y) = \int_0^\pi d\theta \int_{-\infty}^\infty |f|\alpha(f,\theta)\exp[-i2\pi f(x\cos\theta + y\sin\theta)]df$$

$$= \int_0^\pi FBproj(x\cos\theta + y\sin\theta)d\theta, \qquad (2.106)$$

where

$$FBproj(x\cos\theta + y\sin\theta) = \int_{-\infty}^\infty |f|\alpha(f,\theta)\exp[-i2\pi f(x\cos\theta + y\sin\theta)]df. \qquad (2.107)$$

Spectrum $\alpha(f,\vartheta)$ is a cross section of the object's 2D Fourier spectrum along a straight line that goes under angle ϑ, and, according to the projection theorem, it is 1D Fourier spectrum $\alpha_\vartheta(f)$ of object projection $Pr(\vartheta,\xi)$. Therefore, Equation 2.107 describes the reconstruction of object signal $a(x, y)$ by means of accumulating (integration over all angles θ from $\theta = 0$ to $\theta = \pi$) of "back projections" (replication of $FBproj(x\cos\theta + y\sin\theta)$ along straight lines $x\cos\theta + y\sin\theta = \xi$ for all ξ) of its "filtered projections" $FBproj(x\cos\theta + y\sin\theta)$. Filtering of projections defined by Equation 2.107, using as frequency response of the filter ramp-function $|f|$ is called *ramp-filtering*.

Statistical Models of Signals and Transformations

Principles of Statistical Treatment of Signals and Signal Transformations and Basic Definitions

The survey of mathematical models of signal transformations will not be complete without considering statistical approaches to signals and transformations. The need in statistical treatment of a signal arises because any processing is designed to be performed on a certain set of signals rather than on a particular signal and the quality of processing is evaluated on average over this set. Therefore, for the design and optimization of processing, one has to specify (i) a set of signals, or "signal ensemble," (ii) a measure of processing quality of individual signals, and (iii) a method of averaging over the ensemble of the quality measure found for each signal of the ensemble. We will denote signal ensembles as Ω_A and operation of averaging of a numerical measure V over ensemble Ω_A as $AV_{\Omega_A}(V)$. For ensembles of digital signals, the averaging is conventionally performed as weighted summation and for continuous signals it is performed as weighted integration of the measure over the signal ensemble:

$$AV_\Omega(V) = \sum_{\omega \in \Omega} W_\omega V_\omega, AV_\Omega(V) = \int_\Omega W(\omega)V(\omega)\, d\omega, \qquad (2.108)$$

where W_ω and $W(\omega)$ are weight coefficients and, correspondingly, weighting function, ω is an index of the element of ensemble, and V_ω and $V(\omega)$ are numerical measures to be averaged for the cases of digital and continuous signals, correspondingly.

Image ensembles can be specified in a form of databases, such as databases of images of human faces, databases of microscopic images of cells, databases of satellite images of different kinds of terrains and alike, or in the form of mathematical models.

Mathematical models of signal ensembles are usually formulated as probabilistic models. Probabilistic models do not specify individual signals of an ensemble. Rather, they specify only certain signal statistical characteristics, that is, averages over the ensemble of certain signal attributes. In this averaging, it is most natural to associate the contribution of individual signals to the ensemble average, which are embodied in weight coefficients and weighting function, with the rate of occurrence in realizations of individual elements of the ensemble. The mathematical models usually assume that the number of realization of signals is infinitely large, and this rate is called *probability*, or, in continuous case, *probability density*.

The most fundamental notion of probabilistic models is that of a random variable. Quantized (discrete) random variable V is statistically specified by probability distribution $\{P_q\}$ of its values $\{V_q\}$, where q is an integer index, $q = 0,1,...,Q-1$ and Q is the total number of possible quantized values. The probability distribution $\{P_q\}$ is a limit, when the number of realizations of the variable goes to infinity, of the rates of individual values $\{V_q\}$ in a finite set of realizations. The set of these rates is called the *distribution histogram* of the variable and the rates for particular values of $\{V_q\}$ are called *histogram bins*. The number of occurrences of individual values $\{V_q\}$ in a finite set of realizations is called the *cardinality* of the set of these values. The sum of histogram bins from minimal value $V_0 = \min\{V_q\}$ of the variable to a particular value V_q is called the *cumulative histogram*.

Continuous random variables v are specified by the probability $P(V) = \text{Probability}(v < V)$ that the variable has a value lower than V. This probability as a function of V is called the *cumulative distribution function*. It is defined as a limit of the cumulative histogram, when the number of realizations and the number of levels of random variable both go to infinity. Its derivative $p(v) = (dP(V)/dV)|_{V=v}$ is called the probability density. The probability density can be regarded as a limit of histogram bins of random variable, when the number of its quantized levels reaches infinity.

Random variable mean value

$$\bar{V} = \sum_{q=0}^{Q-1} P_q V_q; \quad \bar{V} = \int_{-\infty}^{\infty} p(v)v \, dv, \tag{2.109}$$

variance

$$\overline{V^2} = \sum_{q=0}^{Q-1} P_q (V_q - \bar{V})^2; \quad \overline{V^2} = \int_{-\infty}^{\infty} p(v)(v - \bar{V})^2 \, dv, \tag{2.110}$$

and, generally, n-th centered moments

$$\overline{V^n} = \sum_{q=0}^{Q-1} P_q (V_q - \bar{V})^n; \quad \overline{V^n} = \int_{-\infty}^{\infty} p(v)(v - \bar{V})^n \, dv \tag{2.111}$$

for, correspondingly, quantized and continuous random variables are typical examples of ensemble averaging of signal's attributes, in this case its value, and represent statistical characteristics of random variables complementary to probability distribution and probability density. The square root of variance $\sigma_V = \sqrt{\overline{V^2}}$ is called the *standard deviation* of the random variable.

Statistical characteristics of digital signals alternative to histograms and their moments are *order statistics* and their derivatives. Signal order statistics are defined through a notion of the *variational row*. Variational row $VR_{\{a_k\}}$ of a digital signal $\{a_k\}$, $k = 0,1,\ldots,N-1$ is a sequence of signal samples ordered according to their values:

$$VR_{\{a_k\}} = \left\{ a^{(1)}, a^{(2)}, \ldots, a^{(N)} \right\} : a^{(1)} \le a^{(2)} \le \cdots \le a^{(N)}. \tag{2.112}$$

If signal samples happen to have the same quantized value, they are placed in the variational row one after another in an arbitrary order. Note that coordinate indices $\{k\}$ of signal samples are ignored in the variational row.

Element $a^{(R)}$ that occupies R-th place in the variational row is called signal R-th-*order statistics*. Index R ($R = 1,2,\ldots,N$) of the order statistics is called its *rank*. The rank shows how many signal samples have value lower or equal to that of the element. The very first-order statistics is signal minimum:

$$a^{(1)} = \text{MIN}\{a_k\}. \tag{2.113}$$

The very last-order statistics in the variational row is signal maximum:

$$a^{(N)} = \text{MAX}\{a_k\}. \tag{2.114}$$

If the number of signal samples N is odd, $(N+1)/2$-th signal order statistics that is located exactly in the middle of the variational row is called the signal *median*:

$$a^{((N+1)/2)} = \text{MEDN}\{a_k\}. \tag{2.115}$$

If N is an even number, the median can be defined as

$$(a^{(N/2-1)} + a^{(N/2-1)})/2 = \text{MEDN}\{a_k\}. \tag{2.116}$$

The signal median is an alternative to the signal mean value. As such, it is more robust, in a statistical sense, to possible presence, in the sequence of signal samples of "outliers" that belong to a statistical ensemble different from that of the majority of the signal samples. They may appear due to, for instance, spontaneous faults of the signal sensor. Signal median in such cases will be close to the mean of the signal without outlier samples. Yet another alternative to signal mean robust to the presence of outliers, which take extreme values, is the so-called signal *alpha-trimmed mean*:

$$\bar{a}_\alpha = \frac{1}{N - \alpha} \sum_{r=\alpha}^{N-1-\alpha} a^{(r)}, \tag{2.117}$$

the arithmetic mean of the elements of the variational row, which are minimum α elements apart from its ends.

Signal second moment, or its variance, is a parameter that characterizes the spread of the signal values. An alternative to it and robust to outliers characteristic of signal values spread is signal *quasispread* defined as a difference between order statistics, which are certain number r of samples to the right and to the left from the median:

$$\mathrm{QSPR}_r(a) = a^{\left(\frac{N+1}{2}+r\right)} - a^{\left(\frac{N+1}{2}-r\right)}. \tag{2.118}$$

In image processing, order statistics and ranks can be measured both globally over the entire given image or set of images and over a sliding window. In the latter case, they are called local. Local histograms, ranks, and order statistics are the base of nonlinear filters discussed in the section "Correcting Image Gray-Scale Nonlinear Distortions" in Chapter 8.

If a random variable represents values of a function, this function is called a *random process*. In order to specify interrelation of values of random process at different points, the notions of random processes autocorrelation and mutual correlation functions are introduced. *Autocorrelation function* $CF_v(x_1, x_2)$ of a continuous random process $v(x)$ is defined as an ensemble average of the product of the process values $(v_1; v_2)$ at points x_1 and x_2:

$$CF_v(x_1, x_2) = AV_{\Omega_V}[v(x_1)v(x_2)]. \tag{2.119}$$

Correspondingly, correlation functions of discrete and digital random processes $\{V_k\}$ are defined as

$$CF_V(k,l) = AV_{\Omega_V}[V_k V_l]. \tag{2.120}$$

The correlation function as a statistical characteristic of random processes is a generalization of the variance as a statistical characteristic of random variables. Its value $CF_v(x,x)$ ($CF_V(k,k)$ for discrete processes) is equal to the variance of the process at point x (at k-th sample of the discrete process):

$$CF_v(\mathbf{x},\mathbf{x}) = AV_{\Omega_V}[v^2(\mathbf{x})] = \overline{V^2}(\mathbf{x}); \quad CF_V(k,k) = AV_{\Omega_V}\left[V_k^2\right] = \overline{V_k^2}. \tag{2.121}$$

An important special case of random processes is *homogeneous random processes*, or, in the communication engineering jargon, *stationary random processes*. The correlation function of homogeneous random processes $CF_v(x_1, x_2)$ depends only on the difference of coordinates:

$$CF_v(x_1, x_2) = CF_v(x_1 - x_2) \tag{2.122}$$

and its value $CF_v(0)$ is equal to the process variance:

$$CF_v(0) = \overline{V^2}, \tag{2.123}$$

which in this case does not depend on coordinates due to the homogeneity (stationarity) of the process.

A measure of interrelation of two random processes is the *mutual*, or *cross-correlation, function*:

$$CF_{v_1v_2}(x_1, x_2) = AV_{\Omega_V}[v_1(\mathbf{x}_1)v_2(\mathbf{x}_2)]. \tag{2.124}$$

As we will see in Chapter 7, the cross-correlation function plays an important role in optimal target location.

In the same way as the autocorrelation function of a random process was introduced as an ensemble average of the product of random process realization values in two points, one can consider the ensemble average $AV_{\Omega_v}[v(f_1)v^*(f_2)]$ of the product of values of Fourier spectra $v(f)$ of realizations $v(x)$ of random process taken in two frequencies f_1 and f_2. This "correlation function" of the spectra of realizations of the random process turns out to be the Fourier transform of the process correlation function:

$$AV_{\Omega_v}[v(f_1)v^*(f_2)] = AV_{\Omega_v}\left[\int_{-\infty}^{\infty} v(x_1)\exp(i2\pi f_1 x_1)\,dx_1 \int_{-\infty}^{\infty} v(x_2)\exp(-i2\pi f_2 x_2)\,dx_2\right]$$

$$= \left[\int_{-\infty}^{\infty}\int_{-\infty}^{\infty} AV_{\Omega_v}[v(x_1)v(x_2)]\exp i2\pi(f_1 x_1 - f_2 x_2)\,dx_1\,dx_2\right]$$

$$= \int_{-\infty}^{\infty}\int_{-\infty}^{\infty} CF_v(x_1, x_2)\exp i2\pi(f_1 x_1 - f_2 x_2)\,dx_1\,dx_2. \tag{2.125}$$

If the random process is stationary

$$AV_{\Omega_v}[v(f_1)v^*(f_2)] = \int_{-\infty}^{\infty}\int_{-\infty}^{\infty} CF_v(x_1 - x_2)\exp i2\pi(f_1 x_1 - f_2 x_2)\,dx_1\,dx_2$$

$$= \int_{-\infty}^{\infty}\int_{-\infty}^{\infty} CF_v(\xi)\exp i2\pi(f_1\xi)\exp[i2\pi(f_1 - f_2)x_2]\,d\xi\,dx_2$$

$$= \int_{-\infty}^{\infty} CF_v(\xi)\exp i2\pi(f_1\xi)\,d\xi \int_{-\infty}^{\infty} \exp[i2\pi(f_1 - f_2)x_2]\,dx_2. \tag{2.126}$$

The second integral in Equation 2.125 is, according to Equation 2.72, a Dirac delta-function. Therefore, for stationary random processes

$$AV_{\Omega_v}[v(f_1)v^*(f_2)] = \left[\int_{-\infty}^{\infty} CF_v(\xi)\exp i2\pi(f_1\xi)\,\mathrm{d}\xi \right]\delta(f_1 - f_2) = SD(f_1)\delta(f_1 - f_2).$$

(2.127)

The Fourier transform $SD(f_1)$ of the correlation function of a stationary random process is called its *spectral density* (SD) or *power spectrum*.

By the definition given by Equation 2.127, the spectral density of a stationary random process is also an average, over all realizations of the process $\{v(x)\}$, of squared module of Fourier spectra of the realizations

$$SD(f) = AV_{\Omega_v}\left\{ \left| \int_{-\infty}^{\infty} v(x)\exp(i2\pi fx)\,\mathrm{d}x \right|^2 \right\}.$$

(2.128)

The random process is called uncorrelated if its correlation function is a delta-function, or, respectively, its spectral density is a constant:

$$CF_n(x) = \sigma_n^2\delta(x); \quad SD(f) = \sigma_v^2.$$

(2.129)

By the definition (Equation 2.128), the spectral densities of real homogeneous random processes are nonnegative even functions:

$$SD(f) = AV_{\Omega_v}[v(f)v^*(f)] = AV_{\Omega_v}[v^*(-f)v(-f)] = SD(-f) \geq 0.$$

(2.130)

Therefore, autocorrelation functions, being Fourier transforms of even spectral density functions, are also even functions:

$$CF_v(x) = CF_v(x).$$

(2.131)

Models of Signal Random Interferences

A special case of signal ensembles are those generated by signal random interferences due to factors such as signal sensor noise. The most frequently used models for describing signal interferences are those of additive, multiplicative, Poisson's noise, impulse, and speckle noise models.

Additive Signal-Independent Noise Model

The additive signal-independent noise model (ASIN-model) implies signal distortions described as

$$b(x) = a(x) + n(x),$$

(2.132)

where $a(x)$ is a noiseless signal, $n(x)$ is a random process statistically independent on signal $a(x)$ and referred to as additive noise, and $b(x)$ is a signal contaminated with noise. Given signal $a(x)$, additive noise generates signal ensemble Ω_N that consists of copies of this signal distorted by different realizations of noise.

The assumption of statistical independence implies that statistical averaging over the noise ensemble applied to signal $b(x)$ or to results of its transformations acts only on the noise component $\{n(x)\}$. For instance, for the considered ASIN-model:

$$AV_{\Omega_N}\{b(x)\} = a(x) + AV_{\Omega_N}\{n(x)\}. \qquad (2.133)$$

In this model, mean value $\bar{n} = AV_{\Omega_N}\{n(x)\}$ of noise is usually of no concern and it is natural to assume that $\bar{n} = 0$. With this assumption, the mean value of the signal corrupted with additive zero mean noise is equal to the uncorrupted signal and standard deviation of its fluctuations is equal to the standard deviation of noise:

$$AV_{\Omega_N}\{b(x)\} = a(x); \qquad (2.134)$$

$$\sqrt{AV_{\Omega_N}\left\{\left|b(x) - AV_{\Omega_N}\{b(x)\}\right|^2\right\}} = \sqrt{AV_{\Omega_N}\left\{\left|b(x) - a(x)\right|^2\right\}} = \sqrt{AV_{\Omega_N}\left\{\left|n(x)\right|^2\right\}}.$$
$$(2.135)$$

A typical statistical model of additive noise $n(x)$ is homogeneous zero mean Gaussian random process. It is completely statistically determined by its probability density called normal distribution

$$p(n) = \frac{1}{\sqrt{2\pi}\sigma_n}\exp\left(-\frac{n^2}{2\sigma_n^2}\right), \qquad (2.136)$$

where σ_n is its standard deviation, and by its correlation function $CF_n(\mathbf{x})$.

One of the basic results of the probability theory is the *central limit theorem*, which states that distribution density of sum of statistically independent random values tends to the normal distribution with the growth of the number of those random values. This implies that the normal distribution is a good model of the distribution of many types of noise in image sensors, and, in particular, of *thermal noise* caused by random fluctuations of velocities of electrons in electronic signal sensors.

Uncorrelated random process, whose correlation function is a delta-function and spectral density is a constant, is conventionally called *white noise* by analogy with "white light" that contains all colors with uniform weights.

In a number of applications, highly correlated additive noise should be considered. These correlations manifest themselves as concentrated peaks in noise power spectrum in certain areas of the frequency domain. If noise power spectrum is concentrated only around a few isolated points in the frequency domain, this type of additive noise is called periodical noise, or *moiré noise*.

Multiplicative Noise Model

Multiplicative noise model (MN-model) implies that signal distortions appear due to signal multiplication by a random noise process:

$$b(x) = n(x) \cdot a(x), \tag{2.137}$$

where $n(x)$ is a signal-independent random process with mean $AV_{\Omega_N}\{n(x)\}$ equal to 1. Similar to mean value of signal corrupted by additive zero mean noise, the mean value of a signal corrupted by multiplicative noise is also equal to the noncorrupted signal:

$$AV_{\Omega_N}\{b(x)\} = AV_{\Omega_N}\{n(x)\}a(x) = a(x). \tag{2.138}$$

Standard deviation of the corrupted signal is proportional to the standard deviation of noise times the signal magnitude:

$$\sqrt{AV_{\Omega_N}\{b^2(x)\}} = \sqrt{AV_{\Omega_N}\{n^2(x)\}}\sqrt{a^2(x)} = \sqrt{AV_{\Omega_N}\{n^2(x)\}}\,|a(x)|. \tag{2.139}$$

Therefore, the ratio of corrupted signal standard deviation to its value is equal to the standard deviation of the multiplicative noise and does not depend on the signal. This is a characteristic property of the multiplicative noise.

Poisson Model

The Poisson model describes quantum fluctuations in interaction of radiation with its sensors, such as photo and x-ray-sensitive sensors. For Poisson's model, random variable values n are nonnegative integer numbers that have *Poisson probability distribution*:

$$P(n) = \frac{\exp(-\bar{n})}{n!}\bar{n}^n, \tag{2.140}$$

where $\bar{n}$ is the mean value of n. It is the property of Poisson's probability distribution that standard deviation of n is equal to its mean $\bar{n}$:

$$\sqrt{AV_{\Omega_N}\{(n - \bar{n})^2\}} = \bar{n}. \tag{2.141}$$

When $\bar{n} \to \infty$, the Poisson distribution, according to the Moivre–Laplace theorem, tends to normal distribution with standard deviation $\sqrt{\bar{n}}$:

$$P(n) = \frac{1}{\sqrt{2\pi\bar{n}}} \exp\left[-\frac{(n - \bar{n})^2}{2\bar{n}}\right].$$

(2.142)

An important peculiarity of the Poisson distributions is that the ratio of its standard deviation to mean value tends to zero with growth of the mean value:

$$\lim_{\bar{n}\to\infty}\left(\frac{\sqrt{AV_{\Omega_N}\left|n - AV_{\Omega_N}(n)\right|^2}}{AV_{\Omega_N}}\right) = \lim_{\bar{n}\to\infty}\left(\frac{\sqrt{\bar{n}}}{\bar{n}}\right) = 0,$$

(2.143)

which means that with growth of the mean value $\bar{n}$ of random variable with Poisson's distribution, its fluctuations from its mean value are becoming relatively less noticeable.

Impulse Noise Model

Impulse noise model (ImpN-model) is defined by the following equation:

$$b(x) = [1 - e(x)] \cdot a(x) + e(x)n(x),$$

(2.144)

which implies that random distortions of signal $a(x)$ are caused by two independent random factors: by a binary random process $e(x)$ that assumes values 0 and 1, and by a random variable $n(x)$ that assumes values from the same value range as that of the noiseless signal $a(x)$. ImpN-model is characteristic for signal transmission and storage systems with random faults. The process $e(x)$ describes signal transmission and memory faults and is specified by the probability that $e(x) = 1$ (the *probability of error*, or the probability of signal loss). Statistical properties of noise $n(x)$ can be arbitrary; most frequently, it is uniformly distributed in the signal dynamic range. If random process $n(x)$ assumes only two values equal to the minimum and to the maximum of $a(x)$, it produces impulse noise that in image processing jargon is called the *pepper-and-salt noise* because it is seen as contrast black and white points.

Speckle Noise Model

The phenomenon of *speckle noise* is characteristic for imaging systems such as holographic ones, synthetic aperture radar, and ultrasound imaging systems

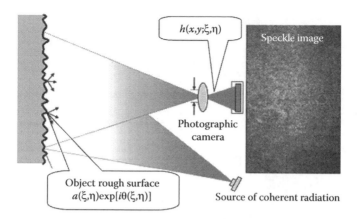

FIGURE 2.12
Speckle formation model.

that form images using coherent radiation. Speckle noise originates from the property of objects to diffusely scatter radiation and is caused by distortions introduced by radiation wavefront sensors. Most frequently, people associate it with the finite *resolving power* of sensors, or, in holographic systems, with limitation of the area over which the wavefront is measured by sensors or recorded on holograms, though other types of distortions must be considered as well.

Consider the diagram in Figure 2.12 that illustrates the formation of speckle noise caused by the limitation of the sensor's resolving power.

Let an object with a rough surface be illuminated by a coherent radiation and $a(\xi, \eta)\exp[i\theta(\xi, \eta)]$ be the complex amplitude of the wavefront on the object surface at the point (ξ, η) in the object surface plane. The magnitude $a(\xi, \eta)$ of the wavefront defines the reflected or transmitted radiation intensity; its phase $\theta(\xi, \eta)$ describes how object surface changes the phase of the illuminating coherent radiation wavefront. For describing the property of objects to diffusely scatter radiation, $\theta(\xi, \eta)$ is considered as a random process. If $\theta(\xi, \eta)$ is a highly spatially correlated process, object surface is specular: it scatters light predominantly in one direction. If process $\theta(\xi, \eta)$ is uncorrelated, radiation is scattered uniformly in all directions. In what follows, we will consider this uniform scattering case.

Assume that the PSF of the coherent imaging system is $h(x, y; \xi, \eta)$. Then, for input object $a(\xi, \eta)\exp[i\theta(\xi, \eta)]$, output image $B(x, y)$ formed by a sensor that is sensitive to squared magnitude of the output wavefront is

$$B(x, y) = |b(x, y)|^2 = \left| \int_{-\infty}^{\infty} \int_{-\infty}^{\infty} a(\xi, \eta)\exp[i\theta(\xi, \eta)]h(x, y; \xi, \eta)\, d\xi\, d\eta \right|^2 . \quad (2.145)$$

For an object with uniformly painted surface such that $a(\xi, \eta) = A_0$

$$B(x,y) = |b(x,y)|^2 = \left| A_0 \int\limits_{-\infty}^{\infty} \int\limits_{-\infty}^{\infty} \exp[i\theta(\xi,\eta)]h(x,y;\xi,\eta)\, d\xi\, d\eta \right|^2 = |b^{re}|^2 + |b^{im}|^2,$$

(2.146)

where

$$b^{re} = A_0 \int\limits_{-\infty}^{\infty} \int\limits_{-\infty}^{\infty} \cos\theta(\xi,\eta)h(x,y;\xi,\eta)\, d\xi\, d\eta;$$

(2.147)

$$b^{im} = A_0 \int\limits_{-\infty}^{\infty} \int\limits_{-\infty}^{\infty} \sin\theta(\xi,\eta)h(x,y;\xi,\eta)\, d\xi\, d\eta$$

are real and imaginary components of the complex-valued process $b(x, y)$. The assumption that $\theta(\xi, \eta)$ is an uncorrelated random process implies that many uncorrelated values of $\cos\theta(\xi, \eta)$ and $\sin\theta(\xi, \eta)$ are observed within the aperture $h(x, y; \xi, \eta)$ of the imaging system. Therefore, the central limit theorem of the probability theory can be used to assert that b^{re} and b^{im} are random processes with normal distribution and zero mean values:

$$\overline{b^{re}} = AV_\theta(b^{re}) = A_0 \int\limits_{-\infty}^{\infty} \int\limits_{-\infty}^{\infty} AV_\theta\{\cos\theta(\xi,\eta)\}h(x,y;\xi,\eta)\, d\xi\, d\eta = 0;$$

(2.148)

$$\overline{b^{im}} = AV_\theta(b^{im}) = A_0 \int\limits_{-\infty}^{\infty} \int\limits_{-\infty}^{\infty} AV_\theta\{\sin\theta(\xi,\eta)\}h(x,y;\xi,\eta)\, d\xi\, d\eta = 0.$$

Let us find the correlation function of b^{re} and b^{im}. For b^{re}

$$CF_{b^{re}}(\mathbf{x}_1,\mathbf{x}_2) = AV_{\Omega_\theta}\{b^{re}(\mathbf{x}_1)[b^{re}(\mathbf{x}_2)]^*\}$$

$$= A_0^2 \int\limits_{-\infty}^{\infty} \int\limits_{-\infty}^{\infty} AV_{\Omega_\theta}\{\cos\theta(\xi_1)\cos\theta(\xi_2)\}h(\mathbf{x}_1;\xi_1)h^*(\mathbf{x}_2;\xi_2)\, d\xi_1\, d\xi_2$$

$$= A_0^2 \int\limits_{-\infty}^{\infty} \int\limits_{-\infty}^{\infty} CF_\theta(\xi_1,\xi_2)h(\mathbf{x}_1;\xi_1)h^*(\mathbf{x}_2;\xi_2)\, d\xi_1\, d\xi_2,$$

(2.149)

where $\mathbf{x} = (x, y)$, $\xi = (\xi, \eta)$, and $CF_\theta(\xi_1, \xi_2)$ is the correlation function of $\cos\theta(\xi, \eta)$. In the above-accepted assumption of applicability of the central limit theorem, $CF_\theta(\xi_1, \xi_2)$ is, with respect to integration with $h(x, y; \xi, \eta)$, a delta-function:

$$CF_\theta(\xi_1,\xi_2) = AV_{\Omega_\theta}(\cos^2\theta)\delta(\xi_1 - \xi_2) = \frac{1}{2}\delta(\xi_1 - \xi_2).$$

(2.150)

Therefore

$$CF_{b^{re}}(\mathbf{x}_1,\mathbf{x}_2) = \frac{1}{2} A_0^2 \int\limits_{-\infty}^{\infty} h(\mathbf{x}_1;\xi) h^*(\mathbf{x}_2;\xi)\, d\xi.$$

(2.151)

In the same way, one can obtain that

$$CF_{b^{im}}(\mathbf{x}_1,\mathbf{x}_2) = CF_{b^{re}}(\mathbf{x}_1,\mathbf{x}_2) = \frac{1}{2} A_0^2 \int\limits_{-\infty}^{\infty} h(\mathbf{x}_1;\xi) h^*(\mathbf{x}_2;\xi)\, d\xi.$$

(2.152)

From Equations 2.151 and 2.152, it follows that variances of orthogonal components b^{re} and b^{im} are

$$\sigma_b^2 = \overline{(b^{re})^2} = \overline{(b^{im})^2} = \frac{1}{2} A_0^2 \int\limits_{-\infty}^{\infty}\int\limits_{-\infty}^{\infty} |\, h(x,y;\xi,\eta)\,|^2\, d\xi\, d\eta.$$

(2.153)

In general, they are functions of coordinates (x, y) in image plane. For space invariant imaging system, for which $h(x,y;\xi,\eta) = h(x - \xi, y - \eta)$, variances are coordinate independent:

$$\sigma_b^2 = \frac{1}{2} A_0^2 \int\limits_{-\infty}^{\infty}\int\limits_{-\infty}^{\infty} |h(x,y)|^2\, dx\, dy$$

(2.154)

Having defined the probability density function of orthogonal components $b^{re}(x, y)$ and $b^{im}(x, y)$ of $b^2(x, y)$, one can now find the probability density of $b^2(x, y)$ as the probability density of sum of squared independent variables b^{re} and b^{im} with normal distribution. To this end, introduce random variables R and ϑ such that

$$b^{re} = R\cos\vartheta; \quad b^{im} = R\sin\vartheta; \quad B(x,y) = |\, b(x,y)\,|^2 = (b^{re})^2 + (b^{im})^2 = R^2. \quad (2.155)$$

Then, the joint probability that b^{re} and b^{im} take values within a rectangle ($db^{re} \times db^{im}$) is

$$p(b^{re}, b^{im}) = \frac{1}{2\pi\sigma_b^2} \exp\left[-\frac{(b^{re})^2 + (b^{im})^2}{2\sigma_b^2}\right] db^{re}\, db^{im}$$

$$= \frac{d\theta}{2\pi} \frac{R}{\sigma_b^2} \exp\left(-\frac{R^2}{2\sigma_b^2}\right) dR = \frac{d\theta}{2\pi} \frac{1}{2\sigma_b^2} \exp\left(-\frac{R^2}{2\sigma_b^2}\right) dR^2. \quad (2.156)$$

This is the joint probability that variables $\{R^2, \vartheta\}$ take values in the range $[R^2 \div R^2 + dR^2, \vartheta \div \vartheta + d\vartheta]$. It follows, therefore, that $\{R^2 = B(x, y)\}$ and ϑ are statistically independent, ϑ is uniformly distributed in the range $\{0, 2\pi\}$ and $B(x, y)$ has an exponential distribution density:

$$P(b) = \frac{1}{2\sigma_b^2} \exp\left(-\frac{B}{2\sigma_b^2}\right), \tag{2.157}$$

where σ_b^2 is defined by Equation 2.154.

From Equations 2.140 and 2.141, one can immediately see that the mean value of $B(x, y)$ is

$$\overline{B(x, y)} = \overline{|b^{re}|^2} + \overline{|b^{im}|^2} = 2\sigma_b^2 = A_0^2 \int_{-\infty}^{\infty} \int_{-\infty}^{\infty} |h(x, y)|^2 \, dx \, dy. \tag{2.158}$$

It is the property of the exponential distribution that its standard deviation is equal to its mean value

$$\sigma_B = \left(\overline{B^2}\right)^{1/2} = A_0^2 \int_{-\infty}^{\infty} \int_{-\infty}^{\infty} |h(x, y)|^2 \, dx \, dy. \tag{2.159}$$

Fluctuations of $b^2(x, y)$ around its mean value are called speckle noise. It is customary to characterize the intensity of speckle noise by the ratio of its standard deviation to mean value called "speckle contrast":

$$SC = \frac{\sigma_B}{B(x, y)}. \tag{2.160}$$

It follows from Equations 2.158 and 2.159 that, for objects that scatter radiation almost uniformly in the space, speckle contrast of their image obtained in coherent imaging systems of a finite resolving power is unity:

$$SC = 1. \tag{2.161}$$

Therefore, the speckle noise originated from the limitation of the imaging system resolving power is multiplicative with respect to the image signal.

Quantifying Signal-Processing Quality

Consider now the basic notions associated with the evaluation of signal-processing quality in numerical terms. The measures of processing quality of individual signals are formulated in the form of *loss functions* (LF) that measure how much an obtained result departs from the desired one. The most frequently used loss functions are, for continuous and discrete signals, respectively

$$LF_p[a(x);\hat{a}(x)] = |a(x) - \hat{a}(x)|^p \quad \text{and} \quad LF_p(a_k;\hat{a}_k) = |a_k - \hat{a}_k|^p, \quad (2.162)$$

where p is a numerical parameter equal to 0, 1, or 2, $a(x)$ is the desired signal, and $\hat{a}(x)$ is its estimate, of which quality is evaluated.

Loss functions defined in this way are functions of signal coordinates (index, for discrete signals). In order to characterize the deviations of one signal as a whole from another signal by one number, an average value of loss functions over the signal extent is used. We will denote this value as $LOSS_p(\mathbf{a};\hat{\mathbf{a}})$. For continuous signals given on interval X and for discrete signals of N samples, they are defined as, correspondingly,

$$LOSS_p(\mathbf{a};\hat{\mathbf{a}}) = \int_X LF_p[a(x);\hat{a}(x)]\,dx = \int_X |a(x) - \hat{a}(x)|^p\,dx;$$

$$LOSS_p(\mathbf{a};\hat{\mathbf{a}}) = \sum_{k=0}^{N-1} LF_p[a_k;\hat{a}_k] = \sum_{k=0}^{N-1} |a_k - \hat{a}_k|^p. \tag{2.163}$$

In mathematical jargon, $LOSS_p(\mathbf{a};\hat{\mathbf{a}})$ is called the "Lp-norm" (*L0-norm* for $p = 0$, *L1-norm* for $p = 1$, and *L2-norm* for $p = 2$). L2 norm is frequently called also the mean square error (MSE), assuming that signal difference $a_k - \hat{a}_k$ is treated as an "error" in representing signal a through $\hat{a}$. Because $|a_k - \hat{a}_k|^0 = \delta(a_k - \hat{a}_k)$, L0-norm computes the rate of nonzero errors, that is, probability of errors.

The processing quality for signal ensembles is evaluated as ensemble average of loss values obtained for individual particular realizations $(a_\omega, \hat{a}_\omega)$:

$$\overline{LOSS_\Omega} = AV_\Omega[LOSS(\mathbf{a};\hat{\mathbf{a}})] = \sum_{\omega \in \Omega} W_\omega LOSS(\mathbf{a}_\omega, \hat{\mathbf{a}}_\omega)$$

$$\overline{LOSS_\Omega} = AV_\Omega[LOSS(\mathbf{a};\hat{\mathbf{a}})] = \int_\Omega W(\omega)LOSS(\mathbf{a}_\omega, \hat{\mathbf{a}}_\omega)\,d\omega \tag{2.164}$$

for discrete and continuous signals, correspondingly.

Basics of Optimal Statistical Parameter Estimation

The measurement of physical parameters of objects is one of the most fundamental tasks in image processing. It is required in many applications.

Typical examples are measuring the number of various objects in the field of view, their orientations, dimensions and coordinates, outlining objects by parameterized curves (so-called "active contours" or "snakes"), and measuring maps of pixel displacement ("motion vectors") in sequences of video frames, to name a few. One of the important special cases is object recognition, when it is required to determine which object from a list of possible objects is observed.

As the measurement devices and algorithms for measuring parameters are designed for the use for multitude of images and imaged objects, the most appropriate approach to their design is, as we already mentioned, a statistical approach that assumes that objects and their parameters to be estimated are, in any particular observation, selected from corresponding ensembles of images and objects and performance of parameter estimation algorithms is evaluated on average over these ensembles.

Let $\mathbf{a}(\rho)$ be an object signal that depends on a certain parameter ρ, scalar or vectorial, which uniquely specifies the signal, and $P_{\Re}(\rho)$ be a parameter probability distribution over an ensemble $\rho \in \Re$ of parameter values (probability of values if they are discrete or probability density if the parameter values are continuous variables). Knowledge of signal $\mathbf{a}(\rho)$ allows one to uniquely determine the parameter ρ. This signal is sensed by a signal sensor, which, in response to it, generates an observed signal that can be used for estimating the parameter. Let it be signal $\mathbf{b}_\omega(\rho)$. Realizations $\{\mathbf{b}_\omega(\rho)\}$ belong to a statistical ensemble $\omega \in \Omega$ generated by the ensemble of parameter values $\Re$ and by factors that specify the relationship between signals $\mathbf{a}(\rho)$ and $\{\mathbf{b}_\omega(\rho)\}$, including factors such as a sensor's noise. The task of parameter estimation is to determine, in the best possible way, the parameter ρ from the observed signal $\mathbf{b}_\omega(\rho)$.

Within the statistical approach, the best estimation operation is the one that minimizes losses, measured as average, over ensembles of observed signals Ω and of parameter values $\Re$, of a certain loss function $LF(\hat{\rho}_\omega, \rho)$, which characterizes losses due to deviations of the parameter estimates from their true values:

$$\hat{\rho}_{opt} = \arg \min_{\mathbf{b}_\omega(\rho) \Rightarrow \hat{\rho}} \{AV_\Omega AV_\Re [LF(\hat{\rho}_\omega, \rho)]\}. \tag{2.165}$$

Let us initially assume that parameter ρ takes discrete values from a set of values $\{\rho_q\}$, numbered by an integer index q. A natural measure of the estimation performance is in this case the probability P_{err} that estimate $\hat{\rho}_\omega$ is not equal to the parameter true value ρ_q (the probability of estimation error). Loss function $LF(\hat{\rho}_\omega, \rho_q)$ is in this case a Kronecker delta: $LF(\hat{\rho}_\omega, \rho) = \delta(\hat{\rho}_\omega - \rho_q)$.

For any given realization $\mathbf{b}_\omega$ of the observed signal, parameter value ρ_q might be expected with a certain probability $P(\rho_q / \mathbf{b}_\omega)$ called its *a posteriori probability*. It is determined by the *a priori probability* $P(\rho_q)$ of the parameter value ρ_q and the conditional probability $P(\mathbf{b}_\omega/\rho_q)$ that object signal $\mathbf{a}(\rho_q)$ generates sensor signal $\mathbf{b}_\omega$:

$$P(\rho_q/\mathbf{b}_\omega) = \frac{P(\mathbf{b}_\omega/\rho_q)P(\rho_q)}{\sum_q P(\mathbf{b}_\omega/\rho_q)}. \tag{2.166}$$

The relationship (Equation 2.166) is called the *Bayes rule*.

The value $1 - P(\rho_q/\mathbf{b}_\omega)$ that complements the *a posteriori* probability to one is the probability that parameter takes values other than ρ_q. Therefore, if the parameter estimation device upon observing signal $\mathbf{b}_\omega$ makes a decision that $\hat{\rho} = \rho_q$, the value $1 - P(\rho_q/\mathbf{b}_\omega)$ is the conditional probability of error:

$$P(Err/\mathbf{b}_\omega) = 1 - P[\rho_q/\mathbf{b}_\omega]. \tag{2.167}$$

On average over the ensemble Ω of realizations of $\{\mathbf{b}_\omega\}$, the probability of error will be

$$AV_\Omega P(Err/\mathbf{b}_\omega) = AV_\Omega\{1 - P(\rho_q/\mathbf{b}_\omega)\} = 1 - AV_\Omega P(\rho_q/\mathbf{b}_\omega), \tag{2.168}$$

where $AV_\Omega(\cdot)$ symbolizes the averaging over the ensemble of observed signals. Because all values involved in the averaging are nonnegative, the probability of parameter estimation error is minimal when, for every particular $\mathbf{b}_\omega$, a parameter estimate $\hat{\rho}$ is selected that has maximal *a posteriori* probability:

$$\hat{\rho}_{MAP} = \arg\max_{\{\rho_q\}} P(\rho_q/\mathbf{b}_\omega) = \arg\max_{\{\rho_q\}} \frac{P(\mathbf{b}_\omega/\rho_q)P(\rho_q)}{\sum_q P(\mathbf{b}_\omega/\rho_q)} \tag{2.169}$$

or, as the denominator does not depend on $\{\rho_q\}$,

$$\hat{\rho}_{MAP} = \arg\max_{\{\rho_q\}} P(\mathbf{b}_\omega/\rho_q)P(\rho_q). \tag{2.170}$$

Such an estimate is called maximum *a posteriori* probability estimate (*MAP-estimate*) and the described approach to optimization of parameter estimation is called the *Bayesian approach*.

MAP-estimates require knowledge of the conditional probability $P\{\mathbf{b}_\omega/\rho_q\}$, which is determined by the relationship between sensor signals $\{\mathbf{b}_\omega\}$ and object signals $\{\mathbf{a}(\rho_q)\}$ and of *a priori* probabilities $\{P(\rho_q)\}$ of parameter values. Estimates

$$\hat{\rho}_{ML} = \arg \max_{\{\rho_q\}} P\{\mathbf{b}_\omega / \rho_q\} \qquad (2.171)$$

that ignore *a priori* probabilities (or, which is equivalent, assume that *a priori* probability distribution is uniform) are called maximum likelihood, or *ML-estimates*.

Equations 2.167 and 2.168 were obtained in the assumption that parameter to be estimated is scalar and takes values from a discrete set of values. It is quite obvious that they can be in a straightforward way extended to vectorial parameters, in which case probabilities involved in the equations are corresponding multivariate probabilities, and to parameters with continuum of values, in which case probabilities should be replaced by the corresponding probability densities.

To conclude this exposure of principles of statistically optimal parameter estimation, consider an important and illustrative special case when the relationship between object signals and observed signal is determined by the above-described model of additive signal-independent uncorrelated Gaussian noise (Equation 2.119):

$$\mathbf{b}_\omega = \mathbf{a}(\rho) + \mathbf{n}_v, \qquad (2.172)$$

where $\mathbf{n}_v$ is a realization of noise from a noise ensemble Δ_N. From this equation, it follows that the conditional probability $P\{\mathbf{b}_\omega / \rho_q\}$ is a probability that noise realization $\mathbf{n}_v$ is equal to the difference $\mathbf{b}_\omega - \mathbf{a}(\rho)$:

$$P\{\mathbf{b}_\omega / \rho_q\} = P\{\mathbf{n}_v = \mathbf{b}_\omega - \mathbf{a}(\rho)\}. \qquad (2.173)$$

For N-component discrete signals $\mathbf{b}_\omega = \{b_k\}$ and $\mathbf{a}(\rho) = \{u_k(\rho)\}$, $k = 0,1,\ldots,$ $N-1$ and Gaussian noise with uncorrelated components, this relationship becomes

$$P\{\mathbf{b}_\omega / \rho_q\} = P\{\mathbf{n}_{v,k} = \mathbf{b}_{\omega,k} - \mathbf{a}_k(\rho)\} = \prod_{k=0}^{N-1}\left[\frac{1}{\sqrt{2\pi}\sigma_n} \exp\left(-\frac{[b_k - a_k(\rho)]^2}{2\sigma_n^2} \right) \right]$$

$$= \left(\frac{1}{\sqrt{2\pi}\sigma_n} \right)^N \exp\left(-\frac{1}{2\sigma_n^2} \sum_{k=0}^{N-1} [b_k - a_k(\rho)]^2 \right). \qquad (2.174)$$

Because the exponential function in Equation 2.174 is a monotonic function of its argument, Equations 2.167 and 2.168 for optimal MAP- and ML-estimates of the parameter are reduced to the equations

$$\hat{\rho}^{(MAP)} = \underset{\{\rho\}}{\arg\min}\left\{\sum_{k=0}^{N-1}[b_k - a_k(\rho)]^2 - 2\sigma_n^2 \ln P(\rho)\right\} \qquad (2.175)$$

$$\hat{\rho}^{(ML)} = \underset{\{\rho\}}{\arg\min}\sum_{k=0}^{N-1}[b_k - a_k(\rho)]^2. \qquad (2.176)$$

The latter equation means that the optimal ML-estimate of the parameter minimizes mean squared difference between components of the observed signal and object signal, or L2-norm of the signal difference. This fundamental result is, in particular, a justification of very frequent use in signal processing of L2-norm as a measure of deviation of one signal from another.

Appendix

Derivation of Equation 2.32

For $\mathbf{T}^{(b)} = \{\vec{\phi}_n^{(b)}\}$ and $\mathbf{T}^{(a)} = \{\vec{\phi}_k^{(a)}\}$ as orthogonal transforms for representation of signal vector a and its linear transformation $\mathbf{b} = L(\mathbf{a})$:

$$\beta_n = \mathbf{T}^{(b)}\mathbf{b} = \mathbf{b} \circ \vec{\phi}_n^{(b)} = L(\mathbf{a}) \circ \vec{\phi}_n^{(b)} = L\left(\sum_k \alpha_k \phi_k^{(a)}\right) \circ \vec{\phi}_n^{(b)}$$

$$= \sum_k \alpha_k \left(L\left(\phi_k^{(a)}\right) \circ \vec{\phi}_n^{(b)}\right) = \sum_k \lambda_{k,n}\alpha_k,$$

where symbol $\circ$ represents scalar product and $\lambda_{k,n} = \left(L\left(\phi_k^{(a)}\right) \circ \vec{\phi}_n^{(b)}\right)$.

Derivation of Equation 2.65

$$\int_{-\infty}^{\infty} \exp(-i\pi\omega^2 x^2)\exp(i2\pi px)\,dx = \int_{-\infty}^{\infty} \exp[-(i\pi\omega^2 x^2 - i2\pi px)]\,dx$$

$$= \int_{-\infty}^{\infty} \exp\left[-\left(\sqrt{i\pi}\omega x - \frac{\sqrt{i\pi}}{\omega}p\right)^2\right]\exp\left(i\pi\frac{p^2}{\omega^2}\right)dx$$

$$= \exp\left(i\pi\frac{p^2}{\omega^2}\right)\int_{-\infty}^{\infty} \exp\left[-(\sqrt{i\pi}\omega x)^2\right]dx$$

$$= \exp\left(i\pi \frac{p}{\omega^2}\right) \frac{1}{\sqrt{i2\pi\omega}} \int_{-\infty}^{\infty} \exp\left(\frac{x^2}{2}\right) dx$$

$$= \exp\left(i\pi \frac{p^2}{\omega^2}\right) \frac{\sqrt{2\pi}}{\sqrt{i2\pi\omega}} = \frac{1}{\sqrt{i\omega}} \exp\left(i\pi \frac{p^2}{\omega^2}\right).$$

Derivations of Equations 2.84 through 2.87

If $a(x) = \int_{-\infty}^{\infty} \alpha(f)\exp(-i2\pi fx)\, dx,$

$$\pm a(-x) = \pm \int_{-\infty}^{\infty} \alpha(f)\exp[-i2\pi f(-x)]\, dx = \pm \int_{-\infty}^{\infty} \alpha(f)\exp[-i2\pi(-f)x]\, dx$$

$$= \pm \int_{-\infty}^{\infty} \alpha(-f)\exp[-i2\pi fx]\, dx.$$

Therefore, from $a(\mathbf{x}) = \pm a(-\mathbf{x})$ it follows that $\alpha(\mathbf{f}) = \pm\alpha(-\mathbf{f})$. Furthermore

$$\pm a^*(x) = \pm \int_{-\infty}^{\infty} \alpha^*(f)\exp(i2\pi f)\, dx = \pm \int_{-\infty}^{\infty} \alpha^*(-f)\exp(-i2\pi fx)\, dx.$$

Therefore, from $a(\mathbf{x}) = \pm a^*(\mathbf{x})$, it follows that $\alpha(\mathbf{f}) = \pm\alpha^*(-\mathbf{f})$. Combining these relationships, also obtain Equations 2.87 and 2.88.

Reference

1. D. Gabor, A new microscopic principle, *Nature*, 161, 777–778, 1948.

3

Image Digitization

This chapter addresses the very first problem of digital imaging, the problem of converting continuous signals that carry information on natural objects into digital signals. We call this conversion "signal digitization."

Principles of Signal Digitization

As it was stated in the section "Primary Definitions" in Chapter 2, signal digitization can be treated in general terms as determination, for each particular signal, of an index of the equivalency cell to which the signal belongs in the signal space, and signal reconstruction can be treated as generating a representative signal of the cell with this index. This is, for instance, what we do when we describe everything in our life with words in speaking or writing. In this case, this is our brain that performs the job of subdividing "signal space" into the "equivalency cells" of notions and selects the word (cell representative signal) that corresponds to what we want to describe.

The volume of our vocabulary is about 10^5–10^6 words. The variety of signals we have to deal with in imaging is immeasurably larger. One can see that from this simple example: the number of digital images of, for instance, TV quality (500×500 *pixels* with 256 gray levels in RGB channels) is $256^{3 \times 500 \times 500}$. No technical device will ever be capable of implementing such a huge look-up table.

A solution of this problem of the digitization complexity is found in a two-stage digitization procedure. At the first stage, called *signal discretization*, continuous signals are converted into a set of real numbers that form *signal discrete representation*. At the second stage, called *scalar (element-wise) quantization*, this set of numbers is, on a number-by-number basis, converted into a set of quantized numbers, which finally results in a digital signal that corresponds to the initial continuous one. In the same way, as written speech is a sequence of letters selected from an alphabet, digital representation of signals is a sequence of samples that can take one of a finite set of quantized values. In terms of the signal space, discretization can be regarded as introducing a coordinate system in the signal space and determining signal coordinates in this coordinate system. Element-wise quantization is then quantization of those coordinates. Equivalence cells of the signal space are therefore approximated in this case by hyper-cubes with edges parallel to the coordinate axis. Note that element-wise quantization can also be considered as an implementation of the general quantization in 1D signal space.

Signal Discretization

Signal Discretization as Expansion over a Set of Basis Functions

In principle, there might be many different ways to convert continuous signals into discrete ones represented by a set of real numbers. However, the entire technological tradition and all technical devices that are used at present for such a conversion implement a method that can be mathematically modeled as computing coefficients representing signals in their expansion over a set of basis functions.

Let $\left\{\varphi_k^{(d)}(x)\right\}$ be a set of basis functions used for signal discretization. Then, coefficients $\{\alpha_k\}$ of signal discrete representations are computed as

$$\alpha_k = \int_X a(x)\varphi_k^{(d)}(x)\,dx. \tag{3.1}$$

This equation is a mathematical formulation of operations performed by signal sensors. Functions $\left\{\varphi_k^{(d)}(x)\right\}$ describe spatial sensitivity of sensors, that is, sensor *PSF*.

Signal discretization with discretization basis function $\left\{\varphi_k^{(d)}(x)\right\}$ assumes that a reciprocal set of reconstructing basis functions $\left\{\varphi_k^{(r)}(x)\right\}$ is defined, with which signal can be reconstructed from its discrete representation coefficients $\{\alpha_k\}$ as follows:

$$a(x) \approx \sum_{k=0}^{N-1} \alpha_k \varphi_k^{(r)}(x), \tag{3.2}$$

where the approximation symbol $\approx$ in Equation 3.2 indicates that signal reconstructed from its discrete representation is, generally, not a precise copy of the initial signal, N is the number of representation coefficients needed for signal reconstruction with a desired accuracy, and functions $\left\{\varphi_k^{(r)}(x)\right\}$ are PSFs of signal reconstruction devices (such as, for instance, computer display), or their *reconstruction aperture functions*.

For different bases, the signal approximation accuracy for a given number of basis functions N used for signal reconstruction might be different. Naturally, the discretization and reconstruction bases that provide better approximation accuracy for a given N are preferable. Therefore, the accuracy of the expansion (3.2) is the first issue one should check when selecting bases for signal discretization and reconstruction. Approaches to the selection of bases optimal in this respect are discussed in the section "Optimality of Bases: Karhunen–Loeve and Related Transform."

The second issue to consider when selecting discretization and reconstruction basis functions is that of complexity of generating and implementing basis functions and computing signal representation coefficients.

In principle, discretization and reconstruction basis functions can be hard-wired in the signal discretization and restoration devices in a form of templates. However, practically this is rarely feasible because the number N of the required template functions is very large (for instance, for images it is of the order of magnitude of 10^5–10^7). This is why these functions are usually produced from a single *"mother" function* by means of its modification with one or another method.

Typical Basis Functions and Classification

Shift (Convolutional) Basis Functions

The simplest method for generating sets of discretization and reconstruction basis functions from a "mother" function $\varphi_M(x)$ is that of translation (coordinate shift):

$$\varphi_k^{(s)}(x) = \varphi_M^{(s)}(x - k\Delta x); \quad \varphi_k^{(r)}(x) = \varphi_M^{(r)}(x - k\Delta x). \tag{3.3}$$

We call functions generated in this way *shift basis functions*. We call the elementary shift interval Δx *the discretization* or *sampling interval*, the discretization using shift basis functions *sampling*, and the positions $\{x = k\Delta x\}$ *sampling points*. The sampling interval is usually the same for the entire range of signal coordinates x, although it need not necessarily be so.

The signal representation coefficients $\{a_k\}$ for sampling basis functions are obtained as

$$a_k = \int_X a(x)\varphi_M^{(s)}(x - k\Delta x). \tag{3.4}$$

They are called *samples*. Correspondingly, signal reconstruction is performed as

$$\tilde{a}(x) = \sum_k a_k \varphi_M^{(r)}(x - k\Delta x). \tag{3.5}$$

Equation 3.5 can be regarded as convolution *conv(.,.)* of signals

$$\tilde{\tilde{a}}(x) = \sum_k \alpha_k \delta(x - k\Delta x) \tag{3.6}$$

with reconstruction basis functions

$$\tilde{a}(x) = \sum_k \alpha_k \varphi^{(r)}(x - k\Delta x) = \int_{-\infty}^{\infty} \left[\sum_{k=0}^{N-1} \alpha_k \delta(\xi - k\Delta x) \right] \varphi_0^{(r)}(x - \xi) d\xi$$

$$= conv\left(\sum_{k=0}^{N-1} \alpha_k \delta(x - k\Delta x), \varphi_0^{(r)}(x) \right). \tag{3.7}$$

This justifies yet another name for shift basis functions, the *convolution basis functions*, and treatment of Equation 3.5 for reconstruction of continuous signals from their samples as *convolution-based interpolation* of the samples.

The most immediate example of sampling functions are functions built from rectangular impulse rect-functions defined by Equation 2.8:

$$\varphi_k^{(s)}(x) = \frac{1}{\Delta x} rect\left(\frac{x - k\Delta x}{\Delta x} \right). \tag{3.8}$$

For rectangular sampling functions, signal representation coefficients are their mean values over sampling intervals:

$$\alpha_k = \frac{1}{\Delta x} \int_X a(x) rect\left(\frac{x - k\Delta x}{\Delta x} \right) dx = \frac{1}{\Delta x} \int_{(k-1/2)\Delta x}^{(k+1/2)\Delta x} a(x) dx = \overline{a(k\Delta x)}. \tag{3.9}$$

By the definition of the rect-function, rectangular sampling functions defined by Equation 3.8 are orthogonal. Therefore, the reciprocal signal reconstruction basis in this case is composed of functions

$$\left\{ \varphi_k^{(r)}(x) = rect\left(\frac{x - k\Delta x}{\Delta x} \right) \right\}. \tag{3.10}$$

With this basis, signals are approximated by piece-wise constant functions

$$a(x) \approx \tilde{a}(x) = \sum_{k=0}^{N-1} \overline{a(k\Delta x)} rect\left(\frac{x - k\Delta x}{\Delta x} \right) \tag{3.11}$$

as it is illustrated in Figure 3.1.

2D shift basis functions are most frequently implemented as a product of 1D functions of corresponding coordinates, say, x and y, that is, as *separable functions*. For rectangular sampling functions, these are

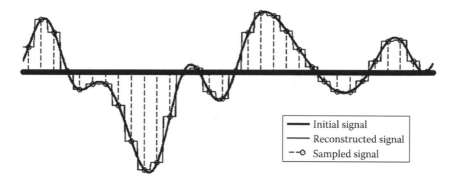

FIGURE 3.1
Continuous signal sampled and reconstructed using rectangular impulse sampling and reconstruction basis functions.

$$\left\{ \varphi_{k,l}^{(s)}(x,y) = \frac{1}{\Delta x \Delta y} rect\left(\frac{x - k\Delta x}{\Delta x} \right) rect\left(\frac{y - l\Delta y}{\Delta y} \right) \right\}, \tag{3.12}$$

$$\left\{ \varphi_{k,l}^{(r)}(x,y) = rect\left(\frac{x - k\Delta x}{\Delta x} \right) rect\left(\frac{y - l\Delta y}{\Delta y} \right) \right\}. \tag{3.13}$$

Sampling points form in this case a rectangular *sampling grid* in Cartesian coordinates (x, y) (see Figure 3.2) and sampling is performed by measuring average signal intensity over rectangular apertures of light-sensitive cells of image sensors shown in Figure 3.3. Samples of images are commonly called *pixels* (from picture elements).

Owing to fabrication technology limitations, some gaps must usually be left between adjacent light-sensitive cells. Because of this, dimensions dx and dy of light-sensitive cells are smaller than corresponding sampling intervals

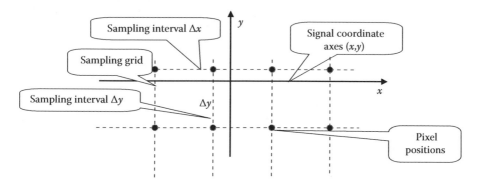

FIGURE 3.2
2D rectangular sampling grid.

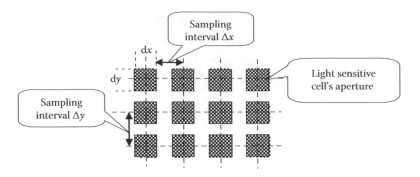

FIGURE 3. 3
Arrangement of light-sensitive elements in image sensor array.

Δx and Δy. Sampling functions that describe such image sensors with rectangular apertures can be described as

$$\left\{ \varphi_{k,l}^{(s)}(x,y) = \frac{1}{dx\,dy} rect\left(\frac{x - k\Delta x}{dx} \right) rect\left(\frac{y - l\Delta y}{dy} \right) \right\}. \tag{3.14}$$

The ratios $(dx/\Delta x, dy/\Delta y)$ of dimensions of light-sensitive cells to sampling intervals are called camera *fill factors*.

Piece-wise constant image approximation in their reconstruction from samples using rectangular reconstruction functions given by Equation 3.13 produces specific visual effects called image *pixilation*. They are illustrated in Figure 3.4.

FIGURE 3.4
Effect of "pixilation" in image reconstruction from sampled data using rectangular reconstruction basis functions: (a) initial image and (b) image reconstructed from sampled initial image using rectangular basis functions.

2D sampling functions produced from 1D rect-functions do not necessarily have to be separable functions with rectangular apertures. Frequently used options are sampling functions with circular apertures:

$$\left\{\varphi_{k,l}^{(s)}(x,y) = \frac{1}{\pi r^2} circ\left(\frac{x - k\Delta x}{r}, \frac{y - l\Delta y}{r}\right)\right\} \tag{3.15}$$

$$\left\{\varphi_{k,l}^{(r)}(x,y) = circ\left(\frac{x - k\Delta x}{r}, \frac{y - l\Delta y}{r}\right)\right\}, \tag{3.16}$$

where *circ-function* $circ(x, y)$ is defined as

$$circ(x,y) = \begin{cases} 1, & \sqrt{x^2 + y^2} \leq 1 \\ 0, & \text{otherwise} \end{cases} \tag{3.17}$$

and r is radius of the circular aperture. Rectangular sampling grid with circular sampling apertures is illustrated in Figure 3.5.

Yet another example of sampling functions is an idealistic mathematical model of *sinc-function* defined by Equation 2.37. For sampling interval Δx, sampling basis functions build on sinc-functions are defined as

$$\varphi_k(x) = \frac{1}{\Delta x} sinc\left(\pi \frac{x - k\Delta x}{\Delta x}\right) \tag{3.18}$$

with sampling performed as

$$\alpha_k = \frac{1}{\Delta x} \int_{-\infty}^{\infty} a(x) sinc\left(\pi \frac{x - k\Delta x}{\Delta x}\right) dx. \tag{3.19}$$

As it follows from Equations 2.104, sinc-functions and rect-functions are dual: they are Fourier transforms of one another.

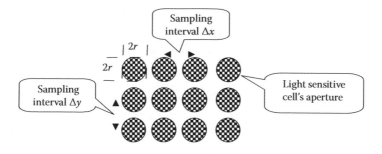

FIGURE 3.5
Rectangular sampling grid with circular sampling apertures.

Using *shift theorem* and *convolution theorem* of the theory of Fourier transform (the section "Imaging Systems and Integral Transforms" in Chapter 2) and Equations 2.74 and 2.75, one can see that sampling sinc-functions are mutually orthogonal:

$$\frac{1}{\Delta x} \int\limits_{-\infty}^{\infty} \text{sinc}\left(\frac{\pi(x - k\Delta x)}{\Delta x}\right) \text{sinc}\left(\frac{\pi(x - l\Delta x)}{\Delta x}\right) dx$$

$$= \frac{1}{\Delta x} \int\limits_{-\infty}^{\infty} \text{sinc}\left(\frac{\pi x}{\Delta x}\right) \text{sinc}\left(\frac{\pi(x - (k - l)\Delta x)}{\Delta x}\right) dx$$

$$= \Delta x \int\limits_{-1/2\Delta x}^{1/2\Delta x} \exp\left[-i2\pi f(k - l)\Delta x\right] df$$

$$= \text{sinc}\left[\pi(k - l)\right] = \begin{cases} 1, & k = l \\ 0, & k \neq l \end{cases} = \delta(k - l). \tag{3.20}$$

Therefore, reconstruction functions for this basis are functions $\{\text{sinc}[\pi(x - k\Delta x)/\Delta x]\}$ and signal reconstruction is performed as

$$a(x) \approx \tilde{a}(x) = \sum_{k=0}^{N-1} \alpha_k \, \text{sinc}\left(\pi \frac{x - k\Delta x}{\Delta x}\right). \tag{3.21}$$

Equation 2.104 also shows that the Fourier spectrum of functions $\{\text{sinc}[\pi(x - k\Delta x)/\Delta x]\}$ is zero outside the frequency interval $[-1/2\Delta x, 1/2\Delta x]$. This implies that signals reconstructed from their samples according to Equation 3.21 belong to the class of so-called *band-limited functions* and that coefficients $\{\alpha_k\}$ of discrete representation of band-limited signals $\tilde{a}(x)$ are equal to their samples taken at equidistant points $\{x = k\Delta x\}$:

$$\alpha_k = \tilde{a}(k\Delta x). \tag{3.22}$$

This is where the name "sampling functions" derives from.

Scale (Multiplicative) Basis Functions

Sinusoidal Basis Functions

Yet another method for generating basis functions from one mother function $\varphi_M(x)$ is scaling the argument of functions:

$$\varphi_k(x) = \varphi_M(kx). \tag{3.23}$$

An immediate example of signal expansion over *scaling basis functions* is signal Fourier series expansion over a finite interval:

$$a(x) = \sum_k \alpha_k \exp(i2\pi kx/X), \tag{3.24}$$

where X is the interval, within which signal is approximated by the Fourier series, and coefficients $\{\alpha_k\}$ are complex numbers computed as

$$\alpha_k = \frac{1}{X} \int_{-X/2}^{X/2} a(x)\exp(-i2\pi kx/X)\,dx. \tag{3.25}$$

They can be regarded as $(1/X)$-normalized samples of the signal Fourier transform spectrum taken at frequencies $\{k/X\}$.

While in reality the number of terms in the expansion (3.24) is always finite, classical Fourier series expansion assumes that it is infinite. The limitation of the number of terms in Fourier series (3.24) is equivalent to dropping all signal components with frequencies higher than $(N-1)/2X$, where N is the number of terms in the Fourier series expansion. This frequency band-limitation can mathematically be described as multiplying signal Fourier spectrum by a rectangular window function $rect((f + (N-1)/2X)/((N-1)/X))$, which is equivalent to the signal convolution with the corresponding sinc-function, Fourier transform of the rectangular function. As a result, sinc-function-originated oscillations may appear at signal edges. As it was already mentioned in the section "Imaging Systems and Integral Transforms" in Chapter 2, these specific distortions in reconstructed signals are called *Gibb's effects*.

The fundamental property of Fourier series signal expansion is that, while it converges to the signal within interval of expansion X, outside this interval Fourier series it converges, owing to the periodicity of basis functions $\{\exp(i2\pi kx/X)\}$, to a periodical function with a period X. Periodical signals are dual to signals composed from shifted delta-functions introduced in Equation 3.6. This duality will be discussed in connection with the *sampling theorem* in the section "Image Sampling."

Walsh Functions

Generating basis functions by means of scaling their argument can also be treated, for the basis of exponential functions $\{\exp(i2\pi kx/X)\}$, as generating by means of multiplying mother function:

$$\exp(i2\pi kx/X) = \prod_{l=1}^{k} \exp(i2\pi x/X). \tag{3.26}$$

There exists yet another family of orthogonal functions built with the same principle, *Walsh functions*. Walsh functions are binary functions that assume only two values, 1 and –1. Walsh functions are generated by multiplication

$$wal_k(x) = \prod_{m=0}^{\infty} \left[rad_{m+1}(x/X) \right]^{k_m^{GC}} \tag{3.27}$$

of clipped sinusoidal functions called the *Rademacher functions*:

$$rad_m(\xi) = sign\left[\sin(\pi 2^m \xi) \right] = \begin{cases} +1, & \sin(2^m \pi \xi) > 0 \\ -1, & \sin(2^m \pi \xi) < 0 \end{cases}, \tag{3.28}$$

where X is interval on which functions are defined and k_m^{GC} is the m-th digit of the so-called *Gray code* of index k. Gray code digits are generated from digits $\{k_m\}$ of ordinary binary representation of the number

$$k = \sum_{m=0}^{\infty} k_m 2^m; \quad k_m = 0, 1 \tag{3.29}$$

according to the following rule:

$$k_m^{GC} = k_m \oplus k_{m+1}, \tag{3.30}$$

where $\oplus$ stands for *modulo 2 addition*, according to which $0 + 0 = 1 + 1 = 0; \ 0 + 1 = 1 + 0 = 1$.

Formula 3.27 reveals the multiplicative nature of Walsh functions. For the purpose of calculating their values in computers, another representation of the Walsh functions (see Appendix) is useful:

$$wal_k(\xi) = \prod_{m=0}^{\infty} \left[(-1)^{\xi_{m+1}} \right]^{k_m^{GC}} = (-1)^{\sum_{m=0}^{\infty} k_m^{GC} \xi_{m+1}}, \tag{3.31}$$

where

$$\xi = x/X = \sum_{m=1}^{\infty} \xi_m 2^{-m} \tag{3.32}$$

with $\{\xi_m = 0, 1\}$ as digits of binary representation of the normalized coordinate ξ.

Walsh functions are orthogonal on the interval [0, X]. As one can see from Equation 3.31, multiplication of two Walsh functions results in another Walsh function with a shifted index or argument:

$$wal_k(\xi)wal_l(\xi) = wal_{k\oplus l}(\xi); \qquad (3.33)$$

$$wal_k(\xi)wal_k(\zeta) = wal_k(\xi \oplus \zeta). \qquad (3.34)$$

The shift, called *dyadic shift*, is determined through the bit-by-bit modulo 2 addition of functions' indices or, respectively, arguments.

One can regard the sum $\sum_{m=0}^{\infty} k_m^{GC}\xi_{m+1}$ in the definition of Walsh as a scalar product of vectors $\left\{k_m^{GC}\right\}$ and $\{\xi_{m+1}\}$, composed from binary digits of function's index and argument. It is in this sense that one can treat the basis of Walsh functions as a scale basis functions. The multiplicative nature of Walsh functions and exponential ones justifies yet another name for these families of bases, *multiplicative bases*.

Walsh functions are akin to complex exponential functions $\{exp(i2\pi kx/X)\}$ in one more respect. This can be seen if (–1) in Equation 3.42 is replaced by $exp(i\pi)$ to obtain

$$wal_k(\xi) = exp\left(i\pi \sum_{m=0}^{\infty} k_m^{GC}\xi_{m+1} \right) \qquad (3.35)$$

and from comparison of plots of sinusoidal and Walsh functions shown in Figure 3.6.

As one can see from the figure, an important parameter that unites both families of functions is the number of function zero crossing which coincides, for these functions, with their index k. For sinusoidal functions, this parameter is associated with the notion of frequency. For Walsh functions, this parameter was termed *sequency* by Harmuth [1].

Sequency-wise ordering of Walsh functions is not accidental. In principle, ordering basis functions can be arbitrary. The only condition one should obey is matching between signal representation coefficients $\{\alpha_k\}$ and the corresponding discretization and reconstruction basis functions $\left\{\phi_k^{(d)}(x)\right\}$ and $\left\{\phi_k^{(r)}(x)\right\}$. However, almost always a certain "natural" ordering of basis functions exists. For instance, shift (convolution) bases are ordered according to successive coordinate shifts. Sinusoidal basis functions are naturally ordered according to their frequency. Sequence-wise ordering of Walsh functions can be regarded natural for the following reason.

One of the most natural orderings of basis functions for signal representation is ordering, for which signal approximation error for a finite number N of terms in signal expansion over the basis decreases with the growth of N. For Walsh functions, it happens that, for many analog signals, sequency-wise

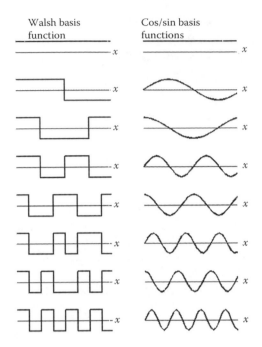

FIGURE 3.6
First eight sinusoidal and Walsh functions.

ordering satisfies this requirement much better than other methods of order-ing. Figure 3.7 illustrates this feature of Walsh function signal spectra.

Graphs on the figure show averaged squared Walsh spectral coefficients (power spectra) of image rows as functions of coefficients' indices for sequence-wise (Walsh) ordering and for Hadamard ordering. In Hadamard ordering, digits $\{k_m\}$ of ordinary binary representation of index k Equation 3.43 rather than index Gray code digits $\{k_m^{GC}\}$ are used for generating the basis functions:

$$walhad_k(\xi) = (-1)^{\sum\limits_{m=0}^{\infty} k_m \xi_{m+1}}. \tag{3.36}$$

Walsh functions are basis functions of the discrete transform called *Walsh transform*, which is discussed in Chapter 4, section "Binary Transforms." In conclusion, note that signal expansion over Walsh function basis approxi-mates signals with piece-wise constant functions just as it is the case for rect-angular impulse basis functions.

Wavelets

A distinctive feature of shift (convolution) basis functions is that they are concentrated in a vicinity of sampling points and, for them, signal

(a)

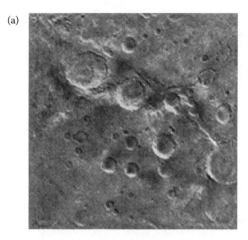

(b)

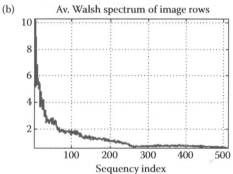

Av. Walsh spectrum of image rows

Sequency index

Av. Hadamard spectrum of image rows

(c)

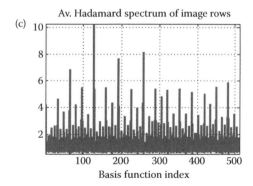

Basis function index

FIGURE 3.7

A test image (a) and averaged power spectra of its rows in Walsh (b) and Hadamard (c) orderings.

representation coefficients (in this case signal samples) depend mostly on signal values in those vicinities; therefore, they carry "local" information about signals. In contrast to them, signal discrete representation for scale (multiplicative) bases is "global": signal representation coefficients depend

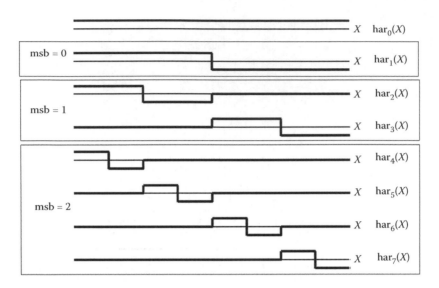

FIGURE 3.8
First eight Haar functions. Boxes unify Haar functions of the same scale.

on the entire signal. Sometimes it is useful to have a combination of these two features in the signal discrete representation. This is achieved with *wavelet basis functions* built using a combination of shift and scaling of a mother function. We have already addressed this feature in the section "Multiresolution Imaging: Wavelet Transforms" in Chapter 2 on multiresolution imaging and indicated that a classic example of such a combination represents basis functions of Haar transform. *Haar functions* defined on interval X are combinations of shifted rectangular impulse functions $\{rect(x - k\Delta x/\Delta x)\}$ and *Rademacher functions* $rad_k(\xi) = \{sign(sin(2^k \pi \xi))\}$:

$$har_k(\xi) = 2^{msb} rad_{msb+1}(\xi) rect\left(2^{msb}\xi - [k] \bmod 2^{msb}\right), \quad \xi = x/X \leq 1, \qquad (3.37)$$

where *msb* is the index of the most significant nonzero digit (bit) in binary representation of k (Equation 3.29) and $[k] \bmod 2^{msb}$ is a residual from division of k by 2^{msb}.

Haar functions are orthogonal on the interval $[0, X]$. Graphs of the first eight Haar functions are shown in Figure 3.8.

The figure clearly reveals the way in which Haar functions are built. One can see from the figure that Haar functions form groups according to the scale indexed by symbol *msb* (outlined by boxes in Figure 3.8) and that functions within each group are generated by shift in coordinates by multiple of $X/2^{msb}$. One can also see that signals reconstructed from their discrete representation over Haar basis functions belong to the same family of piecewise constant functions as signals reconstructed in the base of rectangular impulse basis functions and Walsh functions.

Numerical values of Haar functions can be found at each point by expressing Haar functions, as in the case of Walsh functions, via a binary code of the argument (Equations 3.29 and 3.32):

$$har_k(x) = 2^{msb}(-1)^{\xi_{msb+1}}\delta\left([\xi]_{msb} - [k]\,\mathrm{mod}\,2^{msb}\right). \tag{3.38}$$

Haar functions are basis functions of the discrete transform called *Haar transform,* which we discuss in Chapter 4, section "Binary Transforms."

Currently, numerous wavelet bases are known. Interested readers are referred to numerous monographs on wavelets. As for each scale wavelets are convolutional functions, using wavelets as discretization bases can be treated as multiresolution sampling.

Optimality of Bases: Karhunen–Loeve and Related Transform

As it was mentioned, the selection of basis functions for signal discretization is governed, in particular, by the accuracy of signal approximation with a limited number of terms in their expansion over the bases. In this section, we address the issue of optimization of bases in this respect. Because bases have to be optimally selected not for individual signals but for certain signal ensembles, we will apply statistical treatment of signals outlined in the section "Statistical Models of Signals and Transformations" in Chapter 2 and assume that signals belong to a certain family (ensemble) of signals and that the accuracy of signal representation is evaluated on average over this family.

Let $\{a^{(\omega)}(x)\}$, $\omega \in \Omega_A$ be a set of signals to be represented by their approximations in the form of a finite expansion:

$$a^{(\omega)}(x) \cong \tilde{a}^{(\omega)}(x) = \sum_{k=0}^{N-1} \alpha_k^{(\omega)}\varphi_k^{(r)}(x) \tag{3.39}$$

over orthonormal basis discretization and reconstruction basis functions $\{\varphi_k^{(d)}(x)\};\{\varphi_k^{(r)}(x)\}$ such that

$$\int_X \varphi_k^{(d)}(x)\left[\varphi_k^{(r)}(x)\right]^* dx = \delta(k,l) = 0^{|k-l|}, \tag{3.40}$$

where * stands for a complex conjugate. Let us also assume that the approximation accuracy is evaluated in terms of mean square approximation error (MSE):

$$|\varepsilon(\omega)|^2 = \int_X \left|a^{(\omega)}(x) - \tilde{a}^{(\omega)}(x)\right|^2 dx = \int_X \left|a^{(\omega)}(x) - \sum_{k=0}^{N-1} \alpha_k^{(\omega)}\varphi_k^{(r)}(x)\right|^2 dx. \tag{3.41}$$

Optimal values of coefficients $\left\{\alpha_k^{(\omega)}(\omega)\right\}$ that minimize MSE can be found by equating to zero derivatives of $|\varepsilon(\omega)|^2$ (Equation 3.47) over $\left\{\alpha_k^{(\omega)}(\omega)\right\} = \left\{\alpha_k^{(\omega)re} + i\alpha_k^{(\omega)im}\right\}$:

$$\frac{\partial}{\partial\alpha_k^{(\omega)re}}\int\limits_X\left|a^{(\omega)}(x) - \sum_{k=0}^{N-1}\alpha_k^{(\omega)}\varphi_k^{(r)}(x)\right|^2 dx = 0,$$

$$\frac{\partial}{\partial\alpha_k^{(\omega)im}(\omega)}\int\limits_X\left|a^{(\omega)}(x) - \sum_{k=0}^{N-1}\alpha_k^{(\omega)}\varphi_k^{(r)}(x)\right|^2 dx = 0, \tag{3.42}$$

which implies that signal transform coefficients should be computed as

$$\alpha_k^{(\omega)}(\omega) = \int\limits_X a^{(\omega)}(x)\varphi_k^{*(r)}(x)\,dx. \tag{3.43}$$

As it is shown in Appendix, the minimal MSE is equal in this case to

$$|\varepsilon(\omega)|_{\min}^2 = \int\limits_X|a(x,\omega)|^2\,dx - \sum_{k=0}^{N-1}|\alpha_k(\omega)|^2. \tag{3.44}$$

One can further minimize average MSE over the given set Ω of signals by an appropriate selection of bases functions $\left\{\varphi_k^{(r)}(x)\right\}$ as follows (see Appendix):

$$\left\{\varphi_k(x)\right\}_{opt} = \underset{\{\varphi_k(x)\}}{\arg\min}\left\{AV_\Omega\left(\int\limits_X|a^{(m)}(x)|^2\,dx - \sum_{k=0}^{N-1}|\alpha_k^{(\omega)}|^2\right)\right\}$$

$$= \underset{\{\varphi_k(x)\}}{\arg\max}\left\{\sum_{k=0}^{N-1}\iint\limits_X R_a(x_1,x_2)\varphi_k^*(x_1)\varphi_k(x_2)\,dx_1\,dx_2\right\}, \tag{3.45}$$

where AV_Ω means averaging over the signal set Ω_A. Function $R_a\,(x,y)$

$$R_a(x_1,x_2) = AV_{\Omega_A}\left[a^{(\omega)}(x_1)a^{*(\omega)}(x_2)\right] \tag{3.46}$$

is an autocorrelation function of signals $\{a^{(\omega)}\,(x)\}$. Then, optimal bases functions $\left\{\varphi_k^{(r)}(x)\right\}$ are defined by the equation

$$\left\{\varphi_k^{(r)}(x)\right\} = \underset{\{\varphi_k(x)\}}{\arg\max}\left\{\sum_{k=0}^{N-1}\int_X\left[\int_X R_a(x,\xi)\varphi_k^{(r)}(\xi)d\xi\right]\varphi_k^{*(r)}(x)dx\right\}. \tag{3.47}$$

From Cauchy–Schwarz–Buniakowsky inequality

$$\int_X f_1(x)f_2^*(x)dx \leq \left(\int_X|f_1(x)|^2\,dx\right)^{1/2}\left(\int_X|f_2(x)|^2\,dx\right)^{1/2}, \tag{3.48}$$

which becomes equality when $f_1(x)$ is proportional to $f_2(x)$, it follows that optimal basis functions are *eigen functions* of the integral equation

$$\int_X R_a(x,\xi)\varphi_k(\xi)dy = \lambda_k\varphi_k(x) \tag{3.49}$$

with kernel defined by the autocorrelation function (Equation 3.46) of signals. Scalar coefficients $\{\lambda_k\}$ in this equation are called *eigen values* of the equation. Such bases are called *Karhunen–Loeve bases* [2,3].

Similar reasoning can be applied to integral representation of signals. Let signals $a^{(\omega)}(x)$ be approximated by their integral representation over self-conjugate basis $\varphi(x,f)$:

$$a^{(\omega)}(x) \cong \tilde{a}^{(\omega)}(x) = \int_F \alpha^{(\omega)}(f)\varphi(x,f)df. \tag{3.50}$$

Then mean square approximation error

$$AV_\Omega|\varepsilon(\omega)|^2 = AV_\Omega\left\{\left|a^{(\omega)}(x) - \int_F\alpha^{(\omega)}(f)\varphi(x,f)df\right|^2\right\} \tag{3.51}$$

will be minimal if signal spectrum $\alpha^{(\omega)}(f)$ in this basis is obtained as

$$\alpha^{(\omega)}(f) = \int_X a^{(\omega)}(x)\varphi^*(x,f)dx \tag{3.52}$$

and basis $\varphi(x,f)$ satisfies the integral equation

$$\int_X R_a(x_1,x_2)\varphi(x_2,f)dx_2 = \lambda(f)\varphi^*(x_1,f), \tag{3.53}$$

where $R_a(x_1, x_2)$, as before, is the autocorrelation function (Equation 3.46) of signals $\{a^{(\omega)}(x)\}$. Signal transform with transform kernel defined by Equation 3.53 is called the *Karhunen–Loeve* transform.

An important, though idealized, special case is that of statistically homogeneous ("stationary") signals, whose correlation functions depend only on the difference of its arguments:

$$R_a(x_1, x_2) = R_a(x_1 - x_2). \tag{3.54}$$

In this case, the basis optimality condition of Equation 3.53 takes the form

$$\int_{-\infty}^{\infty} R_a(x_1 - x_2)\varphi(x_2, f)\,dx_2 = \lambda(f)\varphi^*(x_1, f) \tag{3.55}$$

A solution of this equation is an exponential function

$$\varphi(x, f) = \exp(i2\pi f x). \tag{3.56}$$

Two important conclusions follow from this result:

Integral Fourier transform is an MSE optimal Karhunen–Loeve transform for signals with correlation function that depends only on the difference of its arguments.

For signals with power spectrum that decays with frequency, the best, in MSE sense, their approximation is the approximation with *band-limited functions*, whose Fourier spectrum is equal to zero outside a certain bandwidth $[-F,F]$:

$$\tilde{a}^{(\omega)}(x, \omega) = \int_{-F}^{F} \alpha^{(\omega)}(f)\exp(i2\pi f x)\,df. \tag{3.57}$$

Approximation MSE in this case is minimal and is equal to

$$\left[AV_{\Omega_A}\left|\varepsilon(\omega)\right|^2\right]_{\min} = \int_{-\infty}^{-F} AV_{\Omega_A}\left\{\left|\alpha^{(\omega)}(f)\right|^2\right\}df + \int_{F}^{\infty} AV_{\Omega_A}\left\{\left|\alpha^{(\omega)}(f)\right|^2\right\}Ef. \tag{3.58}$$

Karhunen–Loeve transform provides the theoretical lower bound for compact (in terms of the number of terms in signal decomposition) signal discrete representation for MSE criterion of signal approximation.

Consider now a discrete equivalent of the Karhunen–Loeve expansion. Let $a_\omega = \left\{a_k^{(\omega)}\right\}$, $k = 0,1,\ldots,N-1$, $\omega \in \Omega$ be a discrete signal from an ensemble of signals Ω. Find an orthogonal transform $T_N = \left\{\varphi_{k,r}\right\}$

$$\alpha^{\omega} = T_N \alpha^{\omega} = \left\{ \alpha_r^{(\omega)} = \sum_{k=0}^{N-1} a_k^{(\omega)} \varphi_{k,r} \right\} \tag{3.59}$$

that minimizes, on average over the signal ensemble Ω, mean square error

$$\overline{|\varepsilon(\omega)|^2}\Big|_{\Omega} = AV_{\Omega} \left[\sum_{k=0}^{N-1} \left(a_k^{(\omega)} - \tilde{a}_k^{(\omega)} \right)^2 \right] \tag{3.60}$$

between signals $\{a_k^{\omega}\}$ and their approximation by truncated expansion

$$\tilde{\alpha}_{\omega} = \left\{ a_k = \sum_{r=0}^{\tilde{N}-1} \alpha_r^{(\omega)} \varphi_{k,r}^* \right\} \tag{3.61}$$

over $\tilde{N} \leq N$ terms. By virtue of the Parseval's identity, the transform is a solution of the equation

$$T = \underset{\{\varphi_{k,r}\}}{\arg\max} AV_{\Omega} \left[\sum_{r=0}^{\tilde{N}-1} \left| \alpha_r^{(\omega)} \right|^2 \right] = \underset{\{\varphi_{k,r}\}}{\arg\max} AV_{\Omega} \left\{ \sum_{r=0}^{\tilde{N}-1} \sum_{k=0}^{N-1} \sum_{n=0}^{N-1} a_k^{\omega} (a_n^{\omega})^* \varphi_{k,r} (\varphi_{n,r})^* \right\}. \tag{3.62}$$

Denote

$$R_a(k,n) = AV_{\Omega} \left[a_k^{(\omega)} \left(a_n^{(\omega)} \right)^* \right]. \tag{3.63}$$

Then, the transform optimization equation (3.62) becomes

$$T = \underset{\{\varphi_{k,r}\}}{\arg\max} \left\{ \sum_{r=0}^{\tilde{N}-1} \sum_{k=0}^{N-1} \sum_{l=0}^{N-1} R_a(k,n) \varphi_{k,r} (\varphi_{n,r})^* \right\}$$

$$= \underset{\{\varphi_{k,r}\}}{\arg\max} \left\{ \sum_{r=0}^{\tilde{N}-1} \left(\sum_{n=0}^{N-1} (\varphi_{n,r})^* \sum_{k=0}^{N-1} R_a(k,n) \varphi_{k,r} \right) \right\}. \tag{3.64}$$

From the Cauchy–Schwarz–Bunyakowsky inequality

$$\left(\sum_{k=0}^{N-1} a_k b_k \right)^2 \leq \sum_{k=0}^{N-1} |a_k|^2 \cdot \sum_{k=0}^{N-1} |b_k|^2 \tag{3.65}$$

that converts to equality for $\{a_k = \lambda b_k\}$, where λ is an arbitrary constant, it follows that the transform optimization equation becomes

$$\sum_{k=0}^{N-1} R_a(k,n)\varphi_{k,r} = \lambda_r\varphi_{n,r}. \qquad (3.66)$$

The transform whose basis functions satisfy Equation 3.66 is called the *Hottelling transform*. The result of applying this transform is called the signal *principal component decomposition* [4].

If no ensemble averaging is required when evaluating the approximation error, Equation 3.66 becomes

$$\sum_{k=0}^{N-1} a_k(a_n)^*\varphi_{k,r} = \lambda_r\varphi_{n,r}. \qquad (3.67)$$

The signal decomposition with a transform defined by basis functions that are solutions of Equation 3.67 is called *singular value decomposition (SVD)*.

The capability of transforms to concentrate signal energy in few signal decomposition coefficients is called *energy compaction capability*. Theoretically, the Karhunen–Loeve transform and the related Hottelling and SVD transforms have the best energy compaction capability among linear transforms. However, in the practice of image processing, they remain, because of implementation problems, to serve only as a theoretical benchmark and other transforms are used, such as discrete Fourier, discrete cosine, discrete Walsh, and discrete Haar transforms that will be introduced in Chapter 4.

Figure 3.9 generated using MATLAB program EnergyCompact_DFT_DCT_Walsh_Haar_2D_CRC.m provided in Exercises illustrates, for comparison, energy compaction capability of these transforms for four test images. One can see from the plots that, for these test images, discrete cosine transform (DCT) exhibits the best energy compaction capability. In fact, DCT has the edge over other known transforms in most of the image processing applications. We will discuss the reasons for this later in the section "Discrete Cosine and Sine Transforms."

Image Sampling

The Sampling Theorem and Signal Sampling

Representation of images in the form of arrays of image samples taken, usually, in nodes of a uniform rectangular sampling grid is the basic form of image discrete representation. Although imaging devices that produce

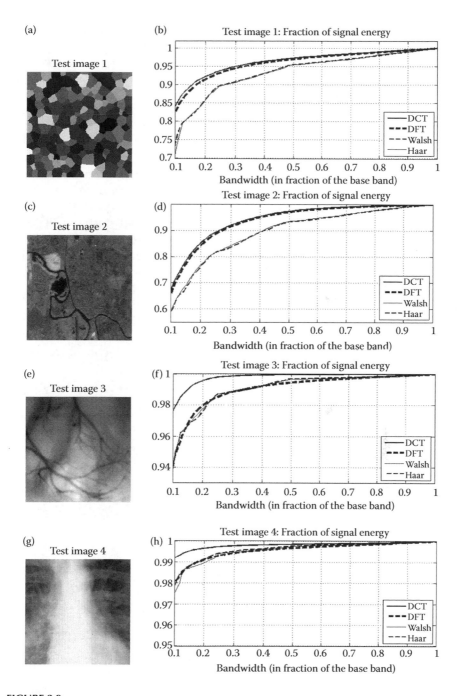

FIGURE 3.9
Comparison of energy compaction capability of discrete transforms for four test images. Plots on the right of each image show the fraction of image signal energy contained within a fraction of image transform coefficients limited by a square in transform domain.

images may work on different principles, for image displays and image processing software, such representation is the standard.

The theoretical foundation of signal sampling is provided by the *sampling theorem*. It was introduced in communication engineering by Vladimir A. Kotelnikov [5] and Claude Shannon [6], although, as a mathematical theorem, it was formulated earlier by J. M. Whittaker [7]. Here is how the sampling theorem is formulated in Shannon's classical paper [6]: *"If a function* x(t) *contains no frequencies higher than B hertz, it is completely determined by giving its ordinates at a series of points spaced 1/(2B) seconds apart"*. It is in this formulation that it is usually known and taught. In reality, however, there are no band-limited signals and precise signal reconstruction from its sampled representation is never possible. Therefore, if sampling is used for signal discretization, one needs to know how to minimize signal distortions due to its sampling.

In what follows, we introduce an approach to formulation of the sampling theorem that directly addresses this problem of minimization of approximation error in signal reconstruction from its sampled representation.

1D Sampling Theorem

Let $\varphi^{(r)}(x)$ be the PSF of signal/image reconstruction/display device, $\varphi^{(s)}(x)$ be the PSF of signal sampling device, and Δx be the signal sampling interval. Then, discrete representation coefficients $\{\alpha_k\}$ of a signal $a(x)$ are its samples:

$$\alpha_k = \int_X a(x)\varphi^s(x - k\Delta x)\,dx \qquad (3.68)$$

and its reconstructed copy is

$$a^{(r)}(x) = \sum_k \alpha_k \varphi^{(r)}(x - k\Delta x). \qquad (3.69)$$

Integration over x in Equation 3.68 and summation over k in Equation 3.69 are performed within signal boundaries. Note that in this formulation, we made a deliberate simplifying assumption that the coordinate of the signal sampling point with index $k = 0$ is $x = 0$ in both signal sampling and reconstruction. In general, there might be arbitrary shifts of the signal sampling grid with respect to the sampling and reconstruction devices coordinate systems. However, in the analysis of signal sampling, this arbitrary shift plays no essential role and can be disregarded. Later, in Chapter 4, we will see that these possible shifts must be taken into account when describing discrete representations of signal transforms other than convolution.

Without violating the generality, one can, in order to simplify further mathematical analysis, rewrite the integration in Equation 3.68 and the summation in Equation 3.69 in infinite limits:

$$\alpha_k = \int_{-\infty}^{\infty} a(x)\varphi^{(s)}(x - k\Delta x)dx \tag{3.70}$$

$$a^{(r)}(x) = \sum_{k=-\infty}^{\infty} \alpha_k \varphi^{(r)}(x - k\Delta x) \tag{3.71}$$

assuming that signal $a(x)$ and its samples $\{\alpha_k\}$ are equal to zero outside the real intervals of support. Now, modify Equations 3.70 and 3.71 in the following way:

$$\alpha_k = \int_{-\infty}^{\infty} a(x)\varphi^{(s)}(x - k\Delta x)dx = \int_{-\infty}^{\infty} \left[\int_{-\infty}^{\infty} a(\xi)\varphi^{(s)}(x - \xi)d\xi \right] \delta(x - k\Delta x)dx$$

$$= \int_{-\infty}^{\infty} a^{(s)}(x)\delta(x - k\Delta x)dx = a^{(s)}(k\Delta x) \tag{3.72}$$

$$a^{(r)}(x) = \sum_{k=-\infty}^{\infty} \alpha_k \varphi^{(r)}(x - k\Delta x) = \int_{-\infty}^{\infty} \left[\sum_{k=-\infty}^{\infty} \alpha_k \delta(\xi - k\Delta x) \right] \varphi^{(r)}(x - \xi)d\xi$$

$$= \int_{-\infty}^{\infty} \left[\sum_{k=-\infty}^{\infty} a^{(s)}(k\Delta x)\delta(\xi - k\Delta x) \right] \varphi^{(r)}(x - \xi)d\xi$$

$$= \int_{-\infty}^{\infty} \tilde{a}^{(s)}(\xi)\varphi^{(r)}(x - \xi)d\xi, \tag{3.73}$$

which introduces "virtual" signals

$$a^{(s)}(x) = \int_{-\infty}^{\infty} a(\xi)\varphi^{(s)}(x - \xi)d\xi \tag{3.74}$$

and

$$\tilde{a}^{(s)}(x) = \sum_{k=-\infty}^{\infty} a^{(s)}(k\Delta x)\delta(x - k\Delta x). \tag{3.75}$$

Further analysis of signal sampling is much simplified in the Fourier transform domain. Compute the Fourier spectra of virtual signals $a^{(s)}(x)$ and $\tilde{a}^{(s)}(x)$.

The spectrum $\alpha^{(s)}(f)$ of signal $a^{(s)}(x)$ (Equation 3.74), whose instantaneous values $\{a^{(s)}(k\Delta x)\}$ form sampled representation of signal $a(x)$, is, by the convolution theorem, a product

$$\alpha^{(s)}(f) = \alpha(f)FR^{(s)}(f) \tag{3.76}$$

of spectrum $\alpha(f)$ of the signal $a(x)$ and the frequency response $FR^{(s)}(f)$ of the sampling device, Fourier transform of its PSF $\varphi^{(s)}(x)$:

$$FR^{(s)}(f) = \int\limits_{-\infty}^{\infty} \varphi^{(s)}(x)\exp(i2\pi fx)\,dx. \tag{3.77}$$

Fourier spectrum of the virtual signal $\tilde{a}^{(s)}(x)$ involved in the reconstruction of signal $a(x)$ from its sampled representation $\{a^{(s)}\,(k\Delta x)\}$ can be found as (see Appendix)

$$\tilde{\alpha}^{(s)}(f) = \int\limits_{-\infty}^{\infty} \tilde{a}^{(s)}(x)\exp(i2\pi fx)\,dx = \int\limits_{-\infty}^{\infty}\left[\sum_{k=-\infty}^{\infty} a^{(s)}(k\Delta x)\delta(x - k\Delta x)\right]\exp(i2\pi fx)\,dx$$

$$= \int\limits_{-\infty}^{\infty} \alpha(p)FR^{(s)}(p)\,dp\Delta x\left\{\sum_{k=-\infty}^{\infty}\frac{1}{\Delta x}\exp\left[i2\pi k\Delta x(f - p)\right]\right\}. \tag{3.78}$$

The sum in the latter expression is a Fourier series expansion on interval $1/\Delta x$ of a function with expansion coefficients, equal to 1, that is, of a Dirac delta (Equation 2.24). Therefore, this series converges to a periodical replication of delta-function with a period $1/\Delta x$. In this way, we arrive at the identity

$$\frac{1}{\Delta x}\sum_{k=-\infty}^{\infty}\exp[i2\pi k\Delta x(f - p)] = \sum_{m=-\infty}^{\infty}\delta\left(f - p + \frac{m}{\Delta x}\right) \tag{3.79}$$

called the Poisson summation formula.

Inserting this formula into Equation 3.77, obtain that the spectrum of the virtual signal $\tilde{\alpha}^{(s)}(f)$ is composed of periodical replicas of the spectrum $\alpha^{(x)}(f)$:

$$\tilde{\alpha}(f) = \Delta x\sum_{m=-\infty}^{\infty}\alpha^{(s)}\left(f + \frac{m}{\Delta x}\right) = \Delta x\sum_{m=-\infty}^{\infty}\alpha\left(f + \frac{m}{\Delta x}\right)FR^{(s)}\left(f + \frac{m}{\Delta x}\right). \tag{3.80}$$

This analysis reveals that coefficients $\{\alpha_k\}$ of discrete representation of signal $a^{(x)}$ over shift basis functions $\{\varphi^{(s)}(x - k\Delta x); \varphi^{(r)}\,(x - k\Delta x)\}$ are instantaneous values $\{a_s\,(k\Delta x)\}$ of a signal $a^{(s)}(x)$ obtained from the signal $a(x)$ by its "prefiltering" in the sampling device (and not those of the signal $a(x)$).

Signal sampling can be interpreted as converting prefiltered input signal $a^{(s)}(x)$ into a virtual signal $\tilde{a}^{(s)}(x)$ (Equation 3.75), whose Fourier spectrum is composed of periodically repeated replicas of initial signal spectrum modified by the frequency response of signal sampling device (Equation 3.80); the replication period $BB = 1/\Delta x$ is defined by the sampling interval and forms; it is called the signal *baseband*.

Signal reconstruction from its sampled representation can be treated as a "postfiltering" of the virtual discrete signal $\tilde{a}^{(s)}(x)$ in the reconstruction device:

$$a^{(r)}(x) = \int_{-\infty}^{\infty} \tilde{a}(\xi)\varphi^{(r)}(x - \xi)d\xi. \tag{3.81}$$

This "postfiltering" can be described in the Fourier transform domain as multiplication

$$\alpha^{(r)}(f) = \tilde{\alpha}^{(s)}(f)FR^{(r)}(f) \tag{3.82}$$

of spectrum $\tilde{\alpha}^{(s)}(f)$ of the virtual signal $\tilde{a}^{s}(x)$ by frequency response $FR'(f)$ of the signal reconstruction device:

$$FR'(f) = \int_{-\infty}^{\infty} \varphi^{(r)}(x)\exp(i2\pi fx)dx. \tag{3.83}$$

The described interpretation of signal sampling and signal reconstruction from its samples is illustrated by flow diagrams in Figures 3.10 and 3.11 and by plots in Figure 3.12.

Now, one can easily see that distortions of the reconstructed signal compared to the initial nonsampled signal are due to the following reasons:

- Modifications of the signal spectrum by its prefiltering in the sampling device (Equation 3.76)
- Penetrating tails of the prefiltered signal spectrum periodical replicas inside the signal baseband, which causes spectra *aliasing*
- Remains of the signal spectrum periodical replicas outside the signal baseband not perfectly filtered out by the "postfiltering" in the reconstruction device

We postpone detailed qualitative and quantitative analysis of sampling artifacts till the sections "Sampling Artifacts: Quantitative Analysis" and "Sampling Artifacts: Qualitative Analysis" and formulate here the major conclusion that, for signals, whose spectral energy does not grow with frequency,

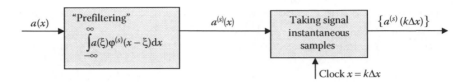

FIGURE 3.10
Flow diagram of signal sampling.

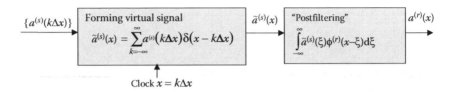

FIGURE 3.11
A mathematical model of devices for restoring continuous signals from their samples.

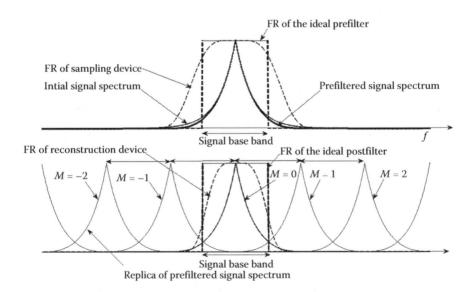

FIGURE 3.12
Fourier domain representation of signal sampling and reconstruction: upper plot—initial signal spectrum, its spectrum after "prefiltering" in the sampling device and frequency responses (FR) of the sampling device and of the ideal low-pass filter that defines the signal baseband; lower plot—periodical replicas of sampled signal spectrum described by Equation 3.80 and frequency responses (FR) of the signal reconstruction device and of the ideal low-pass reconstruction filter.

signal distortions due to sampling can be minimized in terms of signal reconstruction MSE, if signal prefiltering and postfiltering are performed by the
ideal low-pass filters with frequency response

$$FR^{(LP)}(f) = rect\big[(f + 1/2\Delta x)\Delta x\big]. \tag{3.84}$$

PSF of these filters is sinc-function $(\pi x/\Delta x)$ introduced in the section
"Properties of the Integral Fourier Transform" in Chapter 2. Thus, with the
ideal low-pass pre- and postfilters, signal samples are obtained as

$$\alpha_k = a^{(s)}(k\Delta x) = \frac{1}{\Delta x} \int_{-\infty}^{\infty} a(x)\,\mathrm{sinc}\left(\pi\frac{x - k\Delta x}{\Delta x}\right) dx \tag{3.85}$$

and signal reconstruction from its samples is performed as

$$a^{(r)}(x) = \sum_{k=-\infty}^{\infty} a^{(s)}(k\Delta x)\,\mathrm{sinc}\left(\pi\frac{x - k\Delta x}{\Delta x}\right). \tag{3.86}$$

Reconstruction of a continuous signal from its samples according to
Equation 3.86 is called *sinc-interpolation*.

Ideal low-pass filters do not distort the signal spectrum within the baseband and remove all spectral components outside the baseband, thus pre-
venting signal spectrum from aliasing. Signal prefiltering with the ideal
low-pass filter makes the signal band-limited.

In a special case of band-limited signals, the result of sampling are samples $\{a(k\Delta x)\}$ of signals themselves and signals are perfectly (distortion less)
reconstructed from the result of sampling:

$$a(x) = \sum_{k=-\infty}^{\infty} a(k\Delta x)\,\mathrm{sinc}\left(\pi\frac{x - k\Delta x}{\Delta x}\right). \tag{3.87}$$

The above reasoning can be summarized in the following formulation of
the sampling theorem:

For signals with power spectrum that does not grow with frequency, the
MSE of their reconstruction from their sampled representation is minimal
if sampling and reconstruction are performed using ideal low-pass filters
for pre- and postfiltering. Band-limited signals with baseband $BB = 1/\Delta x$ are
precisely reconstructed by sinc-interpolation of their instantaneous values
$a(k\Delta x)$ taken with interval Δx.

Sampling theorem also allows evaluating the volume of signal discrete representation, that is, the number of signal samples N needed to represent signals of length X and baseband F:

$$N = \frac{X}{\Delta x} = \frac{X}{1/F} = XF. \tag{3.88}$$

Signal space-bandwidth product XF is a fundamental parameter that defines the number N of signal degrees of freedom.

Sampling Two-Dimensional and Multidimensional Signals

For sampling 2D signals, 2D discretization and reconstruction basis functions should be used. The simplest way to generate and to implement in image sampling and display devices 2D sampling and reconstruction basis functions is to use separable 2D basis functions that are formed as a product of 1D function:

$$\left\{\varphi_1^{(s)}(x - k\Delta x)\varphi_2^{(s)}(y - k\Delta y)\right\} \quad \text{and} \quad \left\{\varphi_1^{(r)}(x - k\Delta x)\varphi_2^{(r)}(y - k\Delta y)\right\} \tag{3.89}$$

With these sampling and reconstruction functions, image sampling and reconstruction are carried out successively row-wise/column-wise over a rectangular sampling grid in Cartesian coordinates. In the domain of 2D Fourier transform, this corresponds to periodical replication of signal spectrum in Cartesian coordinates as it is shown in Figure 3.13a and c. For sampling intervals $\{\Delta x, \Delta y\}$ along two coordinates $\{x, y\}$, replication periods of spectra replicas are $\{1/\Delta x, 1/\Delta y\}$, respectively. Widening sampling intervals in order to reduce the number of signal samples is possible until spectra replicas do not overlap too much to cause unacceptable spectra aliasing. Therefore, the optimal will be a sampling grid that corresponds to most dense possible packing spectra replicas in the spectral domain. The opportunities for achieving spectra replicas dense packing depend on the shape of the figure that encompasses the area of image spectrum, which contains image spectral components that must not be sacrificed due to sampling.

If image spectrum is bounded by a rectangle, rectangular sampling grid is optimal and the number N of image samples, or the number of image degrees of freedom, will be equal to the product of image area $S_{x,y}$ and spectrum area S_{f_x,f_y}:

$$N = S_{x,y}S_{f_x,f_y}. \tag{3.90}$$

However, if the shape of the figure that bounds image spectrum is not a rectangle, rectangular sampling grid is not optimal. Consider, for instance, the shape of the spectrum shown in Figure 3.13a. If images with such spectrum shape are sampled using square sampling grid (Figure 3.13b), sampling will correspond

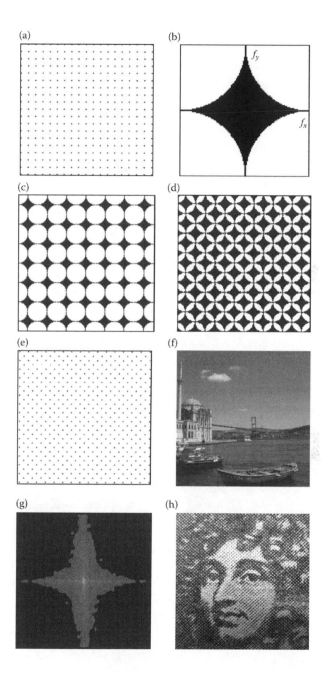

FIGURE 3.13

(a) An example of a figure that encompasses image spectrum; (b) square sampling grid; (c) and (d) its spectrum periodical replications in rectangular and 45°-tilted coordinate systems, correspondingly; (e) 45°-tilted sampling grid; (f) and (g) an example of a natural outdoor scene and its spectrum bounded on a certain intensity level; (h) a magazine printout of a gray-scale image printed using the 45°-tilted sampling grid.

to periodical replication of spectra as shown in Figure 3.13c. However, for this particular spectrum shape, one can achieve substantially denser packing of spectrum replicas, which is shown in Figure 3.13d. To achieve this, image sampling over a square sampling grid in 45°-rotated coordinates should be performed as it is shown in Figure 3.13e. As one can see in Figure 3.13d, the denser periodical pattern of spectrum replicas preserves same sampling intervals over 45°-rotated coordinates as those for the rectangular sampling grid in nonrotated coordinates (Figure 3.13b). Therefore, distances between samples in nonrotated coordinates become larger by $\sqrt{2}$, which results in twofold saving of the number of samples needed to reconstruct images.

This particular example of the image spectrum shape is not occasional. It is characteristic for many images of outdoor scenes, such as that shown in Figure 3.13f, in which horizontally and vertically oriented objects prevail; in their spectra, correspondingly, vertical and horizontal spatial high frequencies have higher intensity than diagonal ones as it is seen in Figure 3.13g. It is not surprising that the human visual system evolved to have higher sensitivity to horizontal and vertical spatial frequencies than to diagonal frequencies.

In viewing printed or displayed images, human eye optics plays the role of a postfilter. Selection of a 45°-rotated rectangular sampling grid that fits the spectra of natural images and human eye spatial filtering secures invisibility of sampling artifacts for less dense sampling grid. The 45°-rotated sampling grid became a standard in the print industry for printing gray-scale images. An example of a magnified fragment of such a printout, in which rotated sampling grid is seen, is shown in Figure 3.13h.

In many applications, images have spectra that decay more or less uniformly in all directions such as, for instance, shown in spectrum of Figure 3.14a of the image presented in Figure 3.7a. For such images with isotropic spectra that can be regarded as bounded by a circle, hexagonal periodical replication of spectra is the densest (compare Figures 3.14b and 3.14c). Hence, hexagonal sampling grid shown in Figure 3.14d will be optimal in this case. Hexagonal sampling grids are frequently used in color displays and print art. Hexagonal arrangement of light-sensitive cells can also be found in compound eyes of insects and in retinas of eyes of vertebrates (Figures 3.14e and 3.14f).

The described examples of optimization of sampling grids by means of rotating and tilting do allow reducing the number of image samples N required for image reconstruction, but this number still exceeds the product $S_{x,y}S_{f_x,f_y}$ of image and its spectrum areas as long as gaps remain empty in periodic replication of image spectra. This product can be regarded as a lower bound for the required number of image samples

$$N \geq S_{x,y}S_{f_x,f_y}. \tag{3.91}$$

This lower bound can, at least in principle, be achieved with *sub-band decomposition sampling*.

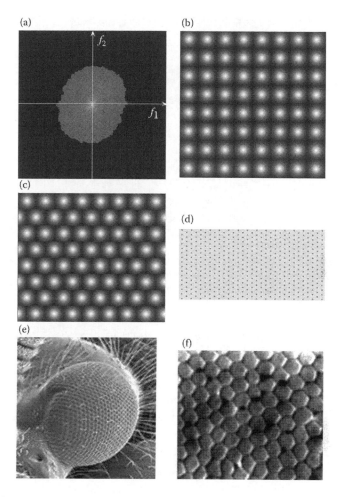

FIGURE 3.14

(a) Bounded on a certain intensity spectrum of the image shown in Figure 3.7; (b) and (c) replication of circularly bounded image spectrum for Cartesian and hexagonal sampling grids: the same number of circular spectra replicas occupy less area, when they are packed in a hexagonal coordinate system than in Cartesian coordinates; (d) hexagonal sampling grid; (e) compound eye of tsycada; (f) hexagonal arrangement of cones in fovea of human eye retina.

In discretization with sub-band decomposition sampling, image $a(x_1, x_2)$ is decomposed into a sum

$$a(x,y) \cong \sum_{k=0}^{K-1} a^{(k)}(x,y) \qquad (3.92)$$

of certain number K of components $\{a^{(k)}(x, y)\}$ with rectangular spectra that together approximate image spectrum as it is shown in Figure 3.15.

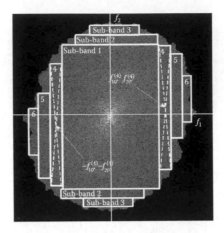

FIGURE 3.15

Image spectrum sub-band decomposition. As an example, the spectrum of the image of Figure 3.7a is shown bounded on a certain intensity level; six sub-bands of this spectrum are presented to illustrate the sub-band decomposition process that can be extended ad infinitum.

For each of the components, rectangular sampling grid with sampling intervals defined by dimensions of the component spectrum will be optimal. Hence, the required number N_k of samples of the k-th component with spectrum $S^{(k)}_{f_x,f_y}$ will be equal to $S_{x,y}S^{(k)}_{f_x,f_y}$, which, for the entire image, amounts to $N = \sum_{k=1}^{K} N_k = S_{x,y} \sum_{k=1}^{K} S^{(k)}_{f_x,f_y}$ samples. As soon as $\lim_{k \to \infty} \sum_{k=1}^{K} S^{(k)}_{f_x,f_y} = S_{f_x,f_y}$, $\lim_{k \to \infty} \sum_{k=1}^{K} a^{(k)}(x_1, x_2) = a(x_1, x_2)$, and the lower bound $S_{x,y}S_{f_x,f_y}$ for N is achieved

$$\lim_{K \to \infty} N = \lim_{K \to \infty} \sum_{k=1}^{K} N_k = S_{x,y}S_{f_x,f_y}. \tag{3.93}$$

Elucidate that the spectra of high-frequency components (components 2, 3, 4, etc. in Figure 3.15) extracted from the image by means of the corresponding ideal bandpass filters must be, before sampling, shifted to the coordinate origin. For this, they must be multiplied (or, in communication engineering jargon, modulated) by a sinusoidal signal of spatial frequencies defined by coordinates of the symmetry center of the sub-band (for instance, for the fourth component in Figure 3.15 it will be $\cos[2\pi(f^{(4)}_{10}x + f^{(4)}_{20}y)]$). The modulated signal then has to be subjected to a low-pass prefiltering using a filter with a flat frequency response within the bandwidth defined by the bandwidth of the shifted sub-band. After that the sub-band component extracted in this way is ready for sampling over a rectangular sampling grid with sampling intervals defined by dimensions of the sub-band. For image reconstruction, each sampled sub-band component is reconstructed using the corresponding 2D low-pass interpolating filter and then, before summation with other components, multiplied by the corresponding "demodulating" sinusoidal signal in order to return the component spectrum in its original position in the signal spectrum.

Sampling Artifacts: Quantitative Analysis

In this section, we will show how one can numerically evaluate sampling artifacts. For the sake of simplicity, a 1D case will be analyzed. Let $\varphi^{(s)}(x)$ and $\varphi^{(r)}(x)$ be PSFs of sampling and reconstruction devices, Δx be the sampling interval, and N be the number of sampled signal samples used for signal reconstruction.

Derive a relationship between initial signal $a(x)$ and signal $a^{(r)}(x)$ reconstructed from its samples that explicitly involves sampling and reconstruction device parameters. From Equations 3.68 and 3.69, it follows that

$$a^{(r)}(x) = \sum_{k=0}^{N-1} \alpha_k \varphi^{(r)}(x - k\Delta x) = \sum_{k=0}^{N-1} \left\{ \int_{-\infty}^{\infty} a(\xi)\varphi^{(s)}(\xi - k\Delta x)\,d\xi \right\} \varphi^{(r)}(x - k\Delta x)$$

$$= \int_{-\infty}^{\infty} a(\xi)\,d\xi \sum_{k=0}^{N-1} \varphi^{(s)}(\xi - k\Delta x)\varphi^{(r)}(x - k\Delta x) = \int_{-\infty}^{\infty} a(\xi)h^{(s\&r)}(x,\xi)\,d\xi, \quad (3.94)$$

where

$$h^{(s\&r)}(x,\xi) = \sum_{k=0}^{N-1} \varphi^{(s)}(\xi - k\Delta x)\varphi^{(r)}(x - k\Delta x) \tag{3.95}$$

is an overall PSF of the sampling and reconstruction devices.

As we saw in the section "The Sampling Theorem and Signal Sampling," it is easier to analyze those procedures in Fourier domain. Find the Fourier spectrum $\alpha^{(r)}(f)$ of the reconstructed signal $a^{(r)}(x)$ in its connection with spectrum $\alpha(f)$ of the initial signal:

$$\alpha^{(r)}(f) = \int_{-\infty}^{\infty} a^{(r)}(x)\exp(i2\pi fx)\,dx = \int_{-\infty}^{\infty} \left\{ \int_{-\infty}^{\infty} a(\xi)h^{(s\&r)}(x,\xi)\,d\xi \right\} \exp(i2\pi fx)\,dx$$

$$= \int_{-\infty}^{\infty}\int_{-\infty}^{\infty} \left\{ \int_{-\infty}^{\infty} \alpha(p)\exp(-i2\pi p\xi)\,dp \right\} h^{(s\&r)}(x,\xi)\exp(i2\pi fx)\,dx\,d\xi$$

$$= \int_{-\infty}^{\infty} \alpha(p)\,dp \int_{-\infty}^{\infty}\int_{-\infty}^{\infty} h^{(s\&r)}(x,\xi)\exp\left[i2\pi(fx - p\xi)\right]dp\,dx\,d\xi$$

$$= \int\limits_{-\infty}^{\infty} \alpha(p)FR^{(s\&r)}(f,p)\mathrm{d}p, \qquad (3.96)$$

where $FR^{(s\&r)}(f, p)$ is the overall frequency response of the sampling and reconstruction procedures, Fourier transform of their overall PSF:

$$FR^{(s\&r)}(f,p) = \int\limits_{-\infty}^{\infty} h^{(s\&r)}(x,\xi)\exp\left[i2\pi(fx - p\xi)\right]\mathrm{d}x\,\mathrm{d}\xi. \qquad (3.97)$$

Substitute Equation 3.95 into Equation 3.97 and obtain:

$$FR^{(s\&r)}(f,p) = \int\limits_{-\infty}^{\infty}\int\limits_{-\infty}^{\infty}\sum_{k=0}^{N-1}\varphi^{(s)}(\xi - k\Delta x)\varphi^{(r)}(x - k\Delta x)\exp\left[i2\pi(fx - p\xi)\right]\mathrm{d}x\,\mathrm{d}\xi$$

$$= FR^{(r)}(f)FR^{*(s)}(p)\Big\{N\,\mathrm{sincd}\left[N;\pi(f - p)N\Delta x\right]$$

$$\times \exp\left[i\pi(f - p)(N - 1)\Delta x\right]\Big\}, \qquad (3.98)$$

where $FR^{(s)}(\cdot)$ and $FR^{(r)}(\cdot)$ are frequency responses of signal sampling and reconstruction devices, defined by Equations 3.77 and 3.83, and

$$\mathrm{sincd}(N;x) = \frac{\sin x}{N\sin(x/N)} \qquad (3.99)$$

is the *discrete sinc-function*, discrete analog of the sinc-function. Note that Equation 3.98 is a special case of Equation 4.24 for overall frequency response of digital filters.

Because of the finite number of signal samples N used for signal reconstruction from samples, signal sampling and reconstruction procedures are not shift invariant and behave differently on signal borders and on signal parts that are far from the borders. Border effects add additional artifacts in the reconstructed signal. When the number of signal samples increases, border effects diminish and therefore can be neglected for some sufficiently large N.

Consider the asymptotic behavior of $FR^{(s\&r)}(f, p)$ when $N \to \infty$. By Poisson's summation formula (Equation 3.79), obtain from Equation 3.102:

$$\lim_{N\to\infty} FR^{(s\&r)}(f,p) = \Delta x FR^{(r)}(f)FR^{*(s)}(p)\sum_{m=-\infty}^{\infty}\delta\left(f - p + \frac{m}{\Delta x}\right). \qquad (3.100)$$

Substitution of Equation 3.100 in Equation 3.96 gives

$$\lim_{N \to \infty} \alpha_r(f) = \int_{-\infty}^{\infty} \alpha(p) \cdot FR^{*(s)}(p) \cdot FR^{(r)}(f) \sum_{m=-\infty}^{\infty} \delta\left(f - p + \frac{m}{\Delta x}\right) dp$$

$$= FR^{(r)}(f) \sum_{m=-\infty}^{\infty} \alpha\left(f + \frac{m}{\Delta x}\right) FR^{*(s)}\left(f + \frac{m}{\Delta x}\right) = FR^{(r)}(f) FR^{*(s)}(f) \alpha(f)$$

$$+ FR^{(r)}(f) \sum_{m=1}^{\infty} \left[\alpha^{(s)}\left(f + \frac{m}{\Delta x}\right) + \alpha^{(s)}\left(f - \frac{m}{\Delta x}\right)\right]. \tag{3.101}$$

Equation 3.101 shows that the signal $a^{(r)}(x)$ reconstructed from samples of signal $a(x)$ consists of two components. The first component (the first term of Equation 3.104) is a copy of signal $a(x)$, modified by its convolution with PSFs of the sampling and reconstruction devices. This component does not contain aliasing terms and specifies deviation of reconstructed signal spectrum from the initial one:

$$\varepsilon_{sp}(f) = \left[1 - FR^{(r)}(f) FR^{*(s)}(f)\right] \alpha(f). \tag{3.102}$$

The second component is an aliasing one. Its spectrum $\varepsilon_{alsng}(f)$ consists of periodical replicas $\{\alpha^{(S)}(f \pm m/\Delta x)\}$ of spectrum $\alpha^{(s)}(f)$ of sampled signal $a^{(s)}(x)$ (Equation 3.76):

$$\varepsilon_{alsng}(f) = FR^{(r)}(f) \sum_{m=1}^{\infty} \left\{\alpha\left(f + \frac{m}{\Delta x}\right) FR^{(s)}\left(f + \frac{m}{\Delta x}\right)\right.$$

$$\left. + \alpha\left(f - \frac{m}{\Delta x}\right) FR^{(s)}\left(f - \frac{m}{\Delta x}\right)\right\}. \tag{3.103}$$

A reasonable measure to characterize this component is its energy computed on average over the signal ensemble

$$\overline{\varepsilon_{alsng}^2} = AV_\Omega \left[\int_{-\infty}^{\infty} \left|\varepsilon_{alsng}(f)\right|^2 df\right], \tag{3.104}$$

which, as it is shown in Appendix, can be computed as

$$\overline{\varepsilon_{alsng}^2} = 2 \int_{-\infty}^{\infty} \left|FR^{(r)}(f)\right|^2 \left\{\sum_{m=1}^{\infty} SD_a\left(f + \frac{m}{\Delta\xi}\right) \left|FR^{(s)}\left(f + \frac{m}{\Delta\xi}\right)\right|^2\right\} df$$

$$+ 2 \int\limits_{-\infty}^{\infty} \left| FR^{(r)}(f) \right|^2 \left\{ \sum_{m=1}^{\infty} SD_a \left(f - \frac{m}{\Delta \xi} \right) \left| FR^{(s)} \left(f - \frac{m}{\Delta \xi} \right) \right|^2 \right\} df. \qquad (3.105)$$

Interpretation of this formula is straightforward from Figure 3.12.

Sampling Artifacts: Qualitative Analysis

Although Equations 3.102 and 3.105 provide certain numerical characterization of signal distortions due to sampling, they do not tell much about their qualitative features, such as, their visual appearance in images or their influence on readability of images and performing other image analysis tasks. In this section, we will address these issues.

One of the most characteristic and clearly visible sampling artifacts is a stroboscopic reduction of frequency of periodical signal components with frequencies that exceed the highest frequency $1/2\Delta x$ of the signal baseband as defined by the sampling interval Δx. If properly prefiltered, such components are removed from the signal. Otherwise their replicated copies will get inside the baseband and appear with reduced frequencies. Specifically, a component with frequency $f > 1/2\Delta x$ appears in the reconstructed signal with frequency $(1/\Delta x - f)$. Figure 3.16 illustrates this phenomenon, called the *strobe effect*, for 1D signals.

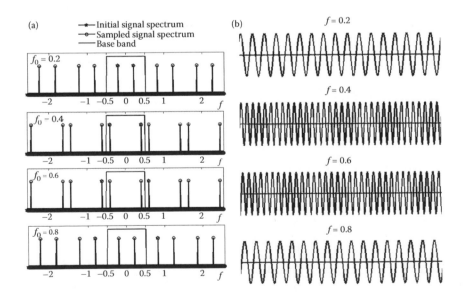

FIGURE 3.16
Strobe effects in sampling periodical signals: (a) from top to bottom, spectra of a sinusoidal signals with frequencies 0.2, 0.4, 0.6, and 0.8 (in fraction of the baseband); (b) corresponding reconstructed signals.

(a)   (b)

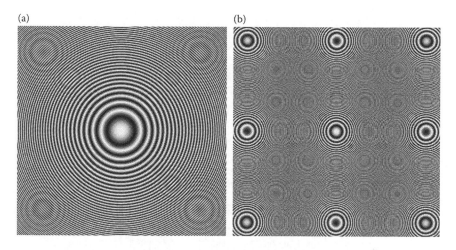

FIGURE 3.17
Strobe effect in sampling a 2D periodical signal $\cos[2\pi\kappa(x^2 + y^2)]$ ((x, y) are vertical and horizontal coordinates centered in the centers of images): (a) signal with local frequencies that do no exceed the sampling rate; (b) signal with a larger value of the parameter κ which results in higher local frequencies that exceed the sampling rate.

One can see from the figure that while signals with frequencies 0.2 and 0.4 fit the baseband and are not affected by sampling, signals with frequencies 0.6 and 0.8 that exceed the baseband border frequency 0.5 are reconstructed as having reduced frequencies, correspondingly, $1 - 0.6 = 0.4$ and $1 - 0.8 = 0.2$. Strobe effects for 2D periodical signals of different frequencies are illustrated in Figure 3.17 generated using MATLAB demo program fringe_aliasing_demo_CRC.m provided in Exercises. In video, strobe effects are observed on moving objects, and in particular, on rotating objects, such as rotating wheels or propellers. On these objects, they cause reducing visible rotation speed up to even its inversion.

While strobe effects appear due to inappropriate prefiltering prior sampling, signal-inappropriate postfiltering at the reconstruction stage results in the appearance of signal ghost high-frequency components from spectrum replicas not removed by the reconstruction postfilter. For periodical signals, these ghost high-frequency components cause low-frequency "beatings" with original high-frequency components as illustrated in Figure 3.18. As beatings form moiré patterns, these aliasing effects are called *moiré effects*. Figure 3.19 illustrates aliasing artifacts in real-life images that contain periodical patterns.

Appropriate pre- and postfiltering may have a profound positive effect on performing image analysis tasks. Figure 3.20, generated using MATLAB program IdealVsNonidealSampling_CRC.m provided in Exercises, illustrates this on an example of their influence on readability of a printed text.

A remarkable demonstration that appropriate postfiltering can even change the visual content of images is given through a painting by Salvador

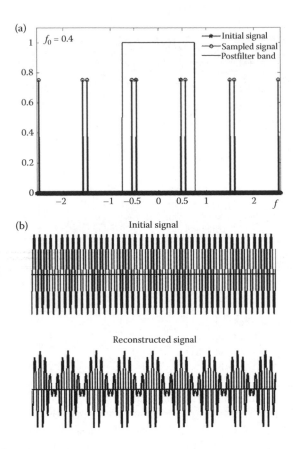

FIGURE 3.18

Moire effects in sampling sinusoidal signals. For sinusoidal signal with frequency 0.4 (in fraction of the signal base band) shown in the upper plot in (b), postfiltering in the band that exceeds the base band ([−0.5, 0.5]) does not suppress in the spectrum of sampled signal (a) the replica with frequency 0.6, which results in the appearance in the reconstructed signal (bottom plot in (b)) of beating between two signals, the original and the ghost.

Dali "Gala Contemplating the Mediterranean Sea which at Twenty Meters becomes a Portrait of Abraham Lincoln" shown in Figure 3.21: pixilated image around figure of Gala converts to the portrait of A. Lincoln when painting is viewed with low resolution from a distant position, which corresponds to low-pass filtering.

Alternative Methods of Discretization in Imaging Devices

In many digital imaging devices, discretization basis functions and restoration basis functions belong to different families of functions. While restoration basis functions implemented in commonly used display and printing

(a)   (b)

(c)   (d)

FIGURE 3.19
Sampling artifacts in an image that contains periodical patterns: (a) initial image; (b) image reconstructed after sampling and reconstruction with an ideal low-pass filter as pre- and post-filters; (c) image reconstructed after sampling without any pre- and postfiltering with rect-function as filter PSF; (d) image reconstructed after sampling with pre- and postfiltering using filter with rect-function as filter PSF. In all cases, sampling interval is 1/70 of the image height.

devices are always shift (convolution) ones, in many modern imaging methods, discretization basis functions are other than shift ones. Most immediate examples are *coded aperture imaging, holographic imaging, computed tomography,* and *magnetic resonance imaging (MRI),* to name a few. Discrete data collected in the discretization process in such devices should be transformed into image samples for image display or printing. This process, which is usually carried out in computers, is called *image reconstruction.*

Coded aperture imaging methods were suggested for nonoptical imaging such as x-ray, gamma-ray, and other nuclear radiation imaging when no optical focusing devices are available. They are an alternative to pinhole cameras that also build images directly without any need of focusing the radiation.

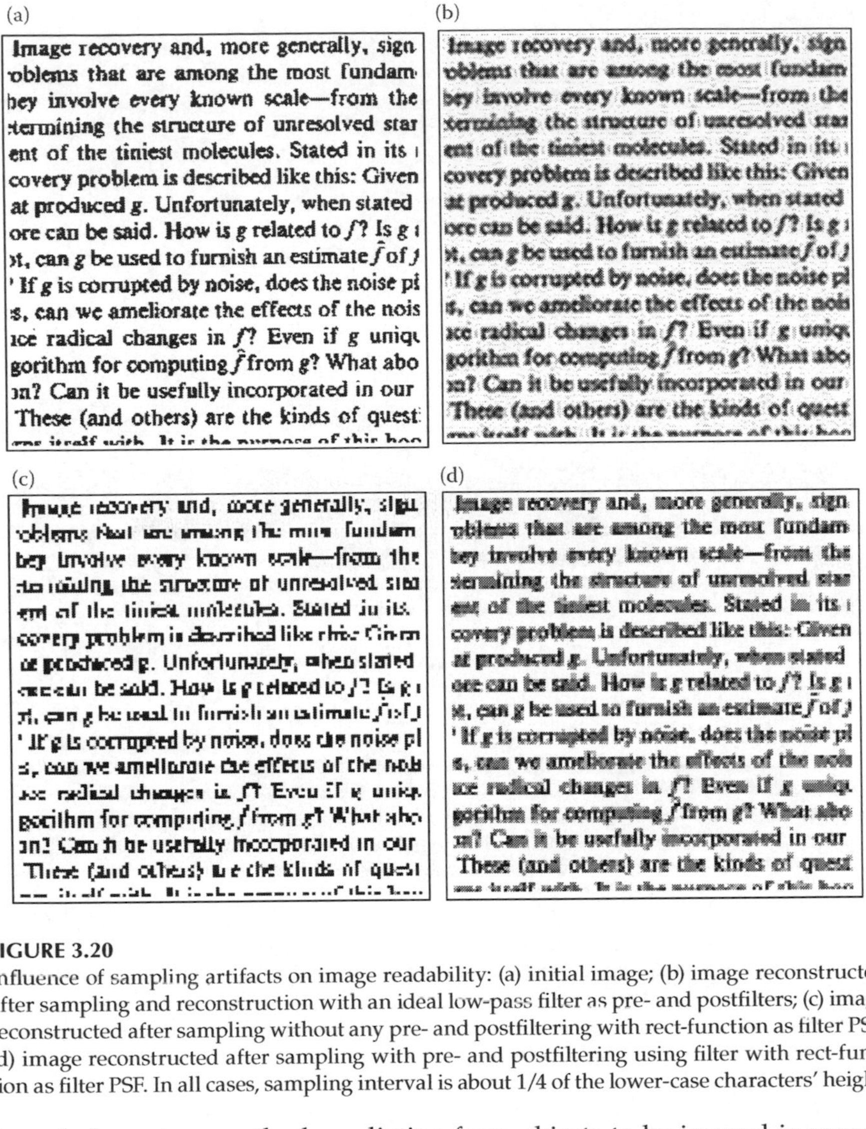

(a)

Image recovery and, more generally, sign
oblems that are among the most fundam·
hey involve every known scale—from the
:termining the structure of unresolved star
ent of the tiniest molecules. Stated in its |
covery problem is described like this: Given
at produced g. Unfortunately, when stated
ore can be said. How is g related to f? Is g |
>t, can g be used to furnish an estimate f of f
' If g is corrupted by noise, does the noise pi
:s, can we ameliorate the effects of the nois
ice radical changes in f? Even if g uniqu
gorithm for computing f from g? What abo
>n? Can it be usefully incorporated in our
These (and others) are the kinds of quest
 are itself with It is the purpose of this hoo

(b)

Image recovery and, more generally, sign
oblems that are among the most fundam
hey involve every known scale—from the
termining the structure of unresolved star
ent of the tiniest molecules. Stated in its |
covery problem is described like this: Given
at produced g. Unfortunately, when stated
ore can be said. How is g related to f? Is g |
>t, can g be used to furnish an estimate f of f
' If g is corrupted by noise, does the noise pi
s, can we ameliorate the effects of the noi
ice radical changes in f? Even if g uniqu
gorithm for computing f from g? What abo
>n? Can it be usefully incorporated in our
These (and others) are the kinds of quest
are itself with It is the purpose of this hoo

(c)

Image recovery and, more generally, sign
oblems that are among the most fundam
hey involve every known scale—from the
termining the structure of unresolved star
ent of the tiniest molecules. Stated in its
covery problem is described like this: Given
at produced g. Unfortunately, when stated
ore can be said. How is g related to f? Is g |
>t, can g be used to furnish an estimate f of f
' If g is corrupted by noise, does the noise pi
s, can we ameliorate the effects of the noi
ice radical changes in f? Even if g uniqu
gorithm for computing f from g? What abo
>n? Can it be usefully incorporated in our
These (and others) are the kinds of quest

(d)

Image recovery and, more generally, sign
oblems that are among the most fundam
hey involve every known scale—from the
termining the structure of unresolved star
ent of the tiniest molecules. Stated in its |
covery problem is described like this: Given
at produced g. Unfortunately, when stated
ore can be said. How is g related to f? Is g |
>t, can g be used to furnish an estimate f of f
' If g is corrupted by noise, does the noise pi
s, can we ameliorate the effects of the noi
ice radical changes in f? Even if g uniqu
gorithm for computing f from g? What abo
>n? Can it be usefully incorporated in our
These (and others) are the kinds of quest
are itself with It is the purpose of this hoo

FIGURE 3.20
Influence of sampling artifacts on image readability: (a) initial image; (b) image reconstructed
after sampling and reconstruction with an ideal low-pass filter as pre- and postfilters; (c) image
reconstructed after sampling without any pre- and postfiltering with rect-function as filter PSF;
(d) image reconstructed after sampling with pre- and postfiltering using filter with rect-function as filter PSF. In all cases, sampling interval is about 1/4 of the lower-case characters' height.

In coded aperture methods, radiation from objects to be imaged is sensed
through binary (transparent/opaque) masks ("coded apertures") such as, for
instance, 2D Walsh function masks shown in Figure 3.22, or by means of
correspondingly wired arrays of parallel sensors. In our terminology, these
masks implement discretization basis functions. In such binary masks, half
of their area is transparent, therefore radiation energy is collected over half
of the image area. Thus, they provide better ratio of signal-to-quantum noise
at sensor's output than pinhole cameras, which collect radiation energy only
within the area of the pinhole. The gain amounts to the order of $\sqrt{N}$, where
N is the number of pixels in reconstructed images.

FIGURE 3.21
A close-up view of the painting by Salvador Dali "Gala Contemplating the Mediterranean Sea which at Twenty Meters becomes a Portrait of Abraham Lincoln" (a) and same painting as seen with low resolution from a distant point (b), which is equivalent to low-pass filtering.

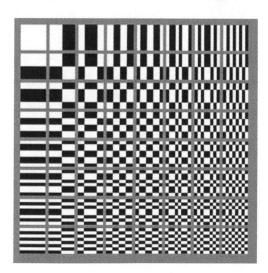

FIGURE 3.22
8×8 array of 2D Walsh basis function for $N = 8$.

In *numerical reconstruction of holograms*, sampled are holograms, that is, integral Fourier, Fresnel, or Kirchhoff transforms of objects under study. Sampling is performed by photographic cameras in special optical set-ups, and the results of sampling are used for image reconstruction in the computer. We will detail this process in the Section "Computer-Generated Display Holography" in Chapter 5.

In *computed tomography*, slices of bodies are imaged, and 2D discretization is separable in a polar coordinate system: linear array of sensors rotates around the body to be imaged and, for a discrete set of observation angles, samples of projections of body slices, that is, of their Radon transforms described in the section "Imaging from Projections and Radon Transform" in Chapter 2 are collected. The obtained discrete data are then used for the reconstruction of slice images in computers. In the section "Digital Image Formation by Means of Numerical Reconstruction of Holograms" in Chapter 5, we will overview corresponding basic algorithms.

In *MRI* or *NMR* (nuclear magnetic resonance) *imaging*, objects to be imaged are placed in a strong and spatially nonhomogeneous magnetic field. When an electromagnetic excitation signal on a radiofrequency is applied to the object, the object reemits the signal but modulates its amplitude and frequency. The intensity of the radiofrequency signal reemitted by different elements of the body volume is proportional to the density of protons in these volume elements and its frequency is determined by the strength of the magnetic field in their coordinates. Because the magnetic field is spatially nonhomogeneous, the frequency of the reemitted signal carries information about the spatial coordinates of the elements. Therefore, for collecting data about distribution of proton density within the body, the reemitted signal is sampled in its Fourier domain and, therefore, the discretization bases functions are sinusoidal ones.

Signal Scalar Quantization

Optimal Quantization: Principles

Scalar (element-wise) quantization is the second stage of signal digitization. It is applied to coefficients of signal discrete representation obtained as the result of signal discretization. Scalar quantization implies that a finite interval first has to be specified in the entire range of the signal representation coefficient values α by defining their minimum α_{min} and maximum α_{max} and then interval $[\alpha_{min}, \alpha_{max}]$ is split into a certain number $Q - 2$ of *quantization intervals* by defining their border values $\{\alpha^{(q)}\}$, $q = 0, 1, \ldots,$ $Q - 2$, Q being the total number of quantization levels. Quantization intervals are indexed by an integer index and for each particular q-th interval, its representative value, or *quantization level* $\aleph^{(q)}$, is chosen. In signal reconstruction,

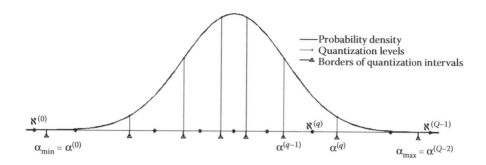

FIGURE 3.23
Signal values probability density and arrangement of signal quantization intervals and quantization levels.

all values within a particular quantization interval are replaced with its representative quantization level. Figure 3.23 illustrates these concepts.

Difference $\varepsilon^{(q)} = \alpha - \aleph^{(q)}$ between true value α and its corresponding quantization level $\aleph^{(q)}$ is called *quantization error*. One should distinguish quantization errors within the selected dynamic range

$$\varepsilon^{(q)} = \alpha - \aleph^{(q)}; \alpha \in \left[\alpha^{(q-1)}, \alpha^{(q)}\right], \quad 1 \le q \le Q - 2 \tag{3.106}$$

from those outside the boundaries of the dynamic range:

$$\varepsilon^{(min)} = \alpha - \aleph^{(0)}; \quad \alpha < \alpha_{min}; \tag{3.107}$$

$$\varepsilon^{(max)} = \alpha - \aleph^{(Q-1)}; \quad \alpha > \alpha_{max}. \tag{3.108}$$

Quantization errors within the dynamic range are limited in the range by the size of the quantization intervals, whereas dynamic range limitation errors may, in principle, be unlimited in value.

The arrangement of quantization intervals and selection of quantization levels is governed by requirements to the accuracy of signal-quantized representation, which are generally formulated in terms of certain constraints imposed on the quantization errors. With the most common approach to formulating the constraints, losses due to quantization errors are evaluated over all possible coefficient values on average according to their probability density. To this goal, loss functions $D_{lr}(\varepsilon^{(q)})$, $D_l(\varepsilon^{(min)})$, $D_r(\varepsilon^{(max)})$ are introduced that measure losses owing to the quantization errors within and outside the dynamic range quantization and precision of signal reconstruction is evaluated, separately for quantization errors within and outside the dynamic range, as

$$\bar{D}_l = \int\limits_{-\infty}^{\alpha_{min}} p(\alpha)D_l(\varepsilon^{(min)})d\alpha, \tag{3.109}$$

$$\bar{D}_r = \int_{\alpha_{max}}^{\infty} p(\alpha)D_r(\varepsilon^{(max)})d\alpha, \tag{3.110}$$

and

$$\bar{D}_{lr} = \sum_{q=1}^{Q-2} \int_{\alpha^{(q-1)}}^{\alpha^{(q)}} p(\alpha)D_{lr}(\varepsilon^{(q)})d\alpha; \quad \alpha^{(0)} = \alpha_{min}; \alpha^{(Q-2)} = \alpha_{max}, \tag{3.111}$$

where $p(\alpha)$ is the probability density of the values under quantization.

Equations 3.109 and 3.110 can be used for determining dynamic range boundaries $\{\alpha_{min}, \alpha_{max}\}$ and their corresponding quantization levels given the constraints to dynamic range limitation error measures $\bar{D}_l$ and $\bar{D}_r$. Equation 3.111 can be used for determining sets of quantization intervals boundaries $\{\alpha^{(q)}\}$ and quantization levels $\{\aleph^{(q)}\}$ that minimize the number of quantization levels Q given the average quantization error measure $\bar{D}_{lr}$, or minimize the average quantization error measure given the number of quantization levels. We call quantizers designed in this way *optimal scalar quantizers*.

Design of Optimal Quantizers

There are two approaches to the design of optimal quantizers, a direct optimization approach and a *compressor–expander (compander)* approach. Direct optimization approach assumes solving the optimization equation

$$\left\{\alpha_{opt}^{(q)}, \aleph_{opt}^{(q)}\right\} = \underset{\{\alpha^{(q)}, \aleph^{(q)}\}}{\arg\min} \left\{ \sum_{q=1}^{Q-2} \int_{\alpha^{(q-1)}}^{\alpha^{(q)}} p(\alpha)D_{lr}\left(\alpha - \aleph^{(q)}\right)d\alpha \right\}. \tag{3.112}$$

The optimization is simplified if loss function $D_{lr}(\varepsilon^{(q)})$ is an even function: $D_{lr}(\varepsilon) = D_{lr}(-\varepsilon)$. In this case, from

$$\frac{\partial}{\partial \alpha^{(q)}}\left\{ \sum_{q=1}^{Q-2} \int_{\alpha^{(q-1)}}^{\alpha^{(q)}} p(\alpha)D_{lr}\left(\alpha - \aleph^{(q)}\right)d\alpha \right\} = p(\alpha^{(q)})D_{lr}\left(\alpha^{(q)} - \aleph^{(q)}\right)$$

$$- p\left(\alpha^{(q)}\right)D_{lr}\left(\alpha^{(q)} - \aleph^{(q-1)}\right) = 0 \tag{3.113}$$

it follows that optimal boundaries of the quantization intervals should be placed halfway between the corresponding quantized values:

$$\alpha_{opt}^{(q)} = \left(\aleph_{opt}^{(q-1)} + \aleph_{opt}^{(q)}\right)/2. \tag{3.114}$$

For the quadratic loss function $D_{lr}(\varepsilon^{(q)}) = (\varepsilon^{(q)})^2$, a further simplification of the optimization search is possible. From

$$\frac{\partial}{\partial \aleph^{(q)}} \left\{ \sum_{q=1}^{Q-2} \int_{\alpha^{(q-1)}}^{\alpha^{(q)}} p(\alpha) D_{lr}\left(\alpha - \aleph^{(q)}\right) d\alpha \right\} = \int_{\alpha^{(q-1)}}^{\alpha^{(q)}} p(\alpha) \frac{\partial}{\partial \aleph^{(q)}} \left(\alpha - \aleph^{(q)}\right)^2 d\alpha$$

$$= -2 \int_{\alpha^{(q-1)}}^{\alpha^{(q)}} p(\alpha)(\alpha - \aleph^{(q)}) d\alpha = 0 \qquad (3.115)$$

it follows that in this case optimal quantization levels are centers of mass of the probability density within the quantization intervals:

$$\aleph_{opt}^{(q)} = \int_{\alpha^{(q-1)}}^{\alpha^{(q)}} \alpha p(\alpha) d\alpha \bigg/ \int_{\alpha^{(q-1)}}^{\alpha^{(q)}} p(\alpha) d\alpha. \qquad (3.116)$$

Such a solution of the quantization optimization problem for the quadratic loss function is called *Max–Lloyd quantization*.

Compressor–expander quantization is a method suited for hardware implementation of nonuniform optimal quantization using nonlinear analog signal amplifiers and readily available uniform quantizers in which signal dynamic range is split into quantization intervals of equal size, and centers of the intervals are used as quantization levels. In order to implement a nonuniform quantization with uniform quantizers, signal, before being sent to the uniform quantizer, has to be subjected to an appropriate nonlinear point-wise transformation. Correspondingly, at the signal reconstruction stage, uniformly quantized values have to be subjected to a nonlinear transformation inverse to the one used for quantization. Usually, the required nonlinear prequantization transformation compresses the signal dynamic range, thus the name "compressor–expander quantization." Flow diagram of the compressor–expander quantization and reconstruction is presented in Figure 3.24.

Optimization of the compressor–expander quantization is achieved by an appropriate selection of the compressive nonlinear point-wise transformation. The optimal transformation transfer function can be found in the following way. Let $w(.)$ be a compression transfer function and $\Delta_{(u)}$ be a uniform quantization interval. Then, for a particular value α to be quantized, quantization interval $\Delta^{(q)}$ that corresponds to uniform quantization of values of function $w(\alpha)$ with quantization interval $\Delta_{(u)}$ will be equal to

$$\Delta^{(q)} = \frac{\Delta_u}{dw(\alpha)/d\alpha}\bigg|_{\alpha = \frac{a^{(q)} - a^{(q-1)}}{2}}, \quad \alpha^{(q-1)} \leq \alpha \leq \alpha^{(q)}. \qquad (3.117)$$

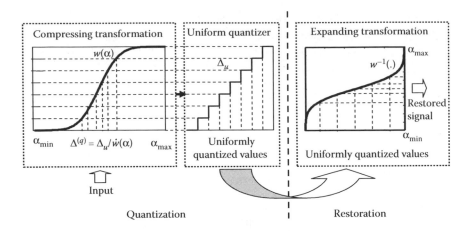

FIGURE 3.24
Schematic diagram of the compressor/expander quantization and restoration.

This interval determines the quantization error for the Δ_u-th quantization interval. In order to enable analytical solution for optimal selection of $w(.)$, modify slightly the quantization error measure incorporated in the optimization Equation 3.112: introduce a virtual quantization interval

$$\Delta_\alpha = \frac{\Delta_u}{dw(\alpha)/d\alpha} = \frac{\Delta_u}{\dot{w}(\alpha)}, \quad \alpha^{(min)} \le \alpha \le \alpha^{(max)}, \tag{3.118}$$

a modified quantization loss function $\tilde{D}_{lr}(\Delta_\alpha)$ as measure of losses due to quantization error in the range of Δ_α and replace integration of quantization loss function over individual quantization intervals and summation over all intervals in Equation 3.112 by integration of the modified loss function $\tilde{D}_{lr}(\Delta_\alpha)$ over the range of quantized values. With these modifications, the optimization Equation 3.112 for optimal transfer function $w(\alpha)$ can be rewritten as

$$w_{opt}(\alpha) = \underset{w(\alpha)}{\arg\min} \left\{ \int_{\alpha_{min}}^{\alpha_{max}} p(\alpha)\tilde{D}_{lr}\left(\frac{\Delta_u}{\dot{w}(\alpha)}\right) d\alpha \right\}. \tag{3.119}$$

This equation can be solved using the Euler–Lagrange equation, which in this case takes the form

$$\frac{\partial}{\partial \dot{w}(\alpha)} \left\{ p(\alpha)\tilde{D}_{lr}\left(\frac{\Delta_u}{\dot{w}(\alpha)}\right) \right\} = \text{const.} \tag{3.120}$$

Consider some special cases. We will begin with threshold quantization loss functions that assume that quantization error is nil if it does not exceed a certain threshold. For *threshold loss functions*, optimal arrangement of quantization intervals does not depend on the probability distribution $p(\alpha)$ because zero average losses can be secured if quantization errors are kept under the threshold level. The following two examples represent a particular practical interest.

Example 3.1: Uniform Threshold Criterion for Absolute Value of Quantization Error

$$\tilde{D}_{lr}(\Delta_\alpha) = \begin{cases} 0, & \Delta_\alpha \le \Delta_{thr} \\ 1, & \text{otherwise} \end{cases}. \tag{3.121}$$

From Equations 3.117 and 3.121, it follows that, for monotonic functions $w(\alpha)$. $\bar{D}_{lr} = 0$ if

$$\dot{w}(\alpha) = \Delta_u / \Delta_{thr}, \tag{3.122}$$

and, therefore, $w_{opt}(\alpha)$ is a linear function

$$\frac{w_{opt}(\alpha) - w_{opt}(\alpha_{min})}{w_{opt}(\alpha_{max}) - w_{opt}(\alpha_{min})} = \frac{\alpha - \alpha_{min}}{\alpha_{max} - \alpha_{min}}. \tag{3.123}$$

Therefore, in this case, optimal quantization is the standard uniform quantization with quantization interval and the number of quantization levels equal to, respectively

$$\Delta_u = \frac{w_{opt}(\alpha_{max}) - w_{opt}(\alpha_{min})}{\alpha_{max} - \alpha_{min}} \Delta_{trh}, \quad Q = \frac{\alpha_{max} - \alpha_{min}}{\Delta_{thr}}. \tag{3.124}$$

Example 3.2: Uniform Threshold Criterion of Relative Quantization Error

$$\bar{D}_{dr}(\Delta_\alpha) = \begin{cases} 0, & \Delta_\alpha \le \Delta_{thr} = \delta_{thr}\alpha \\ 1, & \text{otherwise} \end{cases}. \tag{3.125}$$

In this case, $\bar{D}_{lr} = 0$ if

$$\dot{w}(\alpha) = \Delta_u / \delta_{thr}\alpha, \tag{3.126}$$

and, therefore, uniform quantization in a logarithmic scale is optimal:

$$\frac{w_{opt}(\alpha) - w_{opt}(\alpha_{min})}{w_{opt}(\alpha_{max}) - w_{opt}(\alpha_{min})} = \frac{\ln(\alpha/\alpha_{min})}{\ln(\alpha_{max}/\alpha_{min})} \tag{3.127}$$

with the number of quantization levels defined as

$$
\begin{aligned}
Q &= \frac{w_{opt}(\alpha_{max}) - w_{opt}(\alpha_{min})}{\Delta_u} \\
&= \frac{w_{opt}(\alpha_{max}) - w_{opt}(\alpha_{min})}{\alpha \dot{w}_{opt}(\alpha)\delta_{thr}} \\
&= \frac{w_{opt}(\alpha_{max}) - w_{opt}(\alpha_{min})}{\alpha\left\{\left[w_{opt}(\alpha_{max}) - w_{opt}(\alpha_{min})\right]/\alpha\ln(\alpha_{max}/\alpha_{min})\right\}\delta_{thr}} \\
&= \frac{\ln(\alpha_{max}/\alpha_{min})}{\delta_{thr}}.
\end{aligned}
\tag{3.128}
$$

This particular case is of a special interest in image processing. In images, artifacts of quantization of image pixel gray levels most frequently exhibit themselves in appearance of what is called *"false contours,"* visible boundaries between image patches quantized to different quantization levels. This phenomenon is illustrated in Figure 3.25.

A natural primary requirement for quantization of image gray levels is to secure invisibility of false contours in displayed digital images. The threshold of visibility of patches of constant gray level depends on their contrast with respect to their background and of size of patches. Quantitative estimates can be made using the *Weber–Fechner's law*: the ratio of brightness contrast ΔI_{thr} of a stimulus on the threshold of its visibility to background of brightness I is constant for a wide range of the brightness:

$$
\frac{\Delta I_{thr}}{I} = \delta_{thr} = \text{const.}
\tag{3.129}
$$

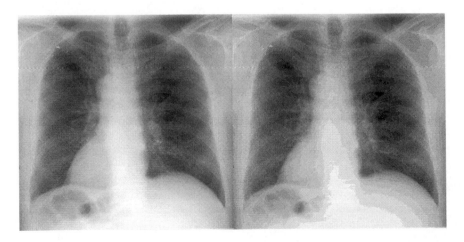

FIGURE 3.25
False contours in image quantization: initial test image (left) and same image quantized uniformly to 16 quantization levels (right), in which false patches and their borders become visible.

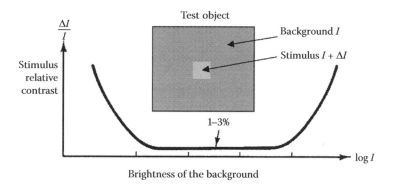

FIGURE 3.26
The Weber–Fechner law on the threshold visibility of a stimulus on a uniform background.

This law is illustrated in Figure 3.26, which shows the dependence of the visibility threshold of stimulus relative contrast on the brightness of stimulus background (in a logarithmic scale).

The value of the threshold contrast δ_{thr} depends on the stimulus size. Its lowest value of 1–3% shown in the figure corresponds to stimuli of sufficiently large angular size (of the order of 10 or more angular resolution elements). From the above example and Weber–Fechner's law, it follows that optimal, for digital image display, quantization of image samples should be uniform in the logarithmic scale. The number Q of required quantization levels for logarithmic quantization can be found from Equation 3.121. For good-quality photographic, TV, and computer displays, dynamic range $\alpha_{max}/\alpha_{min}$ of displayed image brightness is about 100. Using this estimation of the dynamic range and taking visual sensitivity threshold equal to 2%, obtain from Equation 3.121 that the required number of quantization levels Q for which false contours are under visibility threshold is of the order of $\ln 100/0.02 = 2 \cdot 2.43/0.02 = 243$. Remarkably, similar figures are characteristic also for human audio sensibility. This was the main reason why 256 quantization levels (8 bits) were chosen as a standard for gray-scale image, sound, and other analog signal representation and why 8 bits (byte) had become a basic unit in the computer industry.

Consider now examples of "soft" (nonthreshold) criteria.

Example 3.3: Exponential Criterion of Absolute Quantization Error

$$\tilde{D}_{lr}(\Delta_{\alpha}) = \Delta_{\alpha}^{2G} = \left| \Delta_u / \dot{w}(\alpha) \right|^{2G}. \tag{3.130}$$

Substituting Equation 3.130 into Equation 3.120 and solving the resulting differential equation, obtain in this case

$$\frac{w_{opt}(\alpha) - w_{opt}(\alpha_{min})}{w_{opt}(\alpha_{max}) - w_{opt}(\alpha_{min})} = \frac{\int\limits_{\alpha_{min}}^{\alpha} [p(\alpha)]^{1/(2G+1)} d\alpha}{\int\limits_{\alpha_{min}}^{\alpha_{max}} [p(\alpha)]^{1/(2G+1)} d\alpha} = \frac{\int\limits_{\alpha_{min}}^{\alpha} [p(\alpha)]^{P} d\alpha}{\int\limits_{\alpha_{min}}^{\alpha_{max}} [p(\alpha)]^{P} d\alpha}, \tag{3.131}$$

where $P = 1/(2G + 1)$. Thus, the required nonlinear prequantization transformation depends solely on the probability distribution of the quantized values. The meaning of this relationship becomes evident from the expression

$$\Delta_\alpha = \Delta_u / \dot{w}(\alpha) \propto [p(\alpha)]^{-P}, \tag{3.132}$$

which implies that the width of quantization intervals for the various values of α is inversely proportional to their probability densities raised to the corresponding power. For the widely used mean squared quantization error criterion ($G = 1, P = 1/3$)

$$\frac{w(\alpha) - w(\alpha_{min})}{w(\alpha_{max}) - w(\alpha_{min})} = \frac{\int\limits_{\alpha_{min}}^{\alpha} [p(\alpha)]^{1/3} d\alpha}{\int\limits_{\alpha_{min}}^{\alpha_{max}} [p(\alpha)]^{1/3} d\alpha}. \tag{3.133}$$

Image modification using transfer function defined by Equation 3.131 turned out to be very useful for image enhancement. We will discuss this issue in detail in the section "Filter Classification Tables and Particular Examples" in Chapter 8. We call it *p-histogram equalization*. Its special case, when $G = 0; P = 1$, that is

$$\frac{w(\alpha) - w(\alpha_{min})}{w(\alpha_{max}) - w(\alpha_{min})} = \frac{\int\limits_{\alpha_{min}}^{\alpha} p(\alpha) d\alpha}{\int\limits_{\alpha_{min}}^{\alpha_{max}} p(\alpha) d\alpha} \tag{3.134}$$

is commonly called *histogram equalization* because it converts signal probability density, or, for quantized signals, signal histogram into a uniform one. Histogram equalization is a popular image enhancement transformation. For image quantization, this transformation assumes that quantization errors are equally important whatever their values are (see Equation 3.130) and make quantization intervals inversely proportional to the probability density for the level to be quantized (see Equation 3.132).

Sometimes, as in the case of quantizing spectral coefficients of signals in Fourier, Walsh, and other bases, probability density of quantized coefficients of the discrete signal representation is a truncated Gaussian probability density distribution:

$$p(\alpha) \propto \exp\left[-(\alpha - \bar{\alpha})^2/2\sigma_\alpha^2\right], \quad \alpha_{min} \leq \alpha \leq \alpha_{max}. \tag{3.135}$$

Then, for mean square error criterion ($G = 1$), one can get from Equation 3.120 that (see Appendix)

$$\frac{w(\alpha) - w(\alpha_{min})}{w(\alpha_{max}) - w(\alpha_{min})} = \frac{erf\left(\alpha/\sqrt{6}\sigma_\alpha\right) + erf\left(\alpha_{min}/\sqrt{6}\sigma_\alpha\right)}{erf\left(\alpha_{max}/\sqrt{6}\sigma_\alpha\right) + erf\left(\alpha_{min}/\sqrt{6}\sigma_\alpha\right)}, \tag{3.136}$$

where

$$erf(x) = \frac{2}{\sqrt{\pi}} \int_0^x \exp(-\xi^2)d\xi \tag{3.137}$$

is the *error function*. This compression function is shown in Figure 3.27 along with the probability density function.

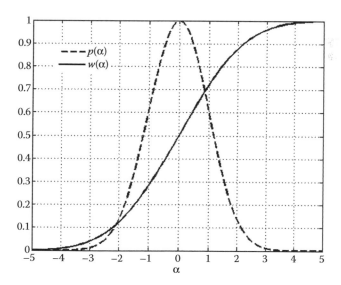

FIGURE 3.27
Optimal compression function as given by Equation 3.136 for quantization variables with normal probability density.

Example 3.4: Exponential Criterion of Relative Quantization Error

$$\tilde{D}_{lr}(\Delta_\alpha) = \left| \Delta_\alpha / \alpha \right|^{2G} = \left(\Delta_u / \dot{w}(\alpha)\alpha \right)^{2G}. \tag{3.138}$$

In this case, solution of the Euler–Lagrange equation (3.120) yields

$$\frac{w_{opt}(\alpha) - w_{opt}(\alpha_{min})}{w_{opt}(\alpha_{max}) - w_{opt}(\alpha_{min})} = \frac{\displaystyle\int_{\alpha_{min}}^{\alpha} \left[p(\alpha)/\alpha^{2G} \right]^{1/(2G+1)} d\alpha}{\displaystyle\int_{\alpha_{min}}^{\alpha_{max}} \left[p(\alpha)/\alpha^{2G} \right]^{1/(2G+1)} d\alpha}. \tag{3.139}$$

If quantized values α are distributed uniformly in the dynamic range $[\alpha_{min}, \alpha_{max}]$, the optimal compression function is

$$\frac{w(\alpha) - w(\alpha_{min})}{w(\alpha_{max}) - w(\alpha_{min})} = \frac{\alpha^{1/(2G+1)} - \alpha_{min}^{1/(2G+1)}}{\alpha_{max}^{1/(2G+1)} - \alpha_{min}^{1/(2G+1)}}. \tag{3.140}$$

We will refer to such type of nonlinear transformations as to "*P-th law quantization*":

$$\frac{w(\alpha) - w(\alpha_{min})}{w(\alpha_{max}) - w(\alpha_{min})} = \frac{\alpha^P - \alpha_{min}^P}{\alpha_{max}^P - \alpha_{min}^P}, \tag{3.141}$$

where the exponent P is a transformation parameter. For the mean square relative quantization error criterion (Equation 3.130), $G = 1$, the optimal is $P = 1/3$:

$$\frac{w_{opt}(\alpha) - w_{opt}(\alpha_{min})}{w_{opt}(\alpha_{max}) - w_{opt}(\alpha_{min})} = \frac{\alpha^{1/3} - \alpha_{min}^{1/3}}{\alpha_{max}^{1/3} - \alpha_{min}^{1/3}}. \tag{3.142}$$

The P-th law compression transformation, thanks to its simple parameterization through a single parameter P, proved to be a useful approximation to optimal compression transformation for quantizing variables with normal probability distribution, in particular, for quantizing of absolute values of DFT and DCT coefficients of images in transform image coding. (We will introduce discrete Fourier and discrete cosine transforms in Chapter 4.) Optimal values of the nonlinearity index P, for which standard deviation of errors in reconstructed images because of quantization of their spectra is minimal, are usually about 0.2–0.3. For instance, Figure 3.28 presents an example of a test image and results of optimization of P-th law quantization of absolute values of its DCT spectral coefficients.

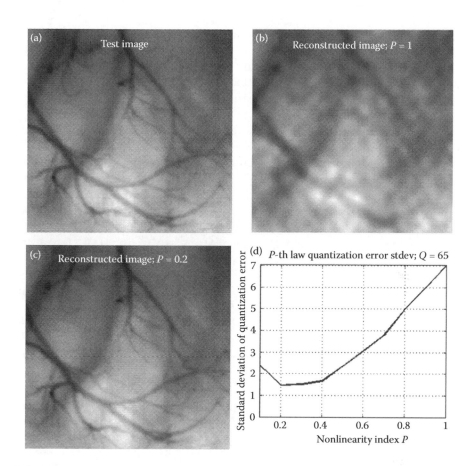

FIGURE 3.28

Optimization of *P*-th law quantization of image DCT spectrum. (a) Initial test image; (b) and (c) images reconstructed from uniformly ($P = 1$) and *P*-th ($P = 0.2$) law quantized spectral coefficients; (d) plot of standard deviation of quantization error as a function on the nonlinearity index *P* for 65 quantization levels, which shows that standard deviation of quantization error is minimal for $P = 0.2$. Note that uniform quantization ($P = 1$) substantially destroys the image.

Quantization in Digital Holography

Numerical reconstruction of electronically recorded hologram and computer-generated holograms, topics united under the name *digital holography*, will be addressed later in Chapter 5. Here, we will discuss some peculiarities of signal quantization in digital holography. First of all, as we will see in the sections "Digital Image Formation by Means of Numerical Reconstruction of Holograms" and "Computer-Generated Display Holography" in Chapter 5, this is quantization in Fourier or Fresnel transform domains. Therefore, in the quantization of holograms, a nonuniform, as, for instance, *P*-th law quantization, is advisable. For numerical reconstruction of images from holograms (see the section "Digital Image Formation by Means of

Numerical Reconstruction of Holograms" in Chapter 5), corresponding correcting expanding transformations should be implemented in hologram preprocessing. For recording computer-generated holograms (see the section "Computer-Generated Display Holography" in Chapter 5), correcting expanding transformation should be implemented in hologram encoding for recording on an optical medium.

In the synthesis of computer-generated display holograms, yet another quantization method, the *pseudorandom diffuser* method, can be employed. The method is aimed at compressing a dynamic range of holograms and makes use of the fact that, for display holograms, reproduction of amplitude is only required in object wavefront, reconstructed from holograms, because human vision senses only the intensity of radiation. For synthesis of the display hologram of an object, object wavefront amplitude and phase should be specified. However, the phase component is irrelevant for visualization. Therefore, it can be selected so as to minimize hologram quantization artifacts. The pseudorandom diffuser method consists of using, for specifying object wavefront phase component, appropriately generated pseudorandom numbers. Note that this imitates, in a certain sense, properties of real object to diffusely scatter light.

The method is mathematically described as follows. Let $\{A_{k,l}^2\}$ be samples, with (k, l) as sample indices, of the given object wavefront intensity and $\{\theta_{k,j}\}$ be an array of pseudorandom numbers taken from the range $[0, 2\pi]$. Then samples of the object wavefront $\tilde{A}_{k,l}$ are defined as:

$$\tilde{A}_{k,l} = A_{k,l} \exp(i\theta_{k,l}), \tag{3.143}$$

which preserves wavefront amplitude and assigns to its samples a pseudorandom phase.

The simplest solution is to use for $\{\theta_{k,j}\}$ binary statistically independent numbers that assume, with equal probabilities, values 0 and π. Such a pseudorandom phase modulation of the object wavefront redistributes wavefront Fourier and Fresnel spectrum energy uniformly (statistically) between all spectral coefficients so that all coefficients have the same dynamic range. Figure 3.29 illustrates this phenomenon of spectrum "uniformization" by means of image pseudorandom phase modulation. Image pseudorandom phase modulation eliminates the need of nonuniform quantization and simplifies the hologram encoding for recording on optical media.

In optical reconstruction of images from computer-generated holograms, image distortions may appear due to quantization for hologram in hologram recording devices. In order to minimize the image reconstruction error, one can optimize the pseudorandom object wavefront phase through an iterative optimization procedure of assignment of the phase. In this process, at each iteration step, a quantization that would be implemented in hologram encoding in applied to the hologram synthesized from the object and iterated object wavefront is reconstructed. Then the amplitude component of the reconstructed

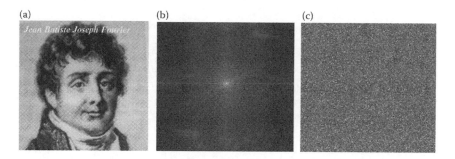

FIGURE 3.29

Image Fourier spectrum "uniformization" by means of pseudorandom phase modulation: (a) test image; (b) Fourier power spectrum of the test image; (c) spectrum of the test image, to which a pseudorandom binary (0 and π) phase component is assigned. Note that, for the display purposes, spectrum (b) is displayed using P-th law transformation ($P = 0.5$), otherwise high-frequency spectral components would be invisible in print.

wavefront is replaced by the given object wavefront amplitude, while the phase component is kept unchanged. The obtained wavefront with this iterated phase component is used for the next iteration. Optimization of the pseudorandom diffuser is especially required for recording *kinoforms*, computer-generated holograms, in which hologram amplitude variations are ignored and are replaced by a constant (see Section "Recording Computer-Generated Holograms on Optical Media" in Chapter 5). Figure 3.30 illustrates that the iterative optimization of pseudorandom phase distribution allows achieving quite good image in image reconstruction from kinoform in spite of ignoring, in hologram recording, its amplitude. In Exercises, a MATLAB program quantization_demo_CRC.m is provided that illustrates issues discussed in this section.

Basics of Image Data Compression

What Is Image Data Compression and Why Do We Need It?

As it was mentioned, image representation in the form of arrays of samples taken in nodes of uniform sampling grids is standard for image displays. This means that any imaging system and image processing should ultimately produce sampled images for displaying.

For sampling, one has to select the sampling interval. For band-limited signals, sampling interval is determined according to the sampling theorem, by signal bandwidth. However, in reality, there are no band-limited signals and sampling interval should be selected on the basis of evaluation of vulnerability of images to sampling artifacts: image blurring due to low-pass prefiltering, strobe, moiré, and other aliasing effects. A good practical rule is to decide how many pixels should be allocated for representing most sharp

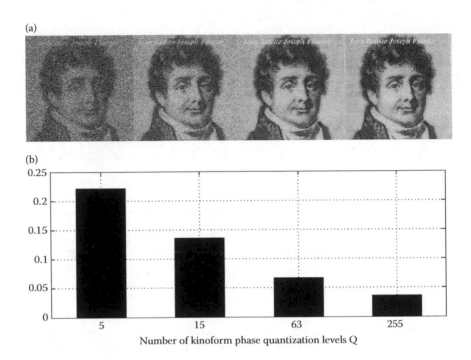

Number of kinoform phase quantization levels Q

FIGURE 3.30
Pseudorandom diffuser optimization for computer generated kinoform of the object given in Figure 3.29a: (a) reconstructed images for (from left to right) 5, 15, 63, and 255 quantization levels of the hologram phase and (b) corresponding standard deviations of image reconstruction normalized error.

image details such as objects' boundaries, or edges. But such details usually occupy a relatively small fraction of image area. Therefore, sampled representation of images is as a rule very redundant in terms of the volume of data needed to produce images because most pixels belong to not edgy areas where image signal is changing much slower and sampling rate selected for edges is, for those areas, excessive. This redundancy amounts to tens and, for some images, even hundreds of times. The same refers to video data represented as sets of video frames that are also highly redundant. In addition to this, scalar quantization of pixels requires excessive number of quantization levels in order to secure invisibility of quantization artifacts, such as false contours, to which most vulnerable are those excessive image samples from nonedgy image areas.

All this means that for image data storage and transmission, sampled image representation can be substantially compressed. The compression is usually a supplemental image processing applied to the "primary" sampled image representation. *Image compression,* or *image coding,* as well as *video compression* are well-established fields covered in many books. In this chapter, we will review only their basic principles beginning with basic facts from information theory.

Signal Rate Distortion Function, Entropy, and Statistical Encoding

The purpose of the signal coding and data compression is encoding signals from a signal ensemble into a stream of binary digits, or bits (binary code) of the least possible length as evaluated on average over the signal ensemble. A fundamental result provided by information theory [7] is that there is a lower bound for the length of binary code required to specify signals from a signal ensemble. This bound is given by the rate distortion function H_ε of the signal ensemble. In terms of signal space and general digitization discussed in the section "Linear Signal Space, Basis Functions and Signal Representation as Expansion over a Set of Basis Functions" in Chapter 2, rate distortion function is defined as minimal entropy of the set of representative signals $\hat{\Omega}_A$ selected to represent signals from the signal ensemble Ω_A with a given accuracy:

$$H_\varepsilon(\Omega_A) = \min_{\Omega_A \Rightarrow \hat{\Omega}_A} \left(-\sum_{\hat{\Omega}_A} P(\hat{A}_k) \log P(\hat{A}_k) \right), \qquad (3.144)$$

where $\hat{A}_k$ is a representative signal for the k-th equivalency cell of the signal space, $P(\hat{A}_k)$ is the probability of the subset of signals that belong to this cell and minimum is sought over all possible mappings of the signal ensemble Ω_A onto the set $\hat{\Omega}_A$ of representative signals provided a given accuracy of the representation. This lower bound can, in principle, be achieved through the general digitization process.

The upper bound of the number of bits per representative signal is determined by the amount N_Ω of the representative signals: it is equal to $\log_2 N_\Omega$, which means that the number of bits per a representative signal H_A sufficient to encode its index lies in the range

$$H_\varepsilon(\Omega_A) \le H_A \le \log_2 N_\Omega \qquad (3.145)$$

the upper bound being achieved when representative signals are equally probable.

As we mentioned, in reality, signal digitization is performed in two steps through discretization and scalar quantization and signals are represented by their quantized representation coefficients over discretization basis functions. The entropy of the obtained ensemble of digital signals, which determines the amount of bits sufficient to specify each of the ensemble signal, will be higher than the above lower bound $H_\varepsilon(\Omega_A)$ and can be computed as

$$H_\varepsilon\left(\hat{\Omega}_A\right) = -\sum_{\hat{\Omega}_A} \sum_r p\left(\left\{\hat{\alpha}_r^{|q_r|}\right\}\right) \log_2 p\left(\left\{\hat{\alpha}_r^{|q_r|}\right\}\right) \ge H_\varepsilon(\Omega_A) \qquad (3.146)$$

where r is index of the signal representation coefficients, $p(\{\hat{\alpha}_r^{|q_r|}\})$ is mutual probability of the sets $\{\hat{\alpha}_r^{|q_r|}\}$ of quantized values of representation coefficients for ensemble $\hat{\Omega}_A$ of signals reconstructed from their quantized representation coefficients.

If representation coefficients can be regarded as mutually statistically independent

$$p\left(\{\hat{\alpha}_r\}\right) = p\left(\{\hat{\alpha}_r^{(q_r)}\}\right) = \prod_{r=0}^{N-1} p\left(\hat{\alpha}_r^{(q_r)}\right), \qquad (3.147)$$

where N is the number of representation coefficients, $p(\hat{\alpha}_r^{(q_r)})$ is the probability of q-th quantization level of coefficient α_r $(q_r = 0,\ldots,Q_r - 1)$, the entropy and, therefore, the quantity of bits per signal increase and can be, taking into account that $\{\sum_{q_r=0}^{Q_r-1} p(\hat{\alpha}_s^{q_r}) = 1\}$, evaluated as

$$H_\varepsilon(\hat{\Omega}_A) \le -\sum_{q_0=0}^{Q_0-1}\cdots\sum_{q_{N-1}}^{Q_{N-1}-1}\left\{\prod_{r=0}^{N-1} p\left(\hat{\alpha}_r^{(q_r)}\right)\right\}\log_2\left\{\prod_{r=0}^{N-1} p\left(\hat{\alpha}_r^{(q_r)}\right)\right\}$$

$$= -\sum_{q_0=0}^{Q_0-1} p\left(\hat{\alpha}_r^{(q_0)}\right)\cdots\sum_{q_{N-1}}^{Q_{N-1}-1} p\left(\hat{\alpha}_r^{(q_{N-1})}\right)\sum_{r=0}^{N-1}\log_2 p\left(\hat{\alpha}_r^{(q_r)}\right)$$

$$= -\sum_{r=0}^{N-1}\sum_{q_r=0}^{Q_r-1} p(\hat{\alpha}_r^q)\,\log_2 p(\hat{\alpha}_r^q). \qquad (3.148)$$

Furthermore, if the probability distribution and the number of quantization levels of quantized coefficients do not depend on their index r, entropy per signal will further increase to entropy $H_{iid}(A)$ of N identically distributed variables

$$H_\varepsilon(\Omega_A) \le H_\varepsilon(\hat{\Omega}_A) \le H_{iid}(A) = -N\sum_{q=0}^{Q-1} p\left(\hat{\alpha}^{(q)}\right)\log_2 p\left(\hat{\alpha}^{(q)}\right). \qquad (3.149)$$

The upper limit to the per-signal entropy is given by the entropy $H_0 = N\log_2 Q$ of uniformly distributed Q level variable:

$$H_\varepsilon(\Omega_A) \le H_\varepsilon(\hat{\Omega}_A) \le -N\sum_{q=0}^{Q-1} p\left(\hat{\alpha}^{(q)}\right)\log_2 p\left(\hat{\alpha}^{(q)}\right) \le N\log_2 Q. \qquad (3.150)$$

In fact, $H_0 = N \log_2 Q$ determines the amount of bits per pixel in sampled image representation, and it usually very substantially exceeds the lower bound. The reason lies in substantial statistical dependence of and nonefficient scalar quantization of pixels. The purpose of image compression is to remove this redundancy and to decrease the number of bits per signal to make it as much as possible close to its lower bound.

Outline of Image Compression Methods

For more than 60 years since the first works on digital image compression, numerous image compression methods have been developed. Figure 3.31 represents the image compression methods classified into two groups.

The major group of the methods outlined Figure 3.31 by a dashed line consists of methods implemented in three steps:

- Decorrelating image-sampled representation
- Scalar quantization of decorrelated data
- Binary statistical encoding of quantized-decorrelated data

Decorrelating image-sampled representation is a processing that converts a set of image or video samples into a set of discrete data that are statistically,

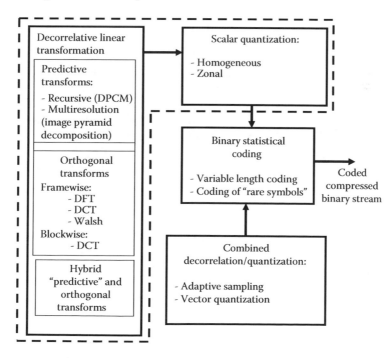

FIGURE 3.31
Classification of digital image compression methods.

that is, regarded as a statistical ensemble, uncorrelated as much as possible. This is achieved by applying to them one or another decorrelating linear transformation from two groups of transformation: predictive transformations and orthogonal transforms.

Predictive transformation computes pixel-wise differences between pixel gray level and its predicted estimate found as a weighted sum over pixels that surround the given one on the sampling grid. These differences are then subjected to scalar quantization. The method in which the prediction and computing differences are performed in course of image row-wise/column-wise scanning is called differential pulse code modulation (*DPCM*). In DPCM, the difference between the current pixel gray level and the average over the nearest to it pixels in its scanning row and in the previous row is quantized and then encoded for transmission or storage. It is one of the earliest methods of image and signal compression. Nowadays, it is used mostly for "interframe" coding of video sequences.

A more efficient decorrelation is achieved when larger and isotropic spatial neighborhood of pixels is involved in the prediction. This is implemented in multiresolution image expansion. The multiresolution image decomposition methods are outlined in section "Discrete Wavelet Transforms and Multiresolution Analysis." In these methods, decorrelated difference signals are obtained, at each resolution (scale) level, by subtracting a low-pass filtered image from its higher-resolution original (see Figure 4.16) and then are optimally quantized and statistically encoded. Depending on the implementation, this method is known under different names: *pyramid coding, sub-band decomposition coding*, and *wavelet coding*.

Decorrelation with orthogonal transforms is an alternative to predictive decorrelation. Redundancy of the images exhibits itself in this case in concentration of most of image energy in a small number of transform coefficients. As it was shown in section "Optimality of Bases: Karhunen-Loewe and Related Transforms," Karhunen–Loewe transform is statistically the best one in terms of energy compaction property. However, because of its high computational complexity, in practice, other transforms, *fast transforms* such as DFT, DCT, and Walsh transforms that can be computed with so-called fast algorithms are used. These discrete transforms are described in details in Chapter 4. Compression methods that assume the use of orthogonal transform for image decorrelation are united by the name *transform image coding*. In transform image coding, it is the transform coefficients that undergo subsequent scalar quantization.

Two versions of transform coding are known: frame-wise and block-wise transform coding. In frame-wise coding, the image frame as a whole is transformed, transform coefficients with very low energy are truncated, and the remaining ones are quantized using described methods of optimal scalar quantization. In block-wise coding, image is split into nonoverlapping blocks of relatively small size and individual blocks are transformed separately. For every block, low-energy block transform coefficients are truncated and the remaining are optimally quantized.

Applying decorrelative transforms block-wise is well suited to spatial image inhomogeneity and transform coefficient quantization can be better optimized if it is performed individually for transform spectra of image blocks. The size of blocks is determined by the degree of image inhomogeneity and, in principle, can vary within an image frame: it should be as large as possible provided image remains to be "homogeneous" within the block. Block transform coding with variable block size is the most efficient implementation of transform image coding. Among block transforms, DCT had proved to be the best one and it is put in the base of modern image and video compression standards JPEG, H.261, H.262, H.263, and H.320.

Depending on implementation issues, predictive and orthogonal transform can be used in a combination. For instance, in video coding standard MPEG, block DCT transform JPEG coding is used for "intraframe" coding while predictive transforms are used for DPCM-coding of dc-components of blocks, when they are encoded one by one in successive order of image scanning, and also for "interframe" coding as a method for *motion compensation*.

As it was already mentioned, data obtained after applying to primary image-sampled representation of a decorrelating transform are subjected to scalar quantization, principles of which were described in the section "Signal Scalar Quantization." The quantization can be homogeneous, that is, applied uniformly to all data, or inhomogeneous, that is, different for different groups of data, or, in terms of data indices, different zones of data indices. For instance, in block image coding using DCT, indices of spectral coefficients of image blocks are split into a number of zones according to their spatial frequency, and low-frequency coefficients, which carry most of the signal energy, are quantized more precisely than middle-frequency coefficients, and the latter are quantized more precisely than high-frequency coefficients.

The third step of the three-step compression is *statistical (entropy) coding*. It is applied to the results of quantization and makes use of nonuniformity of probabilities of different quantization levels for generating their binary codes with the number of bits that are as close as possible to the logarithm of the inverse to their probabilities as it is dictated by the information theoretical lower bound (Equation 3.149). Two families of the methods of statistical coding are used: variable-length coding and coding of "rare symbols." The latter are used when one of the symbols to be encoded, such as symbol that corresponds to zero-quantized prediction error, has an overwhelmingly higher probability than other much more rare symbols. The basics of variable-length coding and coding of "rare symbols" are detailed in Appendix.

The described three-step procedure is called *lossy compression* to emphasize the fact that the original sampled image representation coefficients cannot be precisely restored from the compressed binary codes because the procedure includes quantization of decorrelated data.

When quantization is avoided, precise image restoration from its binary code is possible because decorrelative transforms are reversible. This type of coding is called *loss-less coding*. Data compression in this case is achieved only through statistical (entropy) coding.

An alternative to the three-step procedure represents methods in which decorrelation and quantization are not separated from one another. They can be considered as attempts to implement general digitization by means of splitting signal space into equivalency cells and selection of representative signals of the cells. These methods are not as well developed and widely used as the methods described above and can be exemplified by adaptive sampling and vector quantization methods.

The adaptive sampling method assumes taking samples of images not uniformly as in conventional sampling but selectively only in areas of "importance," the latter being defined by requirements to the image restoration quality. The most known example of adaptive sampling is representing 2D surfaces by their "level lines." Image reconstruction from sets of nonuniform samples is discussed in Chapter 5.

The principle of vector quantization is right that of the general quantization. But due to the complexity limitation, the idea of the general quantization is applied to relatively small image patches or blocks: for each particular image patch or block, a representative "typical block" is found from a "coded book," which is supposed to be prepared in advance by a kind of learning procedure using image data base and image reconstruction quality requirements.

Appendix

Derivation of Equation 3.31

$$rad_m(\xi) = sign\left[\sin(2^m \pi \xi)\right] = sign\left[\sin\left(2^m \pi \sum_{n=-\infty}^{1} \xi_n 2^{-n}\right)\right]$$

$$= sign\left[\sin\left(\pi \sum_{n=-\infty}^{1} \xi_n 2^{m-n}\right)\right] = (-1)^{\xi_m},$$

where $\{\xi_n\}$ are binary digits of binary representation of ξ ($\xi < 1$).

Then

$$wal_k(x) = \prod_{m=0}^{\infty}\left[rad_{m+1}(x)\right]^{k_m^{GC}} = \prod_{m=0}^{\infty}(-1)^{k_m^{GC}\xi_{m+1}} = (-1)^{\sum_{m=0}^{\infty}k_m^{GC}\xi_{m+1}}.$$

Derivation of Equation 3.44

$$\left|\varepsilon(\omega)\right|^2 = \int_X \left|a(x,\omega) - \tilde{a}(x,\omega)\right|^2 dx = \int_X \left|a(x,\omega) - \sum_{k=0}^{N-1} \alpha_k(\omega)\, \varphi_k^{(r)}(x)\right|^2 dx$$

$$= \int_X \left|a(x,\omega)\right|^2 dx - \int_X a^*(x,\omega) \sum_{k=0}^{N-1} \alpha_k(\omega)\, \varphi_k^{(r)}(x) dx$$

$$- \int_X a(x,\omega) \sum_{k=0}^{N-1} \alpha_k^*(\omega) \left[\varphi_k^{(r)}(x)\right]^* dx$$

$$+ \int_X \sum_{k=0}^{N-1} \alpha_k(\omega)\varphi_k^{(r)}(x) \sum_{l=0}^{N-1} \alpha_l^*(\omega) \left[\varphi_l^{(r)}(x)\right]^* dx$$

$$= \int_X \left|a(x,\omega)\right|^2 dx - \sum_{k=0}^{N-1} \alpha_k(\omega) \int_X a^*(x,\omega)\varphi_k^{(r)}(x) dx$$

$$- \sum_{k=0}^{N-1} \alpha_k^*(\omega) \int_X a(x,\omega) \left[\varphi_k^{(r)}(x)\right]^* dx$$

$$+ \sum_{k=0}^{N-1} \sum_{l=0}^{N-1} \alpha_k(\omega)\alpha_l^*(\omega) \int_X \varphi_k^{(r)}(x) \left[\varphi_l^{(r)}(x)\right]^* dx$$

$$= \int_X \left|a(x,\omega)\right|^2 dx - \sum_{k=0}^{N-1} \alpha_k(\omega)\alpha_k^*(\omega)$$

$$- \sum_{k=0}^{N-1} \alpha_k^*(\omega)\alpha_k(\omega) + \sum_{k=0}^{N-1}\sum_{l=0}^{N-1} \alpha_k(\omega)\alpha_l^*(\omega)\delta(k-l)$$

$$= \int_X \left|a(x,\omega)\right|^2 dx - \sum_{k=0}^{N-1} \alpha_k(\omega)\alpha_k^*(\omega)$$

$$- \sum_{k=0}^{N-1} \alpha_k^*(\omega)\alpha_k(\omega) + \sum_{k=0}^{N-1} \alpha_k(\omega)\alpha_k^*(\omega)$$

$$= \int_X \left|a(x,\omega)\right|^2 dx - \sum_{k=0}^{N-1} \left|\alpha_k(\omega)\right|^2$$

Derivation of Equation 3.45

$$
\{\varphi_k(x)\}_{opt} = \underset{\{\varphi_k(x)\}}{\arg\min}\left\{ AV_{\Omega_A}\left(\int_X \left|a^{(\omega)}(x)\right|^2 dx - \sum_{k=0}^{N-1}\left|\alpha_k^{(\omega)}(\omega)\right|^2 \right)\right\}
$$

$$
= \underset{\{\varphi_k(x)\}}{\arg\min}\left\{ AV_{\Omega_A}\left(\int_X \left|a^{(\omega)}(x)\right|^2 dx \right) - AV_{\Omega_A}\left(\sum_{k=0}^{N-1}\left|\alpha_k^{(\omega)}(\omega)\right|^2 \right)\right\}
$$

$$
= \underset{\{\varphi_k(x)\}}{\arg\max}\left\{ AV_{\Omega_A}\left(\sum_{k=0}^{N-1}\left|\alpha_k^{(\omega)}(\omega)\right|^2 \right)\right\}
$$

$$
= \underset{\{\varphi_k(x)\}}{\arg\max}\left\{ \left(\sum_{k=0}^{N-1}\iint_X AV_{\Omega_A}\left[a^{(\omega)}(x_1)a^{*(\omega)}(x_2)\right]\varphi_k^{(r)}(x_1)\varphi_k^{*(r)}(x_2)dx_1 dx_2 \right)\right\}
$$

$$
= \underset{\{\varphi_k(x)\}}{\arg\max}\left\{ \sum_{k=0}^{N-1}\iint_X R_a(x_1,x_2)\varphi_k^{(r)}(x_1)\varphi_k^{*(r)}(x_2)dx_1 dx_2 \right\}
$$

where $R_a(x_1,x_2) = AV_{\Omega_A}\left[a^{(\omega)}(x_1,\omega)a^{*(\omega)}(x_2)\right]$.

Derivation of Equation 3.78

$$
\tilde{\alpha}^{(s)}(f) = \int_{-\infty}^{\infty} \tilde{a}^{(s)}(x)\exp(i2\pi fx)\,dx = \int_{-\infty}^{\infty}\left[\sum_{k=-\infty}^{\infty} a^{(s)}(k\Delta x)\delta(x - k\Delta x)\right]\exp(i2\pi fx)\,dx
$$

$$
= \sum_{k=-\infty}^{\infty} a^{(s)}(k\Delta x)\exp(i2\pi k\Delta xf) = \sum_{k=-\infty}^{\infty}\int_{-\infty}^{\infty}\alpha(p)FR^{(s)}(p)\exp\left[i2\pi k\Delta x(f - p)\right]dp
$$

$$
= \int_{-\infty}^{\infty}\alpha(p)FR^{(s)}(p)\,dp\sum_{k=-\infty}^{\infty}\exp\left[i2\pi k\Delta x(f - p)\right]
$$

$$
= \int_{-\infty}^{\infty}\alpha(p)FR^{(s)}(p)\,dp\Delta x\left\{ \sum_{k=-\infty}^{\infty}\frac{1}{\Delta x}\exp\left[i2\pi k\Delta x(f - p)\right]\right\}.
$$

Derivation of Equation 3.98

$$H^{(s\&r)}(f, p) = \int_{-\infty}^{\infty}\int_{-\infty}^{\infty}\sum_{k=0}^{N-1}\varphi^{(s)}(\xi - k\Delta x)\varphi^{(r)}(x - k\Delta x)\exp\left[i2\pi(fx - p\xi)\right]dx\,d\xi$$

$$= \sum_{k=0}^{N-1}\int_{-\infty}^{\infty}\int_{-\infty}^{\infty}\varphi^{(s)}(\xi - k\Delta x)\varphi^{(r)}(x - k\Delta x)\exp\left[i2\pi(fx - p\xi)\right]dx\,d\xi$$

$$= \sum_{k=0}^{N-1}\int_{-\infty}^{\infty}\varphi^{(s)}(\xi - k\Delta x)\exp(-i2\pi p\xi)d\xi$$

$$\times \int_{-\infty}^{\infty}\varphi^{(r)}(x - k\Delta x)\exp(i2\pi fx)dx$$

$$= \sum_{k=0}^{N-1}\int_{-\infty}^{\infty}\varphi^{(s)}(\tilde{\xi})\exp\left[-i2\pi p(\tilde{\xi} + k\Delta x)\right]d\tilde{\xi}$$

$$\times \int_{-\infty}^{\infty}\varphi^{(r)}(\tilde{x})\exp\left[i2\pi f(\tilde{x} + k\Delta x)\right]d\tilde{x}$$

$$= \sum_{k=0}^{N-1}\exp\left[i2\pi(f - p)k\Delta x\right]\int_{-\infty}^{\infty}\varphi^{(s)}(\tilde{\xi})\exp(-i2\pi p\tilde{\xi})d\tilde{\xi}$$

$$\times \int_{-\infty}^{\infty}\varphi^{(r)}(\tilde{x})\exp(i2\pi f\tilde{x})d\tilde{x}$$

$$= \frac{\exp\left[i2\pi(f - p)N\Delta x\right] - 1}{\exp\left[i2\pi(f - p)\Delta x\right] - 1}FR^{(r)}(f)\left[FR^{(s)}(p)\right]^{*}$$

$$= \left\{\frac{\exp\left[i\pi(f - p)N\Delta x\right] - \exp\left[-i\pi(f - p)N\Delta x\right]}{\exp\left[i\pi(f - p)\Delta x\right] - \exp\left[-i\pi(f - p)\Delta x\right]}\right.$$

$$\left.\times\exp\left[i\pi(f - p)(N - 1)\Delta x\right]\right\}Fr^{(r)}(f)\left[FR^{(s)}(p)\right]^{*}$$

$$= \left\{N\frac{\sin\left[\pi(f - p)N\Delta x\right]}{N\sin\left[\pi(f - p)\Delta x\right]}\exp\left[i\pi(f - p)(N - 1)\Delta x\right]\right\}$$

$$\times FR^{(r)}(f)\left[FR^{(s)}(p)\right]^{*}$$

$$= \left\{N\,\text{sincd}\left[N;\pi(f - p)N\Delta x\right]\exp\left[i\pi(f - p)(N - 1)\Delta x\right]\right\}$$

$$\times FR^{(r)}(f)\left[FR^{(s)}(p)\right]^{*}$$

where $FR^{(s)}(\cdot)$ and $FR^{(r)}(\cdot)$ are frequency responses of signal sampling and reconstruction devices, defined by Equations 3.77 and 3.83 and

$$\text{sincd}(N;x) = \frac{\sin x}{N \sin(x/N)}$$

Derivation of Equation 3.105

$$
\overline{\varepsilon_{alsng}^2} = \left[\int_{-\infty}^{\infty} \left| FR^{(r)}(f) \sum_{m=1}^{\infty} \left\{ \alpha\left(f + \frac{m}{\Delta x} \right) FR^{(s)}\left(f + \frac{m}{\Delta x} \right) \right.\right.\right.
$$

$$
\left.\left.\left. + \alpha\left(f - \frac{m}{\Delta x} \right) FR^{(s)}\left(f - \frac{m}{\Delta x} \right) \right\} \right| df \right]
$$

$$
= AV_{\Omega}\left[\int_{-\infty}^{\infty} \left| FR^{(r)}(f) \sum_{m=1}^{\infty} \left[\alpha^{(s)}\left(f + \frac{m}{\Delta x} \right) + \alpha^{(s)}\left(f - \frac{m}{\Delta x} \right) \right] \right|^2 df \right]
$$

$$
= \int_{-\infty}^{\infty} \left| FR^{(r)}(f) \right|^2 AV_{\Omega}\left\{ \left| \sum_{m=1}^{\infty} \left[\alpha^{(s)}\left(f + \frac{m}{\Delta x} \right) + \alpha^{(s)}\left(f - \frac{m}{\Delta x} \right) \right] \right|^2 \right\} df \qquad \text{(A3.1)}
$$

where we replaced

$$\alpha(f) FR^{(s)}(f) = \alpha^{(s)}(f)$$

Consider

$$
AV_{\Omega}\left\{ \left| \sum_{m=1}^{\infty} \left[\alpha^{(s)}\left(f + \frac{m}{\Delta x} \right) + \alpha^{(s)}\left(f - \frac{m}{\Delta x} \right) \right] \right|^2 \right\}
$$

$$
= AV_{\Omega}\left\{ \left[\sum_{n=1}^{\infty} \sum_{m=1}^{\infty} \left[\alpha^{(s)}\left(f + \frac{m}{\Delta x} \right) + \alpha^{(s)}\left(f - \frac{m}{\Delta x} \right) \right] \right.\right.
$$

$$
\left.\left. \times \left[\alpha^{*(s)}\left(f + \frac{n}{\Delta x} \right) + \alpha^{*(s)}\left(f - \frac{n}{\Delta x} \right) \right] \right] \right\}
$$

$$
= \sum_{n=1}^{\infty} \sum_{m=1}^{\infty} \left\{ AV_{\Omega}\left[\alpha^{(s)}\left(f + \frac{m}{\Delta x} \right) \alpha^{*(s)}\left(f + \frac{n}{\Delta x} \right) \right] \right.
$$

$$
+ AV_{\Omega}\left[\alpha^{(s)}\left(f - \frac{m}{\Delta x} \right) \alpha^{*(s)}\left(f + \frac{n}{\Delta x} \right) \right]
$$

$$
+ AV_{\Omega}\left[\alpha^{(s)}\left(f + \frac{m}{\Delta x} \right) \alpha^{*(s)}\left(f - \frac{n}{\Delta x} \right) \right]
$$

$$
\left. + AV_{\Omega}\left[\alpha^{(s)}\left(f - \frac{m}{\Delta x} \right) \alpha^{*(s)}\left(f - \frac{n}{\Delta x} \right) \right] \right\} \qquad \text{(A3.2)}
$$

Consider now

$$
AV_\Omega\left[\alpha^{(s)}\left(f + \frac{m}{\Delta x}\right)\alpha^{*(s)}\left(f + \frac{n}{\Delta x}\right)\right]
$$

$$
= AV_\Omega\left[\int_{-\infty}^{\infty}\int_{-\infty}^{\infty} a^{(s)}(x)\exp\left[i2\pi\left(f + \frac{m}{\Delta x}\right)x\right]a^{*(s)}(\xi)\exp\left[-i2\pi\left(f + \frac{n}{\Delta x}\right)\xi\right]dx\,d\xi\right]
$$

$$
= \int_{-\infty}^{\infty}\int_{-\infty}^{\infty} AV_\Omega\left[a^{(s)}(x)a^{*(s)}(\xi)\right]\exp\left(i2\pi\left[f(x-\xi) + \frac{mx - n\xi}{\Delta x}\right]\right)dx\,d\xi
$$

$$
= \int_{-\infty}^{\infty}\int_{-\infty}^{\infty} CF_\Omega(x,\xi)\exp\left(i2\pi\left[f(x-\xi) + \frac{mx - n\xi}{\Delta x}\right]\right)dx\,d\xi, \tag{A3.3}
$$

where $CF_\Omega(x, \xi)$ is correlation function of ensemble of signals $\{a^{(s)}(x)\}$. For stationary signals $CF_\Omega(x, \xi) = CF_\Omega(x - \xi)$,

$$
AV_\Omega\left[\alpha^{(s)}\left(f + \frac{m}{\Delta x}\right)\alpha^{*(s)}\left(f + \frac{n}{\Delta x}\right)\right]
$$

$$
= \int_{-\infty}^{\infty}\int_{-\infty}^{\infty} CF_\Omega(x-\xi)\exp\left(i2\pi\left[f(x-\xi) + \frac{mx - n\xi}{\Delta x}\right]\right)dx\,d\xi
$$

$$
= \int_{-\infty}^{\infty}\int_{-\infty}^{\infty} CF_\Omega(\tilde{\xi})\exp\left(i2\pi\left[f\tilde{\xi} + \frac{m(\tilde{\xi} + \xi) - n\xi}{\Delta x}\right]\right)d\tilde{\xi}\,d\xi
$$

$$
= \int_{-\infty}^{\infty} CF_\Omega(\tilde{\xi})\exp\left[i2\pi\left(f + \frac{m}{\Delta x}\right)\tilde{\xi}\right]d\tilde{\xi}\int_{-\infty}^{\infty}\exp\left(i2\pi\frac{m - n}{\Delta x}\xi\right)d\xi \tag{A3.4}
$$

Consider

$$
\int_{-\infty}^{\infty}\exp\left(i2\pi\frac{m - n}{\Delta x}\xi\right)d\xi = \lim_{X\to\infty}\frac{1}{2X}\int_{-X}^{X}\exp\left(i2\pi\frac{m - n}{\Delta x}\xi\right)d\xi
$$

$$
= \lim_{X\to\infty}\frac{1}{2X}\frac{\exp\left(i2\pi(m - n/\Delta x)X\right) - \exp\left(-i2\pi(m - n/\Delta x)X\right)}{i2\pi(m - n/\Delta x)}
$$

$$
= \lim_{X\to\infty}\frac{\sin(2\pi(m - n/\Delta x)X)}{2\pi(m - n/\Delta x)X}
$$

$$
= \lim_{X\to\infty}\text{sinc}\left(2\pi\frac{m - n}{\Delta x}X\right) = \delta(m - n), \tag{A3.5}
$$

where $\delta(m - n)$ is the Kronecker delta defined by Equation 2.4. Therefore

$$AV_\Omega\left[\alpha^{(s)}\left(f + \frac{m}{\Delta x}\right)\alpha^{*(s)}\left(f + \frac{n}{\Delta x}\right)\right]$$

$$= \left[\int_{-\infty}^{\infty} CF_{a^{(s)}}(\tilde{\xi})\exp\left[i2\pi\left(f + \frac{m}{\Delta x}\right)\tilde{\xi}\right]d\tilde{\xi}\right]\delta(m - n)$$

$$= SD_{a^{(s)}}\left(f + \frac{m}{\Delta\xi}\right)\delta(m - n), \qquad (A3.6)$$

where $SD_{a^{(s)}}\left(f + (m/\Delta x)\right)$ is spectral density of the signal ensemble $\{a^{(s)}(x)\}$ on frequency $(f + m/\Delta x)$.

Substitute the right part of Equation A3.6 into each of the four terms of Equation A3.2 with an appropriate replacement of m by $-m$ when needed and obtain

$$AV_\Omega\left\{\left|\sum_{m=1}^{\infty}\left[\alpha^{(s)}\left(f + \frac{m}{\Delta x}\right) + \alpha^{(s)}\left(f - \frac{m}{\Delta x}\right)\right]\right|^2\right\}$$

$$= \sum_{n=1}^{\infty}\sum_{m=1}^{\infty}\left\{SD_{a^{(s)}}\left(f + \frac{m}{\Delta\xi}\right)\delta(m - n) + SD_{a^{(s)}}\left(f + \frac{m}{\Delta\xi}\right)\delta(m + n)\right.$$

$$\left. + SD_{a^{(s)}}\left(f + \frac{m}{\Delta\xi}\right)\delta(m + n) + SD_{a^{(s)}}\left(f + \frac{m}{\Delta\xi}\right)\delta(m - n)\right\}$$

$$= 2\sum_{n=1}^{\infty}\sum_{m=1}^{\infty}SD_{a^{(s)}}\left(f + \frac{m}{\Delta\xi}\right)\delta(m - n) + 2\sum_{n=1}^{\infty}\sum_{m=1}^{\infty}SD_{a^{(s)}}\left(f + \frac{m}{\Delta\xi}\right)\delta(m + n)$$

$$= 2\sum_{m=1}^{\infty}SD_{a^{(s)}}\left(f + \frac{m}{\Delta\xi}\right) + 2\sum_{m=1}^{\infty}SD_{a^{(s)}}\left(f - \frac{m}{\Delta\xi}\right).$$

$$(A3.7)$$

Now, we can use this result to obtain finally

$$\varepsilon_{alsng}^2 = \int_{-\infty}^{\infty}\left|FR^{(r)}(f)\right|^2 AV_\Omega\left\{\left|\sum_{m=1}^{\infty}\left[\alpha^{(s)}\left(f + \frac{m}{\Delta x}\right) + \alpha^{(s)}\left(f - \frac{m}{\Delta x}\right)\right]\right|^2\right\}df$$

$$= \int_{-\infty}^{\infty}\left|FR^{(r)}(f)\right|^2\left\{2\sum_{m=1}^{\infty}SD_{a^{(s)}}\left(f + \frac{m}{\Delta\xi}\right) + 2\sum_{m=1}^{\infty}SD_{a^{(s)}}\left(f - \frac{m}{\Delta\xi}\right)\right\}df \qquad (A3.8)$$

or, with an account for $\alpha(f)FR^{(s)}(f) = \alpha^{(s)}(f)$

$$\overline{\varepsilon_{alsng}^2} = 2\int_{-\infty}^{\infty} |FR^{(r)}(f)|^2 \left\{ \left| \sum_{m=1}^{\infty} SD_a\left(f + \frac{m}{\Delta\xi}\right) FR^{(s)}\left(f + \frac{m}{\Delta\xi}\right) \right|^2 \right.$$

$$\left. + \left| \sum_{m=1}^{\infty} SD_a\left(f - \frac{m}{\Delta\xi}\right) FR^{(s)}\left(f - \frac{m}{\Delta\xi}\right) \right|^2 \right\} df \qquad (A3.9)$$

Derivation of Equation 3.136

For $P = 1$ and $p(\alpha) \propto \exp\left(-\alpha^2/2\sigma_\alpha^2\right)$, compute

$$\int_{\alpha_{min}}^{\alpha} \left[p(\alpha)\right]^{1/(2P+1)} d\alpha = \int_{\alpha_{min}}^{\alpha} \exp(-\alpha^2/6\sigma_\alpha^2) d\alpha$$

$$= \int_0^{\alpha} \exp(-\alpha^2/6\sigma_\alpha^2) d\alpha - \int_{\alpha_{min}}^0 \exp(-\alpha^2/6\sigma_\alpha^2) d\alpha$$

$$= \int_0^{\alpha/\sqrt{6}\sigma_\alpha} \exp(-\tilde{\alpha}^2) d\tilde{\alpha} - \int_{\alpha_{min}/\sqrt{6}\sigma_\alpha}^0 \exp(-\alpha^2/6\sigma_\alpha^2) d\alpha$$

$$= erf\left(\alpha/\sqrt{6}\sigma_\alpha\right) + erf\left(\alpha_{min}/\sqrt{6}\sigma_\alpha\right),$$

where

$$erf(x) = \frac{2}{\sqrt{\pi}} \int_0^x \exp(-\xi^2) d\xi$$

then

$$\frac{w(\alpha) - w(\alpha_{min})}{w(\alpha_{max}) - w(\alpha_{min})} = \frac{erf\left(\alpha/\sqrt{6}\sigma_\alpha\right) + erf\left(\alpha_{min}/\sqrt{6}\sigma_\alpha\right)}{erf\left(\alpha_{max}/\sqrt{6}\sigma_\alpha\right) + erf\left(\alpha_{min}/\sqrt{6}\sigma_\alpha\right)}$$

Basics of Statistical Coding

Two mutually complement methods of binary statistical coding are known: *variable-length coding (VL-coding)* and *coding of rare symbols*. In the data coding jargon, objects of encoding are called symbols. VL-coding is aimed at generating, for each symbol to be coded, a binary code word with number of bits as close to the logarithm of the inverse to its probability. Originally suggested by Shannon [8] and Fano [9], it was later improved by Huffman [10], and the Huffman's coding procedure had become the preferred implementation of VL-coding. VL-Huffman coding is an iterative process, in which, at each iteration, two symbols with the least probabilities are found in the current set of symbols, bits 0 and 1, correspondingly, are assigned to their codes and then these symbols are united to form a new auxiliary symbol, whose probability is the sum of that of the united symbols. The iteration process is repeated with the iterated set of symbols until all symbols are encoded. If probabilities of symbols are integer power of 1/2, such a procedure generates binary codes with number of bits per symbol exactly equal to the binary logarithm of the inverse to its probability, and therefore average number of bits per symbol is equal to the entropy of the symbol ensemble. An example of Huffman-coding is presented in Table A3.1 for eight symbols (A through H). As one can see, when probability of one of the symbols of the ensemble is much higher than 1/2, Huffman coding is inefficient because it does not allow allocating to symbols <1 bit. In such cases, coding of rare symbols is applied.

For coding of rare symbols, either of the two methods, *run length coding* and *coding coordinates of rare symbols*, can be used. In run length coding, the encoder generates from the initial sequence of symbols a new sequence, in which the runs of the most frequent initial symbol are replaced with new auxiliary symbols that designate the run lengths. This new sequence can then be subjected to VL-coding if necessary. Run length coding is an essentially 1D procedure. For coding 2D data, it is applied after 2D data are converted into 1D data by means of one or another scanning procedure; in image compression, a zigzag scanning (see Figure A3.1) is used.

Coding coordinates of rare symbols is an alternative to the run length coding. In this method, positions of rare symbols (others than the most frequent one) in the sequence are found. Then a new sequence of symbols is generated from the initial one in which all occurrences of the most frequent symbol are removed and at each occurrence of rare symbols an auxiliary symbol is added that designates their position in the initial sequence. Coding coordinates of rare symbols is less vulnerable to errors in transmitting the binary code than the run length coding. For the former, transmission errors cause only localized errors in the decoded symbol sequence, while in the latter they result in shifts of the entire decoded symbol sequence after the erroneous one. This property is of special importance in image coding for transmission over noisy channel: in the case of run

TABLE A3.1

An Example of VL-Huffman Coding of Eight Symbols

Iteration	A $P(A) = 0.49$	B $P(B) = 0.27$	C $P(C) = 0.12$	D $P(D) = 0.065$	E $P(E) = 0.031$	F $P(F) = 0.014$	G $P(G) = 0.006$	H $P(H) = 0.004$
1st	—	—	—	—	—	—	0	1
2nd	—	—	—	—	—	0	1 $P(GH) = 0.01$	
3rd	—	—	—	—	0	1 $P(FGH) = 0.024$		
4th	—	—	—	0	1 $P(EFGH) = 0.055$			
5th	—	—	0	1 $P(DEFGH) = 0.12$				
6th	—	0	1 $P(CDEFGH) = 0.24$					
7th	0	1 $P(BCDEFGH) = 0.51$						
Binary code	0	10	110	1110	11110	111110	1111110	1111111

Entropy $H = 1.9554$.

Average number of bits per symbol: 1.959.

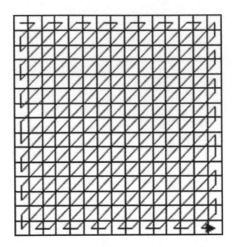

FIGURE A3.1
The principle of zig-zag scanning 2D data on a rectangular sampling grid.

length coding, channel errors may cause substantial deformation of object boundaries.

Exercises

EnergyCompact_DFT_DCT_Walsh_Haar_2D_CRC.m fringe_aliasing_demo_
   CRC.m
IdealVsNonidealSampling_CRC.m
quantization_demo_CRC.m

References

1. H. F. Harmuth, *Transmission of Information by Orthogonal Functions*, Springer Verlag, Berlin, 1970.
2. H. Karhunen, Über Lineare Methoden in der Wahscheinlichkeitsrechnung, *Ann. Acad. Sci. Finn.*, Ser. A.I.37, Helsinki, 1947.
3. M. Loeve, Fonctions Aleatoires de Seconde Ordre, in: P. Levy, Processes Stochastiques et Movement Brownien, Hermann, Paris, France, 1948.
4. H. Hotelling, Analysis of complex of statistical variables into principal components, *J. Educ. Psychol.*, 24: 417, 441, 498–520, 1933.

5. V. A. Kotelnikov, On the carrying capacity of the ether and wire in telecommunications, Material for the First All-Union Conference on Questions of Communication, Izd. Red. Upr. Svyazi RKKA, Moscow, 1933 (Russian). (English translation: http://ict.open.ac.uk/classics/1.pdf.)

6. C. E. Shannon, Communication in the presence of noise, *Proc. Inst. Radio Eng.*, 37(1): 10–21, January 1949. Reprint as classic paper in *Proc. IEEE*, 86(2), 1998.

7. J. M. Whittaker, *Interpolatory Function Theory*, Cambridge University Press, Cambridge, England, 1935.

8. C. Shannon, A mathematical theory of communication, *Bell Syst. Tech. J.*, 27(3): 379–423; (4): 623–656, 1948.

9. R. M. Fano, Technical Report No. 75, The Research Laboratory of Electronics, MIT, March 17, 1949.

10. D. A. Huffman, A method for the construction of minimum redundancy codes, *Proc. IRE*, 40(10): 1098–1101, 1952.

4

Discrete Signal Transformations

Having established methods for digital representation of analog signals, consider how transformations of analog signals must be represented in computers. In the section "Basic Principles of Discrete Representation of Signal Transformations," we formulate the basics principles of discrete representation of signal transformations and in the subsequent sections "Discrete Representation of the Convolution Integral," "Discrete Representation of Fourier Integral Transform," and "Discrete Representation of Fresnel Integral Transform," we apply them to introduce digital convolution as discrete representation of the convolution integral, DFTs as discrete representation of the integral Fourier transform and discrete Fresnel transforms as discrete representation of the integral Fresnel transform. In the section "Discrete Representation of Kirchhoff Integral," we introduce binary Hadamard and Walsh transforms as well as discrete wavelet transforms in their association with signal multiresolution analysis. This chapter is concluded with an extension of discrete transforms to their application in sliding window, which leads to signal "space-frequency" representation.

Basic Principles of Discrete Representation of Signal Transformations

Two principles lie in the base of digital representation of analog signal transformations:

- *Consistency principle* with digital representation of signals
- Mutual correspondence principle between continuous and discrete transformations

The consistency principle requires that digital representation of signal transformations should be parallel to that of signals. The mutual correspondence principle between continuous and digital transformations is said to hold if both act to transform identical input signals into identical output signals. Thus, digital signal transformations should be treated in terms of equivalent analog transformations. This principle is illustrated in Figure 4.1.

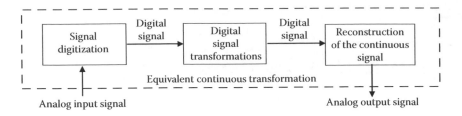

FIGURE 4.1
Mutual correspondence principle between continuous and digital signal transformations.

In Chapter 3, we admitted that signal digitization is carried out in two steps: discretization and element-wise quantization. Let signal $a(x)$ be represented in a digital form by quantized coefficients $\{\alpha_k\}$ of its discrete representation over a chosen discretization basis. In the section "Signal Transformations," two classes of signal transformations were specified, point-wise nonlinear and linear ones. Digital representation of point-wise transformations is based on their definition through transfer functions $\{F_k(\alpha_k)\}$ of the representation coefficients

$$\mathbf{PWTa} = \{F_k(\alpha_k)\}. \tag{4.1}$$

If coefficient values are quantized, functions $\{F_k(\cdot)\}$ can be specified in a tabular form. The volume of the table depends upon the number Q of quantization levels. If this number is too large and especially when numbers in computers are represented in the floating point format, other methods of specifying transformation transfer functions are used. Most commonly, a polynomial approximation of functions is used:

$$F_k(\alpha_k) = \sum_{s=0}^{S-1} c_s \alpha_k^s, \tag{4.2}$$

where $\{c_s\}$ are polynomial weight coefficients. This settles the problem of digital representation of point-wise signal transformations.

Representation of linear transformations in computers is more involved. Let $\{\phi_n^{(d)}(x)\}$, $\{\phi_n^{(r)}(x)\}$, $\{\phi_k^{(d)}(x)\}$, and $\{\phi_k^{(r)}(x)\}$ be, respectively, discretization and reciprocal reconstruction bases for input $a(\xi)$ and output $b(x)$ signals of a linear transformation $b = \mathbf{L}a$ and let $\{\alpha_k\}$ and $\{\beta_k\}$ be the corresponding sets of signal representation coefficients on these bases:

$$a(\xi) = \sum_n \alpha_n \phi_n^{(r)}(\xi); \quad \alpha_n = \int_X a(\xi)\phi_n^{(d)}(\xi)\,d\xi; \tag{4.3}$$

$$b(x) = \mathbf{L}a(\xi) = \sum_k \beta_k \varphi_k^{(r)}(x); \quad \beta_k = \int_X b(x)\phi_k^{(d)}(x)\,\mathrm{d}x. \tag{4.4}$$

Substitute Equation 4.3 into Equation 4.4 to find numerical parameters of the linear transformation specified by its PSF $h(x, \xi)$ that are needed to compute the set of representation coefficients $\{\beta_k\}$ of the linear transformation output signal from those $\{\alpha_n\}$ of its input signal:

$$\beta_k = \int_X b(x)\phi_k^{(d)}(x)\,\mathrm{d}x = \int_X \mathbf{L}a(\xi)\phi_k^{(d)}(x)\,\mathrm{d}x$$

$$= \int_X \mathbf{L}\sum_n \alpha_n \phi_n^{(r)}(\xi)\phi_k^{(d)}(x)\,\mathrm{d}x = \sum_n \alpha_n \int_X \mathbf{L}\left(\phi_n^{(r)}(\xi)\right)\phi_k^{(d)}(x)\,\mathrm{d}x$$

$$= \sum_n \alpha_n \int_X \left[\int_\Xi h(x,\xi)\phi_n^{(r)}(\xi)\,\mathrm{d}\xi\right]\phi_k^{(d)}(x)\,\mathrm{d}x = \sum_n \alpha_n h_{k,n}, \tag{4.5}$$

where

$$\left\{ h_{k,n} = \int_X \left[\int_\Xi h(x,\xi)\phi_n^{(r)}(\xi)\,\mathrm{d}\xi\right]\phi_k^{(d)}(x)\,\mathrm{d}x \right\}. \tag{4.6}$$

Coefficients $\{h_{k,n}\}$ form the discrete representation of the linear transformation with PSF $h(x, \xi)$ for discretization basis functions of output signals $\{\phi_k^{(d)}(x)\}$ and reconstructing basis functions $\{\phi_n^{(r)}(x)\}$ of input signals. We will refer to the set of coefficients $\{h_{k,n}\}$ as the *discrete point spread function* (DPSF) or *discrete impulse response* of the discrete linear transformation defined by the right part of Equation 4.5

$$\beta_k = \sum_n \alpha_n h_{k,n} \tag{4.7}$$

referred to as the digital linear filter.

Equation 4.7 can be rewritten in a compact vector-matrix form as

$$\mathbf{B} = \mathbf{H}\mathbf{A}^t, \tag{4.8}$$

where $\mathbf{B} = \{\beta_k\}$ and is $\mathbf{A} = \{\alpha_n\}$ are vector-columns of the representative coefficients of output and input signals, superscript t denotes the matrix transposition, and $\mathbf{H}$ is the matrix of coefficients $\{h_{k,n}\}$.

In a special case, when bases of the input and output signals are identical, orthogonal, and built from the "eigen" functions of the transformation:

$$\left\{\phi_k^{(r)}(\cdot) = \varphi_k^{(r)}(\cdot)\right\}; \quad \mathbf{L}\left(\varphi_k^{(r)}(\cdot)\right) = \eta_k \varphi_k^{(r)}(\cdot) \tag{4.9}$$

filter matrix **H** degenerates into a diagonal matrix

$$\mathbf{H} = \{\eta_k \delta_{k,n}\}, \tag{4.10}$$

where $\delta_{k,n}$ is the Kronecker delta (Equation 2.4). In this case, the relationship of Equation 4.7 between input and output representation coefficients is reduced to a simple element-wise multiplication:

$$\beta_k = \eta_k \alpha_k. \tag{4.11}$$

A discrete linear transformation described by a diagonal matrix will be referred to as the *scalar filter*. A general transformation described by Equation 4.7 will be referred to as the *vector filter*.

Equation 4.6 provides a way to determine DPSF of a discrete linear transformation that corresponds to a given analog linear transformation for chosen bases of its input and output signals. One can also formulate an inverse problem: given discrete transformation with DPSF $\{h_{k,n}\}$, what will be the PSF of the equivalent analog transformation? To answer this question, substitute Equations 4.3 and 4.7 into Equation 4.4 and obtain

$$b(x) = \sum_k \beta_k \varphi_k^{(r)}(x) = \sum_k \left(\sum_n h_{n,k} \alpha_n\right)\varphi_k^{(r)}(x)$$

$$= \sum_k \left[\sum_n h_{n,k} \int_\Xi a(\xi)\phi_n^{(d)}(\xi)\,d\xi\right]\varphi_k^{(r)}(x)$$

$$= \int_\Xi a(\xi)\left[\sum_k \sum_n h_{k,n}\phi_n^{(d)}(\xi)\varphi_k^{(r)}(x)\right]d\xi \tag{4.12}$$

from which it follows that PSF $h_{eq}(x, \xi)$ of the analog linear transformation equivalent to the discrete transformation with DPSF $\{h_{k,n}\}$ given signal discretization and reconstruction bases is

$$h_{eq}(x,\xi) = \sum_k \sum_n h_{k,n}\phi_n^{(d)}(\xi)\varphi_k^{(r)}(x). \tag{4.13}$$

In conclusion, note that in digital processing, DPSF coefficients of linear transformations are quantized. However, in most cases, they are represented in the floating point format, that is, the number of quantization levels is very high. Because of this, effects of DPSF quantization are usually assumed to be negligible and hence disregarded.

As it was stated in the section "The Sampling Theorem and Signal Sampling," the most widely used method of signal discretization is signal sampling that uses shifted, or convolution, bases. In what follows, we derive discrete representation of the convolution integral and integral Fourier and Fresnel transforms for shifted bases.

Discrete Representation of the Convolution Integral

Digital Convolution

1D convolution integral of a signal $a(x)$ with shift invariant kernel $h(x)$ is defined as (Equation 2.40)

$$b(x) = \int_{-\infty}^{\infty} a(\xi)h(x - \xi)\,d\xi. \tag{4.14}$$

In signal sampling and reconstruction, sampling and reconstruction grids can, in principle, be arbitrarily shifted with respect to corresponding signal coordinate systems. Let signal sampling and reconstruction be performed over sampling grids shifted with respect to the signal coordinate axis by arbitrary shifts, respectively, $u_x^{(s)}$ and $u_x^{(r)}$ (in units of sampling interval Δx) such that sample a_0 is taken at coordinate $x = u_x^{(s)}\Delta x$ and placed for reconstruction in coordinate $x = u_x^{(r)}\Delta x$ (see Figure 4.2).

Then sampling and reconstruction bases can be written as

$$\phi_k^{(s)}(x) = \phi^{(s)}\left(x - \tilde{k}^{(s)}\Delta x\right) \quad \text{and} \quad \phi_k^{(r)}(x) = \phi^{(r)}\left(x - \tilde{k}^{(r)}\Delta x\right), \tag{4.15}$$

where $\tilde{k}^{(s)} = k + u_x^{(s)}$ and $\tilde{k}^{(r)} = k + u_x^{(r)}$. Substituting Equations 4.14 and 4.15 in Equation 4.6, obtain

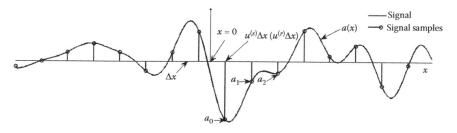

FIGURE 4.2
Signal samples in the signal coordinate system.

$$h_{k,n} = \int_{-\infty}^{\infty} \left\{ \int_{-\infty}^{\infty} h(x-\xi)\phi^{(r)}\left(\xi - \tilde{n}^{(r)}\Delta x\right)d\xi \right\} \phi^{(s)}\left(x - \tilde{k}^{(s)}\Delta x\right)dx$$

$$= \int_{-\infty}^{\infty}\int_{-\infty}^{\infty} h\left[x - \xi - \left(k - n + u_x^{(s)} - u_x^{(r)}\right)\Delta x\right]\phi^{(r)}(\xi)\phi^{(s)}(x)\,dx\,d\xi = h_{k-n}, \qquad (4.16)$$

and Equation 4.7 takes the form

$$b_k = \sum_n a_n h_{k-n}, \qquad (4.17)$$

where the summation is carried out over all signal samples. The definition of the convolution integral assumes that analog signal has infinite extent, which implies that the number of its samples is also infinitely high, that is, that

$$b_k = \sum_{n=-\infty}^{\infty} a_n h_{k-n}. \qquad (4.18)$$

In reality, the number of available signal samples is always finite as is the number of samples of the filter DPSF $\{h_k\}$. Therefore, the summation in Equation 4.18 must be truncated. Because the spatial extent of the convolution kernel $h(x)$ is usually less than that of signals and, consequently, the number $N^{(h)}$ of samples $\{h_n\}$ of the filter DPSF is less than that of signals, it is convenient to rewrite Equation 4.18 in an alternative form as

$$b_k = \sum_{n=-\infty}^{\infty} h_n a_{k-n} \qquad (4.19)$$

and then truncate the summation according to the extent of the filter DPSF:

$$b_k = \sum_{n=0}^{N^{(h)}-1} h_n a_{k-n}. \qquad (4.20)$$

Equation 4.20 is the basic equation for *digital convolution*.

For 2D signals, one can, in a similar way, obtain the following discrete representation of 2D convolution integral:

$$b_{k,l} = \sum_{m,n \in N^h} h_{n,m} a_{k-n,l-m}, \qquad (4.21)$$

where (k, l) are sample indices over corresponding 2D coordinates, and $N^{(h)}$ is area of nonzero samples $h_{n,m}$ in a rectangular sampling grid. The program convolution_demo_CRC.m provided in Exercises illustrates image digital convolution.

Now, find the characteristics of an analog filter equivalent to a given digital one specified by its DPSF $\{h_n\}$ assuming that its analog output signal $b(x)$ is reconstructed using certain limited number, say $N^{(b)}$, of output samples of the digital filter defined by Equation 4.20. To this end, express $b(x)$ as

$$b(x) = \sum_{k=0}^{N^{(b)}-1} b_k \phi^{(r)}\left(x - \tilde{k}^{(r)}\Delta x\right) = \sum_{k=0}^{N^{(b)}-1} \left(\sum_{n=0}^{N^{(h)}-1} h_n a_{k-n}\right) \phi^{(r)}\left(x - \tilde{k}^{(r)}\Delta x\right)$$

$$= \sum_{k=0}^{N^{(b)}-1} \left[\sum_{n=0}^{N^{(h)}-1} h_n \int_{-\infty}^{\infty} a(\xi)\phi^{(s)}\left[\xi - \left(\tilde{k}^{(r)} - \tilde{n}^{(s)}\right)\Delta x\right]d\xi\right] \phi^{(r)}\left(x - \tilde{k}^{(r)}\Delta x\right)$$

$$= \int_{-\infty}^{\infty} a(\xi)\left[\sum_{k=0}^{N^{(b)}-1}\sum_{n=0}^{N^{(h)}-1} h_n \phi^{(r)}\left(x - \tilde{k}^{(r)}\Delta x\right)\phi^{(s)}\left[\xi - \left(\tilde{k}^{(r)} - \tilde{n}^{(s)}\right)\Delta x\right]\right]d\xi, \quad (4.22)$$

where $\tilde{k}^{(r)} = k + u^{(r)}$ and $\tilde{n}^{(s)} = n + u^{(s)}$, as before, are biased sample indices. From Equation 4.22, we can conclude that

$$h_{eq}(x, \xi) = \sum_{k=0}^{N^{(b)}-1}\sum_{n=0}^{N^{(h)}-1} h_n \phi^{(r)}\left(x - \tilde{n}^{(s)}\Delta x\right)\phi^{(s)}\left[\xi - \left(\tilde{k}^{(r)} - \tilde{n}^{(s)}\right)\Delta x\right] \quad (4.23)$$

is the overall PSF (*OPSF*) of the continuous filter equivalent to the digital filter with DPSF $\{h_n\}$.

It is more convenient to characterize the equivalent continuous filter by its overall frequency response (OFR), Fourier transform of its OPSF. As it is shown in Appendix

$$OFR_{eq}(f, p) = \int_{-\infty}^{\infty}\int_{-\infty}^{\infty} h_{eq}(x, \xi)\exp\left[i2\pi(fx - p\xi)\right]dx\,d\xi = \left[\sum_{n=0}^{N^{(x)}-1} h_n \exp(i2\pi p\tilde{n}\Delta x)\right]$$

$$\times \left\{\left[\int_{-\infty}^{\infty}\phi^{(r)}(x)\exp(i2\pi fx)dx\right]\left[\int_{-\infty}^{\infty}\phi^{(s)}(\xi)\exp(-i2\pi p\xi)d\xi\right]\right\}$$

$$\times \left\{\exp\left(i2\pi(f - p)u_x^{(r)}\Delta x\right)\sum_{k=0}^{N^{(b)}-1}\exp\left[i2\pi(f - p)\tilde{k}^{(r)}\Delta x\right]\right\}. \quad (4.24)$$

This expression contains four multiplicands. The first multiplicand

$$CFR(p) = \sum_{n=0}^{N^{(h)}-1} h_n \exp(i2\pi p\tilde{n}\Delta x) \qquad (4.25)$$

is a Fourier series expansion of the digital filter DPSF $\{h_n\}$. It is referred to as *continuous frequency response* of the digital filter. Continuous frequency responses of digital filters are periodical functions of frequency with a period equal to the signal baseband $(1/\Delta x)$.

The second and third multiplicands

$$\Phi^{(r)}(f) = \int_{-\infty}^{\infty} \phi^{(r)}(x)\exp(i2\pi fx)\,dx \qquad (4.26)$$

and

$$\Phi^{(s)}(p) = \int_{-\infty}^{\infty} \phi^{(s)}(x)\exp(i2\pi px)\,dx \qquad (4.27)$$

are, respectively, frequency responses of signal reconstruction device and complex conjugate of the frequency response of signal sampling device. If signal reconstruction and discretization devices are, as required by the sampling theorem, ideal low-pass filters, terms $\Phi^{(r)}(f)$ and $\Phi^{*(s)}(p)$ are rectangular function with support $(1/\Delta x)$ that remove in $OFR_{eq}(f, p)$ all but one period of the digital filter continuous frequency response $CFR(p)$.

The fourth term

$$SV(f,p) = \exp\left[i2\pi(f-p)u^{(r)}\Delta x\right] \sum_{k=0}^{N^{(b)}-1} \exp\left[i2\pi(f-p)\tilde{k}^{(r)}\Delta x\right]$$

$$= \frac{\exp\left[i2\pi N^{(b)}(f-p)\Delta x\right]-1}{\exp\left[i2\pi(f-p)\Delta x\right]-1} \exp\left[i2\pi(f-p)u_x^{(r)}\Delta x\right]$$

$$= N^{(b)} \frac{\sin\left[\pi(f-p)N^{(b)}\Delta x\right]}{N^{(b)}\sin\left[\pi(f-p)\Delta x\right]} \exp\left[i2\pi(f-p)\left(u_x^{(r)}+\frac{N^{(b)}-1}{2}\right)\Delta x\right] \qquad (4.28)$$

depends only on the number of output signal samples. It reflects the fact that the digital filter defined by Equation 4.20 and obtained as a discrete representation of the convolution integral is not spatially invariant owing to the finite number $N^{(b)}$ samples $\{b_k\}$ of the filter output signal involved in the reconstruction of analog output signal $b(x)$. When this number increases, the contribution of border effects into the reconstructed signal diminishes. In the limit, when $N^{(b)} \to \infty$, the filter becomes spatially homogeneous because

$$\lim_{N^{(b)} \to \infty} N_b \frac{\sin\left[\pi(f - p)N^{(b)}\Delta x\right]}{N^{(b)} \sin\left[\pi(f - p)\Delta x\right]} = \delta(f - p). \qquad (4.29)$$

We already addressed the same issue in the section "Sampling Artifacts: Qualitative Analysis" when discussing signal reconstruction from its samples.

Note that phase-factor $\exp[i2\pi(f - p)(u_x^{(r)} + (N^{(b)} - 1/2))\Delta x]$ in Equation 4.28 disappears with a natural selection of the reconstruction shift parameter $u_x^{(r)} = -(N^{(b)} - 1)/2$.

Treatment of Signal Borders in Digital Convolution

Equation 4.20 defines the digital convolution filter output signal samples $\{b_k\}$ only for $k = N^{(h)}$, $N^{(h)} + 1$, $N^{(a)} - 1$, where $N^{(a)}$ is the number of input signal samples. Samples for $k < N^{(h)}$ and $k > N^{(a)} - 1$ are not defined because input signal samples $\{a_n\}$ for $n < 0$ and for $n > N^{(a)}$ are not available. Therefore, in digital filtering, the number of output signal samples is shrinking away unless unavailable input signal samples are additionally defined by one or another method of signal extrapolation. If they are, the output signals within $N^{(h)}/2$ samples at its borders are influenced by the extrapolated values of the input signal; hence, the extrapolation method must be selected to not produce undesirable artifacts. Proper border processing is especially important in image processing because of the "curse of dimensionality." Compare, for instance, cases of an audio signal of 10^6 samples, which corresponds to approximately 1 min of phonation, and an image frame of $10^3 \times 10^3 = 10^6$ pixels. Let filter DPSFs have $N^{(h)} = 100$ for 1D filter and $N^{(h)} = 100 \times 100$ samples, that is, the same 1D extent, for 2D filter. Then, for the audio signal, border artifacts influence only 10^{-4}-th fraction of samples (6 ms of phonation), while for the image signal 10^{-2}-th fraction of pixels is influenced, or one tens on each of the image frame dimensions.

In principle, the task of image extrapolating outside the available frame of image samples is a special case of the problem of image recovery from incomplete set of data. Solving this problem requires formulation of *a priori* knowledge on images or image ensembles and quite sophisticated and computationally expensive methods. Some methods based on the assumption of band-limitedness of images will be discussed in Chapter 7. In practice, more simple methods are conventionally used.

Very frequently, padding signals outside the signal frame by zeros are used. This is, for instance, what is implemented in MATLAB signal and image processing tool boxes. A justification of zero padding may be found in an assumption that signals must be extended by their mean value, which by default, is assumed to be zero. For many signals such as audio signals, the zero mean assumption seems natural. For images, this assumption, as a rule, is not relevant, although in some applications, images of objects do have an empty (zero) background, which can be extended beyond the image frame. Some astronomical images of extraterrestrial objects and tomographic images of body slices exemplify these cases.

In numerical and applied mathematics, another assumption, that of signal periodicity, is quite popular. It originates from the Fourier series expansion of functions and is very well suited with applications of DFT, which, as we will see in the section "Discrete Representation of Fourier Integral Transform," implies, by the definition, that signals are periodical with a period equal to the number of signal samples. In image processing, the periodical image extension beyond its borders, which makes extreme image samples of left and right, or upper and bottom borders immediate neighbors, is definitely not appropriate and frequently results in heavy border artifacts.

The major source of border artifacts associated with the use of the above two signal extrapolation methods is that they tend to introduce discontinuities in the extrapolated signal at borders of the initial signal. An alternative extrapolation method that is completely free of this drawback is signal even extension by means of mirror reflection from its borders. These methods and the way they work in signal convolution are illustrated in Figure 4.3.

In the section "Signal Convolution in the DCT Domain," we will show that signal even extension by mirror reflection translates DFT into DCT, an orthogonal transform with very good energy compaction capability that can replace DFT in all its applications, including Fourier analysis, fast convolution, and signal resampling.

Discrete Representation of Fourier Integral Transform

Discrete Fourier Transforms

In this section, we apply the general approach to discrete representation of linear transformations formulated in the section "Basic Principles of Discrete Representation of Signal Transformations" to integral Fourier transform. Let signal reconstructing and spectrum sampling basis functions be, respectively, $\{\varphi_k^{(r)}(x) = \varphi^{(r)}(x - \tilde{k}\Delta x)\}$ and $\{\varphi_r^{(s)}(f) = \varphi^{(s)}(f - \tilde{r}\Delta f)\}$, where, as before, $\tilde{k} = k + u_x$ and $\tilde{r} = r + v_f$ are biased indices of signal and its spectrum samples and u_x and v_f are, respectively, shifts, in fractions of the discretization

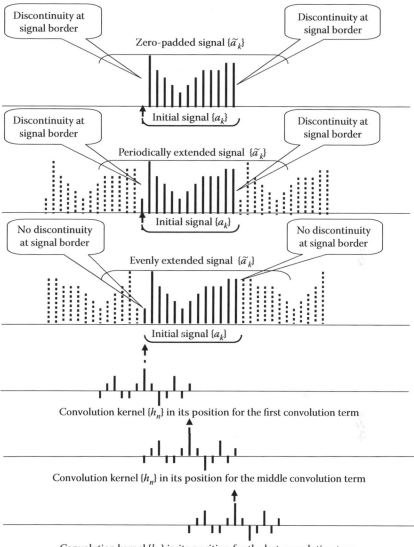

FIGURE 4.3
Signal extrapolation by means of zero padding, periodical extension, and even extension by means of mirror reflection at signal borders and how do they work in signal convolution.

intervals Δx and Δf, of signal and its spectrum samples with respect to the corresponding coordinate system as illustrated in Figure 4.4.

Then, according to Equation 4.6, a discrete representation of the Fourier transform kernel can be obtained as (see Appendix)

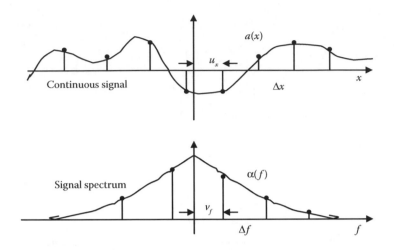

FIGURE 4.4
Geometry of dispositions of signal and its Fourier spectrum samples (stems).

$$h_{k,r} = \int\limits_{-\infty}^{\infty}\int\limits_{-\infty}^{\infty} \exp(i2\pi fx)\phi^{(r)}\left(x - \tilde{k}\right)\phi^{(s)}\left(f - \tilde{r}\Delta f\right)dx\,df$$

$$= \exp\left(i2\pi\tilde{k}\tilde{r}\Delta x\Delta f\right)\int\limits_{-\infty}^{\infty} \Phi^{(r)}\left(f + \tilde{r}\Delta f\right)\phi^{(s)}(f)\exp\left(i2\pi f\tilde{k}\Delta x\right)df, \quad (4.30)$$

where $\Phi^{(r)}(f)$ is the frequency response of the signal reconstruction device. For discrete representation of the Fourier transform, only the term

$$\tilde{h}_{k,r} = \exp\left(i2\pi\tilde{k}\tilde{r}\Delta x\Delta f\right) \quad (4.31)$$

is usually used, and the term

$$\int\limits_{-\infty}^{\infty} \Phi^{(r)}\left(f + \tilde{r}\Delta f\right)\phi^{(s)}(f)\exp\left(i2\pi f\tilde{k}\Delta x\right)df \quad (4.32)$$

is disregarded, though, in principle, it also depends on sample indices k and r. Some justification reasonings for this are discussed in Appendix.

Now, let X be an interval occupied by the signal $a(x)$ and $N = X/\Delta x$ be the number of signal samples within this interval. According to the sampling theory (see the section "The Sampling Theorem and Signal Sampling" in Chapter 3), spectrum sampling interval Δf and extent X of the signal are inverse to each other. Assume, therefore, that $\Delta f = 1/\sigma X$, where σ is a scale parameter. Then, obtain that

$$\Delta x \Delta f = \frac{\Delta x}{\sigma X} = \frac{1}{\sigma N} \tag{4.33}$$

and Equation 4.31 can be rewritten as

$$\tilde{h}_{k,r} = \exp\left(i2\pi \frac{\tilde{k}\tilde{r}}{\sigma N}\right). \tag{4.34}$$

Then, the basic Equation 4.7 of digital filtering takes, for the representation of Fourier transform, the form

$$\alpha_r^{(u,v;\sigma)} \propto \sum_{k=0}^{N-1} a_k \exp\left(i2\pi \frac{\tilde{k}\tilde{r}}{\sigma N}\right) = \sum_{k=0}^{N-1} a_k \exp\left[i2\pi \frac{(k+u_x)(r+v_f)}{\sigma N}\right], \tag{4.35}$$

where superscripts (u_x, v_f) indicate that signal and its Fourier spectrum samples are taken with displacements u_x and v_f with respect to signal and spectrum coordinate systems (Figure 4.4).

To complete the derivation, introduce a normalizing multiplier $1/\sqrt{\sigma N}$ and obtain

$$\alpha_r^{(u,v;\sigma)} = \frac{1}{\sqrt{\sigma N}} \sum_{k=0}^{N-1} a_k \exp\left[i2\pi \frac{(k+u_x)(r+v_f)}{\sigma N}\right]. \tag{4.36}$$

As it is shown in Appendix, this transform has its inverse

$$a_k^{(u,v;\sigma)} = \frac{1}{\sqrt{\sigma N}} \sum_{r=0}^{\sigma N-1} \alpha_r^{(u,v;\sigma)} \exp\left[-i2\pi \frac{(k+u_x)(r+v_f)}{\sigma N}\right]. \tag{4.37}$$

If $\sigma > 1$ and σN is an integer number ($\sigma N \in Z$)

$$a_k^{(u,v;\sigma)} = \begin{cases} a_k, & k = 0,1,...,N-1 \\ 0, & k = N,N+1,...,\sigma N-1 \end{cases}. \tag{4.38}$$

Of especial importance is the case of the *cardinal sampling*, when $\sigma = 1$ and, hence, $\Delta x \Delta f = 1/N$. In this case and for signal and its spectrum sample zero shifts ($u_x = 0$; $v_f = 0$) we arrive at the transformations

$$\alpha_r = \frac{1}{\sqrt{N}} \sum_{k=0}^{N-1} a_k \exp\left(i2\pi \frac{kr}{N}\right) \quad \text{and} \quad a_k = \frac{1}{\sqrt{N}} \sum_{r=0}^{N-1} \alpha_r \exp\left(-i2\pi \frac{kr}{N}\right) \quad (4.39)$$

are known as direct and inverse DFTs. In order to distinguish the general case from this special case, we will refer to transforms defined by Equations 4.36 and 4.37 as *shifted scaled discrete Fourier transforms* SScDFT(u_x, v_f; σ). A version of SScDFT(u_x, v_f; 1) for $\sigma = 1$ will be referred to as *shifted discrete Fourier transforms* SDFT(u,v) and a version of SScDFT(0,0;σ) for $u = v = 0$ will be referred to as *scaled discrete Fourier transforms* ScDFT(σ).

The presence of shift and scale parameters makes the SScDFTs more flexible in simulating integral Fourier transform than the canonical DFT. As we will see in Chapter 6, it enables

- Performing continuous spectrum analysis
- Computing convolution and correlation with subpixel resolution
- Perfect and at the same time flexible and computationally efficient resampling of discrete signals

In all applications, the DFT is the basic transform because for its computation a very efficient fast Fourier transform (FFT) algorithm exists (for the principle of FFT, see Appendix). All the above-introduced generalizations can be computed in one or another way through DFT. In particular, shifted DFTs can be, for any shift parameters, computed through the DFT as follows:

$$\alpha_r^{(u,v)} = \frac{1}{\sqrt{N}} \exp\left(i2\pi \frac{ru_x}{N}\right) \sum_{k=0}^{N-1}\left[a_k \exp\left(i2\pi \frac{kv_f}{N}\right)\right] \exp\left(i2\pi \frac{kr}{N}\right); \quad (4.40)$$

$$\alpha_k^{(u,v)} = \frac{1}{\sqrt{N}} \exp\left(i2\pi \frac{kv_f}{N}\right) \sum_{r=0}^{N-1}\left[\alpha_r \exp\left(i2\pi \frac{ru_x}{N}\right)\right] \exp\left(i2\pi \frac{kr}{N}\right). \quad (4.41)$$

Note that in Equations 4.40 and 4.41, multiplicands $\exp(\pm i2\pi u_x v_f/N)$, which do not depend on sample indices k and r, are omitted as irrelevant and redundant. Equations 4.40 and 4.41 can be used as alternative working definitions of shifted DFT.

Scaled DFTs can be represented as convolution, as it is shown in Appendix, and therefore, can also be computed through DFT using the discrete convolution theorem presented below in Section "Properties of Discrete Fourier Transforms."

2D Discrete Fourier Transforms

In the assumption that 2D signals are defined on a 2D rectangular sampling grid of $N_1 \times N_2$ samples, 1D-shifted scaled DFTs can be straightforwardly generalized to 2D-shifted scaled DFTs, SScDFT($u_1, v_1; u_2, v_2; \sigma_1, \sigma_2$), as

$$\alpha_{r,s}^{(u_1,v_1;u_2,v_2)} = \frac{1}{\sqrt{N_1 N_2}} \sum_{k=0}^{N_1-1} \sum_{l=0}^{N_2-1} a_{k,l} \exp\left[i2\pi \left(\frac{\tilde{k}\tilde{r}}{\sigma_1 N_1} + \frac{\tilde{l}\tilde{s}}{\sigma_2 N_2} \right) \right]; \qquad (4.42)$$

$$a_{k,l} = \frac{1}{\sqrt{N_1 N_2}} \sum_{r=0}^{N_1-1} \sum_{s=0}^{N_2-1} \alpha_{r,s}^{(u_1,v_1;u_2,v_2)} \exp\left[-i2\pi \left(\frac{\tilde{k}\tilde{r}}{\sigma_1 N_1} + \frac{\tilde{l}\tilde{s}}{\sigma_2 N_2} \right) \right], \qquad (4.43)$$

where $\{\tilde{k}, \tilde{l}\}$ and $\{\tilde{r}, \tilde{s}\}$ are biased indices of signal and its spectrum samples:

$$\tilde{k} = k + u_1, k = 0,...,N_1 - 1, \quad \tilde{l} = l + u_2, l = 0,...,N_2 - 1;$$

$$\tilde{r} = r + v_{f_1}, r = 0,...,N_1 - 1, \quad \tilde{s} = s + v_{f_2}, s = 0,...,N_2 - 1; \qquad (4.44)$$

$\{u_1, u_2\}$ and $\{v_{f_1}, v_{f_2}\}$ are corresponding shifts of sampling grids in signal and Fourier transform domains, and $\{\sigma_1, \sigma_2\}$ are scale parameters that define the relationships between signal and its spectrum sampling intervals $\{\Delta x_1, \Delta x_2\}$ and $\{\Delta f_1, \Delta f_2\}$ and the numbers $\{N_1, N_2\}$ of signal and spectrum samples: $\Delta x_1 \Delta f_{x_1} = 1/\sigma_1 N_2$; $\Delta x_2 \Delta f_{x_2} = 1/\sigma_2 N_2$. By the definition, 2D SScDFTs are separable over their indices and therefore can be computed in two separate passes—row-wise and column-wise:

$$\alpha_{r,s}^{(u,v)} = \frac{1}{\sqrt{N_1 N_2}} \sum_{k=0}^{N_1-1} \sum_{l=0}^{N_2-1} a_{k,l} \exp\left[i2\pi \left(\frac{\tilde{k}\tilde{r}}{\sigma_1 N_1} + \frac{\tilde{l}\tilde{s}}{\sigma_2 N_2} \right) \right]$$

$$= \frac{1}{\sqrt{N_1 N_2}} \sum_{k=0}^{N_1-1} \exp\left(i2\pi \frac{\tilde{k}\tilde{r}}{\sigma_1 N_1} \right) \sum_{l=0}^{N_y-1} a_{k,l} \exp\left(i2\pi \frac{\tilde{l}\tilde{s}}{\sigma_2 N_2} \right). \qquad (4.45)$$

Correspondingly, canonical direct and inverse 2D DFTs are defined as

$$\alpha_{r,s} = \frac{1}{\sqrt{N_1 N_2}} \sum_{k=0}^{N_1-1} \sum_{l=0}^{N_2-1} a_{k,l} \exp\left[i2\pi \left(\frac{kr}{N_1} + \frac{ls}{N_2} \right) \right]; \qquad (4.46)$$

$$a_{k,l} = \frac{1}{\sqrt{N_1 N_2}} \sum_{r=0}^{N_1-1} \sum_{s=0}^{N_2-1} \alpha_{r,s} \exp\left[-i2\pi \left(\frac{kr}{N_1} + \frac{ls}{N_2} \right) \right]. \qquad (4.47)$$

A natural generalization of 2D-shifted and -scaled DFTs is 2D *affine DFT* (AffDFT). AffDFT is obtained in the assumption that either signal or its spectrum sampling or reconstruction is carried out in affine transformed, with respect to signal/spectrum coordinate systems (x_1, x_2), coordinates $(\tilde{x}_1, \tilde{x}_2)$:

$$\begin{bmatrix} x_1 \\ x_2 \end{bmatrix} = \begin{bmatrix} A & B \\ C & D \end{bmatrix} \begin{bmatrix} \tilde{x}_1 \\ \tilde{x}_2 \end{bmatrix}. \tag{4.48}$$

where (A, B, C, D) are numerical parameters of the affine transform. With $\sigma_A = 1/N_1 A \Delta \tilde{x}_1 \Delta f_{x_1}$, $\sigma_B = 1/N_2 B \Delta x_2 \Delta f_2$, $\sigma_C = 1/N_1 C \Delta \tilde{x}_2 \Delta f_{x_2}$, and $\sigma_D = 1/N_2 D \Delta \tilde{x}_2 \Delta f_{x_2}$, where $\Delta \tilde{x}_1$, $\Delta \tilde{x}_2$, Δf_{x_1}, and Δf_{x_2} are signal and its Fourier transform sampling intervals in image $(\tilde{x}_1, \tilde{x}_2)$ and Fourier (f_{x_1}, f_{x_2}) planes, *AffDTF* is defined as

$$\alpha_{r,s} = \sum_{k=0}^{N_1-1} \sum_{l=0}^{N_2-1} a_{k,l} \exp\left[i2\pi \left(\frac{\tilde{r}\tilde{k}}{\sigma_A N_1} + \frac{\tilde{s}\tilde{k}}{\sigma_C N_1} + \frac{\tilde{r}\tilde{l}}{\sigma_B N_2} + \frac{\tilde{s}\tilde{l}}{\sigma_D N_2} \right) \right]. \tag{4.49}$$

A special case of affine transforms is rotation. For rotation angle θ

$$\begin{bmatrix} x_1 \\ x_2 \end{bmatrix} = \begin{bmatrix} \cos\theta & \sin\theta \\ -\sin\theta & \cos\theta \end{bmatrix} \begin{bmatrix} \tilde{x}_1 \\ \tilde{x}_2 \end{bmatrix}. \tag{4.50}$$

With $N_1 = N_2 = N$, $\Delta \tilde{x}_1 = \Delta \tilde{x}_2 = \Delta x$, $\Delta f_{x_1} = \Delta f_{x_2} = \Delta f$, and $\Delta x \, \Delta f = 1/N$ (cardinal sampling), 2D-*rotated DFT (RotDFT)* is obtained:

$$\alpha_{r,s}^\theta = \frac{1}{N} \sum_{k=0}^{N-1} \sum_{l=0}^{N-1} a_{k,l} \exp\left[i2\pi \left(\frac{\tilde{k}\cos\theta + \tilde{l}\sin\theta}{N} \tilde{r} - \frac{\tilde{k}\sin\theta - \tilde{l}\cos\theta}{N} \tilde{s} \right) \right]. \tag{4.51}$$

An obvious generalization of RotDFT is *rotated and scaled DFT* (RotScDFT):

$$\alpha_{r,s}^\theta = \frac{1}{\sigma N} \sum_{k=0}^{N-1} \sum_{l=0}^{N-1} a_{k,l} \exp\left[i2\pi \left(\frac{\tilde{k}\cos\theta + \tilde{l}\sin\theta}{\sigma N} \tilde{r} - \frac{\tilde{k}\sin\theta - \tilde{l}\cos\theta}{\sigma N} \tilde{s} \right) \right] \tag{4.52}$$

that assumes 2D signal sampling in θ-rotated and σ-scaled coordinate systems.

Properties of Discrete Fourier Transforms

In this section, we will review basic properties of DFTs. It is instructive to compare them with corresponding properties of the integral Fourier transform discussed in Chapter 2. In order to avoid unnecessary complications in formulae, we will consider here only shifted DFTs. Some useful properties and usage of scaled DFTs will be discussed in Chapter 6.

Invertibility and sincd-Function

Let $\{\alpha_r^{u,v}\}$ be SDFT(u,v) coefficients of a discrete signal $\{a_k\}$, $k = 0,1,\ldots,N-1$. Compute the inverse SDFT with shift parameters (p_x, q_f) of a subset of K spectral samples $\{\alpha_r^{u,v}\}$.

As it is shown in Appendix, in the result, we obtain

$$
\tilde{a}_n^{u/p,v/q} = \frac{1}{\sqrt{N}} \left[\sum_{r=0}^{K-1} \alpha_r^{u,v} \exp\left(-i2\pi \frac{rp_x}{N} \right) \right] \exp\left[-i2\pi \frac{n(r+q_v)}{N} \right]
$$

$$
= \left\{ \sum_{k=0}^{N-1} a_k \exp\left[i\pi k \left(\frac{K-1}{N} + 2v_f \right) \right] \overline{\text{sincd}}\left[K; N; \pi(k-n+u_x-p_x) \right] \right\}
$$

$$
\times \exp\left[-i\pi \left(\frac{K-1}{N} + 2q_f \right) n \right] \exp\left[i\pi \frac{K-1}{N} (u_x - p_x) \right], \qquad (4.53)
$$

where

$$
\overline{\text{sincd}}(K;N;x) = \frac{\sin(\pi(K/N)x)}{N\sin(\pi(x/N))} \qquad (4.54)
$$

is the *general discrete sinc-function*, an extension of the discrete sinc-function introduced in Chapter 3 and defined by Equation 3.61; these two functions are identical when $K = N$.

For integer $x = k - n$ and $K = N$, function $\overline{\text{sincd}}(N;N;k-n)$ is identical to Kronecker delta-function:

$$
\overline{\text{sincd}}[N;N;\pi(k-n)] = \frac{\sin[\pi(k-n)]}{N\sin(\pi(k-n/N))} = \delta(k-n) = \begin{cases} 1, & k = n \\ 0, & k \neq n \end{cases}. \qquad (4.55)
$$

Therefore, when $u_x = p_x$, $v_f = q_f$, and $K = N$

$$
\tilde{a}_n^{u/p,v/q} = \sum_{k=0}^{N-1} a_n \overline{\text{sincd}}[N;N;(k-n)] = \sum_{k=0}^{N-1} a_n \, \text{sincd}[N;(k-n)] = a_n, \qquad (4.56)
$$

which confirms the invertibility of the SDFT. However, the meaning of Equation 4.53 is richer. It shows that the SDFT is invertible in a more general sense. Provided $v_f = q_f = -(K - 1/2)$ and the demodulation of the reconstructed signal by an exponential multiplier $\exp[-i\pi(K-1/N)\,(u_x - p_x)]$,

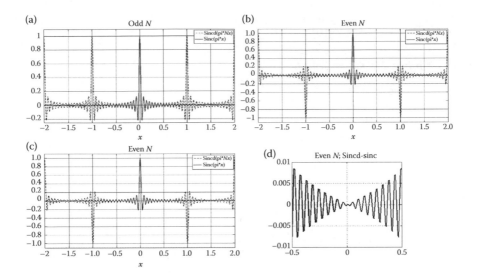

FIGURE 4.5

Discrete sinc- (dash line) along with continuous sinc- (solid line) functions for odd (a) and even (c) number of samples N and differences between sinc- and sincd-functions within the basic interval $N\Delta x$ for odd (b) and even (d) N.

one, according Equation 4.26, can generate sincd-interpolated signal copies shifted with respect to the original signal samples by the $(u_x - p_x)$-th fraction of the sampling interval. In Chapter 6, we will show that such discrete sinc-interpolation is the gold standard for resampling discrete signals.

Discrete sinc-function is a discrete analog of the continuous sinc-function defined by Equation 2.73. Both functions are plotted for comparison in Figure 4.5.

One can see from the figure that continuous and sinc-function and discrete sincd-function are almost identical within the basic interval of N samples for the *discrete sinc-function* and its corresponding interval $N\Delta x$ for the continuous sinc-function; within this interval, the difference between them does not exceed 10^{-2}. Outside this interval, they differ dramatically: while sinc-function continues decaying, sincd-function is a periodical function and the type of its periodicity depends on whether N is odd or even number: $\text{sincd}(N; k + gN) = (-1)^{g(N-1)} \text{sincd}(N;k)$.

Energy Preservation Property

The invertibility of SDFTs implies that SDFTs are orthogonal transforms and, as such, they satisfy Parseval's relationship:

$$\sum_{k=0}^{N-1} \left(a_k^{(u,v)}\right)^2 = \sum_{r=0}^{N-1} \left(\alpha_r^{(u,v)}\right)^2. \tag{4.57}$$

Cyclicity

From Equation 4.37, it follows that, for an integer number g, discrete signal $\{a_k^{u,v}\}$ reconstructed by the inverse SDFT(u_x, v_y) from SDFT(u_x, v_y) spectrum of signal $\{a_k\}$ is defined outside the index interval $(0, N-1)$ as (see Appendix)

$$a_{k+gN}^{(u,v)} = a_k^{(u,v)} \exp(-i2\pi g v_f). \tag{4.58}$$

In particular, when $v_f = 0$, the signal is a periodical function of k:

$$a_{k+gN}^{(u,0)} = a_k^{(u,0)} = a_{(k) \bmod N}^{(u,0)}, \tag{4.59}$$

where $(k) \bmod N$ is a residual of k/N. Similar periodicity property holds for SDFT(u_x, v_f) spectra:

$$\alpha_{r+gN}^{(u,v)} = \alpha_r^{(u,v)} \exp(i2\pi g u_x); \quad \alpha_{r+gN}^{(0,v)} = \alpha_r^{(0,v)} = \alpha_{(r) \bmod N}^{(0,v)}. \tag{4.60}$$

For the canonical DFT, both signal and its spectrum are periodical function of indices:

$$a_k^{(0,0)} = a_{(k) \bmod N}^{(0,0)}; \quad \alpha_r^{(0,0)} = \alpha_{(r) \bmod N}^{(0,0)}. \tag{4.61}$$

Owing to the separability of the DFTs, the same cyclicity feature holds for 2D and multidimensional DFTs for all indices. The cyclicity feature is the main distinction of DTFs from the integral Fourier transform they represent.

Shift Theorem

The shift theorem for the DFTs is completely analogous to that for integral Fourier transform: absolute value of signal SDFT(u_x, v_f) spectrum is invariant to the signal shift; signal shift causes only linear modulation of the phase of the spectrum.

To prove this, let $\alpha_r^{(u,v)}$ be SDFT(u_x, v_f) of signal $\{a_{k+k_0}\}$. Express signal $\{a_{k+k_0}\}$ as inverse SDFT(u_x, v_f):

$$a_{k+k_0} = \frac{1}{\sqrt{N}} \sum_{r=0}^{N-1} \alpha_r^{(u,v)} \exp\left[-i2\pi \frac{(k + k_0 + u_x)(r + v_y)}{N}\right]$$

$$= \frac{1}{\sqrt{N}} \sum_{r=0}^{N-1} \left[\alpha_r^{(u,v)} \exp\left(-i2\pi \frac{r + v_y}{N} k_0\right)\right] \exp\left[-i2\pi \frac{(k + u_x)(r + v_y)}{N}\right]. \tag{4.62}$$

Equation 4.62 implies that SDFT(u_x, v_f) spectrum of the shifted signal $\{a_{k+k_0}\}$ is $\left[\alpha_r^{(u,v)} \exp\left(-i2\pi \frac{r + v_y}{N} k_0\right)\right]$, which proves the theorem. Similarly, shifted signal spectrum $\{\alpha_{r+r_0}^{u,v}\}$ corresponds to phase-modulated signal $a_k^{u,v} \exp\left(i2\pi r_0 \dfrac{k+u_x}{N}\right)$.

Convolution Theorem

The convolution theorem for DFTs is an analog of the convolution theorem for integral Fourier transform (see Chapter 2, Equation 2.82): a product of SDFT spectra of two signals is the SDFT spectrum of the result of convolution of these signals. However, in distinction to arithmetic (aperiodic) convolution, associated with integral Fourier transform, convolution associated with SDFTs is cyclic, by virtue of the cyclicity property of SDFTs.

Let $\{\alpha_r^{(u^a,v^a)}\}$ be SDFT (u_x^a, v_f^a) of a signal $\{a_k\}$ and $\{\beta_r^{(u^b,v^b)}\}$ be SDFT (u_x^b, v_f^b) of a signal $\{b_k\}$, $k,r = 0,1,\ldots,N-1$. Compute the inverse SDFT (u_x^c, v_f^c) of their product and obtain (see Appendix)

$$
c_k^{(u^c,v^c)} = \frac{1}{\sqrt{N}} \sum_{r=0}^{N-1} \alpha_r^{(u^a,v^a)} \beta_r^{(u^b,v^b)} \exp\left(-i2\pi \frac{\tilde{k}^c \tilde{r}^c}{N}\right)
$$

$$
= \frac{1}{\sqrt{N}} \exp\left(-i2\pi \frac{\tilde{k}^c\left(v_f^c - v_f^b\right)}{N}\right) \sum_{n=0}^{N-1} a_n \exp\left(i2\pi \frac{\tilde{n}^a\left(v_f^a - v_f^b\right)}{N}\right) \tilde{b}_{k-n}, \quad (4.63)
$$

where $\tilde{k}^a = k + u_x^a$; $\tilde{k}^b = k + u_x^b$; $\tilde{k}^c = k + u_x^c$; $r^a = r + v_f^a$; $r^b = r + v_f^b$; $r^c = r + v_f^c$; and

$$
\tilde{b}_{k-n} = \frac{1}{\sqrt{N}} \sum_{r=0}^{N-1} \beta_r^{(u^b,v^b)} \exp\left[-i2\pi \frac{\left(k - n + u_x^c - u_x^a\right)\left(r + v_f^b\right)}{N}\right] \quad (4.64)
$$

is signal reconstructed from spectrum $\{\beta_r^{u^b,v^b}\}$ using inverse SDFT $(u_x^c - u_x^a, v_f^b)$. By virtue of the general invertibility property of SDFTs, when it is selected that $v_f^b = -(N-1)/2$

$$
\tilde{b}_m = \sum_{n=0}^{N-1} b_n \operatorname{sincd}\left[N; \pi\left(n - m + u_x^b - u_x^c + u_x^a\right)\right] \quad (4.65)
$$

is a discrete sinc-interpolated copy of signal $\{b_n\}$ shifted by $u_x^b - u_x^c + u_n^a$. The derivation of Equation 4.65 is provided in Appendix.

When $v_f^a = v_f^b = v_f^c = v_f = v$, Equation 4.63 is converted to

$$
c_k^{(u^c,v^c)} = \frac{1}{\sqrt{N}} \sum_{r=0}^{N-1} \alpha_r^{(u^a,v)} \beta_r^{u^b,v} \exp\left[-i2\pi \frac{\left(k + u_x^c\right)\left(r + v_v\right)}{N}\right] = \frac{1}{\sqrt{N}} \sum_{n=0}^{N-1} a_n \tilde{b}_{k-n}. \quad (4.66)
$$

Equation 4.66 represents discrete convolution of signals $\{a_k\}$ and $\{\tilde{b}_k\}$. Being computed through SDFTs, it assumes that the signals are extended outside the index interval $[0, N-1]$ according to the cyclicity property of SDFTs (Equation 4.60).

In a special case of the DFT $(u_a = u_b = u_c = v_a = v_b = v_c = 0)$, it is a *cyclic convolution*:

$$c_{(k)\bmod N} = \frac{1}{\sqrt{N}} \sum_{r=0}^{N-1} \alpha_r^{(0,0)} \beta_r^{(0,0)} \exp\left(-i2\pi \frac{kr}{N}\right) = \frac{1}{\sqrt{N}} \sum_{n=0}^{N-1} a_{(n)\bmod N} b_{(k-n)\bmod N},\qquad(4.67)$$

which regards signals as being periodical with a period of N samples.

In a similar way, considering inverse SDFT (u_x^c, v_f^c) of the product of SDFT (u_x^a, v_f^a) spectrum $\alpha_r^{(u_x^a, v_f^a)}$ of signal $\{a_k\}$ and complex conjugate to SDFT (u_x^b, v_f^b) spectrum $\beta_r^{*(u_x^b, v_f^b)}$ of signal $\{b_k\}$, $k = 0,1,\ldots,N-1$, obtain with selection $v_f^a = v_f^b = v_f^c = -(N-1)/2$ (see Appendix):

$$c_k^{u^c,v^c} = \frac{1}{\sqrt{N}} \sum_{r=0}^{N-1} \alpha_r^{(u^a,v^a)} \beta_r^{*(u^b,v^b)} \exp\left(-i2\pi \frac{\tilde{k}^c \tilde{r}^c}{N}\right) = \sum_{n=0}^{N-1} u_n \tilde{b}_{n-k},\qquad(4.68)$$

where $\tilde{b}_n$ is defined by Equation 4.65.

Equation 4.68 complements Equation 4.66 and describes the correlation of signal $\{a_k\}$ with discrete sinc-interpolated copy of signal $\{b_k\}$. Similar to convolution, in a special case of the DFT $(u_a = u_b = u_c = v_a = v_b = v_c = 0)$ it is a *cyclic correlation*:

$$c_{(k)\bmod N} = \frac{1}{\sqrt{N}} \sum_{r=0}^{N-1} \alpha_r^{(0,0)} \beta_r^{*(0,0)} \exp\left(-i2\pi \frac{kr}{N}\right) = \frac{1}{\sqrt{N}} \sum_{n=0}^{N-1} a_{(n)\bmod N} b_{(n-k)\bmod N}.\qquad(4.69)$$

Symmetry Properties

Symmetry properties of general SDFT(u_x, v_f) are quite intricate. We will consider here several most important special cases with which we deal in this book. First of all, consider symmetry properties of the canonical DFT. As shown in Appendix, for signal $\{a_k\}$ and its DFT spectrum $\{\alpha_r\}$ and their index reversed copies $\{a_{N-k}\}$ and $\{\alpha_{N-r}\}$, the following relationships hold:

$$\{a_{N-k}\} \xrightarrow{\text{DFT}} \{\alpha_{N-r}\}; \quad \{a_k^*\} \xrightarrow{\text{DFT}} \{\alpha_{N-r}^*\}.\qquad(4.70)$$

From Equations 4.70, it follows that

$$\{a_k = \pm a_{N-k}\} \xrightarrow{\text{DFT}} \{\alpha_r = \pm \alpha_{N-r}\};$$

$$\{a_k = \pm a_k^*\} \xrightarrow{\text{DFT}} \{\alpha_r = \pm \alpha_{N-r}^*\}; \quad \alpha_0 = \alpha_0^*; \text{ For even } N, \alpha_{N/2} = \alpha_{N/2}^*,$$

$$\{a_k = a_k^* = \pm a_{N-k}\} \xrightarrow{\text{DFT}} \{\alpha_r = \pm \alpha_{N-r}^* = \pm \alpha_{N-r}\}. \tag{4.71}$$

Reversing signal and its DFT spectrum indices corresponds to reversing sign of signal coordinate and frequency for integral Fourier transform. Equations 4.70 and 4.71 show that this analogy is not complete: the signal and its DFT spectrum samples with index 0 and, for even N, index $N/2$ are not inversed, and index $N/2$, which for odd N is virtual, plays a role of the symmetry center, as do points $x = 0$ and $f = 0$ for integral Fourier transform.

SDFT(1/2,1/2) with both semi-integer shift parameters exhibits more perfect symmetry (see Appendix):

$$\{a_{N-1-k}\} \xrightarrow{\text{SDFT}(1/2,1/2)} \{\alpha_{N-1-r}^{(1/2,1/2)}\}; \tag{4.72}$$

$$\{a_k = \pm a_{N-1-k}\} \xrightarrow{\text{SDFT}(1/2,1/2)} \{\alpha_r^{(1/2,1/2)} = \pm \alpha_{N-1-r}^{(1/2,1/2)}\} \tag{4.73}$$

$$\{a_k = \pm a_k^* = \pm a_{N-1-k}\} \xrightarrow{\text{SDFT}(1/2,1/2)} \{\alpha_r^{(1/2,1/2)} = \mp \alpha_{N-1-r}^{*(1/2,1/2)} = \pm \alpha_{N-1-r}^{(1/2,1/2)}\}. \tag{4.74}$$

For SDFT(u,0) spectra $\{\alpha_r^{(u,0)}\}$, the following properties are useful:

$$\{a_k^*\} \xrightarrow{\text{SDFT}(u,0)} \{\alpha_{N-r}^{*(u,0)} \exp(i2\pi u)\};$$

$$\{a_k = a_k^*\} \xrightarrow{\text{SDFT}(u,0)} \{\alpha_r^{(u,0)} = \alpha_{N-r}^{*(u,0)} \exp(i2\pi u)\}. \tag{4.75}$$

SDFT Spectra of Sinusoidal Signals

SDFT(u_x, v_f) spectrum of a sinusoidal signal $a_k = \cos(2\pi(\rho k/N) + \phi)$ is a combination of two discrete sinc-functions:

$$\alpha_r = \frac{1}{\sqrt{N}} \sum_{k=0}^{N-1} \cos\left(2\pi \frac{\rho k}{N} + \phi\right) \exp\left[i2\pi \frac{(k+u_x)(r+v_f)}{N}\right]$$

$$= \frac{\sqrt{N}}{2} \exp\left(i2\pi \frac{r+v_f}{N} u_x\right) \left\{ \text{sincd}[N; \pi(r+\rho+v_f)] \right.$$

$$\times \exp\left\{i\left[\pi \frac{(N-1)(r+\rho+v_f)}{N} + \phi\right]\right\} + \text{sincd}[\pi(r-\rho+v_f)]$$

$$\left. \times \exp\left\{i\left[\pi \frac{(N-1)(r-\rho+v_f)}{N} - \phi\right]\right\}\right\} \tag{4.76}$$

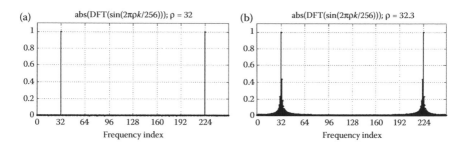

FIGURE 4.6
DFT spectra of sinusoidal signals with 256 samples and with integer (a) and noninteger (b)
frequency parameter ρ.

shifted in index by the frequency index ρ of the signal. In a special case of
DFT $(u_x = 0, v_f = 0)$,

$$\alpha_r = \frac{1}{\sqrt{N}} \sum_{k=0}^{N-1} \cos\left(2\pi \frac{\rho k}{N} + \phi \right) \exp\left(i2\pi \frac{kr}{N} \right)$$

$$= \frac{\sqrt{N}}{2} \left\{ \text{sincd}[N; \pi(r + \rho)] \exp\left\{ i \left[\pi \frac{(N-1)(r+\rho)}{N} + \phi \right] \right\} \right.$$

$$\left. + \text{sincd}[\pi(r - \rho)] \exp\left\{ i \left[\pi \frac{(N-1)(r-\rho)}{N} - \phi \right] \right\} \right\}. \qquad (4.77)$$

For integer ρ, discrete sinc-functions in Equation 4.77 are reduced to two
Kronecker deltas:

$$\alpha_r = \frac{1}{\sqrt{N}} \sum_{k=0}^{N-1} \cos\left(2\pi \frac{\rho k}{N} + \phi \right) \exp\left(i2\pi \frac{kr}{N} \right)$$

$$= \frac{\sqrt{N}}{2} \{ \delta(r + \rho) \exp(i\phi) + \delta(r - \rho) \exp(-i\phi) \}. \qquad (4.78)$$

Examples of DFT spectra of sinusoidal signals with integer and noninteger
frequency parameter ρ are presented in Figure 4.6.

Mutual Correspondence between Signal Frequencies and Indices of Its SDFTs Spectral Coefficients

By definition of SDFTs, index r of SDFT spectral coefficients is linked with
frequency f parameter of signal Fourier spectra by the following relationship:

$$f = r\Delta f = r/N\Delta x; \quad f\Delta x = \tilde{f} = r/N, \tag{4.79}$$

where $\tilde{f}$ is a normalized frequency measured in fractions of the signal baseband $1/\Delta x$.

DFT spectral coefficient $\alpha_0^{(0,0)}$ plays a role of the zero-frequency component of the integral Fourier transform. It is proportional to the signal *dc-component*:

$$\alpha_0 = \sqrt{N}\left(\frac{1}{N}\sum_{k=0}^{N-1} a_k\right). \tag{4.80}$$

For even N

$$\alpha_{N/2} = \frac{1}{\sqrt{N}}\sum_{k=0}^{N-1}(-1)^k a_k \tag{4.81}$$

represents the highest, $f = (N/2)\Delta f = 1/2\Delta x$, signal frequency component. For odd N, coefficients $\alpha_{(N-1)/2} = \alpha_{(N+1)/2}^*$ (for signals with real values) represent the highest signal frequency $f = (N - 1/2)\Delta f = (N - 1)/2N\Delta x$.

In order to maintain similarity between DFT spectrum $\{\alpha_r\}$ of discrete signal $\{a_k\}$ and Fourier spectrum $\alpha(f)$, $f \in [-\infty, \infty]$ of the corresponding continuous signal $a(x)$, it is convenient to cyclically shift sequence $\{\alpha_r\}$ in such a way as to place the dc-component coefficient $\{\alpha_0\}$ in the middle of the sequence. For even and odd N, shifts are $N/2$ and $(N - 1)/2$ correspondingly and spectral coefficients should to be taken, correspondingly, in the following order: $[\alpha_{N/2}, \alpha_{N/2+1}, \ldots, \alpha_{N-1}, \alpha_0, \alpha_1, \ldots, \alpha_{N/2-1}]$ and $[\alpha_{(N+1)/2}, \alpha_{(N+1)/2+1}, \ldots, \alpha_{N-1}, \alpha_0, \alpha_1, \ldots, \alpha_{(N-1)/2}]$. For 2D signals of $N_1 \times N_2$ samples, these cyclical shifts should be taken along the two corresponding indices. In MATLAB, this centering shift is performed by the command, *fftshift*. Figures 4.7 and 4.8 illustrate the spectral coefficient reordering for DFT spectra of 1D and 2D sinusoidal signals.

DFT Spectra of Sparse Signals and Spectrum Zero Padding

Let discrete signal $\{\tilde{a}_{\tilde{k}}\}$ of LN samples ($\tilde{k} = 0,1,\ldots,LN - 1$) be obtained from signal $\{a_k\}$ of N samples ($k = 0,1, \ldots, N - 1$) by placing between its samples $(L - 1)$ zero samples. If index $\tilde{k}$ is represented as a two-component index $\tilde{k} = kL + l$, $k = 0,1, \ldots, N - 1$, $l = 0,1, \ldots, L - 1$, such a signal with sparse samples can be represented as

$$\tilde{a}_{\tilde{k}} = a_k\delta(l), \tag{4.82}$$

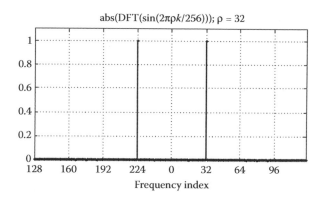

FIGURE 4.7
Spectrum of a sinusoidal signal Figure 4.6a reordered by the MATLAB command "fftshift" for spectrum centering.

where $\delta(\cdot)$ is Kronecker's delta. Compute the DFT of this signal:

$$\tilde{\alpha}_{\tilde{r}} = \frac{1}{\sqrt{LN}} \sum_{k=0}^{LN-1} \tilde{a}_k \exp\left(i2\pi\frac{k\tilde{r}}{LN}\right)$$

$$= \frac{1}{\sqrt{LN}} \sum_{l=0}^{L-1}\sum_{k=0}^{N-1} a_k\delta(l)\exp\left(i2\pi\frac{kL+l}{LN}\tilde{r}\right)$$

$$= \frac{1}{\sqrt{LN}} \sum_{k=0}^{N-1} a_k \exp\left(i2\pi\frac{k\tilde{r}}{N}\right) = \frac{1}{\sqrt{L}}\alpha_{(\tilde{r})\bmod N},\qquad(4.83)$$

where $\{\alpha_{(\tilde{r})\bmod N}\}$, $\tilde{r} = 0,1,\ldots, LN -$ is the DFT of signal $\{a_k\}$. Equation 4.83 shows that placing zeros between signal samples results in a periodical replication of its DFT spectrum with the number of replicas equal to the number of zeros plus one. This property of DFT spectra of sparse signals is illustrated in Figure 4.9a–d. It is an analog of virtual spectrum replication in sampling continuous signals discussed in the section "Sampling Artifacts: Quantitative Analysis."

Let us now multiply spectrum $\{\tilde{\alpha}_r\}$ by a mask $\{\mu_r\}$ that zeros all periods but one (Figure 4.9e). According to the convolution theorem, multiplying signal DFT spectrum by a mask function results in signal cyclic convolution with IDFT of the mask. In order to secure that the convolution kernel is a real-valued function, the mask should maintain spectrum symmetry for real-valued signals (Equation 4.71). Thus, the definition of the mask depends on whether N is an odd or even number. For odd N, the mask should be

$$\mu_{\tilde{r}} = 1 - rect\left(\frac{\tilde{r} - (N+1)/2}{LN - N - 1}\right),\qquad(4.84)$$

where $rect(x) = 1, 0 < x < 1; 0, otherwise$ (Equation 2.45).

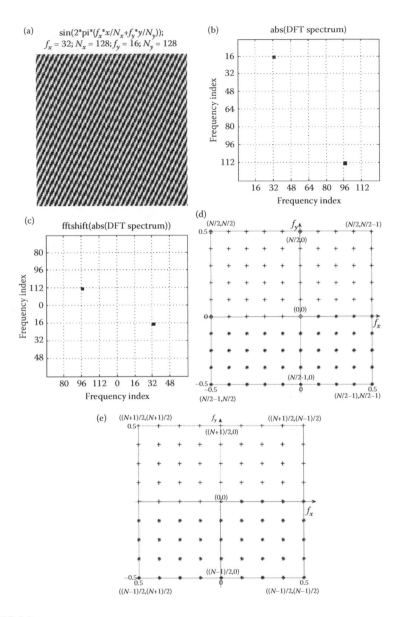

FIGURE 4.8

(a–c) A 2D sinusoidal signal and its DFT spectrum represented as an image in the original index order (b) and in the shifted index order (c), which corresponds to analog Fourier spectrum coordinate system centered at point $(f_x = 0; f_y = 0)$. Two small dark boxes in images (b) and (c) indicate the position of signal spectral components. (d–e) Schematic maps of arrangements of $N \times N$ "fftshifted" 2D DFT spectral coefficients for even (d) and odd (e) number N in their correspondence with frequency (f_x, f_y) coordinates of integral Fourier transform. Numbers in brackets indicate coefficient indices. For real-valued signals, spectral coefficients symmetrical with respect to coordinate origin (0, 0) and marked with crosses and stars are mutually complex conjugate and those marked with circle are real valued.

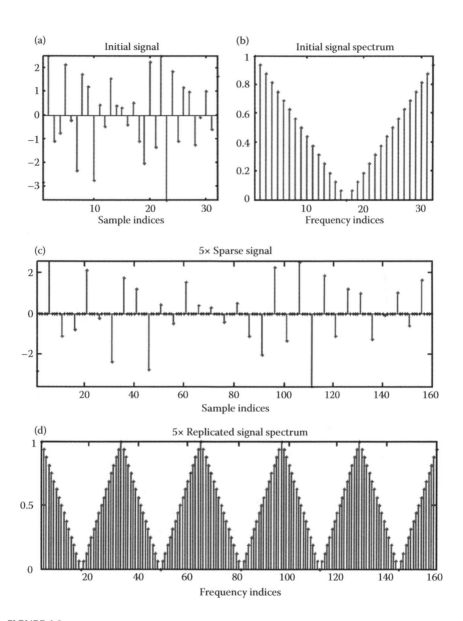

FIGURE 4.9
(a) A test signal; (b) its DFT spectrum; (c) sparse signal obtained from signal (a) by placing four zeros between its samples; (d) DFT spectrum of the obtained sparse signal; (e) zero-padded spectrum of the sparse signal after removing all its periodical components but one; (f) sincd-interpolated signal reconstructed from the zero-padded spectrum.

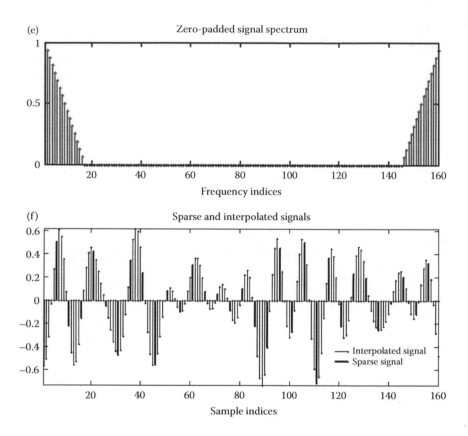

FIGURE 4.9
Continued.

In this case, the inverse DFT of $\{[1 - rect\,((r - (N + 1)/2)/(LN - N - 1))]\alpha_{(r)\bmod N}\}$ produces (see Appendix) a signal

$$\tilde{\tilde{a}}_{\tilde{k}} = \frac{1}{\sqrt{LN}} \sum_{\tilde{r}=0}^{LN-1} \left[1 - rect\,\frac{\tilde{r} - (N + 1)/2}{LN - N - 1}\right]\alpha_{(r)\bmod N} \exp\left(-i2\pi\frac{\tilde{k}\tilde{r}}{LN}\right)$$

$$= \frac{1}{\sqrt{L}} \sum_{n=0}^{N-1} a_n \,\mathrm{sincd}\left[N;\left(\tilde{k} - nL\right)/L\right], \tag{4.85}$$

or

$$\tilde{\tilde{a}}_{Lk+l} = \frac{1}{\sqrt{L}} \sum_{n=0}^{N-1} a_n \mathrm{sincd}\left(N;k - n + \frac{l}{L}\right). \tag{4.86}$$

This equation shows that nonzero samples of signal $\tilde{a}_k$ in signal $\tilde{\tilde{a}}_k$ are equal to $1/\sqrt{L}$-scaled samples of the original signal $\{a_k\}$

$$\tilde{\tilde{a}}_{Lk} = \frac{1}{\sqrt{L}} a_k, \tag{4.87}$$

while its additional zero samples are interpolated with the discrete sinc-function from the initial nonzero ones (see Figure 4.9f).

For even N, maintaining the spectrum complex-conjugated symmetry $\tilde{\alpha}_{\tilde{r}}\mu_{\tilde{r}} = \tilde{\alpha}^*_{LN-\tilde{r}}\mu_{LN-\tilde{r}}$ is possible either with the mask

$$\mu_{\tilde{r}}^{(0)} = 1 - rect\frac{\tilde{r} - N/2}{LN - N}, \tag{4.88}$$

which zeros $N/2$-th component of the original spectrum $\{\alpha_r\}$ or with the mask

$$\mu_{\tilde{r}}^{(2)} = 1 - rect\frac{\tilde{r} - N/2 - 1}{LN - N - 2}, \tag{4.89}$$

which leaves two copies of this component in the extended spectrum. In these two cases one can, similarly to the derivation of Equation 4.86, obtain that inverse DFT (IDFT) of the masked-extended spectrum results in signals:

$$\tilde{\tilde{a}}_{kL+l}^{(0)} = IDFT\left\{\mu_{\tilde{r}}^{(0)}\alpha_{(\tilde{r})\bmod N}\right\} = \frac{1}{\sqrt{L}}\sum_{n=0}^{N-1} a_n \overline{sincd}\left(N - 1; N; k - n + \frac{l}{L}\right) \tag{4.90}$$

or, correspondingly

$$\tilde{\tilde{a}}_k = IDFT\left\{\mu_{\tilde{r}}^{(2)}\alpha_{(\tilde{r})\bmod N}\right\} = \frac{1}{\sqrt{L}}\sum_{n_1=0}^{N-1} a_n \overline{sincd}\left(N + 1; N; k - n + \frac{l}{L}\right), \tag{4.91}$$

where $\overline{sincd}(\cdot;\cdot;\cdot)$ is general discrete sinc-function, defined by Equation 3.61.

Equations 4.86, 4.90, and 4.91 are discrete analogs of reconstruction of continuous signals from their samples discussed in the section "The Sampling Theorem and Signal Sampling" in Chapter 3. Masking functions $\{\mu_r\}$ (Equations 4.84, 4.88, and 4.89) describe discrete *ideal low-pass filters*. If N is an even number, original signal samples are not retained in the interpolated signal because of the special treatment required for the spectral coefficient with index $N/2$. In Chapter 6, we will consider discrete sinc-interpolation in detail. The properties of canonican 1D and 2D DFTs are illustrated by the MATLAB program dft_demo_CRC.m provided in Exercises.

Discrete Cosine and Sine Transforms

There are a number of special cases of SDFTs worthy of a special discussion. All of them are associated with representation of signal and/or spectra that exhibit certain symmetry. The most important special case is that of DCT.

Let, for a signal $\{a_k\}$, $k = 0, 1, \ldots, N - 1$ form an auxiliary signal

$$
\tilde{a}_k = \begin{cases} a_k, & k = 0, 1, \ldots, N - 1 \\ a_{2N-k-1}, & k = N, \ldots, 2N - 1 \end{cases}. \tag{4.92}
$$

It is shown in Appendix that SDFT(1/2,0) of such a symmetrized signal is

$$
\text{SDFT}_{1/2,0}\{\tilde{a}_k\} = \frac{1}{\sqrt{2N}} \sum_{k=0}^{2N-1} \tilde{a}_k \exp\left(i2\pi \frac{k+1/2}{2N} r \right) = \frac{2}{\sqrt{2N}} \sum_{k=0}^{N-1} a_k \cos\left(\pi \frac{k+1/2}{N} r \right). \tag{4.93}
$$

In this way, we arrive at the DCT defined for signal $\{a_k\}$ of N samples as

$$
\alpha_r^{\text{DCT}} = \frac{2}{\sqrt{2N}} \sum_{k=0}^{N-1} a_k \cos\left(\pi \frac{k+1/2}{N} r \right). \tag{4.94}
$$

As one can see directly from Equation 4.94, the DCT signal spectrum is an odd (antisymmetric) sequence if regarded outside its base interval $[0, N - 1]$:

$$
\alpha_r^{\text{DCT}} = -\alpha_{2N-r}^{\text{DCT}}; \quad \alpha_N = 0 \tag{4.95}
$$

while the signal is, by the definition (Equation 4.92), perfectly symmetric (even) on interval $[0, 2, N - 1]$:

$$
a_k = a_{2N-k-1}. \tag{4.96}
$$

From the above derivation of the DCT, it also follows that, for the DCT, signals are regarded as periodical with a period $2N$:

$$
a_k = a_{(k) \bmod 2N}. \tag{4.97}
$$

The inverse DCT can be found as the inverse SDFT(1/2,0) of spectrum with an odd symmetry of Equation 4.95 (see Appendix):

$$
a_k = \frac{1}{\sqrt{2N}} \left[\alpha_0^{\text{DCT}} + 2 \sum_{r=1}^{N-1} \alpha_r^{\text{DCT}} \cos\left(\pi \frac{k+1/2}{N} r \right) \right]. \tag{4.98}
$$

The invertibility of DCT implies that DCT is an orthogonal transform, and, therefore, it satisfies Parseval's relationship:

$$\sum_{k=0}^{N-1} a_k^2 = \sum_{r=0}^{N-1} \left(\alpha_r^{DCT} \right)^2. \tag{4.99}$$

Coefficient α_0^{DCT} is, similarly to the case of DFT, proportional to the signal dc-component:

$$\alpha_0^{DCT} = \sqrt{2N} \left(\frac{1}{N} \sum_{k=0}^{N-1} a_k \right). \tag{4.100}$$

Coefficient

$$\alpha_{N-1}^{DCT} = \frac{2}{\sqrt{2N}} \sum_{k=0}^{N-1} a_k (-1)^k \sin\left(\pi \frac{k+1/2}{N} \right) \tag{4.101}$$

represents the signal's highest frequency. This equation is similar to that for DFT (Equation 4.81) except that signal is multiplied by an "apodization" function {sin $(\pi((k + 1/2)/N))$}.

The DCT was introduced by Ahmed et al. [1]. It has proved to have a very good energy compaction capability and is very frequently considered as an approximation to Karhunen–Loeve transform. From the above derivation, it becomes clear that the good energy compaction capability of the DCT can be attributed to the fact that the DCT is SDFT(1/2, 0) of a signal that is evenly extended (Equation 4.92). This way of signal extension eliminates potential signal discontinuities at the signal borders and in this way the need in high-frequency spectral components to reproduce them. Figure 4.10 illustrates a better energy compaction capability of DCT spectra with respect to that of DFT for a test piece-wise constant image. Plots in Figure 4.10b show energy of DFT and DCT spectral coefficients as functions of a fraction (from 0 to 1) of the signal base band. One can see from the plots that the same fraction of DCT spectral coefficients contains a higher fraction of the total signal energy than that of DFT coefficients.

Being a derivative of SDFTs, DCT can, in principle, be computed using FFT with the same computational complexity of 1/2,0 operations per each of N signal coefficients. There exist dedicated fast transform algorithms of FFT type for computing DCT.

The removal of discontinuities at signal borders and computational efficiency make the DCT attractive in many image processing applications such as image compression (see the section "Basics of Image Data Compression"), image resampling (see Chapter 6), and image perfection and enhancement (see Chapter 8) that involve processing in transform domain. DCT is also a prefect substitute of DFT for the implementation of signal digital convolution with the use of fast transforms. This application will be discussed later in the next section.

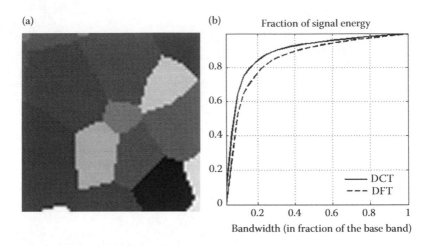

FIGURE 4.10
A test image (a) and energy of its 2D DFT and DCT spectra as functions of fraction of image
base band (b).

The DCT has its complement sine transform, the *discrete cosine–sine transform* (DcST):

$$\alpha_r^{\text{DcST}} = \frac{2}{\sqrt{2N}} \sum_{k=0}^{N-1} a_k \sin\left(\pi \frac{k + 1/2}{N} r \right). \tag{4.102}$$

As shown in Appendix, the DcST is an imaginary part of the SDFT(1/2,0)
of a signal

$$\tilde{\tilde{a}}_k = \begin{cases} a_k, & k = 0, 1, ..., N - 1 \\ -a_{2N-k-1}, & k = N, N + 1, ..., 2N - 1 \end{cases} \tag{4.103}$$

extended to the interval [0, 2N − 1] in an odd (antisymmetric) way:

$$\frac{1}{i\sqrt{2N}} \sum_{k=0}^{2N-1} \tilde{\tilde{a}}_k \exp\left(\pi \frac{k + 1/2}{N} r \right) = \frac{2}{\sqrt{2N}} \sum_{k=0}^{N-1} a_k \sin\left(\pi \frac{k + 1/2}{N} r \right). \tag{4.104}$$

From the definition of DcST, it follows that the DcST spectrum exhibits
even symmetry when being regarded in the interval [0, 2N − 1]:

$$\alpha_r^{\text{DcST}} = \alpha_{2N-r}^{\text{DcST}} \tag{4.105}$$

and that it assumes periodical replication, with a period 2N, of the signal
complemented with its antisymmetrical copy

$$\tilde{a}_k = \tilde{a}_{(k) \bmod 2N}. \tag{4.106}$$

In distinction from the DFT and the DCT, the DcST does not contain a signal dc-component:

$$\alpha_0^{\text{DcST}} = 0 \tag{4.107}$$

and the inverse DcST

$$a_k = \frac{1}{\sqrt{2N}}\left[\alpha_N^{\text{DcST}} + 2\sum_{r=1}^{N-1}\alpha_r^{\text{DcST}}\sin\left(\pi\frac{k+1/2}{N}r\right)\right] \tag{4.108}$$

involves spectral coefficients with indices $\{1, 2, \ldots, N\}$ rather than $\{0, 1, \ldots, N-1\}$ for the DCT.

Direct and inverse DcSTs can be computed through DCT if one changes the ordering of the transform coefficient to a reverse one:

$$\alpha_{N-r}^{\text{DcST}} = \frac{2}{\sqrt{2N}}\sum_{k=0}^{N-1}a_k\sin\left[\pi\frac{(k+1/2)(N-r)}{N}\right]$$

$$= \frac{2}{\sqrt{2N}}\sum_{k=0}^{N-1}a_k\left\{\sin[\pi(k+1/2)]\cos\left(\pi\frac{k+1/2}{N}r\right)\right.$$

$$\left. -\cos[\pi(k+1/2)]\sin\left(\pi\frac{k+1/2}{N}r\right)\right\}$$

$$= \frac{2}{\sqrt{2N}}\sum_{k=0}^{N-1}(-1)^k a_k\cos\left(\pi\frac{k+1/2}{N}r\right) \tag{4.109}$$

$$a_k = \frac{1}{\sqrt{2N}}\left\{\alpha_N^{\text{DcST}} + 2\sum_{r=1}^{N-1}\alpha_{N-r}^{\text{DcST}}\sin\left[\pi\frac{(k+1/2)(N-r)}{N}\right]\right\}$$

$$= \frac{1}{\sqrt{2N}}\left(\alpha_N^{\text{DcST}} + 2(-1)^k\sum_{r=1}^{N-1}\alpha_{N-r}^{\text{DcST}}\cos\left(\pi\frac{k+1/2}{N}r\right)\right). \tag{4.110}$$

2D and multidimensional DCTs and DSTs are defined as separable to 1D transforms on each of the indices. For instance, 2D DCT of a signal $\{a_{k,l}\}$, $k = 0, 1, \ldots, N_1 - 1, l = 0, 1, \ldots, N_2 - 1$ is defined as

$$\alpha_{r,s}^{\text{DCT}} = \frac{2}{\sqrt{N_1 N_2}}\sum_{k=0}^{N_1-1}\sum_{l=0}^{N_2-1}a_{k,l}\cos\left(\pi\frac{k+1/2}{N_1}r\right)\cos\left(\pi\frac{l+1/2}{N_2}s\right) \tag{4.111}$$

that corresponds to fourfold image symmetry illustrated in Figure 4.11.

FIGURE 4.11
Fourfold image symmetry.

Signal Convolution in the DCT Domain

In this section, we will show that applying DFT convolution theorem to signals extended to double length by means of mirror reflection from their borders translates to convolution in the DCT domain.

Let signal $\{\tilde{a}_k\}$ be obtained from signal $\{a_k\}$ of N samples by its extension by means of mirror reflection and periodical replication of the result with a period of $2N$ samples:

$$\tilde{a}_{(k)\bmod 2N} = \begin{cases} a_k, & k = 0, 1, \ldots, N-1 \\ a_{2N-k-1}, & k = N, N+1, \ldots, 2N-1 \end{cases} \tag{4.112}$$

and let $\{\tilde{h}_n\}$ be a convolution kernel $\{h_n\}$ of N samples ($n = 0, 1, \ldots, N-1$) zero-padded to the length $2N$.

$$\tilde{h}_n = \begin{cases} 0, & n = 0, \ldots, \lfloor N/2 \rfloor - 1 \\ h_{n-\lfloor N/2 \rfloor}, & n = \lfloor N/2 \rfloor, \ldots, \lfloor N/2 \rfloor + N - 1, \\ 0, & \lfloor N/2 \rfloor + N, \ldots, 2N - 1 \end{cases} \tag{4.113}$$

where $\lfloor N/2 \rfloor$ is an integer part of $N/2$. Then, the first N samples of cyclic convolution

$$b_{(k)\bmod 2N} = \sum_{n=0}^{2N-1} \tilde{h}_n \tilde{a}_{(k-n+\lfloor N/2 \rfloor)\bmod 2N} \tag{4.114}$$

can be used for computing digital convolution, in which boundary effects characteristic for cyclic convolution are absent, thanks to the signal-mirrored extension as it was illustrated in Figure 4.3.

Consider computing the convolution of such signals by means of inverse DFT of the product of signal and convolution kernel DFT spectra. The DFT spectrum of the extended signal $\{\tilde{a}_k\}$ is

$$\begin{aligned}
\tilde{\alpha}_r &= \frac{1}{\sqrt{2N}} \sum_{k=0}^{2N-1} \tilde{a}_k \exp\left(i2\pi \frac{kr}{2N} \right) \\
&= \left\{ \frac{2}{\sqrt{2N}} \sum_{k=0}^{N-1} a_k \cos\left(\pi \frac{k+1/2}{N} r \right) \right\} \exp\left(-i\pi \frac{r}{2N} \right) \\
&= \alpha_r^{(DCT)} \exp\left(-i\pi \frac{r}{2N} \right),
\end{aligned} \tag{4.115}$$

where $\alpha_r^{(DCT)}$ is DCT coefficients of the initial signal $\{a_k\}$. Therefore, the DFT spectrum of the signal extended by the "mirror reflection" can be computed via DCT using the fast DCT algorithm.

For computing convolution, the signal spectrum defined by Equation 4.115 should be multiplied by DFT coefficients of the zero-padded convolution kernel:

$$\tilde{\eta}_r = \frac{1}{\sqrt{2N}} \sum_{n=0}^{2N-1} \tilde{h}_n \exp\left(i2\pi \frac{nr}{2N} \right) \tag{4.116}$$

and then the inverse DFT of the product should be computed for the first N samples:

$$\begin{aligned}
b_k &= \frac{1}{\sqrt{2N}} \sum_{r=0}^{2N-1} \alpha_r^{(DCT)} \exp\left(-i\pi \frac{r}{2N} \right) \tilde{\eta}_r \exp\left(-i2\pi \frac{kr}{2N} \right) \\
&= \frac{1}{\sqrt{2N}} \sum_{r=0}^{2N-1} \alpha_r^{(DCT)} \tilde{\eta}_r \exp\left(-i2\pi \frac{k+1/2}{2N} r \right).
\end{aligned} \tag{4.117}$$

Splitting the sum in Equation 4.117 into two addends and changing the index r of summation in the second addend to $2N - r$, obtain (see Appendix):

$$b_k = \frac{1}{\sqrt{2N}} \sum_{r=0}^{2N-1} \alpha_r^{(DCT)} \tilde{\eta}_r \exp\left(-i2\pi \frac{k+1/2}{2N} r \right)$$

$$= \frac{1}{\sqrt{2N}} \left\{ \sum_{r=0}^{N-1} \alpha_r^{(DCT)} \tilde{\eta}_r \exp\left(-i2\pi \frac{k+1/2}{2N} r \right) - \sum_{r=1}^{N} \alpha_{2N-r}^{(DCT)} \tilde{\eta}_{2N-r} \exp\left(i2\pi \frac{k+1/2}{2N} r \right) \right\}.$$

(4.118)

From the properties of DFT and DCT spectra (Equations 4.71 and 4.95), it follows that $\alpha_N^{DCT} = 0$; $\alpha_r^{DCT} = -\alpha_{2N-r}^{DCT}$, and $\{\tilde{\eta}_r = \tilde{\eta}_{2N-r}^*\}$. Using them in Equation 4.118, obtain

$$b_k = \frac{1}{\sqrt{2N}} \left\{ \sum_{r=0}^{N-1} \alpha_r^{(DCT)} \tilde{\eta}_r \exp\left(-i2\pi \frac{k+1/2}{2N} r \right) + \sum_{r=1}^{N-1} \alpha_r^{(DCT)} \tilde{\eta}_r^* \exp\left(i2\pi \frac{k+1/2}{2N} r \right) \right.$$

$$= \frac{1}{\sqrt{2N}} \left\{ \alpha_0^{(DCT)} \tilde{\eta}_0 + 2 \sum_{r=1}^{N-1} \alpha_r^{(DCT)} \tilde{\eta}_0^{re} \cos\left(\pi \frac{k+1/2}{N} r \right) \right.$$

$$\left. - 2 \sum_{r=1}^{N-1} \alpha_r^{(DCT)} \tilde{\eta}_r^{im} \sin\left(\pi \frac{k+1/2}{N} r \right) \right\}.$$

(4.119)

The first two terms of this expression represent inverse DCT of the product $\{\alpha_r^{(DCT)} \tilde{\eta}_r^{re}\}$, while the third term is the discrete cosine/sine transform (DcST) of the product $\{\alpha_r^{(DCT)} \tilde{\eta}_r^{im}\}$. As shown in the section "Discrete Cosine and Sine Transforms" (Equation 4.110), inverse DcST can be converted into DCT by means of changing the summation index r in DcST by $N - r$:

$$\sum_{r=1}^{N-1} \alpha_{N-r}^{(DCT)} \tilde{\eta}_{N-r}^{im} \sin\left[\pi \frac{(k+1/2)(N-r)}{N} \right] = (-1)^k \sum_{r=1}^{N-1} \alpha_{N-r}^{(DCT)} \tilde{\eta}_{N-r}^{im} \cos\left(\pi \frac{k+1/2}{N} r \right).$$

(4.120)

Substitute this expression into Equation 4.119 and obtain the final formula for computing, through DCT, digital convolution with substantially reduced boundary effects:

$$b_k = \frac{1}{\sqrt{2N}} \left\{ \alpha_0^{(DCT)} \tilde{\eta}_0 + 2 \sum_{r=1}^{N-1} \alpha_r^{(DCT)} \tilde{\eta}_r^{re} \cos\left(\pi \frac{k+1/2}{N} r \right) \right.$$

$$\left. - 2(-1)^k \sum_{r=1}^{N-1} \alpha_{N-r}^{(DCT)} \tilde{\eta}_{N-r}^{im} \cos\left(\pi \frac{k+1/2}{N} r \right) \right\}. \tag{4.121}$$

The MATLAB programs conv_evenextension_demo_CRC and dct_vs_dft_conv_demo_CRC.m provided in Exercises illustrate the principle of signal convolution in the DCT domain and compare border effects in the implementations of digital convolution in DFT and DCT domains.

DFTs and Discrete Frequency Response of Digital Filter

In this section, we will show that frequency responses of digital filters, which are their main characteristic, can be expressed through DFT coefficients of their PSFs. Consider the continuous frequency response (Equation 4.25) of a digital filter defined by its DPSF $\{h_n\}$:

$$CFR(f) = \sum_{n=0}^{N-1} h_n \exp\left[i2\pi f \Delta x \left(n + u_x^{(s)} \right) \right]. \tag{4.122}$$

Let $\{\eta_r^{(u,0)}\}$ be a set of SDFT(u,0) coefficients of the PSF $\{h_n\}$:

$$h_n = \sum_{r=0}^{N-1} \eta_r^{(u,0)} \exp\left[-i2\pi \frac{(n+u_x)r}{N} \right]$$

$$= \frac{1}{\sqrt{N}} \sum_{r=0}^{N-1} \left[\eta_r^{(u,0)} \exp\left(-i2\pi \frac{u_x r}{N} \right) \right] \exp\left(-i2\pi \frac{nr}{N} \right). \tag{4.123}$$

Then obtain (see Appendix)

$$CFR(f) = \frac{1}{\sqrt{N}} \sum_{n=0}^{N-1} \sum_{r=0}^{N-1} \eta_r \exp\left\{ i2\pi \left[\left(f\Delta x - \frac{r}{N} \right) n + \left(f\Delta x u_x^{(s)} - \frac{u_x r}{N} \right) \right] \right\}$$

$$\propto \sum_{r=0}^{N-1} \eta_r \frac{\sin\left[\pi N \left(f\Delta x - \frac{r}{N} \right) \right]}{N \sin\left[\pi \left(f\Delta x - \frac{r}{N} \right) \right]} \exp\left[i2\pi \left(u_x^{(s)} + \frac{N-1}{2} \right) f\Delta x \right]$$

$$\times \exp\left[-i2\pi \left(u_x + \frac{N-1}{2} \right) \frac{r}{N} \right] \tag{4.124}$$

or, with setting $u_x = u_x^{(s)} = u$

$$CFR(f) = \sum_{r=0}^{N-1} \eta_r \exp\left[i2\pi\left(u + \frac{N-1}{2}\right)\left(f\Delta x - \frac{r}{N}\right)\right] \mathrm{sincd}\left(N; f\Delta x - \frac{r}{N}\right), \quad (4.125)$$

where $\mathrm{sincd}(\cdot;\cdot)$ is the discrete sinc-function defined by Equation 3.61 and the cardinal sampling relationship $\Delta f = 1/N\Delta x$ between signal and its Fourier spectrum sampling intervals Δx and Δf is assumed.

Equation 4.125 implies that, at points $f = r/N\Delta x = r\Delta f$, $r = 0, \ldots, N-1$ within one period of the periodicity of $CFR(f)$, its values are proportional to $SDFT(u,0)$ coefficients $\{\eta_r\}$ of the filter PSF $\{h_n\}$. Between these sampling points, $CFR\_DF(f)$ is interpolated from samples $\{\eta_r \exp[-i2\pi(u + (N-1/2))$ $(r/N)]\}$ with the discrete sinc-interpolation function. The phase multiplier $\exp[-i2\pi(u + (N-1/2))(r/N)]$ affects only values of $CFR(f)$ between its samples. It can be eliminated by the selection of shift parameters $u = u_x = u_x^s = -(N-1)/2$.

In what follows, we will neglect this influence, disregard the phase term $\exp(i2\pi f \Delta x u_x^{(s)})$ of $CFR(f)$ associated with the positioning of the signal sampling grid with respect to signal coordinate system, assume using canonical DFT for the implementation of digital convolution, and call discrete Fourier transform coefficients $\{\eta_r\}$ of the filter PSF $\{h_n\}$

$$\eta_r = DFR(r) = \frac{1}{\sqrt{N}} \sum_{n=0}^{N-1} h_n \exp\left(i2\pi \frac{nr}{N}\right) \quad (4.126)$$

discrete frequency response of the digital filter.

Figure 4.12 illustrates discrete and continuous frequency responses of a digital filter with PSF [–1, 1] that is frequently used for numerical differentiation.

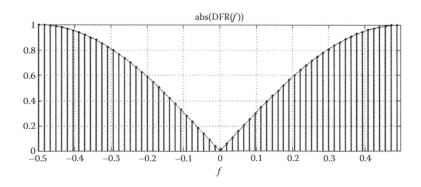

FIGURE 4.12
Discrete (stems) and continuous (solid line) frequency responses of a digital filter that is frequently used for numerical evaluation of signal derivatives.

The graph in the figure shows absolute value of the frequency response in the signal baseband $[-1/2\Delta x \div 1/2\Delta x]$; stems on the curve indicate samples $\{\eta_r\}$ of the continuous frequency response, which represent the filter discrete frequency response.

In Chapter 6, we will show how the notion of digital filter discrete frequency response can be used for the design of discrete filters, and, specifically, of perfect discrete interpolation, differentiation, and integration filters.

Discrete Representation of Fresnel Integral Transform

Canonical Discrete Fresnel Transform and Its Versions

In the section "Imaging in Transform Domain and Diffraction Integrals," we introduced a Fresnel approximation to Kirchhoff integral that describes free space propagation of waves with wavelength λ over distance Z from plane (x, y) to plane (f_x, f_y) and introduced 2D *integral Fresnel transform*

$$\alpha\left(\tilde{f}_x, \tilde{f}_y\right) = \int_{-\infty}^{\infty}\int_{-\infty}^{\infty} a\left(\tilde{x}, \tilde{y}\right)\exp\left\{-i\pi\left[\left(\tilde{x} - \tilde{f}_x\right)^2 + \left(\tilde{y} - \tilde{f}_y\right)^2\right]\right\}dx\,dy \quad (4.127)$$

as defined in dimensionless coordinates

$$\tilde{x} = x/\sqrt{\lambda Z}, \quad \tilde{y} = y/\sqrt{\lambda Z}, \quad \text{and} \quad \tilde{f}_x = f_x/\sqrt{\lambda Z}, \quad \tilde{f}_y = f_y/\sqrt{\lambda Z}. \quad (4.128)$$

Consider its 1D version

$$\alpha(f) = \int_{-\infty}^{\infty} a(x)\exp[-i\pi(x - f)^2]dx. \quad (4.129)$$

According to Equation 4.6, the discrete representation of the Fresnel transform kernel $\exp[-i\pi(x - f)^2]$ is

$$h_{k,r} = \int_{-\infty}^{\infty}\int_{-\infty}^{\infty} \exp[-i\pi(x-f)^2]\phi^{(r)}[x-(k+u_x)]\phi^{(s)}[f-(r+v_f)\Delta f]dx\,df$$

$$= \int_{-\infty}^{\infty} \phi^{(s)}(f)df \int_{-\infty}^{\infty} \exp\left\{-i\pi[x-f+(k+u_x)\Delta x-(r+v_f)\Delta f]^2\right\}\phi^{(r)}(x)dx, \quad (4.130)$$

where, as before, $\phi_k^{(r)}(\cdot)$ and $\phi_r^{(s)}(\cdot)$ are signal reconstruction and spectrum sampling basis functions, k and r are integer indices of signal and its Fresnel spectrum samples, u_x and v_f are, respectively, shifts, in fractions of sampling intervals Δx and Δf, of signal and its spectrum samples with respect to the corresponding coordinate system.

Introduce, temporarily, a variable

$$s_{kr} = (k + u_x)\Delta x - (r + v_f)\Delta f. \tag{4.131}$$

Then

$$h_{k,r} = \int\limits_{-\infty}^{\infty} \phi^{(s)}(f)df \int\limits_{-\infty}^{\infty} \exp[-i\pi(x - f + s_{rk})^2]\phi^{(r)}(x)dx$$

$$= \exp\left(-i\pi s_{rk}^2\right)\int\limits_{-\infty}^{\infty} \phi^s(f)\exp(-i\pi f^2)\exp(i2\pi f s_{rk})df$$

$$\times \int\limits_{-\infty}^{\infty} \phi^{(r)}(x)\exp(-i\pi x^2)\exp[i2\pi x(f - s_{rk})]dx$$

$$= \exp\left(-i\pi s_{rk}^2\right)\int\limits_{-\infty}^{\infty} \phi^{(s)}(f)\exp(-i\pi f^2)\tilde{\Phi}^{(r)}(f - s_{rk})\exp(i2\pi f s_{rk})df, \tag{4.132}$$

where

$$\tilde{\Phi}^{(r)}(f) = \int\limits_{-\infty}^{\infty} \phi^{(r)}(x)\exp(-i\pi x^2)\exp(i2\pi xf)dx \tag{4.133}$$

is the Fourier transform of signal reconstruction basis function $\phi^{(r)}(x)$ modulated by a chirp-function. For $\phi^{(r)}(x)$ is a function more or less compactly concentrated around point $x = 0$ in an interval of about Δx, $\exp(-i\pi x^2) \cong 1$ within this interval and one can regard $\tilde{\Phi}^{(r)}(f)$ as an approximation to frequency response of a hypothetical signal reconstruction device assumed in the signal discrete representation. With similar reservations as those made for discrete representation of Fourier integral, for discrete representation of Fresnel integral, only the term

$$\tilde{h}_{k,r} = \exp\left(-i\pi s_{rk}^2\right) = \exp\left\{-i\pi[(k + u_x)\Delta x - (r + v_f)\Delta f]^2\right\} \tag{4.134}$$

of Equation 4.132 is used.

To complete the derivation, introduce dimensionless variables

$$\mu = (\Delta x/\Delta f)^{1/2} \tag{4.135}$$

and

$$w = u_x\mu - v_f/\mu. \tag{4.136}$$

At this point, one should choose a relationship between sampling intervals Δx and Δf in signal and Fresnel transform domains.

One of the options is to choose the same relationship (Equation 4.33) that was used for DFT:

$$\Delta x\Delta f = 1/\sigma N, \tag{4.137}$$

where σ is a scale parameter. Another option will be discussed later in the section "Convolutional Discrete Fresnel and Angular Spectrum Propagation Transforms." With the choice of Equation 4.137, obtain

$$\begin{aligned}
\tilde{h}_{k,r} &= \exp\{-i\pi[(k + u_x)\Delta x - (r + v_f)\Delta f]^2\} \\
&= \exp\left\{-i\pi\frac{[(k + u_x)\Delta x - (r + v_f)\Delta f]^2}{\sigma N\Delta x\Delta f}\right\} \\
&= \exp\left[-i\pi\frac{(k\mu - r/\mu + w)^2}{\sigma N}\right]
\end{aligned} \tag{4.138}$$

and the basic Equation 4.5 of digital filtering takes, for the representation of Fresnel transform, the form

$$\alpha_r = \frac{1}{\sqrt{N}} \sum_{k=0}^{N-1} a_k \exp\left[-i\pi\frac{(k\mu - r/\mu + w)^2}{\sigma N}\right], \tag{4.139}$$

where multiplier $1/\sqrt{N}$ is introduced, similar to DFTs, for normalization purposes. We will refer to it as *shifted scaled discrete Fresnel transform* (ShScDFrT).

With an account for coordinate normalization (Equation 4.128) and the cardinal sampling relationship $\Delta x = \lambda Z/N\Delta f$, parameter μ of ShScDFrT can be expressed as

$$\mu^2 = \frac{\Delta x}{\Delta f} = \frac{\lambda Z}{N\Delta f}\frac{1}{\Delta f} = \frac{\lambda Z}{N\Delta f^2} = \frac{1}{N\tilde{\Delta f}^2}. \tag{4.140}$$

This representation reveals the physical meaning of the parameter μ^2 as the distance parameter in the Fresnel approximation of Kirchhof's integral

(Equation 2.42), given the wavelength λ, the number of wavefront samples N, and wavefront sampling interval Δf. Parameter w is a combined shift parameter of sampling grid shifts in signal and Fresnel transform domains.

When $\sigma = 1$ (the case of the cardinal sampling), we arrive at *shifted discrete Fresnel transform* (SDFrT):

$$\alpha_r = \frac{1}{\sqrt{N}} \sum_{k=0}^{N-1} a_k \exp\left[-i\pi \frac{(k\mu - r/\mu + w)^2}{N}\right]. \tag{4.141}$$

Its special case for $w = 0$

$$\alpha_r = \frac{1}{\sqrt{N}} \sum_{k=0}^{N-1} a_k \exp\left[-i\pi \frac{(k\mu - r/\mu)^2}{N}\right] \tag{4.142}$$

is the canonical discrete Fresnel transform (CDFrT).

SDFrT (μ, w) is quite obviously connected with SDFT(0, $-w\mu$):

$$\alpha_r = \frac{1}{\sqrt{N}} \left\{\sum_{k=0}^{N-1}\left[a_k \exp\left(-i\pi \frac{k^2\mu^2}{N}\right)\right] \exp\left[i2\pi \frac{k(r - w\mu)}{N}\right]\right\} \exp\left[-i\pi \frac{(r - w\mu)^2}{N\mu^2}\right] \tag{4.143}$$

and can be computed through it as SDFT(0, $-w\mu$) of the chirp-modulated signal with subsequent chirp-demodulation of the result. This computing method can, in particular, be used for numerical reconstruction of holograms recorded in a near diffraction zone. We will discuss the subject later in Chapter 5 (the section "Digital Image Formation by Means of Numerical Reconstruction of Holograms"). In this application, it is called the *Fourier reconstruction algorithm.*

From the relationship (Equation 4.143), one can conclude that SDFrT is invertible and that inverse SDFrT is defined as

$$a_k = \frac{1}{\sqrt{N}} \sum_{r=0}^{N-1} \alpha_r^{(\mu,w)} \exp\left[i\pi \frac{(k\mu - r/\mu + w)^2}{N}\right]. \tag{4.144}$$

Because shift parameter w is a combination of shifts in signal and spectral domains, shift in the signal domain causes a corresponding shift in the transform domain, which, however, depends on the distance parameter μ according to Equation 4.136. One can break this interdependence if, in the definition of the discrete representation of integral Fresnel transform, a symmetry condition is imposed:

$$\alpha_r = \alpha_{N-r} \tag{4.145}$$

of the transform

$$\alpha_r = \exp\left[-i\pi \frac{(r/\mu - w)^2}{N}\right] \tag{4.146}$$

of a point source $\delta(k)$. This symmetry condition is satisfied when

$$w = \frac{N}{2\mu} \tag{4.147}$$

and SDFrT for such shift parameter takes the form

$$\alpha_r = \frac{1}{\sqrt{N}} \sum_{k=0}^{N-1} a_k \exp\left\{-i\pi \frac{[k\mu - (r - N/2)/\mu]^2}{N}\right\}. \tag{4.148}$$

The transform defined by Equation 4.148 allows keeping position of objects reconstructed from Fresnel holograms invariant to distance. We will call this version of the discrete Fresnel transform *focal plane invariant discrete Fresnel transform* (FPIDFrT). We will illustrate the use of FPIDFrT later in the section "Numerical Algorithms for Hologram Reconstruction," when discussing algorithms for numerical reconstruction of holograms.

Invertibility of Discrete Fresnel Transforms and frincd-Function

As it was already mentioned, the invertibility of the SDFrT follows immediately from its connection with SDFT (Equation 4.143). In order to perfectly invert SDFrT, one should know focusing and shift parameters μ, w of the direct SDFrT. In real applications, such as numerical reconstruction of holograms (see Chapter 5), these parameters are not necessarily exactly known. Therefore, it is very instructive to verify what signal will be reconstructed from the DFrT spectrum of a signal using inverse DFrT with focusing and shift parameters other than those used in direct DFrT ("out-of-focus restoration").

If, for a discrete signal $\{a_k\}$, $k = 0, 1, \ldots, N-1$, one computes its SDFrT spectrum with parameters (μ_+, w_+) and then inverses SDFrT with parameters (μ_-, w_-) of the result of the direct transform, the following restored signal will be obtained (see Appendix):

$$a_k^{(\mu^\pm, w^\pm)} = \frac{1}{\sqrt{N}} \sum_{r=0}^{N-1} \left\{ \frac{1}{\sqrt{N}} \sum_{n=0}^{N-1} a_n \exp\left[-i\pi \frac{(n\mu_+ - r/\mu_+ + w_+)^2}{N}\right] \right\}$$

$$\times \exp\left[i\pi \frac{(k\mu_- - r/\mu_- + w_-)^2}{N}\right]$$

$$= \exp\left[i\pi \frac{(k\mu_- + w_-)^2}{N}\right] \sum_{n=0}^{N-1} a_n \exp\left[-i\pi \frac{(n\mu_+ + w_+)^2}{N}\right] \text{frincd}(N;q;k-n-w_\pm),$$

(4.149)

where

$$q = 1/\mu_+^2 - 1/\mu_-^2; \quad w_\pm = w_+/\mu_+ - w_-/\mu_-$$

(4.150)

and frincd(N; q; $k - n - w_\pm$) is a *frincd-function* defined as

$$\text{frincd}(N;q;x) = \frac{1}{N} \sum_{r=0}^{N-1} \exp\left(i\pi \frac{q}{N} r^2\right) \exp\left(-i2\pi \frac{xr}{N}\right).$$

(4.151)

Frincd-function is an analog of sincd-function of the DFT and is identical to it, to the accuracy of an unessential phase factor, when $q = 0$:

$$\text{frincd}(N;0,x) = \frac{1}{N} \sum_{r=0}^{N-1} \exp\left(-i2\pi \frac{xr}{N}\right) = \frac{1}{N} \frac{\exp(-i2\pi x) - 1}{\exp\left(-i2\pi \frac{x}{N}\right) - 1}$$

$$= \frac{\sin(\pi x)}{N \sin\left(\frac{\pi x}{N}\right)} \exp\left(-i\pi \frac{N-1}{N} x\right) = \text{sincd}(N;\pi x) \exp\left(-i\pi \frac{N-1}{N} x\right).$$

(4.152)

Therefore, when $q = 0$ ($\mu_+ = \mu_-$), and $w_\pm = 0$ ($w_+ = w_-$) from Equation 4.149, perfect signal restoration is achieved:

$$a_k = \exp\left[-i\pi \frac{(k\mu + w)^2}{N}\right] \sum_{n=0}^{N-1} a_n \exp\left[i\pi \frac{(n\mu + w)^2}{N}\right]$$

$$\times \text{sincd}[N;\pi(k-n)] \exp\left[i\pi \frac{N-1}{N}(k-n)\right]$$

$$= \exp\left[-i\pi \frac{(k\mu + w)^2}{N}\right] \sum_{n=0}^{N-1} a_n \exp\left[i\pi \frac{(n\mu + w)^2}{N}\right] \delta(k-n) = a_k.$$

(4.153)

From Equation 4.151, one can see that frincd-function is a periodical function with a period of N samples:

$$\text{frincd}(N;q,x) = \text{frincd}(N;q,(x)_{\text{mod} N})$$

(4.154)

and that

$$(\text{frincd}(N;q,x))^* = \text{frincd}(N;-q,N-x). \tag{4.155}$$

Figure 4.13 illustrates the behavior of frincd-function for different values of the focusing parameter q. As one can see from the plots, the absolute value of frincd-function has the shape of a rectangular impulse with oscillations on the borders and of width of about qN samples.

Figure 4.14 presents the frincd-function for focusing parameter q in the range (0–2) as an image, in which columns are frincd-function values shown in gray scale as functions of the sample index r. The image vividly

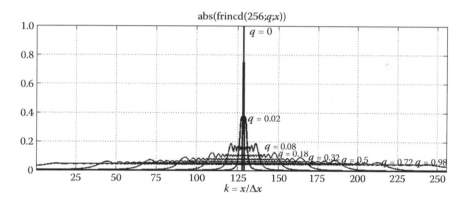

FIGURE 4.13
Plots of absolute values of function frincd($N;q;x$) for $N = 256$ and different values of "focusing" parameter q. (The function is shown centered around the middle point of the range of its argument.)

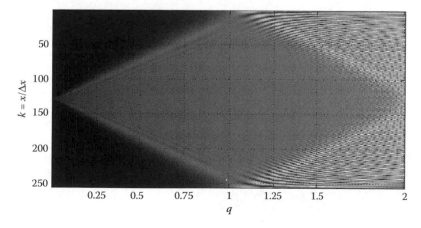

FIGURE 4.14
Frincd-function frincd(256; q; $x/\Delta x$) presented as an image (black is 0, white is 1) for $0 \le q \le 2$.

demonstrates that beginning $q = 1$, frincd-function dramatically changes its behavior with rise of q. As we will see later in Chapter 5 (the section "Numerical Algorithms for Hologram Reconstruction"), this is a manifestation of appearance of the aliasing phenomenon.

According to the definition (Equation 4.151), frincd-function is the DFT of the chirp-function.

DFT is a discrete representation of integral Fourier transform, for which the following relationship between chirp-function and its Fourier transform holds (see the derivation of Equation 2.57 in Appendix to Chapter 2):

$$\int_{-\infty}^{\infty} \exp(i\pi\sigma^2 f^2)\exp(-i2\pi fx)df = \frac{\sqrt{i}}{\sigma}\exp\left(-i\pi\frac{x^2}{\sigma^2}\right).   \tag{4.156}$$

Therefore, one can expect that function frincd(N; q; r) and chirp-functions are also linked by an approximative relationship:

$$\text{frincd}(N;q;k) = \frac{1}{N}\sum_{r=0}^{N-1}\exp\left(i\pi\frac{qr^2}{N}\right)\exp\left(-i2\pi\frac{rk}{N}\right) \cong \frac{\sqrt{i}}{\sqrt{Nq}}\exp\left(-i\pi\frac{k^2}{qN}\right)$$

$$\tag{4.157}$$

obtained from Equation 4.59 by replacements: $f = r\Delta f$; $x = k\Delta x$; $\Delta x\Delta f = 1/N$; $\sigma^2 = qN$. However, taking into account the rectangular shape of the amplitude of the frincd-function, this can be true only within the effective width $(1 - q)$ $N/2 \div (1 + q)N/2$ of the frincd-function, where most of its energy is concentrated. Figure 4.15 shows plots of absolute values and the phase of the ratio of the left and right parts of Equation 4.60 (ratio of frincd-function and chirp-function) for different q, which clearly demonstrate that a chirp-function is a reasonable approximation to the frincd-function within the index interval $[(1 - q)N/2 \div (1 + q)N/2]$, and that even within this approximation interval, noticeable distinctions between them remain in the form of oscillations at the borders of this interval.

Convolutional Discrete Fresnel and Angular Spectrum Propagation Transforms

The cardinal sampling relationship $\Delta x\Delta f = 1/N$ (or $\Delta x\Delta f = \lambda Z/N$ in physical, not normalized, units) put in the base of the derivation of the above versions of discrete Fresnel transforms is relevant for the far diffraction zone. When distance parameter $\mu^2 = \Delta x/\Delta f = \lambda Z/N\Delta f^2$ is <1, the phase of the chirp-function $\exp[-i\pi((k\mu - r/\mu + w)^2/N)]$ involved in the definition of DFrT (Equation 4.141) is changing, for cardinal sampling, too fast with index r, which causes aliasing artifacts. This motivates choosing for $\mu^2 \le 1$ an alternative sampling convention:

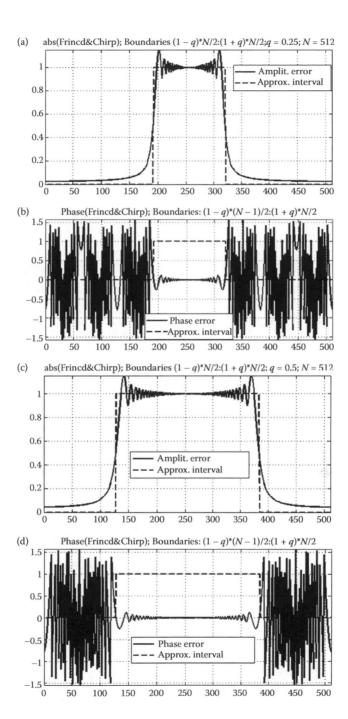

FIGURE 4.15
Amplitudes (a, c, e, g) and phases (b, d, f, h) of the error of approximation of chirp-function by frincd-function for $q = 0.25$ (a, b); 0.5 (c, d); 0.75 (e, f); 0.99(g, h).

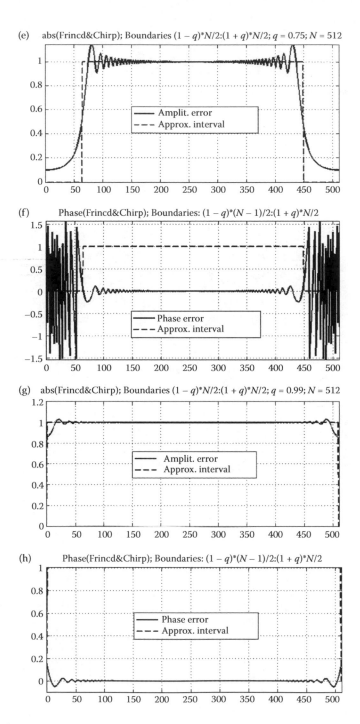

FIGURE 4.15
Continued.

$$\Delta x = \Delta f, \tag{4.158}$$

which obviously must hold on the object plane ($Z = 0$; $\mu^2 = 0$). Then obtain for $\tilde{h}_{k,r}$:

$$\tilde{h}_{k,r} = \exp\left\{-i\pi\left[(k + u_x)\Delta x - (r + v_f)\Delta f\right]^2\right\}$$

$$= \exp\left[-i\pi(k - r + \overline{w})^2\Delta f^2\right] = \exp\left[-i\pi\frac{(k - r + \overline{w})^2}{\mu^2 N}\right], \tag{4.159}$$

where

$$\overline{w} = u_x - v_f. \tag{4.160}$$

This kernel defines a transform:

$$\alpha_r = \frac{1}{\sqrt{N}}\sum_{k=0}^{N-1} a_k \exp\left[-i\pi\frac{(k - r - \overline{w})^2}{\mu^2 N}\right], \tag{4.161}$$

which we call *convolutional discrete Fresnel transform* (ConvDFrT) since it represents a digital convolution of signals with a chirp-function.

Yet another modification of the near zone diffraction integral transform introduced in the section "Imaging Systems and Integral Transforms" in Chapter 2 is the angular spectrum propagation transform (Equation 2.56), which in original physical coordinates can be written as

$$\alpha_{Fr}(f) = \int_{-\infty}^{\infty}\left\{\int_{-\infty}^{\infty} a(x)\exp(i2\pi px)dx\right\}\exp(i\pi\lambda Z p^2)\exp(-i2\pi fp)dp. \tag{4.162}$$

In the discretization of this transform, the following assumptions are made:

- The Fourier transform integral over x is replaced by DFT over a set of N signal samples $\{a_k\}$ at sampling points $\tilde{k}\Delta x\left(\tilde{k} = k + u_x\right)$

$$\int_{-\infty}^{\infty} a(x)\exp(i2\pi px)dx \Rightarrow \sum_{k=0}^{N-1} a_k \exp\left(i2\pi\frac{\tilde{k}s}{N}\right)$$

- Chirp-function $\exp(i\pi\lambda Z p^2)$ is replaced by its samples $\exp[i\pi\lambda Z (s\Delta x)^2]$, where $\{s\}$ are sample indices and Δx is the signal sampling interval.

- λZ is replaced by its expression through parameter $\mu^2 = \lambda Z / N\Delta f^2$ and cardinal relationship $\Delta x = 1/N\Delta f$ between sampling intervals in signal and Fourier transform domains is applied, by virtue of which

$$\exp[i\pi\lambda Z(s\Delta x)^2] = \exp[i\pi\mu^2 N\Delta f^2 (s\Delta x)^2] = \exp\left(i\pi\frac{\mu^2 s^2}{N}\right)$$

- Inverse Fourier transform integral over p is replaced by inverse DFT over index s.

As a result, the *discrete angular spectrum propagation transform* (DASPT) is obtained:

$$\alpha_r = \frac{1}{N}\sum_{s=0}^{N-1}\left[\sum_{k=0}^{N-1} a_k \exp\left(i2\pi\frac{\tilde{k}s}{N}\right)\right]\exp\left(i\pi\frac{\mu^2 s^2}{N}\right)\exp\left(-i2\pi\frac{\tilde{r}s}{N}\right)$$

$$= \frac{1}{N}\sum_{k=0}^{N-1} a_k \left[\sum_{s=0}^{N-1} \exp\left(i\pi\frac{\mu^2 s^2}{N}\right)\exp\left(-i2\pi\frac{r-k-w}{N}s\right)\right], \qquad (4.163)$$

or

$$\alpha_r = \sum_{k=0}^{N-1} a_k \operatorname{frincd}(N;\mu^2;r-k-w), \qquad (4.164)$$

where $w = u_x - v_f$ and $\operatorname{frincd}(\cdot;\cdot;\cdot)$ is the frincd-function defined by Equation 4.151.

DASPT, similar to DFTs and SDFrTs, is an orthogonal transform with inverse DASPT defined as

$$a_k = \frac{1}{N}\sum_{s=0}^{N-1}\left[\sum_{r=0}^{N-1} \alpha_r \exp\left(-i2\pi\frac{k-r+w}{N}s\right)\right]\exp\left(-i\pi\frac{\mu^2 s^2}{N}\right)$$

$$= \sum_{r=0}^{N-1} \alpha_r \operatorname{frincd}^*(N;\mu^2;k-r+w). \qquad (4.165)$$

When $\mu^2 = 0$ (which means $Z = 0$) and $w = 0$, DASPT reduces to the identical transform: in this case, the object plane and the transform planes coincide.

Two-Dimensional Discrete Fresnel Transforms

2D and, generally, multidimensional discrete Fresnel transforms are usually defined as separable transforms over each of the coordinate indices in the assumption of signal and its Fresnel spectrum sampling over rectangular sampling grids. Table 4.1 summarizes the 2D discrete Fresnel transforms.

TABLE 4.1

2D Discrete Fresnel Transforms

2D canonical DFrT

$$\alpha_{r,s} = \frac{1}{\sqrt{N_1 N_2}} \sum_{l=0}^{N_2-1} \sum_{k=0}^{N_1-1} a_{k,r} \exp\left[-i\pi \frac{(k\mu_1 - r/\mu_1)^2}{N_1}\right] \exp\left[-i\pi \frac{(l\mu_2 - s/\mu_2)^2}{N_2}\right]$$

2D-shifted and -scaled DFrT

$$\alpha_{r,s} = \frac{1}{\sqrt{N_1 N_2}} \sum_{l=0}^{N_2-1} \sum_{k=0}^{N_1-1} a_{k,l} \exp\left[-i\pi \frac{(k\mu_1 - r/\mu_1 + w_1)^2}{\sigma_1 N_1}\right] \exp\left[-i\pi \frac{(l\mu_2 - s/\mu_2 + w_2)^2}{\sigma_2 N_2}\right]$$

2D focal plane invariant DFrT

$$\alpha_{r,s} = \frac{1}{\sqrt{N_1 N_2}} \sum_{l=0}^{N_2-1} \sum_{k=0}^{N_1-1} a_{k,l} \exp\left\{-i\pi \frac{[k\mu - (r - N_1/2)/\mu_1]^2}{N}\right\} \exp\left\{-i\pi \frac{[l\mu_2 - (s - N_2/2)/\mu_2]^2}{N}\right\}$$

2D convolutional DFrT

$$\alpha_{r,s} = \frac{1}{\sqrt{N_1 N_2}} \sum_{l=0}^{N_2-1} \sum_{k=0}^{N_1-1} a_{k,l} \exp\left[-i\pi \frac{(k - r - \bar{w}_1)^2}{\mu_1^2 N_1}\right] \exp\left[-i\pi \frac{(l - s - \bar{w}_2)^2}{\mu_2^2 N_2}\right]$$

2D DASPT

$$\alpha_{r,s} = \sum_{l=0}^{N_2-1} \sum_{k=0}^{N_1-1} a_{k,l}\, \mathrm{frincd}\left(N_1; \mu_1^2; k - r + w_1\right) \mathrm{frincd}\left(N_2; \mu_2^2; l - s + w_2\right)$$

Discrete Representation of Kirchhoff Integral

Using the above-described transform discretization principles and assuming, as in the case of the convolutional Fresnel transform, identical sampling intervals Δx and Δf of the signal and its transform, one can obtain the following discrete representations of 2D Kirchhoff integral (Equation 2.40):

$$\alpha_{r,s} = \sum_{k=0}^{N-1}\sum_{l=0}^{N-1} a_{k,l} KRT^{(2D)}\left(\tilde{k}-\tilde{r};\tilde{l}-\tilde{s}\right), \qquad (4.166)$$

where $\tilde{k},\tilde{r};\tilde{l},\tilde{s}$ are shifted indices (as in Equation 4.44),

$$KRS_{\tilde{z},\mu}^{(2D)}(m,n) = \frac{\exp\left[i2\pi\left(\left(\tilde{z}^2\sqrt{1+m^2/\tilde{z}^2+n^2/\tilde{z}^2}\right)/\mu^2 N\right)\right]}{1+m^2/\tilde{z}^2+n^2/\tilde{z}^2}; \qquad (4.167)$$

$$\tilde{z} = Z/\Delta f; \quad \mu^2 = \frac{\lambda Z}{N\Delta f^2} = \frac{\lambda \tilde{z}}{N\Delta f}. \qquad (4.168)$$

We refer to this transform as the 2D *discrete Kirchhoff transform* (DKT). When $\tilde{z} \to 0$, DKT degenerates into the identical transform. When $\tilde{z} \to \infty$, DKT reduces to discrete Fresnel transform. In distinction to 2D DFTs and DFrTs, 2D DKT is an inseparable transform.

As one can see from Equation 4.166, DKT is a digital convolution. Therefore, it can be computed through DFT using FFT as

$$\{\alpha_r\} = IFFT\left\{FFT[\{a_{k,l}\}]\cdot FFT\left[KRS_{\tilde{z},\mu}^{2D}(n,m)\right]\right\}. \qquad (4.169)$$

From this representation, the inverse DKT can be computed as

$$\{\tilde{a}_k\} = IFFT\left\{FFT\{\alpha_r\}\cdot\frac{1}{FFT\left[KRS_{\tilde{z},\mu}^{2D}(n)\right]}\right\}, \qquad (4.170)$$

where FFT[·] is the operator of DFT implemented through FFT and · is the element-wise product of elements of arrays.

Hadamard, Walsh, and Wavelet Transforms

The above-described discrete Fourier and Fresnel transforms originated from corresponding imaging integral transforms. In this section, we consider digital transforms that represent important tools of computational imaging, though they do not have direct natural analogs.

Binary Transforms

Basis functions of DFT and DCT transforms are sinusoidal functions that assume values in the range [–1 ÷ 1]. Computation of their transform coefficients involves multiplication operations that are usually more time consuming than operations of addition. This motivated search for transforms with binary basis functions that assume only two values and due to this do not require multiplication operations for computing transform coefficients. In the section "Sampling Artifacts: Quantitative Analysis," we already introduced notions of Walsh and Haar transforms, whose basis functions are binary functions. Now, we continue the discussion on these transforms and provide additional insight into their origin and properties.

Hadamard and Walsh Transforms

Let the number of signal samples N be an integer power n of 2: $N = 2^n$. Represent signal and transform domain indices k and r on base 2 through their binary digits $\{k_m\}$ and $\{r_m\}$, $m = 0, 1, \ldots, n-1$:

$$k = \sum_{m=0}^{n-1} k_m 2^m, \quad r = \sum_{m=0}^{n-1} r_m 2^m \tag{4.171}$$

and consider n-dimensional DFT of a signal $\{a_k\} = \{a_{\{k_m\}}\}$ over binary indices $\{k_m\}$. Because binary indices assume only two values, 0 and 1, the number of samples in each of the n dimensions is equal to 2, and n-dimensional DFT of signal $\{a_k\} = \{a_{\{k_m\}}\}$ will take the form

$$\alpha_r = \alpha_{\{r_m\}} = \frac{1}{\sqrt{2^n}} \sum_{k_{n-1}=0}^{1} \sum_{k_{n-2}=0}^{1} \cdots \sum_{k_0=0}^{1} a_{\{k_m\}} \exp\left(i2\pi \sum_{m=0}^{n-1} \frac{k_m r_m}{2} \right)$$

$$= \frac{1}{\sqrt{2^n}} \sum_{k_{n-1}=0}^{1} \sum_{k_{n-2}=0}^{1} \cdots \sum_{k_0=0}^{1} a_{\{k_m\}} \left[\exp(i\pi) \right]^{\sum_{m=0}^{n-1} k_m r_m}$$

$$= \frac{1}{\sqrt{2^n}} \sum_{k_{n-1}=0}^{1} \sum_{k_{n-2}=0}^{1} \cdots \sum_{k_0=0}^{1} a_{\{k_m\}} (-1)^{\sum_{m=0}^{n-1} k_m r_m}. \tag{4.172}$$

In this way, we arrive at the transform

$$\alpha_r = \frac{1}{\sqrt{2^n}} \sum_{k_{n-1}=0}^{1} \sum_{k_{n-2}=0}^{1} \cdots \sum_{k_0} a_{\{k_m\}} (-1)^{\sum_{m=0}^{n-1} k_m r_m} \tag{4.173}$$

called the *Hadamard transform*. According to the definition of the Hadamard transform, 1D Hadamard transform is an n-dimensional DFT over two samples in each dimension. From this, it follows that inverse Hadamard transform is identical to the direct one.

The 1D Hadamard transform is, obviously, separable over its n-dimensions:

$$\alpha_r = \frac{1}{\sqrt{2^n}} \sum_{k_{n-1}=0}^{1} (-1)^{k_{n-1}r_{n-1}} \sum_{k_{n-2}=0}^{1} (-1)^{k_{n-2}r_{n-2}} \dots \sum_{k_0}^{1} a_{\{k_m\}}^{k_0 r_0}. \tag{4.174}$$

Owing to this, its computation can be decomposed into n stages with N addition/subtraction operates per stage. Thus, the total computational complexity of the transform is $Nn = N \log_2 N$ operations or $\log_2 N$ operations per transform coefficient. This method of computing Hadamard transform is called the *fast Hadamard transform* algorithm.

The basis functions of Hadamard transform

$$\phi_k(r) = (-1)^{\sum_{m=0}^{n-1} k_m r_m} = \prod_{m=0}^{n-1} (-1)^{k_m r_m} \tag{4.175}$$

are ordered in the same way as exponential basis functions of DFT are: according to the natural ascending order of their index. As we already indicated and demonstrated in section "Typical basis functions and classification" (Chapter 3), sequency-wise ordering of basis functions of Hadamard transform according to their gray code better corresponds to the principle of ordering according to transform coefficients' energy. The transform with such ordering of the basis functions, called *Walsh transform* or *Walsh–Hadamard transform*

$$\alpha_r = \frac{1}{\sqrt{2^n}} \sum_{k_{n-1}=0}^{1} \sum_{k_{n-2}=0}^{1} \dots \sum_{k_0=0}^{1} a_{\{k_m\}} (-1)^{\sum_{m=0}^{n-1} k_m r_m^{GC}} \tag{4.176}$$

can also be computed using fast Hadamard transform algorithm, though it must be complemented with permutation of transform coefficients according to their gray code ordering. Similarly to Hadamard transform, inverse Walsh transform is identical to the direct one.

Haar Transform

As it was indicated in the section "Sampling Artifacts: Quantitative Analysis," Haar transform basis functions are generated by windowing periodical binary sign-alternating functions (*Rademacher functions*) taken at dyadic

TABLE 4.2

Examples of *msb*, (r)mod 2^{msb} and $k/2^{n-msb}$ in Binary Representation of Numbers

k	0	1	2	3	4	5	6	7
Binary Code	000	001	010	011	100	101	110	111
msb		0		0		1		2
(r)mod 2^{msb}	–	–	0	1	00	01	10	11
$k/2^{n-msb}$; $msb = 1$	0	0	0	0	1	1	1	1
$k/2^{n-msb}$; $msb = 2$	00	00	01	01	10	10	11	11

(integer power of 2) scales with a shifted rectangular impulse function, whose extent and dyadic shift intervals are coordinated with the scale. A formal definition of discrete Haar basis functions can be derived from Equation 3.95:

$$har_k(r) = 2^{msb}(-1)^{k_{n-1-msb}}\, \delta\!\left((r)\bmod 2^{msb} - \left\lfloor \frac{k}{2^{msb-1}} \right\rfloor\right), \qquad (4.177)$$

with *msb* as the index of the most significant nonzero digit (bit) in binary representation of r (Equation 4.171), (r) mod 2^{msb} being modulo 2^{msb} value of r, a residual from division of r by 2^{msb} and $\lfloor k/2^{n-msb} \rfloor$ being an integer part of $k/2^{n-msb}$. Table 4.2 provides examples of *msb*, (r)mod 2^{msb} and $k/2^{n-msb}$ for $k = 0, 1, \ldots, 7$.

Plots of first eight Haar functions are shown in Figure 3.8. From the figure, one can see that Haar functions of $N = 2^n$ samples form n groups in scale (index *msb*) and that nonzero fragments of functions within each group are generated by a coordinate shift with a shift interval specific for each scale. Correspondingly, computing Haar transform can be decomposed into n scale groups. Computation in the k-th group involves 2^{n-k} addition/subtraction operations, thus resulting in total $\sum_{k=0}^{n-1} 2^{n-k} = 2^n \sum_{k=0}^{n-1} 2^{-k} = 2^n((2^{-n} - 1)/(2^{-1} - 1)) = 2(N - 1)$ addition/subtraction operations. Such a method of computing Haar transform is called *fast Haar transform* algorithm. Fast Walsh, fast Hadamard, and fast Haar transform algorithms are special cases a large class of fast transform algorithms, to which belong fast Fourier and fast cosine transforms.

Discrete Wavelet Transforms and Multiresolution Analysis

As already mentioned in the section "Sampling Artifacts: Quantitative Analysis," Haar transform is a representative of a large family of transforms called *wavelet transforms*, whose basis functions are built on the principle of combining scaling and shifting of a mother function. The design principle of discrete wavelet transforms can be explained using a signal flow diagram shown in Figure 4.16 for *dyadic wavelets*, in which scales and shifts change as integer powers of 2.

According to this diagram, direct and inverse wavelet transforms consist of several scale levels. The maximal number of levels for signals of $N = 2^n$ samples is $n = \log_2 N$. On each scale level s, $s = 1,\ldots,n$ of the direct transform of a signal $\{a_k\}$, signal $\{\tilde{a}_k^{(s-1)}\}$, ($k = 0, 1, \ldots, 2^{n-s+1}$) from the previous level (for the very first level $\{\tilde{a}_k^{(0)} = a_k\}$) is subjected to low-pass filtering

$$\tilde{a}_k^{(s)} = \sum_{n=0}^{N_h^{(s)}-1} LP_n^{(s)} a_{k-n}^{(s-1)},$$ (4.178)

where $N_h^{(s)}$ is the number of samples of the low-pass filter PSF $\{LP_n^{(s)}\}$. The resulting signal $\tilde{a}_k^{(s)}$ is downsampled two times to the half the number of samples producing a downsampled signal $\tilde{\tilde{a}}_k^{(s)} = \tilde{a}_{2k}^{(s)}$, ($k = 0, 1, \ldots, 2^{n-s} - 1$), which serves as input for the next level. This downsampled signal $\tilde{\tilde{a}}_k^{(s)}$ is also upsampled (interpolated) back to the full length

$$\bar{\tilde{a}}_k^{(s)} = \sum_{n=0}^{2^{N-s-1}-1} \text{INT}_{k-2n}^{(s)} \tilde{\tilde{a}}_n^{(s-1)}$$ (4.179)

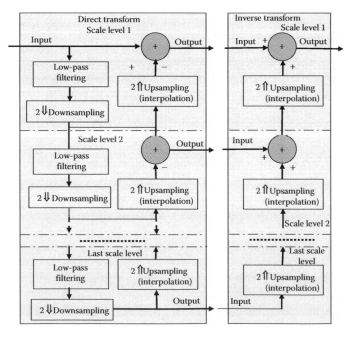

FIGURE 4.16
Flow diagrams of direct and inverse discrete wavelet transforms.

with an interpolation kernel $\left\{INT_k^{(s)}\right\}$ and subtracted from the input signal of this level $\left\{\tilde{a}_k^{(s-1)}\right\}$ to produce the difference signal $\left\{d_k^s = \tilde{a}_k^{(s-1)} - \overline{\tilde{a}}_k^{(s)}\right\}$ which serves, for all levels except for the very last one, as the output signal of the level. At the last n-th level, the level output signal is the downsampled signal $\tilde{\tilde{a}}_k^{(n-1)}$. Thus, the result of the transform is a set of $n-1$ difference signals $\left\{d_k^s\right\}$, obtained at scale levels $s = 1, 2, \ldots, n-1$ and a downsampled signal $\tilde{\tilde{a}}_k^{(n-1)}$ obtained at the last scale level. As a result of the successive downsamplings, these signals have different resolutions with respect to the initial signal, from full resolution, for level $s = 1$, to lower resolution for higher scales. This is why this type of signal decomposition is frequently called *multiresolution analysis* and the result of the decomposition is called *image pyramid*. It is illustrated in Figure 4.17.

Although the maximal number of scales is $n = \log_2 N$, the described signal decomposition does not necessarily perform up to this level; it can be stopped at any intermediate stage.

The inverse transform shown in the right part of the flow diagram in Figure 4.16 consists of successive upsampling (interpolation) of the output signals of

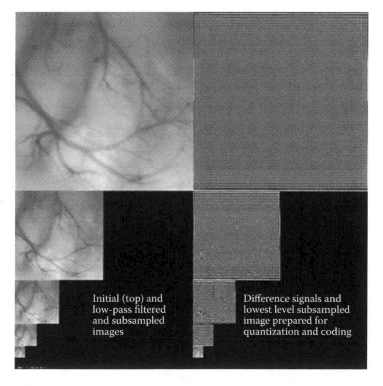

FIGURE 4.17
Image multiresolution decomposition (image pyramid).

corresponding scale levels beginning from the last one and to the level $s = 1$ and adding the results to the output of the previous level. The interpolation is performed using the same interpolation kernel $\left\{INT_k^{(s)}\right\}$ that is used on the corresponding scale level of the direct transform.

Low-pass filtering PSFs $\left\{LP_n^{(s)}\right\}$ and interpolation kernels $\left\{INT_k^{(s)}\right\}$ determine the type of the wavelet transform. Many different types of wavelet transforms designed and tested for different applications are known at present.

Wavelet transforms being linear transforms can be treated in terms of signal DFT spectra. In these terms, they perform signal "sub-band" decomposition: according to the convolution theorem, signal DFT spectrum, at each scale level, is multiplied by the frequency response of the high-pass filter,

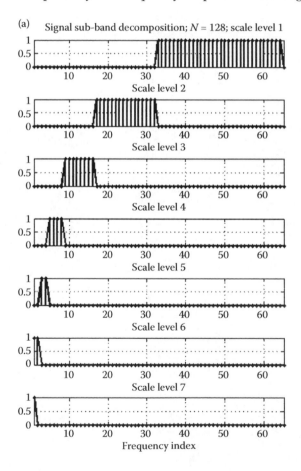

FIGURE 4.18

An illustrative example of dyadic wavelet sub-band decomposition using a trapezoidal frequency response of the high-pass filter for signals of $N = 128$ samples (a) and sub-bands that correspond to Haar transform (b) for the same number of signal samples.

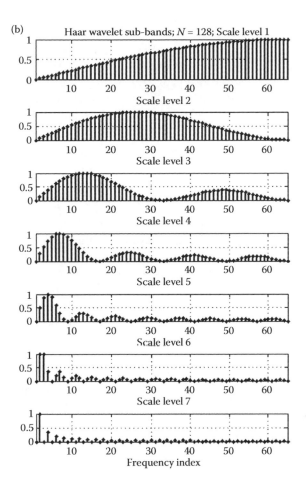

FIGURE 4.18
Continued.

which complements the low-pass filter at this scale and extracts the signal's corresponding sub-band. According to the definition of dyadic wavelets, bandwidth of the sub-band of scale s is proportional to $N/2^s$. Figure 4.18a illustrates this process for signals with $N = 128$ using, as an example, high-pass filters with a trapezoidal frequency response. Figure 4.18b presents for comparison sub-bands that correspond to Haar transform for the same number of signal samples. Note that for Haar transform, low-pass filtering PSF $\left\{ \mathrm{LP}_n^{(s)} \right\}$ and interpolation kernels $\left\{ \mathrm{INT}_k^{(s)} \right\}$ are on each scale rectangular impulses of two samples [1,1].

The computational complexity of wavelet transforms can be evaluated as follows. At scale level s, low-pass filtered and interpolated are $N_s = 2^{n-s}$ signal samples. Let low-pass filtering and interpolation require N_{LP} and N_{LP} multiplication/addition operations per sample, respectively. Then the total number of operations is $(N_{LP} + N_{INT})\sum_{s=0}^{n-1} 2^{n-s} = 2(N_{LP} + N_{INT})(N - 1)$, which amounts to $O(N_{LP} + N_{INT})$ operations per signal sample no matter how many samples the signal contains. Compare this with $O(n = \log_2 N)$ operations per sample of signal of N samples required for transforms, such as DFT and Walsh–Hadamard transforms that feature "FFT-type" algorithms such as the above-described fast Hadamard transform.

Discrete Sliding Window Transforms and "Time-Frequency" Signal Representation

All the described transforms can be applied "globally" to the entire set of available signal samples or "locally" to individual signal fragments. In the latter case, the signal is split into nonoverlapping or overlapping fragments, and transformation is performed over each fragment individually. When transforms are applied to signal fragments within windows that overlap to such a degree that their central samples (pixels) are adjacent in the process of signal (image) scanning over the sampling grid, *"sliding window" transform domain processing* takes place.

Consider, for the sake of simplicity, the 1D case. Sliding window application of transforms to fragment $\{a_{k-n+\lfloor N_w/2 \rfloor}\}$ of signal $\{a_k\}$ centered at the signal k-th sample, where $\lfloor \cdot \rfloor$ is an integer part of the argument, can mathematically be described as

$$\alpha_r^{(k)} = \sum_{n=0}^{N_w-1} w_n a_{k-n+\lfloor N_w/2 \rfloor} \phi_r(n), \qquad (4.180)$$

where $\{w_n\}$ are weight coefficients of a *window function*, which extracts signal fragments of N_w samples and weight them, and $\{\phi_r(n)\}$ are the transform basis functions. For 1D signals of N samples, a set of local transform coefficients $\{\alpha_r^{(k)}\}$ are 2D arrays of $N \times N_w$ samples. When signals are time sequences and their local Fourier analysis using DFT or DCT is performed, such signal representation is called the *time-frequency representation*. For 2D signals, sliding window application of 2D DFT or DCT transforms produces their 4D space–space frequency representation.

Figures 4.19 and 4.20 illustrate the sliding window DCT domain "time-frequency" representation of a test signal that consists of eight pieces of sinusoids with different frequencies (Figure 4.19) and of a fragment of a real speech signal (Figure 4.20).

In Figure 4.19b, one can clearly see bars, which indicate frequencies of the corresponding pieces of the signal. In Figure 4.20b, pieces of the signal with low and high frequencies and signatures of different sounds are easily distinguishable.

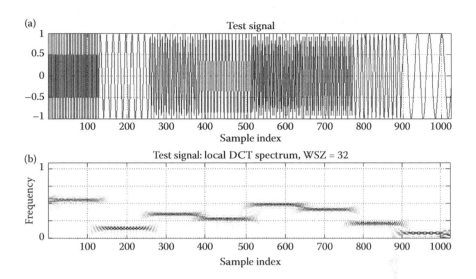

FIGURE 4.19

A test sinusoidal signal with eight fragments of different frequency (a) and its time-frequency representation by local DCT spectral analysis in sliding window of 32 samples (b). Intensities of spectral components are displayed in gray scale (dark—high intensity, bright—low intensity). Frequencies are shown in the normalized scale from 0 (zero frequency) to 1 (the highest frequency of the base band). Note that high-frequency signal fragments are plotted in (a) with aliasing due to the limited resolving power of the printer.

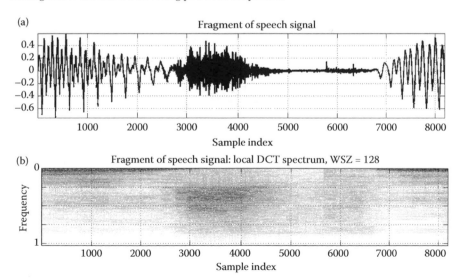

FIGURE 4.20

A fragment of a speech signal (a) and its time-frequency representation by local DCT spectral analysis in sliding window of 128 samples (b). Intensities of spectral components are displayed in gray scale (dark—high intensity, bright—low intensity). Frequencies are shown in the normalized scale from 0 (zero frequency) to 1 (the highest frequency of the base band).

The sliding window transform domain signal representation finds many applications in digital image and video processing. Its use for local adaptive image and video denoising, deblurring, and enhancement is discussed in Chapter 8.

Similar to sliding window transforms of continuous signals treated in terms of their Fourier spectra (see the section "Transforms in Sliding Window (Windowed Transforms) and Signal Sub-band Decomposition," in Chapter 2), local application of discrete transforms can also be regarded, in terms of signal DFT spectra, as signal sub-band decomposition. For fixed spectral index r, the set of signal local transform coefficients $\{\alpha_r^{(k)} = \sum_{n=0}^{N_w-1} w_n a_{k-n+\lfloor N_w/2 \rfloor} \phi_r(n)\}$ of signal $\{a_k\}$ of N samples in the k-th position of window are sequences of N samples. Consider their DFTs over index k:

$$A_s^{(r)} = \frac{1}{\sqrt{N}} \sum_{k=0}^{N-1} \alpha_r^{(k)} \exp\left(i2\pi \frac{ks}{N}\right) = \frac{1}{\sqrt{N}} \sum_{k=0}^{N-1} \left[\sum_{n=0}^{N_w-1} w_n a_{k-n+\lfloor N_w/2 \rfloor} \phi_r(n)\right] \exp\left(i2\pi \frac{ks}{N}\right).$$

(4.181)

Let $\{A_s\}$ be the DFT spectrum of the entire signal $\{a_k\}$, such that

$$a_k = \frac{1}{\sqrt{N}} \sum_{s=0}^{N-1} A_s \exp\left(-i2\pi \frac{ks}{N}\right).$$

(4.182)

Replace $\{a_{k-n+\lfloor N_w/2 \rfloor}\}$ in Equation 4.181 by its expression (Equation 4.182) through its DFT spectrum and obtain, as it is shown in Appendix, that

$$A_s^{(r)} = \frac{1}{\sqrt{N}} \sum_{k=0}^{N-1} \left[\sum_{n=0}^{N_w-1} w_n a_{k-n+\lfloor N_w/2 \rfloor} \phi_r(n)\right] \exp\left(i2\pi \frac{ks}{N}\right) \propto A_s \Phi_r^{(w)}(s),$$

(4.183)

where

$$\Phi_r^{(w)}(s) = \frac{1}{\sqrt{N_w}} \sum_{n=0}^{N-1} w_n \phi_r(n) \exp\left(i2\pi \frac{n}{N}s\right)$$

(4.184)

is DFT spectrum of the r-th basis function weighted with window weight coefficients $\{w_n\}$.

Equation 4.183 implies that the DFT spectrum, over index k, of the sequence of the r-th local spectral coefficients of signal $\{a_k\}$ is the signal DFT spectrum masked by the DFT spectrum of weighted r-th basis function, and, therefore, the sequence $\alpha_r^{(k)}$, in terms of index k, of r-th local transform coefficients represents an r-th sub-band of the signal. In other words, the local application of transforms in a sliding window can be treated as signal sub-band decomposition by means of *band-pass filters* with frequency responses defined by spectra of transform basis functions weighted by the sliding window weight coefficients. In this respect, it is akin to wavelet signal decomposition discussed in the previous section. The only difference between them is that in the sliding window transform domain processing, signal sub-bands are

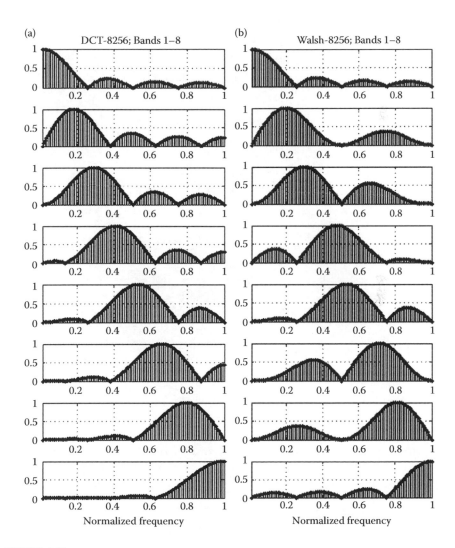

FIGURE 4.21

Frequency responses of band-pass filters that correspond to sliding window DCT (a), Walsh (b), and Haar (c) (window size 8).

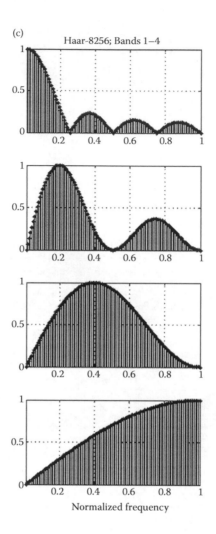

FIGURE 4.21
Continued.

arranged uniformly over the frequency (transform coefficient index *r*) axis and all have the same "bandwidth," while for wavelet decomposition signal sub-bands are arranged in a logarithmic scale with exponentially growing bandwidths.

Figure 4.21 shows, for comparison, frequency responses of band-pass filters that correspond to sliding window DCT, Walsh and Haar processing with uniform weight coefficients ($\{w_n = 1\}$). Note that the very first band-pass filters in all cases are low-pass filters that compute signal local mean in the filter window.

Appendix

Derivation of Equation 4.24

$$OFR_{eq}(f,p) = \int\limits_{-\infty}^{\infty}\int\limits_{-\infty}^{\infty} h_{eq}(x,\xi)\exp\left[i2\pi(fx - p\xi)\right]dx\,d\xi$$

$$= \int\limits_{-\infty}^{\infty}\int\limits_{-\infty}^{\infty}\sum_{k=0}^{N_b-1}\sum_{n=0}^{N^{(h)}-1} h_n\phi^{(r)}(x - \tilde{n}\Delta x)\phi^{(s)}\left[\xi - \left(\tilde{k}^{(r)} - \tilde{n}^{(s)}\right)\Delta x\right]$$

$$\times \exp\left[i2\pi(fx - p\xi)\right]dx\,d\xi$$

$$\sum_{k=0}^{N^{(b)}-1}\sum_{n=0}^{N^{(h)}-1} h_n \int\limits_{-\infty}^{\infty}\int\limits_{-\infty}^{\infty} \varphi^{(s)}(\xi)\varphi^{(r)}(x)\exp\left\{i2\pi\left[f\left(x + \tilde{k}^{(r)}\Delta x\right) - p\left(\xi + \left(\tilde{k}^{(r)} - \tilde{n}^{(s)}\right)\Delta x\right)\right]\right\}dx\,d\xi$$

$$= \left[\sum_{n-0}^{N^{(h)}-1} h_n\exp(i2\pi p\tilde{n}^{(s)}\Delta x)\right]\left\{\sum_{k=0}^{N_b-1}\exp\left[i2\pi(f-p)\tilde{k}^{(r)}\Delta x\right]\right\}$$

$$\times\left\{\exp(i2\pi fu^{(r)}\Delta x)\int\limits_{-\infty}^{\infty}\phi^{(r)}(x)\exp(i2\pi fx)\,dx\right\}$$

$$\times\left\{\exp(-i2\pi pu^{(r)}\Delta x)\int\limits_{-\infty}^{\infty}\phi^{(s)}(\xi)\exp(-i2\pi p\xi)\,d\xi\right\}.$$

Derivation of Equation 4.30

$$h_{k,r} = \int\limits_{-i}^{\infty}\int\limits_{-\infty}^{\infty}\exp(i2\pi fx)\varphi^{(r)}(x - \tilde{k})\varphi^{(s)}(f - \tilde{r}\Delta f)\,dx\,df$$

$$= \int\limits_{-\infty}^{\infty}\varphi^{(s)}(f)df\int\limits_{-\infty}^{\infty}\exp\left\{i2\pi(f + \tilde{r}\Delta f)(x + \tilde{k}\Delta x)\right\}\varphi^{(r)}(x)\,dx$$

$$= \exp(i2\pi\tilde{k}\tilde{r}\Delta x\Delta f)\int\limits_{-\infty}^{\infty}\varphi^{(s)}(f)\exp(i2\pi f\tilde{k}\Delta x)df\int\limits_{-\infty}^{\infty}\exp\left[i2\pi(f + \tilde{r}\Delta f)x\right]\varphi^{(s)}(x)\,dx$$

$$= \exp(i2\pi\tilde{k}\tilde{r}\Delta x\Delta f)\int\limits_{-\infty}^{\infty}\Phi^{(r)}(f + \tilde{r}\Delta f)\phi^{(s)}(f)\exp(i2\pi f\tilde{k}\Delta x)df.$$

Reasonings Regarding Equation 4.31

The spectrum sampling function $\phi^{(s)}(f)$ is a function compactly concentrated in an interval of about Δf around point $f = 0$. Within this interval, the frequency response of the signal reconstruction device $\Phi^{(r)}(f + \tilde{r}\Delta f)$ can be approximated by a constant because it is a Fourier transform of the reconstruction device PSF $\phi^{(r)}(x)$, which is compactly concentrated around point $x = 0$ in an interval of about Δx. Then Equation 4.30c can be approximated as

$$\int_{-\infty}^{\infty} \Phi^{(r)}(f + \tilde{r}\Delta f)\phi^{(s)}(f)\exp(i2\pi f\tilde{k}\Delta x)df \cong \int_{-\infty}^{\infty} \phi^{(s)}(f)\exp\{i2\pi f\tilde{k}\Delta x\}df = \Phi^{(s)}(\tilde{k}\Delta x),$$

where $\Phi^{(s)}(\bullet)$ is the frequency response of the spectrum sampling function $\phi^{(s)}(f)$ that, in its turn and for the same reason of the compactness of $\phi^{(s)}(f)$, can also be approximated by a constant for argument values $\tilde{k}\Delta x$.

Derivation of Equations 4.37 and 4.38

Denote $\tilde{k} = k + u_x$, $\tilde{n} = n + u_x$, and $\tilde{r} = r + u_f$. Then, for Equation 4.37, obtain

$$a_k^\sigma = \frac{1}{\sqrt{\sigma N}} \sum_{r=0}^{\sigma N-1} \alpha_r \exp\left(-i2\pi \frac{\tilde{k}\tilde{r}}{\sigma N}\right) = \frac{1}{\sigma N} \sum_{r=0}^{\sigma N-1}\left[\sum_{n=0}^{N-1} a_k \exp\left(i2\pi \frac{\tilde{n}\tilde{r}}{\sigma N}\right)\right]\exp\left(-i2\pi \frac{\tilde{k}\tilde{r}}{\sigma N}\right)$$

$$= \frac{1}{\sigma N} \sum_{n=0}^{N-1} a_k \left[\sum_{r=0}^{\sigma N-1} \exp\left(i2\pi \frac{\tilde{n}-\tilde{k}}{\sigma N}\tilde{r}\right)\right]$$

$$= \frac{1}{\sigma N} \sum_{n=0}^{N-1} a_k \exp\left(i2\pi \frac{n-k}{\sigma N} u_f\right)\frac{\exp[i2\pi(n-k)]-1}{\exp(i2\pi((n-k)/\sigma N))-1}$$

$$= \frac{1}{\sigma N} \sum_{n=0}^{N-1} a_n \frac{\sin[\pi(n-k)]}{\sin(\pi((n-k)/\sigma N))}\exp\left[i\pi\left(v_f+\frac{\sigma N-1}{2}\right)\frac{n-k}{\sigma N}\right]$$

$$= \frac{1}{\sigma N} \sum_{n=0}^{N-1} a_n \operatorname{sincd}[\sigma N; \pi(n-k)]\exp\left[i\pi\left(v_f+\frac{\sigma N-1}{2}\right)\frac{n-k}{\sigma N}\right]$$

$$= \begin{cases} a_n, & k = 0,1,...,N-1 \\ 0, & k = N, N = 1,...,\sigma N-1 \end{cases}$$

as $\operatorname{sincd}[\sigma N; \pi(n - k)] = \delta(n - k)$ for $k = 0,1,...,N - 1$ and for $k = N, N = 1,...,$ $\sigma N - 1$, $\sin[\pi(n - k)] = 0$ and $\sin\left(\pi \dfrac{n - k}{\sigma N}\right) \neq 0$.

Principle of Fast Fourier Transform Algorithm

FFTs are algorithms for fast computation of the DFT. The principle of the FFT can be easily understood if one compares 1D and 2D DFT. Let $\{a_n^{(1)}\}$ and $\{a_{k,l}^{(2)}\}$ be 1D and 2D arrays with the same number $N = N_1 N_2$ of samples: $n = 0$, 1, ..., $N_1 N_2 - 1$; $k = 0, 1,..., N_1 - 1$; $l = 0, 1, ..., N_2 - 1$. Direct computing of the 1D DFT of array $\{a_n^{(1)}\}$

$$\alpha_r = \frac{1}{\sqrt{N}} \sum_{k=0}^{N-1} a_k^{(1)} \exp\left(i2\pi \frac{kr}{N}\right)$$

requires $N^2 = N_1^2 N_2^2$ operations with complex numbers while computing the 2D DFT of array $\{a_{k,l}^{(2)}\}$

$$\alpha_{r,s} = \frac{1}{\sqrt{N}} \sum_{k=0}^{N_1-1} \sum_{l=0}^{N_2-1} a_{k,l}^{(2)} \exp\left[i2\pi\left(\frac{kr}{N_1} + \frac{ls}{N_2}\right)\right]$$

$$= \frac{1}{\sqrt{N}} \sum_{k=0}^{N_1-1} \exp\left(i2\pi \frac{kr}{N_1}\right) \sum_{l=0}^{N_2-1} a_n \exp\left(i2\pi \frac{ls}{N_2}\right)$$

requires only $N_1^2 N_2 + N_1 N_2^2 = N_1 N_2 (N_1 + N_2)$ operations thanks to the separability of the 2D DFT to two 1D DFTs. Therefore, one can accelerate computing the DFT by representing it in a separable multidimensional form. One can do this if the size of the array is a composite number. Let, as in the above example, $N = N_1 N_2$. Represent indices k and r of signal and its 1D transform samples as 2D ones:

$$k = k_2 N_1 + k_1; \quad k_1 = 0,1,..., N_1 - 1; \quad k_2 = 0,1,..., N_2 - 1;$$

$$r = r_1 N_2 + r_2; \quad r_1 = 0,1,..., N_1 - 1; \quad r_2 = 0,1,..., N_2 - 1.$$

Then, the 1D DFT can be split into two successive 1D DFTs:

$$\alpha_r = \alpha_{r_1,r_2} = \frac{1}{\sqrt{N}} \sum_{k=0}^{N-1} a_k^{(1)} \exp\left(i2\pi \frac{kr}{N}\right)$$

$$= \frac{1}{\sqrt{N_1 N_2}} \sum_{k_2=0}^{N_2-1} \sum_{k_1=0}^{N_1-1} a_{k_1,k_2}^{(1)} \exp\left[i2\pi \frac{(k_2 N_1 + k_1)(r_1 N_2 + r_2)}{N_2 N_1}\right]$$

$$= \frac{1}{\sqrt{N}} \sum_{k_1=0}^{N_1-1} \exp\left(i2\pi \frac{k_1 r_1}{N_1}\right)\left[\exp\left(i2\pi \frac{k_1 r_2}{N_1 N_2}\right) \sum_{k_2=0}^{N_2-1} a_{k_1,k_2}^{(1)} \exp\left(i2\pi \frac{k_2 r_2}{N_2}\right)\right]$$

$$= \mathrm{DFT}_{N_1} \left\{ \exp\left(i2\pi \frac{k_1 r_2}{N} \right) \cdot \mathrm{DFT}_{N_2} \left\{ a^{(1)}_{k_1, k_2} \right\} \right\}$$

over two subsets of N_1 and N_2 samples, which, as in the above example of 2D transform, require only $N_1^2 N_2 + N_1 N_2^2 = N_1 N_2 (N_1 + N_2)$ operations instead of $N^2 = N_1^2 N_2^2$.

Obviously, the larger the number of factors of N, the higher is the dimensionality to which 1D DFT can be decomposed in this way, and the higher is the computational complexity reduction. For n factors of $N = N_0 \cdot N_1, \ldots,$ N_{n-1}, the DFT can be computed with only $N \cdot (N_0 + N_1 + \cdots + N_{n-1})$ operations with complex numbers instead of $N^2 = N(N_0 \cdot N_1, \ldots, N_{n-1})$ operations required for the direct computation. For $N = 2^n$, the computational complexity of the DFT is $O(N \log_2 N)$. FFTs for signal size that is a power of 2 are called *radix-2 FFT*. They are most widespread in the software packages for signal and image processing.

FFT algorithms are a special case of a broad class of fast transform algorithms that have similar structure and whose computational complexity is $O(N \log N)$ or lower.

Representation of Scaled DFT as Convolution

Consider first the inverse scaled DFT defined by Equation 4.37. With denotations $\tilde{k} = k + u_x$ and $\tilde{r} = r + v_f$, Equation 4.37 can be written as

$$\alpha^{(u,v;\sigma)}_{N-r} = \frac{1}{\sqrt{\sigma N}} \sum_{k=0}^{N-1} a^{(u,v;\sigma)}_k \exp\left(i2\pi \frac{N + 2v_f - \tilde{r}}{\sigma N} \tilde{k} \right)$$

$$= \frac{1}{\sqrt{\sigma N}} \sum_{k=0}^{N-1} a^{(u,v;\sigma)}_k \exp\left(i2\pi \frac{N + 2v_f}{\sigma N} \tilde{k} \right) \exp\left(-i2\pi \frac{\tilde{k}\tilde{r}}{\sigma N} \right)$$

$$= \frac{\exp\left(-i2\pi \dfrac{\tilde{r}^2}{\sigma N} \right)}{\sqrt{\sigma N}} \sum_{k=0}^{N-1} \left\{ a^{(u,v;\sigma)}_k \exp\left(i2\pi \frac{N + 2v_f}{\sigma N} \tilde{k} \right) \exp\left(-i2\pi \frac{\tilde{k}^2}{\sigma N} \right) \right.$$

$$\left. \times \exp\left[i2\pi \left(\frac{(\tilde{k} - \tilde{r})^2}{\sigma N} \right) \right] \right\}.$$

The sum in this equation is the digital convolution of the signal spectrum $\left\{ \alpha^{(u,v;\sigma)}_r \right\}$ modulated by a *chirp-function* $\exp(-i2\pi(\tilde{r}^2/\sigma N))$ and a complex conjugate to this function. In a similar way, direct scaled DFT defined by Equation 4.36 can be, for spectral coefficient, taken in a reverse order, written as

$$\alpha_{N-r}^{(u,v;\sigma)} = \frac{1}{\sqrt{\sigma N}} \sum_{k=0}^{N-1} a_k^{(u,v;\sigma)} \exp\left(i2\pi \frac{N + 2v_f - \tilde{r}}{\sigma N}\tilde{k}\right)$$

$$= \frac{1}{\sqrt{\sigma N}} \sum_{k=0}^{N-1} a_k^{(u,v;\sigma)} \exp\left(i2\pi \frac{N + 2v_f}{\sigma N}\tilde{k}\right) \exp\left(-i2\pi \frac{\tilde{k}\tilde{r}}{\sigma N}\right)$$

$$\times \frac{\exp(-i2\pi(\tilde{r}^2/\sigma N))}{\sqrt{\sigma N}} \sum_{r=0}^{N-1} \left\{ a_k^{(u,v;\sigma)} \exp\left(i2\pi \frac{N + 2v_f}{\sigma N}\tilde{k}\right) \exp\left(-i2\pi \frac{\tilde{k}^2}{\sigma N}\right) \right\}$$

$$\times \exp\left[i2\pi\left(\frac{(\tilde{k} - \tilde{r})^2}{\sigma N}\right)\right],$$

which is also a digital convolution of a modulated signal and a chirp-function.

Derivation of Equation 4.53

In the derivation, we will use simplified definitions of SDFT as given by Equations 4.40 and 4.41.

$$\tilde{a}_n^{u/p,v/q} = \frac{1}{\sqrt{N}} \left[\sum_{r=0}^{K-1} \alpha_r^{u,v} \exp\left(-i2\pi \frac{rp_x}{N}\right) \right] \exp\left[-i2\pi \frac{n(r+q_v)}{N}\right]$$

$$= \frac{1}{\sqrt{N}} \sum_{r=0}^{K-1} \left\{ \frac{1}{\sqrt{N}} \sum_{k=0}^{N-1} a_k \exp\left(i2\pi \frac{kv_f}{N}\right) \exp\left[i2\pi \frac{(k+u_x)r}{N}\right] \right\}$$

$$\times \exp\left(-i2\pi \frac{rp_x}{N}\right) \exp\left[-i2\pi \frac{n(r+q_f)}{N}\right]$$

$$= \frac{1}{N} \sum_{k=0}^{N-1} a_k \exp\left(i2\pi \frac{kv_f - nq_f}{N}\right) \sum_{r=0}^{K-1} \exp\left[i2\pi \frac{(k-n+u_x-p_x)r}{N}\right]$$

$$= \frac{1}{N} \sum_{k=0}^{N-1} a_k \exp\left(i2\pi \frac{kv_f - nq_f}{N}\right) \frac{\exp[i2\pi(k-n+u_x-p_x)K/N]-1}{\exp[i2\pi(k-n+u_x-p_x)/N]-1}$$

$$= \frac{1}{N} \sum_{k=0}^{N} a_k \frac{\exp\{((i\pi K[(k+u_x)-(n+p_x)])/N)\} - \exp\{-((i\pi K[(k+u_x)-(n+p_x)])/N)\}}{\exp[i\pi(((k+u_x)-(n+p_x))/N)] - \exp[-i\pi(((k+u_x)-(n+p_x))/N)]}$$

$$\times \exp\left\{i\pi \frac{K-1}{N}[(k+u_x)-(n+p_x)]\right\} \exp\left(i2\pi \frac{kv_f - nq_f}{N}\right)$$

$$= \sum_{k=0}^{N} a_k \frac{\sin\{\pi(K/N)[(k-n)+(u_x-p_x)]\}}{N\sin\{\pi(([(k-n)+(u_x-p_x)])/N)\}}$$

$$\times \exp\left[i\pi\frac{(K-1)(k-n+u_x-p_x)}{N}\right]\exp\left(i2\pi\frac{kv_f - nq_f}{N}\right)$$

$$= \left\{\sum_{k=0}^{N-1} a_k \exp\left[i\pi k\left(\frac{K-1}{N}+2v_f\right)\right]\overline{\mathrm{sincd}}[K;N;(k-n+u_x-p_x)]\right\}$$

$$\times \exp\left[-i\pi\left(\frac{K-1}{N}+2q_f\right)n\right]\exp\left[i\pi\frac{K-1}{N}(u_x-p_x)\right].$$

Derivation of Equations 4.58 and 4.60

By definition of SDFTs:

$$a_{k+gN} = \frac{1}{\sqrt{N}}\sum_{r=0}^{N-1}\alpha_r \exp\left[-i2\pi\frac{(k+gN+u_x)(r+v_f)}{N}\right]$$

$$= \frac{1}{\sqrt{N}}\sum_{r=0}^{N-1}\alpha_r \exp\left[-i2\pi\frac{gN(r+v_f)}{N}\right]\exp\left[-i2\pi\frac{(k+u_x)(r+v_f)}{N}\right]$$

$$= \frac{\exp(-i2\pi g v_f)}{\sqrt{N}}\sum_{r=0}^{N-1}\alpha_r \exp\left[-i2\pi\frac{(k+u_x)(r+v_f)}{N}\right]$$

$$= a_k \exp(-i2\pi g v_f).$$

$$\alpha_{r+gN} = \frac{1}{\sqrt{N}}\sum_{k=0}^{Nc-1}a_k \exp\left[i2\pi\frac{(k+u_x)(r+gN+v_f)}{N}\right]$$

$$= \frac{1}{\sqrt{N}}\sum_{k=0}^{N-1}a_k \exp\left[i2\pi\frac{gN(k+u_x)}{N}\right]\exp\left[i2\pi\frac{(k+u_x)(r+v_f)}{N}\right]$$

$$= \frac{\exp(i2\pi g u_x)}{\sqrt{N}}\sum_{k=0}^{N-1}a_k \exp\left[i2\pi\frac{(k+u_x)(r+v_f)}{N}\right]$$

$$= \alpha_r \exp(i2\pi g u_x).$$

Derivation of Equation 4.63

Denoting $\tilde{k}^a = k + u_x^a;\, \tilde{k}^b = k + u_x^b;\, \tilde{k}^c = k + u_x^c;\; r^a = r + v_f^a;\, r^b = r + v_f^b\; r^c = r + v_f^c$, obtain

$$c_k^{u^c,v^c} = \frac{1}{\sqrt{N}}\sum_{r=0}^{N-1} \alpha_r^{u^a,v^a}\, \beta_r^{u^b,v^b}\, \exp\!\left(-i2\pi\frac{\tilde{k}^c \tilde{r}^c}{N}\right)$$

$$= \frac{1}{\sqrt{N}}\sum_{r=0}^{N-1}\left\{\frac{1}{\sqrt{N}}\sum_{n=0}^{N-1} a_n \exp\!\left(i2\pi\frac{\tilde{n}^a \tilde{r}^a}{N}\right)\right\}\beta_r^{u^b,v^b}\, \exp\!\left(-i2\pi\frac{\tilde{k}^c \tilde{r}^c}{N}\right)$$

$$= \frac{1}{\sqrt{N}}\sum_{n=0}^{N-1} a_n\left\{\frac{1}{\sqrt{N}}\sum_{r=0}^{N-1}\beta_r^{u^b,v^b}\, \exp\!\left(i2\pi\frac{\tilde{n}^a \tilde{r}^a - \tilde{k}^c \tilde{r}^c}{N}\right)\right\}$$

$$= \frac{1}{\sqrt{N}}\sum_{n=0}^{N-1} a_n\left\{\frac{1}{\sqrt{N}}\sum_{r=0}^{N-1}\beta_r^{u^b,v^b}\, \exp\!\left(i2\pi\frac{\tilde{n}^a\left(r+v_f^a\right) - \tilde{k}^c\left(r+v_f^c\right)}{N}\right)\right\}$$

$$= \frac{1}{\sqrt{N}}\sum_{n=0}^{N-1} a_n\left\{\frac{1}{\sqrt{N}}\sum_{r=0}^{N-1}\beta_r^{u^b,v^b}\, \exp\!\left(i2\pi\frac{\tilde{n}^a\left(r+v_f^b+v_f^a-v_f^b\right) - \tilde{k}^c\left(r+v_f^b+v_f^c-v_f^b\right)}{N}\right)\right\}$$

$$= \frac{1}{\sqrt{N}}\sum_{n=0}^{N-1} a_n\left\{\frac{1}{\sqrt{N}}\sum_{r=0}^{N-1}\beta_r^{u^b,v^b}\, \exp\!\left(-i2\pi\frac{\tilde{k}^c\left(r+v_f^b+v_f^c-v_f^b\right) - \tilde{n}^a\left(r+v_f^b+v_f^a-v_f^b\right)}{N}\right)\right\}$$

$$= \frac{1}{\sqrt{N}}\sum_{n=0}^{N-1} a_n\left\{\frac{1}{\sqrt{N}}\sum_{r=0}^{N-1}\beta_r^{u^b,v^b}\, \exp\!\left(-i2\pi\frac{\left(\tilde{k}^c-\tilde{n}^a\right)\left(r+v_f^b\right) + \tilde{k}^c\left(v_f^c-v_f^b\right) - \tilde{n}^a\left(v_f^a-v_f^b\right)}{N}\right)\right\}$$

$$= \frac{1}{\sqrt{N}}\sum_{n=0}^{N-1} a_n\,\frac{1}{\sqrt{N}}\sum_{r=0}^{N-1}\beta_r^{u^b,v^b}\, \exp\!\left[-i2\pi\frac{\left(k-n+u_x^c-u_x^a\right)\left(r+v_f^b\right)}{N}\right]$$

$$\times \exp\!\left(-i2\pi\frac{\tilde{k}^c\left(v_f^c-v_f^b\right) - \tilde{n}^a\left(v_f^a-v_f^b\right)}{N}\right)$$

$$= \frac{1}{\sqrt{N}}\exp\!\left(-i2\pi\frac{\tilde{k}^c\left(v_f^c-v_f^b\right)}{N}\right)\sum_{n=0}^{N-1} a_n \exp\!\left(i2\pi\frac{\tilde{n}^a\left(v_f^a-v_f^b\right)}{N}\right)\tilde{b}_{k-n},$$

where

$$\tilde{b}_{k-n} = \frac{1}{\sqrt{N}}\sum_{r=0}^{N-1}\beta_r^{u^b,v^b}\, \exp\!\left[-i2\pi\frac{\left(k-n+u_x^c-u_x^a\right)\left(r+v_f^b\right)}{N}\right].$$

Derivation of Equation 4.65

$$\tilde{b}_m = \frac{1}{\sqrt{N}} \sum_{r=0}^{N-1} \beta_r^{u^b, v^b} \exp\left[-i2\pi \frac{\left(m + u_x^c - u_n^a\right)\left(r + v_f^b\right)}{N}\right]$$

$$= \frac{1}{\sqrt{N}} \sum_{r=0}^{N-1} \left\{ \frac{1}{\sqrt{N}} \sum_{n=0}^{N-1} b_n \exp\left[i2\pi \frac{\left(n + u_x^b\right)\left(r + v_f^b\right)}{N}\right]\right\}$$

$$\times \exp\left[-i2\pi \frac{\left(m + u_x^c - u_n^a\right)\left(r + v_f^b\right)}{N}\right]$$

$$= \sum_{n=0}^{N-1} b_n \left\{ \frac{1}{N} \sum_{r=0}^{N-1} \exp\left[i2\pi \frac{\left(n + u_x^b\right)\left(r + v_f^b\right)}{N}\right] \exp\left[-i2\pi \frac{\left(m + u_x^c - u_n^a\right)\left(r + v_f^b\right)}{N}\right]\right\}$$

$$= \sum_{n=0}^{N-1} b_n \left\{ \frac{1}{N} \sum_{r=0}^{N-1} \exp\left[i2\pi \frac{\left(n - m + u_x^b - u_x^c + u_n^a\right)}{N}\left(r + v_f^b\right)\right]\right\}$$

$$= \sum_{n=0}^{N-1} b_n \exp\left[i2\pi \frac{\left(n - m + u_x^b - u_x^c + u_n^a\right)v_f^b}{N}\right]$$

$$\times \left\{ \frac{1}{N} \sum_{r=0}^{N-1} \exp\left[i2\pi \frac{\left(n - m + u_x^b - u_x^c + u_n^a\right)}{N}r\right]\right\}$$

$$= \sum_{n=0}^{N-1} b_n \exp\left[i2\pi \frac{\left(n - m + u_x^b - u_x^c + u_n^a\right)v_f^b}{N}\right]$$

$$\times \left\{ \frac{1}{N} \frac{\exp\left[i2\pi\left(n - m + u_x^b - u_x^c + u_n^a\right)\right] - 1}{\exp\left[i2\pi\left(\left(n - m + u_x^b - u_x^c + u_n^a\right)/N\right)r\right] - 1}\right\}$$

$$= \sum_{n=0}^{N-1} b_n \exp\left[i2\pi \frac{\left(n - m + u_x^b - u_x^c + u_n^a\right)v_f^b}{N}\right]$$

$$\times \left\{ \frac{1}{N} \frac{\exp\left[i\pi\left(n - m + u_x^b - u_x^c + u_n^a\right)\right] - \exp\left[-i\pi\left(n - m + u_x^b - u_x^c + u_n^a\right)\right]}{\exp\left[i\pi\left(\left(n - m + u_x^b - u_x^c + u_n^a\right)/N\right)r\right] - \exp\left[-i\pi \frac{\left(n - m + u_x^b - u_x^c + u_n^a\right)}{N}r\right]}\right\}$$

$$\times \exp\left[i2\pi \frac{N-1}{N}\left(n - m + u_x^b - u_x^c + u_n^a\right)\right]$$

$$= \sum_{n=0}^{N-1} b_n \exp\left[i\pi \frac{\left(n - m + u_x^b - u_x^c + u_n^a\right)\left(2v_f^b + N - 1\right)}{N}\right]$$

$$\times \frac{\sin\left[\pi\left(n - m + u_x^b - u_x^c + u_n^a\right)\right]}{N \sin\left[\left(\pi\left(n - m + u_x^b - u_x^c + u_n^a\right)/N\right)\right]}$$

$$= \sum_{n=0}^{N-1} b_n \exp\left[i\pi \frac{\left(n - m + u_x^b - u_x^c + u_n^a\right)\left(2v_f^b + N - 1\right)}{N}\right]$$

$$\times \mathrm{sincd}\left[N; \pi\left(n - m + u_x^b - u_x^c + u_n^a\right)\right].$$

When $v_f^b = -(N-1)/2$,

$$\tilde{b}_m = \sum_{n=0}^{N-1} b_n\, \mathrm{sincd}\left[N; \pi\left(n - m + u_x^b - u_x^c + u_n^a\right)\right].$$

Derivation of Equation 4.68

Denoting $\tilde{k}^a = k + u_x^a$; $\tilde{k}^b = k + u_x^b$; $\tilde{k}^c = k + u_x^c$; $r^a = r + v_f^a$; $r^b = r + v_f^b$; $r^c = r + v_f^c$, obtain

$$c_k^{u^c,v^c} = \frac{1}{\sqrt{N}} \sum_{r=0}^{N-1} \alpha_r^{\left(u_x^a,v_f^a\right)} \beta_r^{*\left(u^b,v^b\right)} \exp\left(-i2\pi \frac{\tilde{k}^c \tilde{r}^c}{N}\right)$$

$$= \frac{1}{\sqrt{N}} \sum_{r=0}^{N-1} \left\{ \frac{1}{\sqrt{N}} \sum_{n=0}^{N-1} a_n \exp\left(i2\pi \frac{\tilde{n}^a \tilde{r}^a}{N}\right)\right\} \beta_r^{*\left(u^b,v^b\right)} \exp\left(-i2\pi \frac{\tilde{k}^c \tilde{r}^c}{N}\right)$$

$$= \frac{1}{\sqrt{N}} \sum_{n=0}^{N-1} a_n \left\{ \frac{1}{\sqrt{N}} \sum_{r=0}^{N-1} \beta_r^{*\left(u^b,v^b\right)} \exp\left(i2\pi \frac{\tilde{n}^a \tilde{r}^a - \tilde{k}^c \tilde{r}^c}{N}\right)\right\}$$

$$= \frac{1}{\sqrt{N}} \sum_{n=0}^{N-1} a_n \left\{ \frac{1}{\sqrt{N}} \sum_{r=0}^{N-1} \beta_r^{\left(u^b,v^b\right)} \exp\left(i2\pi \frac{\tilde{k}^c\left(r + v_f^c\right) - \tilde{n}^a\left(r + v_f^a\right)}{N}\right)\right\}^*$$

$$= \frac{1}{\sqrt{N}} \sum_{n=0}^{N-1} a_n \left\{ \frac{1}{\sqrt{N}} \sum_{r=0}^{N-1} \beta_r^{\left(u^b,v^b\right)}\right.$$

$$\times \exp\left(i2\pi \frac{\tilde{k}^c\left(r + v_f^b + v_f^c - v_f^b\right) - \tilde{n}^a\left(r + v_f^b + v_f^a - v_f^b\right)}{N}\right)\Biggr\}^*$$

$$= \frac{1}{\sqrt{N}} \sum_{n=0}^{N-1} a_n \left\{ \frac{1}{\sqrt{N}} \sum_{r=0}^{N-1} \beta_r^{\left(u^b, v^b\right)} \right.$$

$$\left. \times \exp\left(i2\pi \frac{\left(\tilde{k}^c - \tilde{n}^a\right)\left(r + v_f^b\right) + \tilde{k}^c\left(v_f^c - v_f^b\right) - \tilde{n}^a\left(v_f^a - v_f^b\right)}{N}\right)\right\}^*$$

$$= \frac{1}{\sqrt{N}} \sum_{n=0}^{N-1} a_n \left\{ \frac{1}{\sqrt{N}} \sum_{r=0}^{N-1} \beta_r^{\left(u^b, v^b\right)} \exp\left[i2\pi \frac{\left(k - n + u_x^c - u_x^a\right)\left(r + v_f^b\right)}{N}\right] \right.$$

$$\left. \times \exp\left(i2\pi \frac{\tilde{k}^c\left(v_f^c - v_f^b\right) - \tilde{n}^a\left(v_f^a - v_f^b\right)}{N}\right)\right\}^*$$

$$= \frac{1}{\sqrt{N}} \exp\left(-i2\pi \frac{\tilde{k}^c\left(v_f^c - v_f^b\right)}{N}\right) \sum_{n=0}^{N-1} a_n \exp\left(i2\pi \frac{\tilde{n}^a\left(v_f^a - v_f^b\right)}{N}\right) \tilde{b}_{n-k},$$

where

$$\tilde{b}_m = \frac{1}{\sqrt{N}} \left\{ \sum_{r=0}^{N-1} \beta_r^{u^b, v^b} \exp\left[-i2\pi \frac{\left(m - u_x^c + u_x^a\right)\left(r + v_f^b\right)}{N}\right] \right\}^*; \tilde{b}_m$$

$$= \frac{1}{N} \left\{ \sum_{r=0}^{N-1} \left\{ \sum_{n=0}^{N-1} b_n \left[i2\pi \frac{\left(n + u_x^b\right)\left(r + v_f^b\right)}{N}\right] \right\} \exp\left[-i2\pi \frac{\left(m - u_x^c + u_x^a\right)\left(r + v_f^b\right)}{N}\right] \right\}^*$$

$$= \frac{1}{N} \sum_{r=0}^{N-1} \sum_{n=0}^{N-1} b_n \exp\left[i2\pi \frac{\left(m - u_x^c + u_x^a - n - u_x^b\right)\left(r + v_f^b\right)}{N}\right]$$

$$= \frac{1}{N} \sum_{m=0}^{N-1} b_m \exp\left(i2\pi \frac{m - u_x^c + u_x^a - n - u_x^b}{N} v_f^b\right)$$

$$\times \sum_{r=0}^{N-1} \exp\left(i2\pi \frac{m - u_x^c + u_x^a - n - u_x^b}{N} r\right)$$

$$= \frac{1}{N} \sum_{m=0}^{N-1} b_m \exp\left(i2\pi \frac{m - u_x^c + u_x^a - n - u_x^b}{N} v_f^b \right)$$

$$\times \frac{\exp\left[i2\pi \left(m - u_x^c + u_x^a - n - u_x^b \right) \right] - 1}{\exp\left(i2\pi \left(\left(m - u_x^c + u_x^a - n - u_x^b \right)/N \right) \right) - 1}$$

$$= \sum_{m=0}^{N-1} \left\{ b_m \exp\left(i2\pi \frac{m - u_x^c + u_x^a - n - u_x^b}{N} v_f^b \right) \right.$$

$$\times \frac{\sin\left[\pi \left(m - u_x^c + u_x^a - n - u_x^b \right) \right]}{N \sin\left(\pi \left(\left(m - u_x^c + u_x^a - n - u_x^b \right)/N \right) \right)}$$

$$\left. \times \exp\left[i\pi \frac{N-1}{N} \left(m - u_x^c + u_x^a - n - u_x^b \right) \right] \right\}$$

$$= \sum_{m=0}^{N-1} b_m \frac{\sin\left[\pi \left(m - u_x^c + u_x^a - n - u_x^b \right) \right]}{N \sin\left(\pi \left(m - u_x^c + u_x^a - n - u_x^b \right)/N \right)}$$

$$\times \exp\left[i\pi \frac{(N-1)\left(m - u_x^c + u_x^a - n - u_x^b \right) + 2 v_f^b}{N} \right].$$

Select $v_f^b = -(N-1)/2$, and obtain

$$\tilde{b}_{n-k} = \sum_{m=0}^{N-1} b_m \frac{\sin\left[\pi \left(m - u_x^c + u_x^a - n - u_x^b \right) \right]}{N \sin\left(\pi \frac{m - u_x^c + u_x^a - n - u_x^b}{N} \right)}$$

$$= \sum_{m=0}^{N-1} b_m \operatorname{sincd}\left[N; \pi \left(n - m + u_x^c - u_x^a + u_x^b \right) \right].$$

Derivation of Equation 4.70

Given signal $\{a_k\}$ and its DFT spectrum $\{\alpha_r\}$, $k, r = 0, 1, \ldots, N-1$:

$$a_k = \frac{1}{\sqrt{N}} \sum_{r=0}^{N-1} \alpha_r \exp\left(-i2\pi \frac{kr}{N} \right)$$

express reversed signal $\{a_{N-k}\}$ through its DFT spectrum:

$$a_{N-k} = \frac{1}{\sqrt{N}} \sum_{r=0}^{N-1} \alpha_r \exp\left(-i2\pi \frac{N-k}{N} r\right) = \frac{1}{\sqrt{N}} \sum_{r=0}^{N-1} \alpha_r \exp\left(i2\pi \frac{k}{N} r\right)$$

$$= \frac{1}{\sqrt{N}} \sum_{r=0}^{N-1} \alpha_{N-r} \exp\left(i2\pi \frac{N-r}{N} k\right) = \frac{1}{\sqrt{N}} \sum_{r=0}^{N-1} \alpha_{N-r} \exp\left(-i2\pi \frac{kr}{N}\right),$$

which implies that the DFT spectrum of the reversed signal is the reversed spectrum $\{\alpha_{N-r}\}$ of the initial signal. Furthermore,

$$a_k^* = \frac{1}{\sqrt{N}} \sum_{r=0}^{N-1} \alpha_r^* \exp\left(i2\pi \frac{kr}{N}\right) = \frac{1}{\sqrt{N}} \sum_{r=0}^{N-1} \alpha_{N-r}^* \exp\left(i2\pi \frac{N-r}{N} k\right)$$

$$= \frac{1}{\sqrt{N}} \sum_{r=0}^{N-1} \alpha_{N-r}^* \exp\left(-i2\pi \frac{kr}{N}\right),$$

which implies that signal complex conjugation converts its DFT spectrum $\{\alpha_r\}$ into complex conjugated and reversed one $\left\{\alpha_{N-r}^*\right\}$.

Derivation of Equations 4.72 and 4.74

Let

$$a_k = \frac{1}{\sqrt{N}} \sum_{r=0}^{N-1} \alpha_r^{(1/2,1/2)} \exp\left[-i2\pi \frac{(k+1/2)(r+1/2)}{N}\right].$$

Then

$$a_{N-1-k} = \frac{1}{\sqrt{N}} \sum_{r=0}^{N-1} \alpha_r^{(1/2,1/2)} \exp\left[-i2\pi \frac{(N-1-k+1/2)(r+1/2)}{N}\right]$$

$$= -\frac{1}{\sqrt{N}} \sum_{r=0}^{N-1} \alpha_r^{(1/2,1/2)} \exp\left[i2\pi \frac{(k+1/2)(r+1/2)}{N}\right]$$

$$= -\frac{1}{\sqrt{N}} \sum_{r=0}^{N-1} \alpha_{N-1-r}^{(1/2,1/2)} \exp\left[i2\pi \frac{(k+1/2)(N-1-r+1/2)}{N}\right]$$

$$= \frac{1}{\sqrt{N}} \sum_{r=0}^{N-1} \alpha_{N-1-r}^{(1/2,1/2)} \exp\left[-i2\pi \frac{(k+1/2)(r+1/2)}{N}\right].$$

which implies that reversing the order of signal samples reverses the order of its SDFT(1/2, 1/2) spectral coefficients.

Accordingly

$$
a_k^* = \frac{1}{\sqrt{N}} \sum_{r=0}^{N-1} \alpha_r^{*(1/2,1/2)} \exp\left[i2\pi \frac{(k+1/2)(r+1/2)}{N} \right]
$$

$$
= \frac{1}{\sqrt{N}} \sum_{r=0}^{N-1} \alpha_{N-1-r}^{*(1/2,1/2)} \exp\left[i2\pi \frac{(k+1/2)(N-1-r+1/2)}{N} \right]
$$

$$
= \frac{1}{\sqrt{N}} \sum_{r=0}^{N-1} \alpha_{N-1-r}^{*(1/2,1/2)} \exp\left[i2\pi(k+1/2) \right] \exp\left[-i2\pi \frac{(k+1/2)(r+1/2)}{N} \right].
$$

$$
= \frac{1}{\sqrt{N}} \sum_{r=0}^{N-1} \left(-\alpha_{N-1-r}^{*(1/2,1/2)} \right) \exp\left[-i2\pi \frac{(k+1/2)(r+1/2)}{N} \right].
$$

Derivation of Equation 4.75

For SDFT(u,0):

$$
a_k = \frac{1}{\sqrt{N}} \sum_{r=0}^{N-1} \alpha_r \exp\left(-i2\pi \frac{k+u}{N} r \right)
$$

$$
a_k^* = \frac{1}{\sqrt{N}} \sum_{r=0}^{N-1} \alpha_r^* \exp\left(i2\pi \frac{k+u}{N} r \right) = \frac{1}{\sqrt{N}} \sum_{r=0}^{N-1} \alpha_{N-r}^* \exp\left[i2\pi \frac{k+u}{N} (N-r) \right]
$$

$$
= \frac{1}{\sqrt{N}} \sum_{r=0}^{N-1} \left[\alpha_{N-r}^* \exp(i2\pi u) \right] \exp\left(-i2\pi \frac{k+u}{N} r \right)
$$

from which Equation 4.75 follows.

Derivation of Equation 4.76

$$
\alpha_r = \frac{1}{\sqrt{N}} \sum_{k=0}^{N-1} \cos\left(2\pi \frac{\rho k}{N} + \varphi \right) \exp\left[i2\pi \frac{(k+u_x)(r+v_f)}{N} \right]
$$

$$
= \frac{\exp\left(i2\pi \dfrac{r+v_f}{N} u_x \right)}{\sqrt{N}} \sum_{k=0}^{N-1} \cos\left(2\pi \frac{\rho k}{N} + \phi \right) \exp\left(i2\pi \frac{r+v_f}{N} k \right)
$$

$$= \frac{\exp\left(i2\pi \dfrac{r+v_f}{N} u_x\right)}{2\sqrt{N}} \sum_{k=0}^{N-1}\left[\exp\left(i2\pi \frac{\rho k}{N}+i\phi\right)+\exp\left(-i2\pi \frac{\rho k}{N}-i\phi\right)\right]$$

$$\times \exp\left(i2\pi \frac{r+v_f}{N} k\right)$$

$$= \frac{\exp\left(i2\pi \dfrac{r+v_f}{N} u_x\right)}{2\sqrt{N}}\left\{\exp(i\phi)\sum_{k=0}^{N-1}\exp\left(i2\pi \frac{\rho k}{N}\right)\exp\left(i2\pi \frac{r+v_f}{N} k\right)\right.$$

$$\left.+\exp(-i\phi)\sum_{k=0}^{N-1}\exp\left(-i2\pi \frac{\rho k}{N}\right)\exp\left(i2\pi \frac{r+v_f}{N} k\right)\right\}$$

$$= \frac{\exp\left(i2\pi \dfrac{r+v_f}{N} u_x\right)}{2\sqrt{N}}\left\{\exp(i\phi)\sum_{k=0}^{N-1}\exp\left(i2\pi \frac{r+\rho+v_f}{N} k\right)\right.$$

$$\left.+\exp(-i\phi)\sum_{k=0}^{N-1}\exp\left(i2\pi \frac{r-\rho+v_f}{N} k\right)\right\}$$

$$= \frac{\exp\left(i2\pi \dfrac{r+v_f}{N} u_x\right)}{2\sqrt{N}}\left[\exp(i\phi)\frac{\exp\left(i2\pi \dfrac{r+\rho+v_f}{N} N\right)-1}{\exp\left(i2\pi \dfrac{r+\rho+v_f}{N}\right)-1}\right.$$

$$\left.+\exp(-i\phi)\frac{\exp\left(i2\pi \dfrac{r-\rho+v_f}{N} N\right)-1}{\exp\left(i2\pi \dfrac{r-\rho+v_f}{N}\right)-1}\right] = \frac{\exp\left(i2\pi \dfrac{r+v_f}{N} u_x\right)}{2\sqrt{N}}$$

$$\times\left\{\frac{\exp[i\pi(r+\rho+v_f)]-\exp[-i\pi(r+\rho+v_f)]}{\exp\left(i\pi \dfrac{r+\rho+v_f}{N}\right)-\exp\left(-i\pi \dfrac{r+\rho+v_f}{N}\right)}\exp\left[i\pi \frac{N-1}{N}(r+\rho+v_f)+i\phi\right]\right.$$

$$\left.+\frac{\exp[i\pi(r-\rho+v_f)]-\exp[-i\pi(r-\rho+v_f)]}{\exp\left(i\pi \dfrac{r-\rho+v_f}{N}\right)-\exp\left(-i\pi \dfrac{r-\rho+v_f}{N}\right)}\exp\left[i\pi \frac{N-1}{N}(r-\rho+v_f)-i\phi\right]\right\}$$

$$
= \frac{\exp\left(i2\pi\dfrac{r+v_f}{N}u_x\right)}{2\sqrt{N}}\left\{\frac{\sin[\pi(r+\rho+v_f)]}{\sin\left(\pi\dfrac{r+\rho+v_f}{N}\right)}\exp\left[i\pi\frac{N-1}{N}(r+\rho+v_f)+i\phi\right]\right.
$$

$$
\left. +\frac{\sin[\pi(r-\rho+v_f)]}{\sin\left(\pi\dfrac{r-\rho+v_f}{N}\right)}\exp\left[i\pi\frac{N-1}{N}(r-\rho+v_f)-i\phi\right]\right\}
$$

$$
= \frac{\sqrt{N}}{2}\exp\left(i2\pi\frac{r+v_f}{N}u_x\right)\left\{\operatorname{sincd}[N;\pi(r+\rho+v_f)]\right.
$$

$$
\times\exp\left\{i\left[\pi\frac{(N-1)(r+\rho+v_f)}{N}+\phi\right]\right\}
$$

$$
\left. +\operatorname{sincd}[\pi(r-\rho+v_f)]\exp\left\{i\left[\pi\frac{(N-1)(r-\rho+v_f)}{N}-\phi\right]\right\}\right\}.
$$

Derivation of Equation 4.85

$$
\tilde{\tilde{a}}_{\tilde{k}} = \frac{1}{\sqrt{LN}}\sum_{\tilde{r}=0}^{LN-1}\left[1-\operatorname{rect}\frac{\tilde{r}-(N+1)/2}{LN-N-1}\right]\alpha_{(r)\bmod N}\exp\left(-i2\pi\frac{\tilde{k}\tilde{r}}{LN}\right)
$$

$$
= \frac{1}{\sqrt{LN}}\left[\sum_{\tilde{r}=0}^{(N-1)/2}\alpha_r\exp\left(-i2\pi\frac{\tilde{k}\tilde{r}}{LN}\right)+\sum_{\tilde{r}=LN-(N-1)/2}^{LN-1}\alpha_{(\tilde{r})\bmod N}\exp\left(-i2\pi\frac{\tilde{k}\tilde{r}}{LN}\right)\right]
$$

$$
= \frac{1}{\sqrt{LN}}\left\{\sum_{\tilde{r}=0}^{(N-1)/2}\left[\frac{1}{\sqrt{N}}\sum_{n=0}^{N-1}a_n\exp\left(i2\pi\frac{n\tilde{r}}{N}\right)\right]\exp\left(-i2\pi\frac{\tilde{k}\tilde{r}}{LN}\right)\right.
$$

$$
\left. +\sum_{\tilde{r}=LN-(N-1)/2}^{LN-1}\left[\frac{1}{\sqrt{N}}\sum_{n=0}^{N-1}a_n\exp\left(i2\pi\frac{n\tilde{r}}{N}\right)\right]\exp\left(-i2\pi\frac{\tilde{k}\tilde{r}}{LN}\right)\right\}
$$

$$
= \frac{1}{N\sqrt{L}}\left\{\sum_{n=0}^{N-1}a_n\left[\sum_{\tilde{r}=0}^{(N-1)/2}\exp\left(-i2\pi\frac{\tilde{k}-Ln}{LN}\tilde{r}\right)\right.\right.
$$

$$
\left.\left. +\sum_{\tilde{r}=LN-(N-1)/2}^{LN-1}\exp\left(-i2\pi\frac{\tilde{k}-Ln}{LN}\tilde{r}\right)\right]\right\}
$$

$$= \frac{1}{N\sqrt{L}} \sum_{n=0}^{N-1} a_n \left\{ \frac{\exp\left[-i2\pi \dfrac{(\tilde{k}-Ln)(N+1)}{2LN}\right]-1}{\exp\left(-i2\pi \dfrac{\tilde{k}-Ln}{LN}\right)-1} \right.$$

$$\left. + \frac{\exp\left[-i2\pi \dfrac{LN(\tilde{k}-Ln)}{LN}\right]-\exp\left[-i2\pi \dfrac{(\tilde{k}-Ln)(LN-(N-1)/2)}{LN}\right]}{\exp\left(-i2\pi \dfrac{\tilde{k}-Ln}{LN}\right)-1} \right\}$$

$$= \frac{1}{N\sqrt{L}} \sum_{n=0}^{N-1} a_n \frac{\exp\left[-i2\pi \dfrac{(\tilde{k}-Ln)(N+1)}{2LN}\right]-\exp\left[-i2\pi \dfrac{(\tilde{k}-Ln)(N-1)}{2LN}\right]}{\exp\left(-i2\pi \dfrac{\tilde{k}-Ln}{LN}\right)-1}$$

$$= \frac{1}{\sqrt{L}} \sum_{n=0}^{N-1} a_n \frac{\sin\left(\pi N \dfrac{\tilde{k}-Ln}{LN}\right)}{N\sin\left(2\pi \dfrac{\tilde{k}-Ln}{LN}\right)} \frac{1}{\sqrt{L}} \sum_{n=0}^{N-1} a_n \, \mathrm{sincd}\left[N;(\tilde{k}-nL)/L\right].$$

Rotated and Scaled DFTs as Digital Convolution

Equation 4.91 can be rewritten as follows:

$$\alpha_{r,s}^{\theta} = \frac{1}{\sigma N} \sum_{k=0}^{N-1} \sum_{l=0}^{N-1} a_{k,l} \exp\left[i2\pi\left(\frac{\tilde{k}\cos\theta + \tilde{l}\sin\theta}{\sigma N}\tilde{r} - \frac{\tilde{k}\sin\theta - \tilde{l}\cos\theta}{\sigma N}\tilde{s}\right)\right]$$

$$= \frac{1}{\sigma N} \sum_{k=0}^{N-1} \sum_{l=0}^{N-1} a_{k,l} \exp\left[i2\pi\left(\frac{\tilde{k}\tilde{r} + \tilde{l}\tilde{s}}{\sigma N}\cos\theta + \frac{\tilde{l}\tilde{r} - \tilde{k}\tilde{s}}{\sigma N}\sin\theta\right)\right].$$

Using in this equation identities

$$2\left(\tilde{k}\tilde{r} + \tilde{l}\tilde{s}\right) = \tilde{r}^2 + \tilde{k}^2 - \tilde{s}^2 - \tilde{l}^2 - \left(\tilde{r} - \tilde{k}\right)^2 + \left(\tilde{s} + \tilde{l}\right)^2 ; 2\left(\tilde{k}\tilde{r} + \tilde{l}\tilde{s}\right)$$

$$= \tilde{r}^2 + \tilde{k}^2 - \tilde{s}^2 - \tilde{l}^2 - \left(\tilde{r} - \tilde{k}\right)^2 + \left(\tilde{s} + \tilde{l}\right)^2 ,$$

obtain

$$\alpha_{r,s}^{\theta} = \frac{1}{\sigma N}\sum_{r=0}^{N-1}\sum_{s=0}^{N-1}a_{k,l}\exp\left[i2\pi\left(\frac{\tilde{k}\tilde{r}+\tilde{l}\tilde{s}}{\sigma N}\cos\theta + \frac{\tilde{l}\tilde{r}-\tilde{k}\tilde{s}}{\sigma N}\sin\theta\right)\right]$$

$$= \frac{1}{\sigma N_a}\sum_{k=0}^{N-1}\sum_{l=0}^{N-1}a_{k,l}\exp\left[i\pi\frac{\tilde{r}^2 + \tilde{k}^2 - \tilde{s}^2 - \tilde{l}^2 - \left(\tilde{r}-\tilde{k}\right)^2 + \left(\tilde{s}+\tilde{l}\right)^2}{\sigma N}\cos\theta\right]$$

$$\times\exp\left[-i2\pi\frac{2\tilde{k}\tilde{l} - 2\tilde{r}\tilde{s} + 2\left(\tilde{r}-\tilde{k}\right)\left(\tilde{s}+\tilde{l}\right)}{\sigma N}\sin\theta\right]$$

$$= \frac{\exp\left[-i\pi\dfrac{\left(\tilde{r}^2 - \tilde{s}^2\right)\cos\theta + 2\tilde{r}\tilde{s}\sin\theta}{\sigma N}\right]}{\sigma N}$$

$$\times\sum_{k=0}^{N-1}\sum_{l=0}^{N-1}\left\{\left(a_{k,l}\exp\left[-i\pi\frac{\left(\tilde{k}^2 - \tilde{l}^2\right)\cos\theta - 2\tilde{k}\tilde{l}\sin\theta}{\sigma N}\right]\right)\right.$$

$$\left.\times\exp\left[i\pi\frac{\left(\tilde{s}+\tilde{l}\right)^2\cos\theta - \left(\tilde{r}-\tilde{k}\right)^2\cos\theta - 2\left(\tilde{r}-\tilde{k}\right)\left(\tilde{s}+\tilde{l}\right)\sin\theta}{\sigma N}\right]\right\},$$

which is a 2D chirp-function-modulated 2D convolution, with output coordinates $\tilde{r}, N - \tilde{s}$, of 2D chirp-function $\exp[-i\pi((\tilde{k}^2 - \tilde{l}^2)\cos\theta - 2\tilde{k}\tilde{l}\sin\theta)/(\sigma N)]$ and 2D chirp-function-modulated signal.

Derivation of Equation 4.93

$$\mathbf{SDFT}_{1/2,0}\{\tilde{a}_k\} = \frac{1}{2N}\sum_{k=0}^{2N-1}\tilde{a}_k\exp\left(i2\pi\frac{k+1/2}{2N}r\right)$$

$$= \frac{1}{\sqrt{2N}}\left\{\sum_{k=0}^{N-1}\tilde{a}_k\exp\left(i2\pi\frac{k+1/2}{2N}r\right) + \sum_{k=N}^{2N-1}\tilde{a}_k\exp\left(i2\pi\frac{k+1/2}{2N}r\right)\right\}$$

$$= \frac{1}{\sqrt{2N}}\left\{\sum_{k=0}^{N-1}a_k\exp\left(i2\pi\frac{k+1/2}{2N}r\right) + \sum_{k=N}^{2N-1}a_{2N-k-1}\exp\left(i2\pi\frac{k+1/2}{2N}r\right)\right\}$$

$$
= \frac{1}{\sqrt{2N}} \left\{ \sum_{k=0}^{N-1} a_k \exp\left(i2\pi \frac{k+1/2}{2N} r \right) + \sum_{k=0}^{N-1} a_k \exp\left(i2\pi \frac{2N-k-1/2}{2N} r \right) \right\}
$$

$$
= \frac{1}{\sqrt{2N}} \left\{ \sum_{k=0}^{N-1} a_k \exp\left(i2\pi \frac{k+1/2}{2N} r \right) + \sum_{k=0}^{N-1} a_k \exp\left(i2\pi \frac{k+1/2}{2N} r \right) \right\}
$$

$$
= \frac{2}{\sqrt{2N}} \sum_{k=0}^{N-1} a_k \cos\left(\pi \frac{k+1/2}{N} r \right).
$$

Derivation of Equation 4.98

$$
\tilde{a}_k = \frac{1}{\sqrt{2N}} \sum_{r=0}^{2N-1} \alpha_r^{DCT} \exp\left(-i2\pi \frac{k+1/2}{2N} r \right)
$$

$$
= \frac{1}{\sqrt{2N}} \left\{ \alpha_0^{DCT} + \sum_{r=1}^{N-1} \alpha_r^{DCT} \exp\left(-i2\pi \frac{k+1/2}{2N} r \right) + \sum_{r=N+1}^{2N-1} \alpha_r^{DCT} \exp\left(-i2\pi \frac{k+1/2}{2N} r \right) \right.
$$

$$
= \frac{1}{\sqrt{2N}} \left\{ \alpha_0^{DCT} + \sum_{r=1}^{N-1} \alpha_r^{DCT} \exp\left(-i2\pi \frac{k+1/2}{2N} r \right) \right.
$$

$$
\left. + \sum_{r=1}^{N-1} \alpha_r^{DCT} \exp\left[-i2\pi \frac{(k+1/2)(2N-r)}{2N} \right] \right]
$$

$$
= \frac{1}{\sqrt{2N}} \left\{ \alpha_0^{DCT} + \sum_{r=1}^{N-1} \alpha_r^{DCT} \exp\left(-i2\pi \frac{k+1/2}{2N} r \right) \right.
$$

$$
\left. + \sum_{r=1}^{N-1} \alpha_r^{DCT} \exp\left(-i2\pi \frac{k+1/2}{2N} r \right) = \frac{1}{\sqrt{2N}} \left(\alpha_0^{DCT} + 2\sum_{r=1}^{N-1} \alpha_r^{DCT} \cos\left(\pi \frac{k+1/2}{N} r \right) \right).
$$

Derivation of Equation 4.104

$$
\frac{1}{i\sqrt{2N}} \sum_{k=0}^{2N-1} \tilde{a}_k \exp\left(i2\pi \frac{k+1/2}{2N} r \right)
$$

$$
= \frac{1}{i\sqrt{2N}} \left\{ \sum_{k=0}^{N-1} \tilde{a}_k \exp\left(i2\pi \frac{k+1/2}{2N} r \right) + \sum_{k=N}^{2N-1} \tilde{a}_k \exp\left(i2\pi \frac{k+1/2}{2N} r \right) \right\}
$$

$$
= \frac{1}{i\sqrt{2N}} \left\{ \sum_{k=0}^{N-1} a_k \exp\left(i2\pi \frac{k+1/2}{2N} r \right) - \sum_{k=N}^{2N-1} a_{2N-k-1} \exp\left(i2\pi \frac{k+1/2}{2N} r \right) \right\}
$$

$$= \frac{1}{i\sqrt{2N}} \left\{ \sum_{k=0}^{N-1} a_k \exp\left(i2\pi \frac{k+1/2}{2N} r \right) - \sum_{k=0}^{N-1} a_k \exp\left(i2\pi \frac{2N-k-1/2}{2N} r \right) \right\}$$

$$= \frac{1}{i\sqrt{2N}} \left\{ \sum_{k=0}^{N-1} a_k \exp\left(i2\pi \frac{k+1/2}{2N} r \right) - \sum_{k=0}^{N-1} a_k \exp\left(-i2\pi \frac{k+1/2}{2N} r \right) \right\}$$

$$= \frac{2}{\sqrt{2N}} \sum_{k=0}^{N-1} a_k \sin\left(\pi \frac{k+1/2}{N} r \right).$$

Derivation of Equation 4.118

$$b_k = \frac{1}{\sqrt{2N}} \sum_{r=0}^{2N-1} \alpha_r^{(DCT)} \tilde{n}_r \exp\left(-i2\pi \frac{k+1/2}{2N} r \right)$$

$$= \frac{1}{\sqrt{2N}} \left\{ \sum_{r=0}^{N-1} \alpha_r^{(DCT)} \tilde{\eta}_r \exp\left(-i2\pi \frac{k+1/2}{2N} r \right) + \sum_{r=N}^{2N-1} \alpha_r^{(DCT)} \tilde{\eta}_r \exp\left(-i2\pi \frac{k+1/2}{2N} r \right) \right\}$$

$$= \frac{1}{\sqrt{2N}} \left\{ \sum_{r=0}^{N-1} \alpha_r^{(DCT)} \tilde{\eta}_r \exp\left(-i2\pi \frac{k+1/2}{2N} r \right) + \sum_{r=1}^{N} \alpha_{2N-r}^{(DCT)} \tilde{\eta}_{2N-r} \exp\left[-i2\pi \frac{(k+1/2)(2N-r)}{2N} \right] \right\}$$

$$= \frac{1}{\sqrt{2N}} \left\{ \sum_{r=0}^{N-1} \alpha_r^{(DCT)} \tilde{\eta}_r \exp\left(-i2\pi \frac{k+1/2}{2N} r \right) \right.$$

$$\left. + \sum_{r=1}^{N} \alpha_{2N-r}^{(DCT)} \tilde{\eta}_{2N-r} \exp[-i2\pi(k+1/2)] \exp\left(i2\pi \frac{k+1/2}{2N} r \right) \right\}$$

$$= \frac{1}{\sqrt{2N}} \left\{ \sum_{r=0}^{N-1} \alpha_r^{(DCT)} \tilde{\eta}_r \exp\left(-i2\pi \frac{k+1/2}{2N} r \right) - \sum_{r=1}^{N} \alpha_{2N-r}^{(DCT)} \tilde{\eta}_{2N-r} \exp\left(i2\pi \frac{k+1/2}{2N} r \right) \right\}.$$

Derivation of Equation 4.124

$$CFR(f) = \frac{1}{\sqrt{N}} \sum_{n=0}^{N-1} \sum_{r=0}^{N-1} \eta_r \exp\left\{ i2\pi \left[\left(f\Delta x - \frac{r}{N} \right) n + \left(f\Delta x u_x^{(s)} - \frac{u_x r}{N} \right) \right] \right\}$$

$$= \frac{1}{\sqrt{N}} \sum_{r=0}^{N-1} \eta_r \exp\left[i2\pi \left(f\Delta x u_x^{(s)} - \frac{u_x r}{N} \right) \right] \sum_{n=0}^{N-1} \exp\left[i2\pi \left(f\Delta x - \frac{r}{N} \right) n \right]$$

$$= \frac{1}{\sqrt{N}} \sum_{r=0}^{N-1} \eta_r \frac{\exp\left[i2\pi N \left(f\Delta x - \frac{r}{N} \right) \right] - 1}{\exp\left[i2\pi \left(f\Delta x - \frac{r}{N} \right) \right] - 1} \exp\left[i2\pi \left(f\Delta x u_x^{(s)} - \frac{u_x r}{N} \right) \right]$$

$$= \sqrt{N} \sum_{r=0}^{N-1} \eta_r \frac{\sin\left[\pi N\left(f\Delta x - \frac{r}{N}\right)\right]}{N \sin\left[\pi\left(f\Delta x - \frac{r}{N}\right)\right]} \exp\left[i\pi(N-1)\left(f\Delta x - \frac{r}{N}\right)\right]$$

$$\times \exp\left[i2\pi\left(f\Delta x u_x^{(s)} - \frac{u_x r}{N}\right)\right]$$

$$= \sqrt{N} \sum_{r=0}^{N-1} \eta_r \frac{\sin\left[\pi N\left(f\Delta x - \frac{r}{N}\right)\right]}{N \sin\left[\pi\left(f\Delta x - \frac{r}{N}\right)\right]} \exp\left[i2\pi\left(u_x^{(s)} + \frac{N-1}{2}\right)f\Delta x\right]$$

$$\times \exp\left[-i2\pi\left(u_x + \frac{N-1}{2}\right)\frac{r}{N}\right].$$

Derivation of Equation 4.149

$$a_k^{(\mu^\pm, w^\pm)} = \frac{1}{\sqrt{N}} \sum_{r=0}^{N-1} \left\{ \frac{1}{\sqrt{N}} \sum_{n=0}^{N-1} a_n \exp\left[-i\pi \frac{(n\mu_+ - r/\mu_+ + w_+)^2}{N}\right] \right\}$$

$$\times \exp\left[i\pi \frac{(k\mu_- - r/\mu_- + w_-)^2}{N}\right]$$

$$= \frac{1}{\sqrt{N}} \sum_{r=0}^{N-1} \left\{ \frac{1}{\sqrt{N}} \sum_{n=0}^{N-1} a_n \exp\left[-i\pi \frac{(n\mu_+ + w_+)^2 + r^2/\mu_+^2 - 2r(n\mu_+ + w_+)/\mu_+}{N}\right] \right\}$$

$$\times \exp\left[i\pi \frac{(k\mu_- + w_-)^2 + r^2/\mu_-^2 - 2r(k\mu_- + w_-)/\mu_-}{N}\right]$$

$$= \frac{1}{N} \exp\left[i\pi \frac{(k\mu_- + w_-)^2}{N}\right] \sum_{n=0}^{N-1} a_n \exp\left[-i\pi \frac{(n\mu_+ + w_+)^2}{N}\right]$$

$$\times \sum_{r=0}^{N-1} \exp\left[i\pi \frac{\left(1/\mu_-^2 - 1/\mu_+^2\right)}{N} r^2\right] \exp\left(-i2\pi \frac{k - n - w_+/\mu_+ + w_-/\mu_-}{N} r\right)$$

$$= \exp\left[i\pi \frac{(k\mu_- + w_-)^2}{N}\right] \sum_{n=0}^{N-1} a_n \exp\left[-i\pi \frac{(n\mu_+ + w_+)^2}{N}\right] \mathrm{frincd}(N; q; k - n - w_\pm).$$

Derivation of Equation 4.183

$$
A_s^{(r)} = \frac{1}{\sqrt{N}} \sum_{k=0}^{N-1} \left[\sum_{n=0}^{N_w-1} w_n a_{k-n+\lfloor N_w/2 \rfloor} \phi_r(n) \right] \exp\left(i2\pi \frac{ks}{N} \right)
$$

$$
= \frac{1}{\sqrt{N}} \sum_{k=0}^{N-1} \left\{ \sum_{n=0}^{N_w-1} w_n \phi_r(n) \left[\frac{1}{\sqrt{N}} \sum_{t=0}^{N-1} A_t \exp\left(-i2\pi \frac{k-n+\lfloor N_w/2 \rfloor}{N} t \right) \right] \right\} \exp\left(i2\pi \frac{ks}{N} \right)
$$

$$
= \frac{1}{N} \sum_{k=0}^{N-1} \left\{ \sum_{n=0}^{N_w-1} w_n \phi_r(n) \left[\sum_{t=0}^{N-1} A_t \exp\left(i2\pi \frac{n-\lfloor N_w/2 \rfloor}{N} t \right) \exp\left(-i2\pi \frac{kt}{N} \right) \right] \right\} \exp\left(i2\pi \frac{ks}{N} \right)
$$

$$
= \frac{1}{N} \sum_{t=0}^{N-1} A_t \left\{ \sum_{n=0}^{N_w-1} w_n \phi_r(n) \left(i2\pi \frac{n-\lfloor N_w/2 \rfloor}{N} t \right) \left[\sum_{k=0}^{N-1} \exp\left(-i2\pi \frac{t-s}{N} k \right) \right] \right\}
$$

$$
= \sqrt{N_w} \sum_{t=0}^{N-1} A_t \Phi_r^{(w)}(t) \left\{ \frac{1}{N} \frac{\exp[-i2\pi(t-s)] - 1}{\exp\left(-i2\pi \frac{t-s}{N} \right) - 1} \right\}
$$

$$
= \sqrt{N_w} \sum_{t=0}^{N-1} A_t \Phi_r^{(w)}(t) \delta(t-s) \propto A_s \Phi_r^{(w)}(s),
$$

where

$$
\Phi_r^{(w)}(t) = \frac{1}{\sqrt{N_w}} \sum_{n=0}^{N_w-1} w_n \phi_r(n) \exp\left(i2\pi \frac{n}{N} t \right) = \frac{1}{\sqrt{N_w}} \sum_{n=0}^{N-1} w_n \phi_r(n) \exp\left(i2\pi \frac{n}{N} t \right)
$$

Exercises

convolution_demo_CRC.m
conv_evenextension_demo_CRC.m
dft_demo_CRC.m
dct_vs_dft_conv_demo_CRC.m

Reference

1. N. Ahmed, T. Natarajan, and K. R. Rao, Discrete cosine transform, *IEEE Trans. Computers*, C-23, 90–93, 1974.

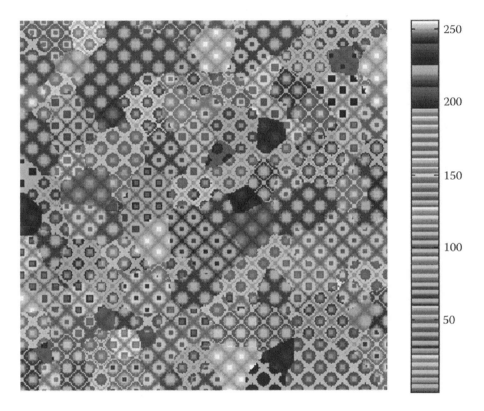

FIGURE 8.35
Pseudocolor representation of the test image of Figure 8.32a, which contains an invisible low-contrast periodical texture. Color bar to the right shows color coding table for image gray levels.

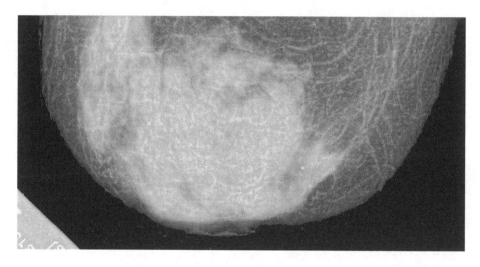

FIGURE 8.36
Colorization of enhanced versions of mammogram shown in Figure 8.33: the lower right version is shown in red, the upper right version is shown in green, and the initial image (upper left) is shown in blue.

5

Digital Image Formation and Computational Imaging

This chapter is devoted to methods of image formation from numerical data, or computational imaging. Apparently, the earliest example of computational imaging is computed tomography. In computed tomography, discrete numerical data of sampling object slice projections obtained under a set of angles are converted into a set of samples of slice image over a regular square sampling grid suited to conventional image displays. This conversion is mainly a matter of image data resampling supplemented with an appropriate transform, DFT, or convolution. We will demonstrate this in Chapter 6. In this chapter, we discuss several other examples of computational imaging: image recovery from sparse nonuniformly sampled data, digital image formation by means of numerical reconstruction of holograms, hybrid digital and analog image formation by means of computer-generated holograms, and, in conclusion, will show that in the conventional imaging scheme, where images are created by optics that focuses radiation into images, optics can, in principle, be replaced by a computer.

Image Recovery from Sparse or Nonuniformly Sampled Data

Formulation of the Task

There are many applications, where, in contrast to the common practice of uniform sampling, sampled data are collected in an irregular manner. Here are a few typical instances:

- Samples are taken not where the regular sampling grid dictates to take them but where it is feasible because of technical or other limitations.
- The pattern of sample disposition is dictated by physical principles of the work of the measuring device (as, e.g., in interferometry or moiré technique, where samples are taken along level lines).
- Sampling device positioning is jittering due to camera or object vibrations or due to other irregularities such as in imaging through a turbulent medium.

- Some samples of the regular sampling grid are lost or unavailable due to losses in communication channels.

Because image display devices as well as computer software for processing sampling data assume using a regular uniform sampling grid, one needs in all these cases to convert irregularly sampled images to regularly sampled ones. Generally, the corresponding regular sampling grid may contain more samples than available because coordinates of positions of available samples might be known with a "subpixel" accuracy, that is, with the accuracy (in units of image size) better than $1/K$, where K is the number of available pixels. Therefore, one can regard available K samples as being sparsely placed at nodes of a denser sampling grid with total amount of nodes $N > K$ determined by the accuracy of specifying positions of the available samples.

In this section, we present a general framework for recovery of discrete signals from a given set of their arbitrarily taken samples. We treat this problem as an approximation task in the assumption that continuous signals are represented in computers by their $K < N$ samples taken at some of, say, N nodes of a regular uniform sampling grid; $K < N$, and it is believed that if all N samples were known, they would be sufficient for representing the continuous signal. The goal of the processing is generating, out of this incomplete set of K samples, a complete set of N signal samples in such a way as to secure the most accurate, in terms of MSE, approximation of the discrete signal, which would be obtained if the continuous signal it is intended to represent were densely sampled at all N positions.

The mathematical foundation of the framework is provided by the discrete sampling theorem for "band-limited" discrete signals that have only few nonzero coefficients in their representation over a certain orthogonal basis. This theorem is introduced in the section "Discrete Sampling Theorem." In the section "Algorithms for Signal Recovery from Sparse Sampled Data," we describe algorithms for signal minimum MSE recovery from such sparse sampled data. In the section "Analysis of Transforms," properties of transforms, which are specifically relevant for signal recovery from sparse data, are analyzed. In the section "Application Examples," we address application issues and illustrate the discrete sampling theorem-based methodology of discrete signal recovery on examples of image superresolution from multiple frames and image recovery from sparse projection data.

Discrete Sampling Theorem

Let $\mathbf{A}_N$ be a vector of N samples $\{a_k\}_{k=0,...,N-1}$ of a discrete signal, Φ_N be an $N \times N$ transform matrix,

$$\Phi_N = \{\phi_r(k)\}_{r=0,1,...,N-1} \tag{5.1}$$

composed of orthonormal basis functions $\phi_r(k)$, and Γ_N be a vector of signal transform coefficients $\{\gamma_r\}_{r=0,\ldots,N-1}$ such that

$$\mathbf{A}_N = \Phi_N \Gamma_N = \left\{ \sum_{r=0}^{N-1} \gamma_r \phi_r(k) \right\}_{k=0,1,\ldots,N-1} . \qquad (5.2)$$

Assume that available are only $K < N$ signal samples $\{a_{\tilde{k}}\}_{\tilde{k} \in \tilde{K}}$, where $\tilde{K}$ is a K-size subset of indices $\{0,1,\ldots, N-1\}$. These available K signal samples define a system of K equations:

$$\left\{ a_{\tilde{k}} = \sum_{r=0}^{N-1} \gamma_r \phi_r(\tilde{k}) \right\}_{\tilde{k} \in \tilde{K}} \qquad (5.3)$$

for K signal transform coefficients $\{\gamma_r\}$ of certain K indices r.

Select a subset $\tilde{R}$ of K transform coefficients indices $\{\tilde{r} \in \tilde{R}\}$ and define a "KofN"-band-limited approximation $\hat{A}_N^{BL}$ to the signal A_N as

$$\hat{A}_N^{BL} = \left\{ \hat{a}_k = \sum_{\tilde{r} \in \tilde{R}} \gamma_{\tilde{r}} \phi_{\tilde{r}}(k) \right\}. \qquad (5.4)$$

Rewrite this equation in a more general form that involves all transform coefficients:

$$\hat{A}_N^{BL} = \left\{ \hat{a}_k = \sum_{r=0}^{N-1} \tilde{\gamma}_r \phi_r(k) \right\} \qquad (5.5)$$

assuming that all transform coefficients with indices $r \notin \tilde{R}$ are set to zero:

$$\tilde{\gamma}_r = \begin{cases} \gamma_r, & r \in \tilde{R} \\ 0, & \text{otherwise.} \end{cases} \qquad (5.6)$$

Then the vector $\tilde{\mathbf{A}}_K$ of available signal samples $\{a_{\tilde{k}}\}$ can be expressed in terms of the basis functions $\{\phi_r(k)\}$ of transform Φ_N as

$$\tilde{\mathbf{A}}_K = \text{KofN}_\Phi \cdot \tilde{\Gamma}_K = \left\{ a_{\tilde{k}} = \sum_{\tilde{r} \in \tilde{R}} \gamma_{\tilde{r}} \phi_{\tilde{r}}(\tilde{k}) \right\}, \qquad (5.7)$$

where $K \times K$ subtransform matrix KofN_Φ is composed of samples $\phi_{\tilde{r}}(\tilde{k})$ of the basis functions with indices $\{\tilde{r} \in \tilde{\mathbf{R}}\}$ for signal sample indices $\tilde{k} \in \tilde{K}$, and $\tilde{\Gamma}_K$ is a vector composed of the corresponding subset $\{\gamma_{\tilde{r}}\}$ of signal nonzero transform coefficients $\{\gamma_r\}$. This subset of the coefficients can be found by means of inverting matrix KofN_Φ as

$$\tilde{\Gamma}_K = \{\tilde{\gamma}_r\} = \mathrm{KofN}_\Phi^{-1} \cdot \tilde{\mathbf{A}}_K, \tag{5.8}$$

provided matrix KofN_Φ^{-1} inverse to the matrix KofN_Φ exists, which, in general, is conditioned, for a specific transform, by positions $\tilde{k} \in \tilde{K}$ of available signal samples and by the selection of the subset $\{\tilde{R}\}$ of transform basis functions that correspond to nonzero transform coefficients.

By virtue of the Parceval's relationship for orthonormal transforms, the band-limited signal $\hat{A}_N^{BL}$ approximates the complete signal A_N with MSE:

$$MSE = \left\| A_N - \hat{A}_N \right\|^2 = \sum_{k=0}^{N-1} \left| a_k - \hat{a}_k \right|^2 = \sum_{r \notin R} \left| \gamma_r \right|^2. \tag{5.9}$$

This error can be minimized by an appropriate selection of the K basis functions of the subtransform KofN_Φ. In order to do so, one must know the energy compaction ordering of basis functions of the transform Φ_N. If, in addition, one knows, for a class of signals, a transform that features the best energy compaction in the smallest number of transform coefficients, one can, by selecting this transform, secure the best minimum MSE band-limited approximation of the signal $\{a_k\}$ for the given subset $\{\tilde{a}_k\}$ of its samples.

In this way, we arrive at the discrete sampling theorem that can be formulated in the following two statements:

Statement 1. For any discrete signal of N samples defined by its $K \le N$ sparse and not necessarily regularly arranged samples, its band-limited, in terms of certain transform Φ_N, approximation defined by Equation 5.5 can be obtained with MSE defined by Equation 5.9 provided positions of the samples secure the existence of the matrix KofN_Φ^{-1} inverse to the subtransform matrix KofN_Φ that corresponds to the band-limitation. The approximation error can be minimized by using a transform with the best energy compaction capability.

Statement 2. Any signal of N samples that is known to have only $K \le N$ nonzero transform coefficients for certain transform Φ_N (Φ_N—transform "band-limited" signal) can be precisely recovered from exactly K samples provided positions of the samples secure the existence of the matrix KofN_Φ^{-1} inverse to the subtransform matrix KofN_Φ that corresponds to the band-limitation.

In this formulation, the discrete sampling theorem is applicable to signals of any dimensionality. It also does not require any assumption regarding compactness of nonzero signal spectral coefficients in the transform domain. The signal dimensionality affects only the formulation of the signal band-limitedness. For 2D images and transforms such as discrete Fourier, discrete cosine, and Walsh transforms, the most simple is compact "low-pass" band-limitedness by a rectangle or circle sector.

Algorithms for Signal Recovery from Sparse Sampled Data

For optimal signal/image band-limited approximation from sampled data, one has to first make a choice of

- The number N of samples to be recovered
- The transform, which promises the best approximation
- The type of band-limitation (e.g., low-pass, band-pass, high-pass, the shape of the figure in the transform domain that is supposed to contain nonzero transform coefficients)

The choice of the transform must be made on the basis of the transform energy compaction capability. As for the type of band-limitation, it is the issue of *a priori* knowledge on the class of images at hand. If the best, for a particular image or set of images, transform and the type of band-limitation are not certain, one can select several transforms and types of band-limitations and then, from obtained approximation results, select the one that has the highest energy. The number N of samples to be recovered is also a matter of *a priori* belief of how many samples of a regular uniform sampling grid would be enough to represent the images to the end user.

Implementation of signal recovery/approximation from sparse nonuniformly sampled data according to Equation 5.8 requires matrix inversion, which is, generally, a very computationally demanding procedure. In applications, in which one can be satisfied with signal reconstruction with a certain limited accuracy, one can apply for the reconstruction a simple iterative reconstruction algorithm of the Gerchberg–Papoulis [1,2] type. The flow diagram of the algorithm is shown in Figure 5.1.

In this algorithm, the initial guess is generated from available sparse signal samples on a dense sampling grid of N samples, supplemented with a guess of the rest of the samples, for which, for instance, zeros, signal mean value or random numbers can be used. Then, at each iteration, the signal is subjected to the selected transform, the obtained transform coefficients are zeroed according to the band-limitation assumption and inverse transformed, after which the next iteration of the restored signal is generated by means of restoring available signal samples in their known positions.

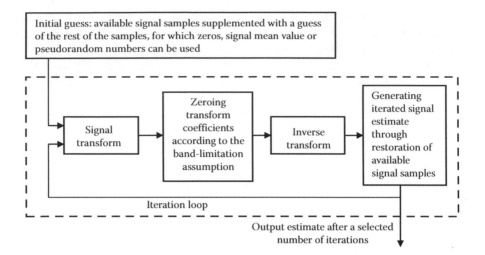

FIGURE 5.1
Flow diagram of the iterative procedure of signal band-limited recovery from sparse samples.

Analysis of Transforms

The applicability of a particular transform for band-limited image approximation depends first of all on whether **KofN** matrix for this transform is invertible, that is, whether available signal samples are compatible with band-limitation type selected for this transform. In what follows, we address the invertibility conditions most widely used in applications DFT, DCT, Walsh, and wavelet transforms.

Discrete Fourier Transform

Consider the $\mathbf{KofN}_{\mathrm{DFT}}^{\mathrm{LP}}$-trimmed DFT_N matrix:

$$\mathbf{KofN}_{\mathrm{DFT}}^{\mathrm{LP}} = \left\{ \exp\left(i2\pi \frac{\tilde{k}\tilde{r}_{LP}}{N} \right) \right\} \tag{5.10}$$

that corresponds to DFT **KofN**-low-pass band-limited signal. Due to complex conjugate symmetry of DFT or real signals, K has to be an odd number, and the set of frequency domain indices of $\mathbf{KofN}_{\mathrm{DFT}}$-low-pass band-limited signals in Equation 5.10 is defined as

$$\tilde{r}_{\mathrm{LP}} \in \tilde{R}_{\mathrm{LP}} = \left\{ [0,1,\ldots,(K-1)/2, N-(K-1)/2,\ldots,N-1] \right\}. \tag{5.11}$$

For such a case, the following theorems hold:

Theorem 5.1

KofN-low-pass DFT band-limited signals of **N** samples with only **K** nonzero low-frequency DFT coefficients can be precisely recovered from exactly **K** of their samples taken in arbitrary positions on a regular uniform sampling grid.

Proof. As it follows from Equations 5.3 through 5.8, the theorem is proven if matrix $\mathbf{KofN}_{DFT}^{LP}$ is invertible. A matrix is invertible if its determinant is nonzero. In order to check whether the determinant of the matrix $\mathbf{KofN}_{DFT}^{LP}$ is nonzero, permute the order of columns of the matrix as follows:

$$\tilde{\tilde{r}} \in \tilde{\tilde{R}} = \{[N - (K-1)/2, \ldots, N-1, 0, 1, \ldots, (K-1)/2]\} \tag{5.12}$$

and obtain the matrix

$$\mathbf{KofN}_{DFT}^{DFTsh} = \left\{\exp\left[i2\pi\frac{\tilde{k}\tilde{r}}{N}\right]\right\} = \left\{\exp\left[i2\pi\frac{N-(K-1)/2}{N}\tilde{k}\right]\delta\left(\tilde{k}-\tilde{\tilde{r}}\right)\right\}$$

$$\times\left\{\exp\left(i2\pi\frac{\tilde{k}\tilde{\tilde{r}}}{N}\right)\right\}, \tag{5.13}$$

where

$$\tilde{\tilde{r}} \in \tilde{\tilde{R}} = \{[0, \ldots, K-1]\}. \tag{5.14}$$

The first matrix $\{\exp[i2\pi((N-(K-1)/2)/N)\tilde{k}]\delta(\tilde{k}-\tilde{\tilde{r}})\}$ in this product of matrices is a diagonal matrix, which is obviously invertible. The second one $\{\exp(i2\pi\tilde{k}\tilde{\tilde{r}}/N)\}$ is a version of Vandermonde matrices, which are also known to have nonzero determinant if, as in our case, its ratios for each row are distinct [3]. As the permutation of the matrix columns does not change the absolute value of its determinant, Equation 5.13 implies that the determinant of $\mathbf{KofN}$-trimmed DFT_N matrix $\mathbf{KofN}_{DFT}^{LP}$ of Equation 5.10 is also nonzero for an arbitrary set $\hat{\mathbf{K}} = \{\hat{k}\}$ of positions of K available signal samples.

For DFT **KofN**-high-pass band-limited signals, for which

$$\mathbf{KofN}_{DFT}^{HP} = \left\{\exp\left(i2\pi\frac{\tilde{k}\tilde{r}_{HP}}{N}\right)\right\}, \tag{5.15}$$

where

$$\tilde{r}_{HP} \in \tilde{R}_{HP} = \{[(N-K+1)/2, (N-K+3)/2, \ldots, (N+K-1)/2]\} \tag{5.16}$$

a similar theorem holds:

Theorem 5.2

KofN-high-pass DFT band-limited signals of **N** samples with only **K** nonzero high-frequency DFT coefficients can be precisely recovered from exactly **K** of their arbitrarily taken samples.

Theorems 5.1 and 5.2 can be extended to a more general case of signal DFT band-limitation, when indices $\{\tilde{r}\}$ of nonzero DFT spectral coefficients form arithmetic progressions with common difference other than one such as, for instance

$$\tilde{r}_{mLP} \in \tilde{R}_{mLP} = \left\{0, m, \ldots, m\frac{(K-1)}{2}, N - m\frac{(K-1)}{2}, \ldots, N - m\frac{(K-1)}{2} + \frac{(K+1)}{2}\right\}.$$

(5.17)

Plots in Figure 5.2a and b illustrate examples of exact reconstruction of a DFT-"band-limited" signal (solid line) by matrix inversion for two cases: (a) all available signal samples are randomly placed within signal support and (b) available signal samples form a compact group.

Note that, as experiments show, the convergence of the iterative algorithm heavily depends on the realization of sample positions and, for some realizations of positions of available samples, it might be quite slow.

Discrete Cosine Transform

As it was shown in the section "Properties of Discrete Fourier Transforms" in Chapter 4, *N*-point DCT of a signal is equivalent to 2*N*-point-shifted discrete Fourier transform (SDFT) with shift parameters (1/2,0) of 2*N* samples signal obtained from the initial one by its mirror reflection. **KofN**-trimmed matrix of SDFT(1/2,0) over 2*N* samples

$$\mathbf{KofN}_{\text{SDFT}} = \left\{\exp\left(i2\pi\frac{(\tilde{k}+1/2)\tilde{r}}{2N}\right)\right\}$$

(5.18)

can be represented as a product

$$\mathbf{KofN}_{\text{SDFT}} = \left\{\exp\left(i2\pi\frac{\tilde{k}\tilde{r}}{2N}\right)\left\{\exp\left(i\pi\frac{\tilde{r}}{2N}\right)\delta(k-r)\right\}\right\}$$

$$= \mathbf{KofN}_{\text{DFT}}\left\{\exp\left(i\pi\frac{\tilde{r}}{2N}\right)\delta(k-r)\right\}$$

(5.19)

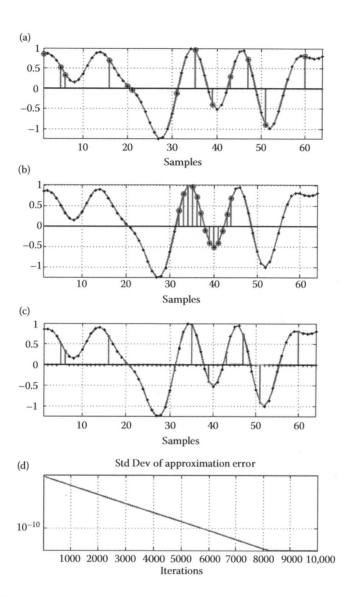

FIGURE 5.2

Restoration of a DFT low-pass band-limited signal by means of matrix inversion for the cases of random (a) and compactly placed signal samples (b) and that by using the iterative algorithm (c). Plot (d) shows standard deviation of signal restoration error as a function of the number of iterations. The experiment was conducted for test signal length of 64 samples and bandwidth of 13 frequency samples (~1/5 of the signal base band). In all plots, the original signal is represented in solid line obtained, for display purposes, by linear interpolation of its samples; available samples are represented by stems and samples reconstructed by matrix inversion are represented by dots.

of a 2N-point DFT matrix and a diagonal matrix $\{\exp(i\pi \tilde{r}/2N)\delta(k - r)\}$. The latter one is invertible and the invertibility of KofN-trimmed DFT_{2N} matrix $KofN_{DFT}$ can be proved, for the above-described band-limitations, in the same way as it has been done above for the DFT case. Therefore, theorems similar to those for DFT hold for DCT as well.

These theorems also hold for 2D DFT and DCT transforms provided band-limitation conditions are separable. The case of nonseparable band-limitation is more involved and requires further study.

We illustrate the above reasoning by some simulation examples. Figures 5.3 and 5.4 illustrate precise restoration from sparse data of images band-limited in DCT domain by a square (separable band-limitation) and by a 90° circle sector (a pie piece, inseparable band-limitation).

The image presented in Figure 5.3a is a 64×64 pixel test image low-pass band-limited in DCT domain by a square of 9×9 samples (Figure 5.3b). It has only $9 \times 9 = 81$ nonzero DCT spectral components out of the 64×64 ones. This image was sampled at 82 "random" positions obtained from the standard MATLAB pseudorandom number generator. One can see from the figure that the iterative algorithm provides quite accurate restoration of the initial image, though precise restoration may require quite a large number of iterations. An important peculiarity of the iterative restoration process is that the convergence of iteration is very nonuniform within the image area. Usually, the restoration error is rapidly becoming very small almost everywhere in the image, and only in some parts, where sample density happens to be low, restoration errors remain to be substantial and decay quite slowly.

Image band-limitation by a square is separable over image coordinates and, as it was shown earlier, it does not impose any limitation on the positions of sparse samples. It is, however, not isotropic. In the case of the isotropic band-limitation in the DCT domain by a circle sector (a pie piece), the situation is quite different. Experiments show that the speed of convergence of the iterative algorithm substantially drops in this case. Many more iterations are needed to make the overall standard deviation of the restoration error low enough, though, again, the restoration error remains to be substantial only in limited areas of the image. The convergence speed of the iterative algorithm in the case of isotropic circle sector band-limitation can be substantially improved if the number of available image samples exceeds the number of nonzero DCT spectral coefficients, that is, if it is redundant from the point of view of the discrete sampling theorem. This is illustrated in Figure 5.4.

The image presented in Figure 5.4a is a 64×64 pixels test image low-pass band-limited in the DCT domain by a circle sector. It has 73 nonzero DCT spectral components out of 64×64, all located within a circle sector shown in white in Figure 5.4b. In distinction from the image of Figure 5.3a, this one was sampled at 93 "random" positions. The redundancy $93/73 = 1.27$ in the number of samples with respect to the number of nonzero spectral coefficients is approximately equal to the ratio of the area of a square to the area of

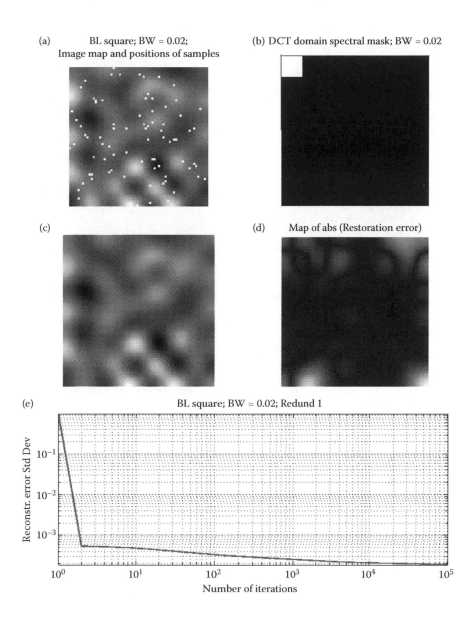

FIGURE 5.3
Recovery of an image band-limited by a square in DCT domain: (a) initial image with 82 "randomly" placed samples in positions shown by white dots; (b) the shape of the image spectrum in DCT domain; (c) image restored by the iterative algorithm after 100,000 iterations; (d) iterative algorithm restoration error (white: large errors; black—small errors); (e) restoration error standard deviation versus the number of iterations for the iterative algorithm.

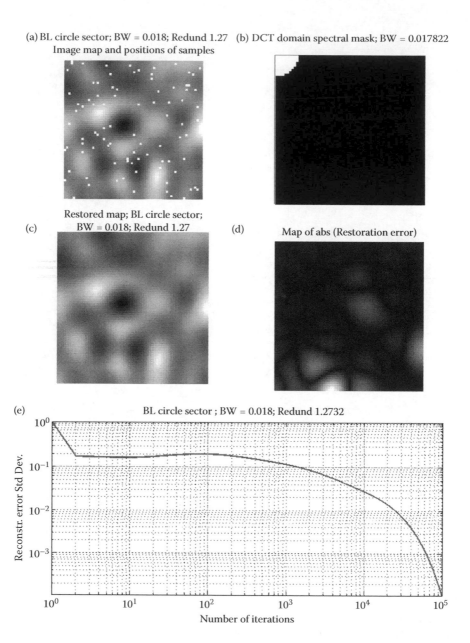

(a) BL circle sector; BW = 0.018; Redund 1.27   (b) DCT domain spectral mask; BW = 0.017822
Image map and positions of samples

Restored map; BL circle sector;
(c)   BW = 0.018; Redund 1.27   (d)   Map of abs (Restoration error)

(e)   BL circle sector ; BW = 0.018; Redund 1.2732

FIGURE 5.4
Recovery of an image band-limited in DCT domain by a circle sector: (a) initial image with 93 "randomly" placed samples in positions shown by white dots; (b) the shape of the image spectrum in DCT domain (upper left corner: lowest frequencies; bottom right corner: highest frequencies); (c) image restored by the iterative algorithm after 100,000 iterations; (d) restoration error found as a difference between initial test and recovered images (white—large errors; black—small errors); (e) restoration error standard deviation versus the number of iterations.

the circle sector inscribed into this square. As one can see from Figure 5.4e, with such a redundancy, iterative restoration converges quite fast and the overall restoration error after 100,000 iterations is comparable to that for the separable band-limitation by a square illustrated in Figure 5.3. Once again, one can see that the convergence of the iterative algorithm is nonuniform over the image and relatively large restoration errors occur only in a small area of the image where the density of available samples happens to be low.

In some applications, there is a natural and substantial redundancy in the number of available image samples with respect to its bandwidth. One of such cases is illustrated in Figure 5.5 by an example of image recovery from its level lines.

A test 256×256 pixels image, shown in Figure 5.5, is band-limited in the DCT domain by a circle sector and contains 302 nonzero spectral coefficients. The image was sampled in 6644 samples on a set of its level lines (eight levels), which resulted in a 22-fold redundancy with respect to the image spectrum 2D bandwidth. As one can see from the figure, such a redundancy accelerated the convergence of the iterative algorithm very substantially and enabled, after a few tens of iterations, restoration with quite low restoration error.

Wavelets and Other Bases

The main peculiarity of wavelet bases is that their basis functions are most naturally ordered in terms of two parameters: scale and position within the scale. Scale index is analogous to the frequency index for DFT. Position index tells only of the shift of the same basis function within the signal extent on each scale. Therefore, band-limitation for DFT translates to scale limitation for wavelets. Limitation in terms of position is trivial: it simply means that some parts of the signal are not relevant. Commonly, discrete wavelets are designed for signals whose length is an integer power of 2 ($N = 2^n$). For such signals, there are $s \leq n$ scales and possible "band-limitations."

The simplest special case of wavelet bases is Haar basis. Signals with $N = 2^n$ samples and with only K lower index nonzero Haar transform (the transform coefficients with indices $\{K,..., N-1\}$ are zero) are ($\tilde{s} = (\lfloor \log_2(K-1) \rfloor + 1)$)- "band-limited," where $\lfloor x \rfloor$ is an integer part of x. Such signals are piece-wise constant within intervals between basis function zero-crossings. The shortest intervals of the signal constancy contain $2^{n-\tilde{s}}$ samples.

As one can see in Figure 5.6, right column plots, where the first eight basis functions of Haar transform are presented, for any two samples located on the same interval, all Haar basis functions on this and lower scales have the same value. Therefore, having more than one sample per constant interval will not change the rank of the matrix **KofN**. The condition for perfect reconstruction is, therefore, to have at least one sample on each of those intervals.

For other wavelets as well as for other bases, a general necessary, sufficient, and easily verified condition for the invertibility of **KofN**-trimmed

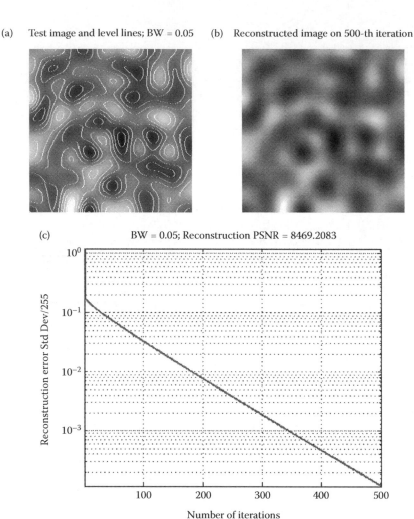

(a)   Test image and level lines; BW = 0.05   (b)   Reconstructed image on 500-th iteration

(c)   BW = 0.05; Reconstruction PSNR = 8469.2083

FIGURE 5.5
Recovery of an image band-limited in DCT domain by a circle sector from its level lines: (a) initial image with level lines (shown by white lines); (b) image restored by the iterative algorithm after 500 iterations; (c) restoration error standard deviation versus the number of iterations.

transform submatrix is yet to be found. Standard linear algebra procedures for determining matrix rank can be used for testing the invertibility of the matrix in each particular case.

For Walsh basis functions, the function index corresponds to the "sequency," or to the number of zero crossings of the basis function. As it was already mentioned in the section "Typical Basis Functions and Classification" in Chapter 3, the sequency carries a certain analogy to the signal frequency. Basis functions ordering according to their sequency, which is a characteristic for Walsh transform, preserves, for many real-life signals, the property

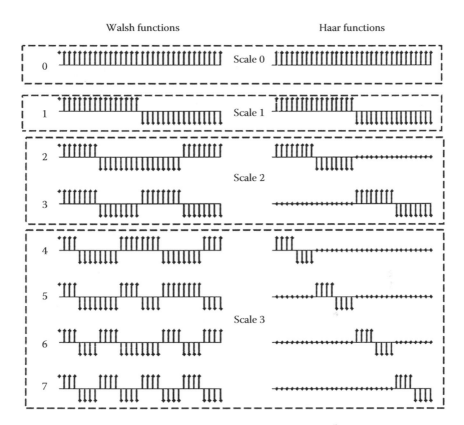

FIGURE 5.6
First eight Walsh and Haar basis functions grouped according the scale parameter.

of more or less regular decaying transform coefficients' energy with their index. Therefore, for Walsh transform, the notion of low-pass band-limited signal approximation, similar to that described for DFT, can be used. On the other hand, as one can see in Figure 5.6, left column plots, Walsh basis functions, similar to Haar basis function, can be characterized by the scale index, which specifies the shortest interval of signal constancy. Signals with $N = 2^n$ samples and band-limitation of K Walsh transform coefficients have shortest intervals of signal constancy of $2^{n-\tilde{s}}$ samples, where $\tilde{s} = \left(\lfloor \log_2(K-1) \rfloor + 1 \right)$. A necessary condition for perfect reconstruction is to have K signal samples taken on different intervals. Unlike the Haar transform case, not all the intervals are needed to be sampled, but only K intervals out of the total number of intervals. For a special case of K equal to a power of 2, there are K intervals, each of which has to be sampled to secure perfect reconstruction. This is the case when the reconstruction condition for the Walsh transform is identical to that for the Haar transform.

Figures 5.7 and 5.8 illustrate recovery of a 1D signal and an image "band-limited" in the Haar transform domain.

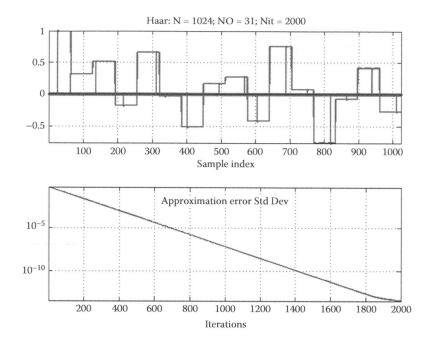

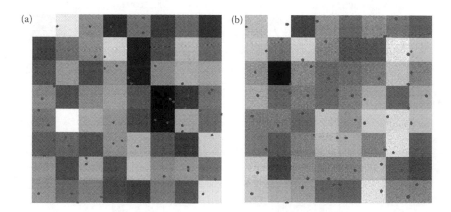

FIGURE 5.7

An illustrative example of restoration of Haar transform 1D band-limited signal using the iterative algorithm and plot of approximation error as function of the number of iterations. Initial signal is shown by solid line linearly interpolated, for display purposes, from its samples; available sparse signal samples are shown by stems, reconstructed samples are shown by light points on top of the initial signal line.

FIGURE 5.8

Two cases of sparse sampling of an image band-limited in Haar transform: (a) not recoverable case; (b) recoverable case. Sample positions are marked with dots; image size is 64×64 pixels; band-limitation is 8×8 samples (scale 3).

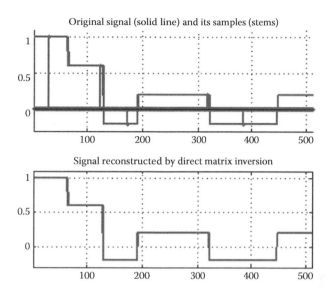

FIGURE 5.9
An example of Walsh band-limited signal recovery by means of matrix inversion.

In Figure 5.8, two examples of image recovery are shown: arrangement of sparse samples, for which signal recovery is possible (Figure 5.8a) and that for which signal is not recoverable from the same number of samples (Figure 5.8b).

An example of the perfect reconstruction of a Walsh transform domain "band-limited" signal of $N = 512$ and of band-limitation $K = 5$ is illustrated in Figure 5.9.

In this example, the resulted $\mathbf{KofN}_{\text{Walsh}}$ matrix is

$$\mathbf{KofN}_{\text{Walsh}}\big|_{K=5} = \begin{bmatrix} 1 & -1 & 1 & -1 & -1 \\ 1 & -1 & -1 & 1 & 1 \\ 1 & 1 & 1 & 1 & 1 \\ 1 & 1 & -1 & -1 & -1 \\ 1 & 1 & 1 & 1 & -1 \end{bmatrix}$$

and its rank equals to 5. One should note that, in this particular example, perfect reconstruction in the Haar transform domain is not possible since one of the shortest intervals of the signal constancy contains no samples.

Selection of Transform for Image Band-Limited Approximation

As it has been already mentioned, the accuracy, in terms of mean square approximation error, of signal band-limited approximation is determined by the energy compaction capability of the transforms. Theoretically (see the section "Optimality of Bases: Karhunen–Loeve and Related Transform"

DCT, Walsh, and Haar transform band-limited images and spectral mask

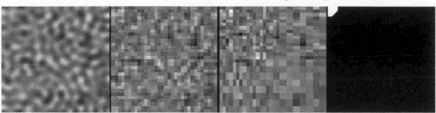

FIGURE 5.10
Examples of images with uniform spectra in (left to right) DCT, Walsh, and Haar transforms
band-limited by a circle sector with radius 0.1 of the base band (the very right image; upper left
corner: lowest spatial frequencies, lower right corner: highest spatial frequencies).

in Chapter 3), optimal transforms are Karhunen–Loeve transform and its
discrete version Hotelling transform. However, the practical value of these
transforms is quite limited because of high computational complexity of generating transform basis functions and computing image representation coefficients. In practice, transforms such as DFT, DCT, Walsh, Haar, and wavelet
transforms that feature fast transform algorithms are preferred for signal
representation, restoration, and analysis because they combine quite good
energy compaction capability and low computational complexity. Their efficiency for different classes of signals might, in principle, be different, and
experimental evaluation of transform energy compaction capability for particular applications is required. In Figures 5.10 and 5.11, some illustrative
data for such an evaluation are presented.

Shown in Figure 5.10, images are examples of pseudorandom images band-
limited in domains of DCT, Walsh, and Haar transform by a circle sector
with radius of 0.1 of the base band size. They are generated by means of
corresponding domain low-pass filtering of arrays of uncorrelated pseudorandom numbers.

Images presented in Figure 5.11 illustrate an approximation error of four test
images approximated by their copies low-pass band-limited by a square of the
size of 0.3 of the base band. A common characteristic feature of these images is
that they represent edges of objects in the images, components of crucial importance for object detection, localization, and recognition (see Chapter 7). One
can also see in this comparison results that the approximation error standard
deviation for DCT band-limitation is the lowest for all of the shown test images.

Application Examples

Image Superresolution from Multiple Differently Sampled Video Frames

One of the potential applications of the above signal recovery technique is
image superresolution from multiple video frames with chaotic pixel displacements due to atmospheric turbulence, camera instability, or similar

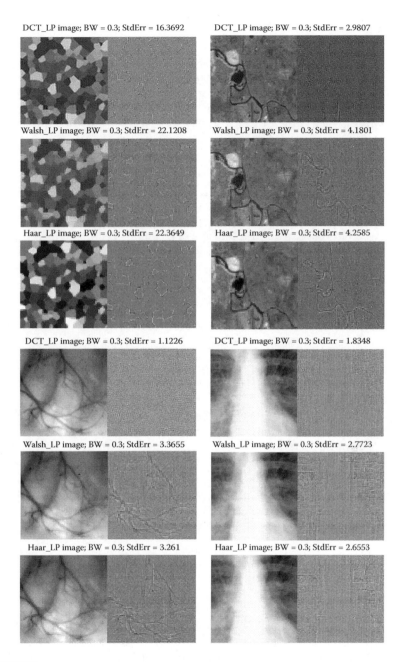

FIGURE 5.11

Low-pass band-limited approximations to test images for band-limitation in a form of a square of 0.3 of the base band in DCT, Walsh, and Haar transform domains and corresponding approximation errors (differences between original and band-limited images; shown to the right of each corresponding image). Error standard deviations (StdErr) are indicated in image headers in units of image dynamic range [0–255].

random factors. Here is, in brief, how this can be done. Using methods of target location (see Chapter 7), one can, by means of registration of pixel neighborhoods in frames of video sequence that contain the same scene, determine, for each image frame in sequences of turbulent video frames and with a subpixel accuracy, pixel displacements caused by the random acquisition factors. Having obtained these data, a synthetic fused image can be generated by placing pixels of the available turbulent video frames in their proper positions on the correspondingly denser sampling grid according to their found displacements. In this process, some pixel positions on the denser sampling grid will remain unoccupied, especially when a limited number of image frames is fused. These missing pixels can then be restored using the above-described iterative band-limited reconstruction algorithm for image recovery from sparse samples. This is illustrated in Figure 5.12 which shows one low-resolution frame (a), an image fused from 15 frames (b), and a result of iterative interpolation (c) achieved after 50 iterations. Image band-limitation was set in this particular experiment twice of the base band of raw low-resolution images.

Image Reconstruction from Sparse Projections in Computed Tomography

The discussed sparse data recovery methods can find application also in tomographic imaging, where it frequently happens that a substantial part of slices, which surrounds the body slice, is known to be an empty field. This means that slice projections (sinograms) are Radon transform band-limited functions. Therefore, whatever number of projections is available, a certain number of additional projections, commensurable, according to the discrete sampling theorem, with the relative size of the slice empty zone, can be obtained and the corresponding resolution increase in the reconstructed images can be achieved using the described iterative band-limited reconstruction algorithm. Figure 5.13 illustrates such superresolution through recovery of missing half of projections achieved using the fact that around 50% of the image area of the head slice is an empty space.

Discrete Sampling Theorem and "Compressive Sensing"

The described methods for image recovery from sparse samples by means of their band-limited approximation in certain transform domain require explicit formulation of the desired band-limitation in the selected transform domain. While for 1D signal, this formulation is quite simple and requires, for most frequently used low-pass band-limitation, specification of only one parameter, signal bandwidth, in 2D case formulation of signal band-limitation requires specification of a 2D shape of signal band-limited spectrum. The simplest shapes, rectangle and circle sector ones, may not be appropriate enough for images at hand. In cases, when the character of spectrum band-limitation is not known or not certain *a priori*, image recovery from sparse

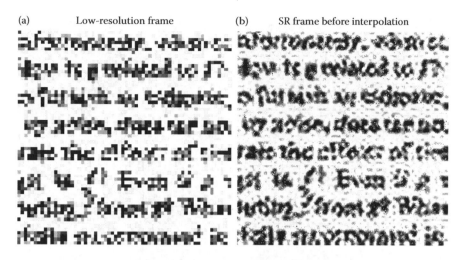

FIGURE 5.12
Iterative image interpolation in the superresolution process: (a) a low-resolution frame; (b) image fused by elastic image registration from 15 frames; (c) a result of iterative interpolation of image (b) after 50 iterations. (Adapted from J. W. Goodman and R. W. Lawrence, *Applied Physics Letters*, 11(3), 77–79, 1967.)

samples can be achieved using an approach, for which the name "compressive sensing" was coined [4].

The compressive sensing approach also assumes obtaining band-limited, in certain selected transform domain, approximation of images but it does not require explicit formulation of the signal band-limitation and achieves signal recovering from an incomplete set of available samples by means of minimization of $L1$ norm $\|\alpha\|_{L1} = \sum_{r=0}^{N-1} |\alpha_r|$ of signal transform coefficients $\{\alpha_r\}$ for the selected transform conditioned by preservation in the recovered

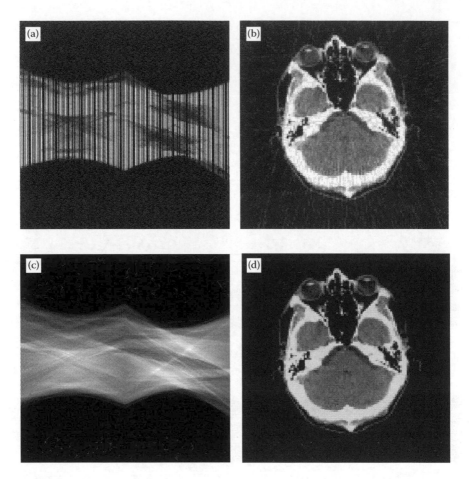

FIGURE 5.13
Superresolution in computed tomography: (a) a set of initial projections supplemented with the same number of presumably lost projections to double the number of projections; initial guesses of the supplemented projections are set to zero (shown in black); (b) image reconstructed from initially available projections; (c) result of iterative restoration of missing projections; (d) image reconstructed from the restored double set of projections.

signal of its available samples. The price one should pay for the uncertainty regarding band-limitation is that the number of required signal samples M must be in this case redundant with respect to the given number K nonzero spectral coefficients: according to [4], $M/K = O(\log N)$.

Minimization of the $L1$ norm replaces minimization of $L0$ norm $\|\alpha\|_{L0} = \sum_{r=0}^{N-1} |\alpha_r|^0$, which would lead to the minimum number of signal transform nonzero coefficients. Justification of this replacement is that for minimization of $L1$ norm, standard numerical optimization algorithms, such as linear programming, are available, whereas minimization of $L0$ norm is computationally problematic, and that the two optimizations are practically equivalent in many applications.

Digital Image Formation by Means of Numerical Reconstruction of Holograms

Introduction

The first experiments in numerical reconstruction of optical holograms date back to the 1960s through the 1970s [5–7]. At that time, scanning devices that could be used for digitizing holograms had low resolution, which required optical magnification of photographic recordings of holograms to fit them to the resolution of scanning devices. Figure 5.14 reproduces the results reported in [5] and [6].

Then, in the 1990s, with the advent of digital photographic cameras, it had become possible to perform direct digitizing optical hologram in the process of hologram recording. In first experiments with CCD cameras [8], holograms were recorded in the Leith–Upatnieks off-axis scheme [9]. Later, *phase-shifting method* for recording holograms was suggested [10] that enabled on-axis hologram recording scheme more efficient in terms of the use of resolution of digital cameras. Since that time, numerous projects in digital recording and numerical reconstruction of optical holograms have been initiated and implemented, especially in the field of optical holographic microscopy.

Principles of Hologram Electronic Recording

As it was already mentioned, two methods of electronic recording of optical holograms using digital photographic cameras are known: "off-axis" and "on-axis" methods.

The schematic diagram of the *"off-axis" method* for electronic recording of holograms is presented in Figure 5.15.

As it is always in holography, the illumination laser beam is split into two beams: object beam, which illuminates the object, and reference beam, which is directed to the recording camera to interfere with the light scattered by the object. Both beams have presumably a plane wavefront. The interfering object and reference wavefronts are sensed and sampled by a photosensor array of the camera placed in a plane called the *hologram plane*. In the off-axis recording, a spatial angle between the reference and object beams is introduced that must exceed the angular size of the object, under which it is seen from the hologram plane.

Consider a mathematical model of "off-axis" recorded holograms. At each point of the hologram plane (f_x, f_y), photosensitive array of the digital camera

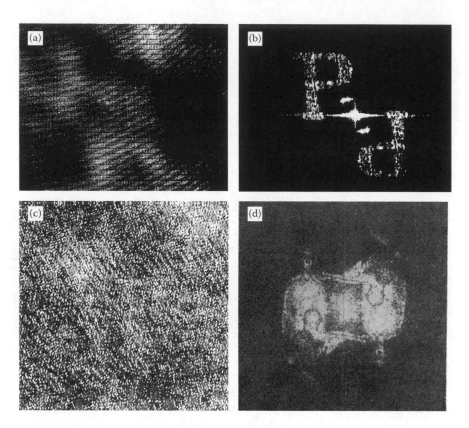

FIGURE 5.14

First holograms numerically reconstructed in computers: (a) a test optical Fourier hologram electronically recorded using vidicon TV camera to 256 × 256 pixels quantized to 8 gray levels; (b) image numerically reconstructed from this hologram on a computer PDP-6; (c) an optical photographically recorded Fourier hologram optically magnified with magnification factor 20 and scanned to 512 × 512 pixels using electromechanical scanner with resolution 0.2 mm and quantized to 64 gray levels in a logarithmic scale; (d) two conjugate images reconstructed from this hologram on a computer Minsk-22 (http://www.computer-museum.ru/english/minsk0.htm). (Images (a) and (b) are adapted from J. W. Goodman, R. W. Lawrence, *Applied Physics Letters*, 11(3), 77–79, 1967; images (c) and (d) are adapted from L. P. Yaroslavskii, N. S. Merzlyakov, *Methods of Digital Holography*, Consultance Bureau, NY, 1980. (English translation from Russian, In: *Methods of Digital Holography*, Editors L. P. Yaroslavskii, N. S. Merzlyakov, Moscow, Izdatel'stvo Nauka, 1977. 192 p.))

measures samples of intensity of the sum of the reference beam $R(f_x, f_y)$ and the object beam $\alpha_{obj}(f_x, f_y)$ reflected or transmitted by the object:

$$H(f_x, f_y) = \left|\alpha_{obj}(f_x, f_y) + R(f_x, f_y)\right|^2 = \alpha_{obj}(f_x, f_y)R^*(f_x, f_y)$$

$$+ \alpha_{obj}^*(f_x, f_y)R(f_x, f_y) + \left|\alpha_{obj}(f_x, f_y)\right|^2 + \left|R(f_x, f_y)\right|^2, \quad (5.20)$$

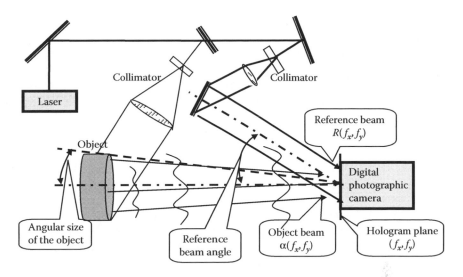

FIGURE 5.15
Schematic diagram of the "off-axis" method for electronic hologram recording.

where asterisk denotes complex conjugation. Numerical reconstruction of the hologram consists in applying to samples of the recorded hologram a discrete transform that implements wave propagation from the hologram plane back to object.

Equation 5.20 can be modified to a form:

$$H(f_x, f_y) = \left|\alpha_{\text{obj}}(f_x, f_y)\right|\left|R(f_x, f_y)\right| \exp\left[i(\theta_{\text{obj}}(f_x, f_y) - \theta_{\text{Rb}}(f_x, f_y))\right]$$

$$+ \left|\alpha_{\text{obj}}(f_x, f_y)\right|\left|R(f_x, f_y)\right| \exp\left[i(\theta_{\text{obj}}(f_x, f_y) - \theta_{\text{Rb}}(f_x, f_y))\right] + \left|\alpha(f_x, f_y)\right|^2$$

$$+ \left|R(f_x, f_y)\right|^2 = +2\left|\alpha(f_x, f_y)\right|\left|R(f_x, f_y)\right| \cos\left[\theta_{\text{Obj}}(f_x, f_y) - \theta_{\text{Rb}}(f_x, f_y)\right],$$

(5.21)

where $\alpha(f_x, f_y) = \left|\alpha(f_x, f_y)\right| \exp(i\theta_{\text{obj}})$ and $R(f_x, f_y) = \left|R(f_x, f_y)\right| \exp(i\theta_{\text{Rb}})$ denote complex amplitudes (amplitudes and phases) of the object and reference beams, respectively, at point (f_x, f_y) of the hologram plane. Equation 5.21 explicitly shows that the recorded hologram signal contains a term with amplitude- and phase-modulated spatial carrier (the last term in Equation 5.21) that carries information on the object beam. In the holograms shown in Figure 5.14a and c, one can clearly see periodical patterns formed by this spatial carrier. It is also seen in the results of hologram reconstruction shown in Figure 5.13b and d that reconstructed images contain a bright spot in the center of reconstructed images. This spot, the so-called *zero-order diffraction term*, is generated by the first two terms in Equation 5.21, while the second term, the spatial carrier one, produces two reconstructed images, direct and conjugated ones. The angular distance between centers of these images is determined by the reference beam angle, which must exceed the image angular size in order to prevent overlapping direct and conjugate images. Because of this, the highest spatial frequency of the recorded hologram is at least twice as that of the object wavefront. Therefore, the hologram recording device must have at least twice as much resolution cells for recording and sampling holograms compared to that required for recording of only the *mathematical hologram* (the first term on the right part of Equation 5.20), which will reconstruct the object image without the conjugate image.

The schematic diagram of the *"on-axis" method* for electronic recording of holograms is shown in Figure 5.16. In this method, also known as *phase-shifting holography* [10], object and reference beams are colinear and several exposures of holograms of the object are carried out with shifting, at each exposure, the phase of the reference beam plane wavefront. In order to compute from the recording results the mathematical hologram, they are combined in the computer in a certain way, which depends on the number of exposures and phase shifts of the reference beam in each exposure.

In what follows, we show that at least three exposures are required in this method. Let $\theta_{Rb}^{(k)}$ be a phase shift of the reference beam in the k-th hologram exposure, $k = 1,\ldots,K$. Then

$$H_k(f_x, f_y) = \left| \alpha(f_x, f_y) + \left| R(f_x, f_y) \right| \exp\left(i\theta_{Rb}^{(k)} \right) \right|^2 = \left| \alpha(f_x, f_y) \right|^2 + \left| R(f_x, f_y) \right|^2$$

$$+ \alpha(f_x, f_y) \left| R(f_x, f_y) \right| \exp\left(-i\theta_{Rb}^{(k)} \right) + \alpha^*(f_x, f_y) \left| R(f_x, f_y) \right| \exp\left(i\theta_{Rb}^{(k)} \right)$$

$$(5.22)$$

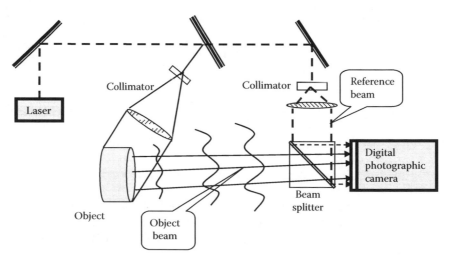

FIGURE 5.16
Schematic diagram of the "on-axis" method for electronic recording of holograms.

is a hologram recorded in the k-th exposure. For separating the mathematical hologram term containing $\alpha(f_x, f_y)$, K recorded at each exposure holograms $\{H_k\}$ are summed up with the same phase shifts as those used in their recording:

$$\bar{H} = \frac{1}{K}\sum_{k=1}^{K} H_n \exp\left(i\theta_{Rb}^{(k)}\right) = \alpha(f_x, f_y)\left|R(f_x, f_y)\right| + \alpha^*(f_x, f_y)\left|R(f_x, f_y)\right|$$

$$\times \sum_{k=1}^{K} \exp\left(i2\theta_{Rb}^{(k)}\right) + \left[\left|\alpha(f_x, f_y)\right|^2 + \left|R(f_x, f_y)\right|\right]\sum_{k=1}^{K}\exp\left(i\theta_{Rb}^{(k)}\right).$$

(5.23)

In this equation, the first term is the record of the wavefront propagated from the object times intensity of the plane reference beam, which is assumed to be a constant; other terms are interfering terms that should be eliminated. In order to secure this, phases $\left\{\theta_{Rb}^{(k)}\right\}$ must be solutions of equations:

$$\begin{cases} \displaystyle\sum_{n=1}^{N} \exp\left(i\theta_{Rb,n}\right) = 0 \\ \displaystyle\sum_{n=1}^{N} \exp\left(i2\theta_{Rb,n}\right) = 0 \end{cases}.$$

(5.24)

It is convenient to use phase shifts $\{\theta_{Rb,n}\}$ that form an arithmetic progression

$$\theta_{Rb,n} = (k-1)\theta_0, \quad k = 1, \ldots, K.$$

(5.25)

Then Equations 5.24 is converted to equations

$$\begin{cases} \displaystyle\sum_{k=1}^{K} \exp\left(i\theta_{Rb}^{(k)}\right) = \sum_{k=1}^{K}\exp\left[i(k-1)\theta_0\right] = \dfrac{\exp\left(iK\theta_0\right)-1}{\exp\left(i\theta_0\right)-1} = 0 \\ \displaystyle\sum_{n=1}^{N} \exp\left(i2\theta_{Rb,n}\right) = \sum_{k=0}^{K-1}\exp\left[i2(k-1)\theta_0\right] = \dfrac{\exp\left(i2K\theta_0\right)-1}{\exp\left(i2\theta_0\right)-1} = 0 \end{cases}$$

(5.26)

The solution of Equations 5.26 is $\theta_0 = (2\pi/K)$ for any integer $N \geq 3$. Note that although for $K = 2$ ($\theta_0 = \pi$) the first equation holds, the second equation does not:s

$$\frac{\exp\left(i2\theta_0\right)-1}{\exp\left(i\theta_0\right)-1}\frac{\exp\left(i2\theta_0\right)+1}{\exp\left(i\theta_0\right)+1} = \frac{\exp\left(i2\theta_0\right)-1}{\exp\left(i2\theta_0\right)-1}\left[\exp\left(i2\theta_0\right)+1\right] = \exp(i2\pi)+1 = 2.$$

(5.27)

Consider an example. In four exposures ($K = 4$, $\theta_0 = (\pi/2)$), we will have from Equation 5.22

$$H_1 = |\alpha|^2 + |R|^2 + \alpha R^* + \alpha^* R; \quad H_2 = |\alpha|^2 + |R|^2 - i\alpha R^* + i\alpha^* R;$$

$$H_3 = |\alpha|^2 + |R|^2 - \alpha R^* + \alpha^* R; \quad H_4 = |\alpha|^2 + |R|^2 + i\alpha R^* - i\alpha^* R. \tag{5.28}$$

Then the mathematical hologram can be computed as

$$\alpha R = \left(H_1 + iH_2 - H_3 - iH_4 \right)/4. \tag{5.29}$$

To this hologram, a discrete transform that implements wave propagation from the hologram plane back to object is to be applied for reconstruction the object wavefront.

Numerical Algorithms for Hologram Reconstruction

As it is shown in the section "Imaging in Transform Domain and Diffraction Integrals," for holograms recorded in far diffraction zone (*Fourier holograms*), wavefronts in object and hologram planes are linked through integral Fourier transform. Therefore, for reconstruction of Fourier holograms, DFT introduced in the section "Discrete Representation of Fourier Integral Transform" are used in their FFT algorithmic implementation. Thanks to FFTs, the computational complexity of reconstruction of Fourier hologram of N samples is of the order $O(\log_2 N)$ per sample. Note that in cases of reconstruction of holograms of the same object recorded with different wavelength, such as in case of color holograms, reconstructed images will be scaled according to the wavelength. In order to compensate for the scaling and obtain images of the same size for different wavelengths, SDFT (the section "Discrete Fourier Transforms") can be used. Images shown in Figure 5.14b and c are examples of reconstruction of holograms recorded in the far diffraction zone.

For holograms recorded in near diffraction zone, wavefronts in object and hologram planes are linked through Fresnel integral transform and, for especially small distances between the object and the hologram plane, by Kirchhoff's integral (the section "Imaging in Transform Domain and Diffraction Integrals"). Correspondingly, for reconstruction of such holograms, discrete Fresnel transforms listed in Table 4.1 and DKT computed via FFT (Equations 4.169 and 4.170) should be used depending on the value of the focusing parameter $\mu^2 = \lambda Z/N\Delta f^2 = \lambda ZN/S_H^2$, which connects the wavelength of the object illumination λ, distance between the object and hologram planes Z, the number of hologram samples N, the pitch of the hologram recording camera Δf, and the hologram physical size $S_H = \Delta f N$ (see Section "Imaging in Transform Domain and Diffraction Integrals").

Specifically, canonical discrete Fresnel transform (CDFrT) implemented through shifted DFT and its versions described in the section "Canonical Discrete Fresnel Transform and Its Versions" provide aliasing-free reconstruction for distances that are large enough to secure that $\mu^2 \geq 1$, while the convolutional discrete Fresnel transform and discrete angular spectrum propagation transform secure aliasing-free reconstruction for closer distances when $\mu^2 \leq 1$. When $\mu^2 = 1$, all these transforms produce identical results. These two applicability zones are sketched in Figure 5.17.

Figure 5.18 illustrates the reconstruction of an off-axis near zone hologram using discrete Fresnel and focal plane invariant discrete Fresnel transforms (see also the MATLAB demo program FocPlaneVar_Invar_reconstr_illustr_ CRC.m provided in Exercises). One can see from the figure how focal plane invariant discrete Fresnel transform secures a stable position of the reconstructed image for different object distances.

Figure 5.19 generated by the MATLAB program fourier_vers_conv_demo_ CRC.m provided in Exercises illustrates using canonical DFrT and convolution DFrT for numerical reconstruction at different distances of an on-axis near zone hologram recorded in four exposures by the phase-shifting method. One can see the appearance of aliasing artifacts, when canonical DFrT is used for image reconstruction out of the zone of its applicability $\mu^2 \geq 1$, and that at distance, which corresponds to $\mu^2 = 1$, both transforms produce identical results.

The central images in each row in the figure are reconstructed using canonical DFrT from a hologram cropped to μ^2-th fraction of its size. As one can see in the figure, this hologram cropping eliminates reconstruction artifacts for the canonical DFrT reconstruction.

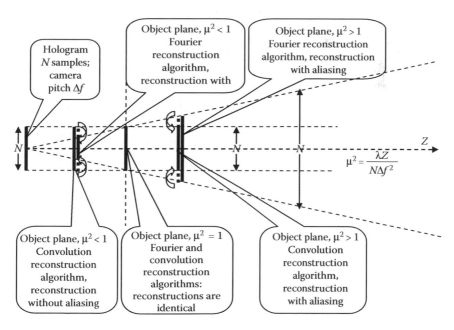

FIGURE 5.17
Zones of applicability of different versions of discrete Fresnel transform.

Focal plane variant (left) and invariant (right) Fresnel hologram reconstruction; $\mu^2 = 0.44$

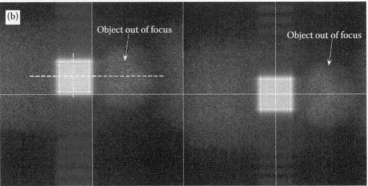

Focal plane variant (left) and invariant (right) Fresnel hologram reconstruction; $\mu^2 = 0.36$

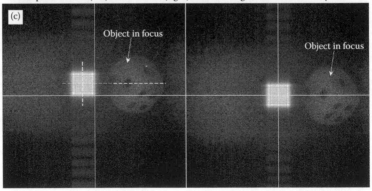

FIGURE 5.18
Reconstruction of an off-axis optical near zone hologram (a) on two different depth using canonical discrete Fresnel transform (left images (b), (c)) and focal plane invariant discrete Fresnel transform (right images (b), (c)). One can clearly see that the displacement of the center of symmetry of the reconstructed image is observed, when canonical DFrT is used for reconstruction on different depths, and that the reconstructed image position is kept stable, when the focal plane invariant discrete Fresnel transform is used. Bright square spots in the images are zero-order diffraction images from dc-component of the hologram. They represent amplitude of the discrete frincd-function, 1D cross sections of which are shown in Figures 4.13 and 4.14.

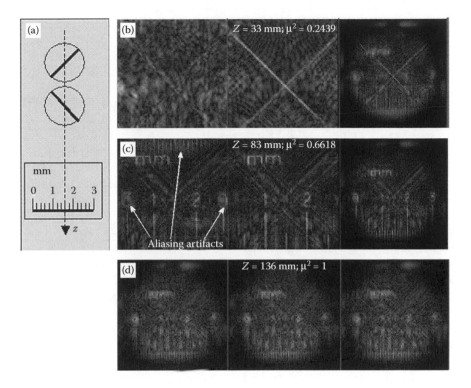

FIGURE 5.19
Comparison of reconstruction of near diffraction zone test hologram of an object (a) that consists of two hairs and a ruler placed at different distances. (b)–(d) In each row, that show images reconstructed for different value of the focusing parameter μ^2, the left image is reconstructed using canonical DFrT, the right image is reconstructed using convolutional DFrT, and the central image is reconstructed using canonical DFrT from the hologram cropped to μ^2-th fraction of its size.

Hologram Pre- and Postprocessing

In the section "Principles of Hologram Electronic Recording," we described idealized models of hologram recording. In reality, optical set-ups and hologram recording devices always have certain deficiencies, which impair the quality of the reconstructed object amplitude and phase. For compensating of this impairment, the preprocessing of the hologram before applying to them, reconstruction algorithms and postprocessing of the reconstruction results are usually required.

Hologram preprocessing is aimed at correcting distortions of hologram in the process of recording. One of the major sources of hologram and reconstructed wavefront distortions is imperfectness of the reference beam. Ideally, the reference beam used for recording holograms must have perfectly planar wavefront, or at least, the shape of the reference wavefront, that is, its amplitude and phase as functions of coordinates in the hologram plane must be precisely known. In reality, reference beams are never precisely planar due to various technical limitations in constructing and adjustment of optical set-ups for hologram recording. Not precisely planar reference beams cause amplitude and phase modulation of the object wavefront, which results in aberrations in the reconstruction of object amplitude and phase profiles. These aberrations can be "inverse filtered" (see Chapter 8), if the amplitude and phase profiles of the reference beam are known. One method to achieve this is recording, in the same optical set-up but without the object, of an additional reference hologram. The principle of the "inverse filtering" is as follows. Having at hand two holograms, one recorded with the object and another one recorded, using the same reference beam, without the object, one can back propagate to the object plane wavefronts reconstructed from both holograms. Back propagation of the result of reconstruction of the initial hologram will reconstruct a product $\alpha_{obj}(f_x, f_y)R^*(f_x, f_y)$ of the object wavefront and reference beam wavefront described by the first term in Equation 5.20. Back propagation of the result of reconstruction of the reference hologram will reconstruct wavefront $R(f_x, f_y)$ of the reference beam at the hologram plane. Dividing the former by the complex conjugate to the latter can obtain pure object wavefront $\alpha_{obj}(f_x, f_y)$, which will reconstruct amplitude and phase profiles of the object, which will be free from aberration caused by a nonplanar reference beam. Of course, this method cannot allow for aberrations caused by mechanical vibrations or other instabilities of the optical set-up and additional hologram rectification might be required. This issue is currently a subject of intensive research, and interested readers are referred to relevant publications.

Yet another example of hologram pre-processing is eliminating zero-order diffraction term that is characteristic for reconstruction of off-axis holograms. This can be done by means of one or another methods of hologram high-pass pre-filtering (see also Chapter 8). One of simple methods is described in [7].

Among other sources of hologram distortion, one can mention imperfections of hologram recording devices that frequently introduce nonlinear distortions and random interferences into digitized holograms. Some particular examples of preprocessing of holograms for removing periodical noise and correcting nonlinear distortions and recording noise in holograms are given in the sections "MMSE-Optimal Linear Filters for Image Restoration" and "Correcting Image Gray-Scale Nonlinear Distortions" in Chapter 8.

Reconstructed wavefront postprocessing is aimed at further correcting distortions of hologram and other distortions, such as speckle noise, that may appear in the reconstruction results. The basics of methods for perfecting results of hologram reconstruction are discussed in Chapter 8.

Point Spread Functions of Numerical Reconstruction of Holograms
General Formulation

In numerical reconstruction of holograms, samples of the object wavefront are reconstructed out of samples of its recorded hologram using the discrete diffraction transforms. This process can be treated as sampling the object wavefront by a sampling system that consists of the Hologram recording and sampling device and a computer, in which the object wavefront samples are numerically reconstructed from samples of the hologram.

Signal sampling is a linear transformation. As such, it is fully specified by its PSF, which establishes a link between an object signal $a(x)$ and its samples $\{a_k\}$:

$$a_k = \int_X a(x)PSF(x,k)\,dx. \tag{5.30}$$

According to the sampling theory (see the section "Image Sampling"), for a given sampling interval Δx, the PSF of the ideal sampling device is a sinc-function:

$$PSF(x,k) = \text{sinc}\left[\pi(x - k\Delta x)/\Delta x\right] = \frac{\sin\left[\pi(x - k\Delta x)/\Delta x\right]}{\pi(x - k\Delta x)/\Delta x}. \tag{5.31}$$

In this section, we consider how PSFs of different reconstruction algorithms depend on algorithm parameters and on physical parameters of holograms and their sampling devices. For the sake of simplicity, we will consider 1D holograms and transforms. Corresponding 2D results are straightforward in the conventional assumption of the separability of sampling and transforms.

Let, in numerical reconstruction of holograms, samples $\{a_k\}$ of the object wavefront be obtained through a transformation

$$a_k = \sum_{r=0}^{N-1} \alpha_r DRK(r,k) \tag{5.32}$$

of available hologram samples $\{\alpha_r\}$ with a certain discrete reconstruction kernel $DRK\,(r, k)$, which corresponds to the type of the hologram. Also let samples $\{\alpha_r\}$ of hologram $\alpha(f)$ that are measured by a hologram recording and sampling device be

$$\alpha_r = \int_{-\infty}^{\infty} \alpha(f)\phi_f^{(s)}(f - \tilde{r}\Delta f)\,df, \tag{5.33}$$

where $\left\{\phi_f^{(s)}(.)\right\}$ is a PSF of the hologram sampling device, Δf is a hologram sampling interval, $\tilde{r} = r + v^{(s)}$, r is an integer index of hologram samples, and v is a shift, in units of the hologram sampling interval, of the hologram sampling grid with respect to the hologram coordinate system; these sampling parameters are analogous to those illustrated, for signal and its Fourier spectrum sampling, in Figure 4.4.

The hologram signal $\alpha(f)$ is linked with object wavefront $a(x)$ through a diffraction integral

$$\alpha(f) = \int_{-\infty}^{\infty} a(x)WPK(x,f)dx, \tag{5.34}$$

where $WPK(x,f)$ is a wave propagation kernel. Therefore, one can rewrite Equation 5.33 as

$$\alpha_r = \int_{-\infty}^{\infty}\left[\int_{-\infty}^{\infty} a(x)WPK(x,f)\,dx\right]\varphi_f^{(s)}\left(f - \tilde{r}\Delta f\right)df$$

$$= \int_{-\infty}^{\infty} a(x)\,dx \int_{-\infty}^{\infty} WPK(x,f)\varphi_f^{(s)}\left(f - \tilde{r}\Delta f\right)df. \tag{5.35}$$

Now insert Equation 5.35 into Equation 5.32 and establish a link between the object wavefront $a(x)$ and its samples $\{a_k\}$ reconstructed from the sampled hologram:

$$a_k = \sum_{r=0}^{N-1}\left[\int_{-\infty}^{\infty} a(x)dx \int_{-\infty}^{\infty} WPK(x,f)\phi_f^{(s)}\left(f - \tilde{r}\Delta f\right)df\right]DRK(r,k)$$

$$= \int_{-\infty}^{\infty} a(x)dx\left[\int_{-\infty}^{\infty} WPK(x,f)df\sum_{r=0}^{N-1}DRK(r,k)\phi_f^{(s)}\left(f - \tilde{r}\Delta f\right)\right] = \int_{-\infty}^{\infty} a(x)OPSF(x,k)dx.$$

$$\tag{5.36}$$

where function

$$OPSF(x,k) = \int_{-\infty}^{\infty} WPK(x,f)df\sum_{r=0}^{N-1}DRK(r,k)\phi_f^{(s)}\left(f - \tilde{r}\Delta f\right) \tag{5.37}$$

can be treated as an overall point spread function (OPSF) of numerical reconstruction of holograms using discrete reconstruction kernel $DRK(r,k)$. As one can see from Equation 5.37, OPSF depends on all factors involved in the process of sampling and reconstruction of holograms: wave propagation kernel $WPK\,(.,\,.)$, discrete reconstruction kernel $DRK\,(.,\,.)$, PSF of the hologram sampling device $\phi_f^{(s)}(.)$, and sampling interval Δf.

For further analysis, it is convenient to replace the PSF of the hologram sampling device by its Fourier transform, or its frequency response $\Phi_f^{(s)}(.)$:

$$\phi_f^{(s)}(f - \tilde{r}\Delta f) = \int_{-\infty}^{\infty} \Phi_f^{(s)}(\xi)\exp\left[i2\pi(f - \tilde{r}\Delta f)\xi\right]d\xi$$

$$= \int_{-\infty}^{\infty} \Phi_f^{(s)}(\xi)\exp(-i2\pi\tilde{r}\Delta f\xi)\exp(i2\pi f\xi)\,d\xi. \qquad (5.38)$$

Then obtain

$$\alpha_r = \int_{-\infty}^{\infty}\left[\int_{-\infty}^{\infty} a(x)WPK(x,f)dx\right]\phi_f^{(s)}(f - \tilde{r}\Delta f)df$$

$$= \int_{-\infty}^{\infty} a(x)dx \int_{-\infty}^{\infty} WPK(x,f)\phi_f^{(s)}(f - \tilde{r}\Delta f)df. \qquad (5.39)$$

Introduce function

$$\overline{PSF}(x,\xi;k) = \left[\int_{-\infty}^{\infty} WPK(x,f)\exp(i2\pi f\xi)df\right]\left[\sum_{r=0}^{N-1} DRK(r,k)\exp(-i2\pi\tilde{r}\Delta f\xi)\right]$$

$$= \overline{WPK}(x,\xi) \cdot \overline{DRK}(\xi,k), \qquad (5.40)$$

where

$$\overline{WPK}(x,\xi) = \int_{-\infty}^{\infty} WPK(x,f)\exp(i2\pi f\xi)\,df \qquad (5.41)$$

is the Fourier transform of the wave propagation kernel $WPK(.,\,.)$ and

$$\overline{DRK}(\xi,k) = \sum_{r=0}^{N-1} DRK(r,k)\exp(-i2\pi r\Delta f\xi) \qquad (5.42)$$

is a Fourier series expansion with samples of the discrete reconstruction kernel $DRK(r, k)$ as expansion coefficients.

The function $\overline{PSF}(x,\xi;k)$ does not depend on physical parameters of the hologram sampling device. It depends solely on wave propagation and discrete reconstruction kernels. We will call this function *PSF of sampled hologram reconstruction*. OPSF of the numerical reconstruction of holograms $OPSF(x, k)$ and PSF of sampled hologram reconstruction $\overline{PSF}(x,\xi;k)$ are linked through the integral transform

$$OPSF(x,k) = \int\limits_{-\infty}^{\infty} \Phi_f^{(s)}(\xi)\overline{PSF}(x,\xi;k)\,d\xi \tag{5.43}$$

with frequency response of the hologram sampling device as a transform kernel.

Point Spread Function of Numerical Reconstruction of Holograms Recorded in Far Diffraction Zone (Fourier Holograms)

Consider the PSF of numerical reconstruction of Fourier holograms. For far diffraction zone, wave propagation kernel is that of the integral Fourier transform:

$$WPK(x,f) = \exp\left(-i2\pi\frac{xf}{\lambda Z}\right), \tag{5.44}$$

where λ is object illumination wavelength and Z is the distance between object and hologram planes. Its Fourier transform $\overline{WPK}(x,\xi)$ is a delta-function:

$$\overline{WPK}(x,\xi) = \int\limits_{-\infty}^{\infty} WPK(x,f)\exp(i2\pi f\xi)\,df = \int\limits_{-\infty}^{\infty} \exp\left[-i2\pi f\left(\frac{x}{\lambda Z} - \xi\right)\right]df$$

$$= \delta\left(\frac{x}{\lambda Z} - \xi\right). \tag{5.45}$$

Assume that shifted DFT with discrete reconstruction kernel

$$DRK(r,k) = \exp\left[i2\pi\frac{(k+u)(r+v)}{N}\right] \tag{5.46}$$

defined in the section "Discrete Fourier Transforms" in Chapter 4 is used for the numerical reconstruction of Fourier holograms. As shown in Appendix, Fourier series expansion over this discrete reconstruction kernel is

$$
\overline{DRK}(\xi,k) = \sum_{r=0}^{N-1} DRK(r,k)\exp\left(-i2\pi\tilde{r}\Delta f\xi\right)
$$

$$
= \sum_{r=0}^{N-1} \exp\left[[i2\pi]\frac{(k+u)(r+v)}{N}\right]\exp\left[-i2\pi\left(r+v^{(s)}\right)\Delta f\xi\right]
$$

$$
= \exp\left[-i2\pi\left(v^{(s)}+\frac{N-1}{2}\right)\Delta f\xi\right]\exp\left\{i2\pi\frac{(k+u)}{N}\left(v+\frac{N-1}{2}\right)\right\}
$$

$$
\times \frac{\sin\left[\pi N\left(\Delta f\xi-\frac{k+u}{N}\right)\right]}{\sin\left[\pi\left(\Delta f\xi-\frac{k+u}{N}\right)\right]}. \tag{5.47}
$$

In order to eliminate pure phase exponential multiplicands in Equation 5.47, one can choose shift parameters $V^{(s)}$ of sampling and v of the reconstruction transform as

$$
v^{(s)} = v = -\frac{N-1}{2}. \tag{5.48}
$$

With these shift parameters

$$
\overline{DRK}(\xi,k) = \frac{\sin\left[\pi\left(\Delta f\xi-\frac{k+u}{N}\right)N\right]}{\sin\left[\pi\left(\Delta f\xi-\frac{k+u}{N}\right)\right]} = N\,\mathrm{sincd}\left[N;\pi\left(\Delta f\xi-\frac{k+u}{N}\right)N\right]
$$

$$
\tag{5.49}
$$

and

$$
\overline{PSF}(x,\xi;k) = \overline{WPK}(x,\xi)\overline{DRK}(\xi,k)N\,\mathrm{sincd}\left[N;\pi\left(\Delta f\xi-\frac{k+u}{N}\right)N\right]\delta\left(\frac{x}{\lambda Z}-\xi\right).
$$

$$
\tag{5.50}
$$

Then obtain finally that the OPSF of numerical reconstruction of Fourier hologram is

$$OPSF(x,k) = \int_{-\infty}^{\infty} \Phi_f^{(s)}(\xi)\overline{PSF}(x,\xi;k)\,d\xi$$

$$= N\operatorname{sincd}\left[N;\pi\left(\frac{\Delta fx}{\lambda Z} - \frac{k+u}{N}\right)N\right]\int_{-\infty}^{\infty}\Phi_f^{(s)}(\xi)\delta\left(\frac{x}{\lambda Z} - \xi\right)d\xi$$

$$= N\operatorname{sincd}\left[N;\pi\left(\frac{\Delta fx}{\lambda Z} - \frac{k+u}{N}\right)N\right]\Phi_f^{(s)}\left(\frac{x}{\lambda Z}\right)$$

$$= N\operatorname{sincd}[N;\pi(x - (k+u)\Delta x)/\Delta x]\Phi_f^{(s)}\left(\frac{x}{\lambda Z}\right), \qquad (5.51)$$

where

$$\Delta x = \lambda Z/N\Delta f = \lambda Z/S_H \qquad (5.52)$$

is an efficient object wavefront sampling interval and

$$S_H = N\Delta f \qquad (5.53)$$

is the physical size of the sampled hologram.

Formula 5.51 has a clear physical interpretation illustrated in Figure 5.20.

The figure shows two multiplicands defining OPSFs of numerical reconstruction of holograms digitally recorded in the far diffraction zone: discrete sinc-function (solid line) and frequency responses of the ideal sampling device (shown by dots) and of digital cameras (dotted lines) for two different *camera fill-factors*, or the ratio of the size of camera-sensitive elements to the interpixel distance (*camera pitch*).

As it was mentioned, the PSF of the ideal signal sampling device is a sinc-function and its frequency response is a rect-function. Provided the hologram sampling device is such an ideal sampler with frequency response

$$\Phi_f^{(s)}\left(\frac{x}{\lambda Z}\right) = \operatorname{rect}\left(\frac{x + \lambda z/2\Delta f}{\Delta f/\lambda Z}\right) = \begin{cases} 1, & -\lambda Z/2\Delta f \le x \le \lambda Z/2\Delta f \\ 0, & \text{otherwise,} \end{cases} \qquad (5.54)$$

the OPSF of numerical reconstruction of the Fourier hologram is a discrete sinc-function, and object wavefront samples are measured within the spatial interval defined by the spread $[-\lambda Z/2\Delta f \le x \le \lambda Z/2\Delta f]$ of the hologram sampling

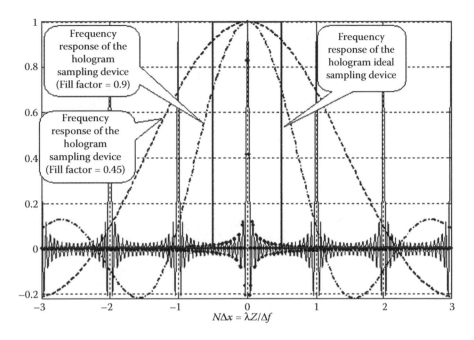

FIGURE 5.20
Components of OPSF of numerical reconstruction of holograms digitally recorded in far diffraction zone. Thin gray solid line represents the discrete sinc-function. Bold dots are samples of the continuous sinc-function, the ideal sampling PSF, to compare it with the discrete sinc-function. Bold rectangle is the frequency response of the hologram ideal sampling device. Dotted lines are frequency responses of real hologram sampling devices, digital cameras with fill factors 0.9 and 0.45.

device frequency response. Therefore, with the ideal hologram sampler, numerical reconstruction of Fourier hologram is almost ideal object wavefront sampling, as the discrete sinc-function approximates continuous sinc-function within the interval $-\lambda Z/2\Delta f \le x \le \lambda Z/2\Delta f$ relatively close, the approximation being the closer, the larger is the number of hologram samples N.

In reality, hologram sampling devices are, of course, not ideal samplers, and their frequency responses are not rectangular functions. They are rather not uniform within the basic object extent interval $-\lambda Z/2\Delta f \le x \le \lambda Z/2\Delta f$ and decay not abruptly outside this interval but quite gradually. As a consequence, each of the object samples is a combination of the sample, measured by the main lobe of the discrete sinc-function within the basic object extend interval and samples collected by other lobes of the discrete sinc-function outside the basic interval, which may cause aliasing artifacts. It is one source of the measurement errors. In particular, for diffusely reflecting objects, it may result in an additional speckle noise in the reconstructed object image. One can avoid this distortion if, in the process of making object hologram, the object is illuminated strictly only within the basic interval $[-\lambda Z/2\Delta f \le x \le \lambda Z/2\Delta f]$ defined by the hologram sampling interval (camera pitch).

The second source of reconstruction errors is associated with nonuniformity of the hologram sampler frequency response within the basic interval. These errors can be compensated by multiplying the reconstruction results by a function inverse to the frequency response of the hologram sampler $\Phi_f^{(s)}(x/\lambda Z)$.

One can also see from Equations 5.51 and 5.52 that the resolving power of numerical reconstruction of Fourier hologram is determined by the distance Δx between zeros of the discrete sinc-function, which is equal to $\lambda Z/N\Delta f = \lambda Z/S_H$. Due to this finite resolving power, one can also expect, for diffuse objects, a certain amount of speckle noise in the reconstructed image (see the section "Speckle Noise Model").

Point Spread Function of Numerical Reconstruction of Holograms Recorded in Near Diffraction Zone (Fresnel Holograms)

For near diffraction zone, wave propagation kernel is, to the accuracy of an irrelevant constant multiplicand

$$WPK(x, f) = \exp\left[i\pi \frac{(x - f)^2}{\lambda Z} \right]. \tag{5.55}$$

Its Fourier transform is

$$\overline{WPK}(x, \xi) = \int_{-\infty}^{\infty} \exp\left[i\pi \frac{(x - f)^2}{\lambda Z} \right] \exp(i2\pi f \xi) df$$

$$= \exp(i2\pi x \xi) \int_{-\infty}^{\infty} \exp\left(i\pi \frac{p^2}{\lambda Z} \right) \exp(-i2\pi p \xi) df \tag{5.56}$$

or, with an account of the relationship between chirp-function and its Fourier spectrum (Equation 4.156),

$$\overline{WPK}(x, \xi) = \exp(i2\pi x \xi) \exp(-i\pi \lambda Z \xi^2) \tag{5.57}$$

again to the accuracy of an irrelevant constant multiplicand omitted. In what follows, we will separately consider the PSF for hologram reconstruction using two complementing discrete representations of Fresnel integral transform, canonical and convolutional discrete Fresnel transforms described in the sections "Canonical Discrete Fresnel Transform and Its Versions" and "Convolutional Discrete Fresnel and Angular Spectrum Propagation

Transforms." For simplicity's sake, zero shifts of sampling grids will be assumed in both hologram and object wavefront domains.

Fourier Reconstruction Algorithm

In the *Fourier reconstruction algorithm,* implemented with canonical DFrT, discrete reconstruction kernel is, with zero shifts of sampling grids in hologram and object wavefront domains

$$DRK(r,k) = \left[\exp\left(-i\pi\frac{k^2\mu^2}{N}\right)\exp\left(i2\pi\frac{kr}{N}\right)\right]\exp\left(-i\pi\frac{r^2}{\mu^2 N}\right). \qquad (5.58)$$

For this kernel, the Fourier series expansion defined by Equation 5.42 is

$$\overline{DRK}(\xi,k) = \exp\left(-i\pi\frac{k^2\mu^2}{N}\right)\sum_{r=0}^{N-1}\exp\left(-i\pi\frac{r^2}{\mu^2 N}\right)\exp\left(i2\pi\frac{kr}{N}\right)\exp\left(-i2\pi r\Delta f^{(s)}\xi\right)$$

$$= \exp\left(-i\pi\frac{k^2\mu^2}{N}\right)\sum_{r=0}^{N-1}\exp\left(-i\pi\frac{r^2}{\mu^2 N}\right)\exp\left[i2\pi r\left(\frac{k}{N}-\Delta f^{(s)}\xi\right)\right]$$

$$= \exp\left(-i\pi\frac{k^2\mu^2}{N}\right)\text{frincd}\left(N;-1/\mu^2;\Delta f\xi-\frac{k}{N}\right). \qquad (5.59)$$

Then, obtain that the PSF of sampled hologram reconstruction is

$$\overline{PSF}(x,\xi;k) = \overline{WPK}(x,\xi)\cdot\overline{DRK}(\xi,k)$$

$$\propto \exp(i2\pi x\xi)\exp(-i\pi\lambda Z\xi^2)\exp\left(-i\pi\frac{k^2\mu^2}{N}\right)\text{frincd}\left(N;-1/\mu^2;\Delta f\xi-\frac{k}{N}\right)$$

$$\qquad (5.60)$$

and that the OPSF of numerical reconstruction of Fresnel holograms with Fourier reconstruction algorithm is

$$OPSF(x,k) = \exp\left(-i\pi\frac{k^2\mu^2}{N}\right)\int_{-\infty}^{\infty}\Phi_f^{(s)}(\xi)\,\exp(-i\pi\lambda Z\xi^2)$$

$$\times\,\text{frincd}\left(N;-1/\mu^2;\Delta f\xi-\frac{k}{N}\right)\exp(i2\pi x\xi)\,d\xi. \qquad (5.61)$$

Equation 5.60 is much more involved for an analytical treatment than the corresponding equation for OPSF of numerical reconstruction of Fourier holograms and, in general, requires numerical methods for analysis. In order to facilitate its interpretation, rewrite Equation 5.60 using Equation 5.58 for $\overline{DRK}(\xi, k)$:

$$
OPSF(x,k) = \exp\left[i\pi \left(\frac{x^2}{\lambda Z} - \frac{k^2 \mu^2}{N} \right) \right] \int_{-\infty}^{\infty} \Phi_f^{(s)}(\xi) \sum_{r=0}^{N-1} \exp\left(-i\pi \frac{r^2}{\mu^2 N} \right)
$$

$$
\times \exp\left[i2\pi r \left(\frac{k}{N} - \Delta f \xi \right) \right] \exp\left[-i\pi \frac{(x - \lambda Z \xi)^2}{\lambda Z} \right] \exp\left[i\pi \left(\frac{x^2}{\lambda Z} - \frac{k^2 \mu^2}{N} \right) \right]
$$

$$
\times \sum_{r=0}^{N-1} \exp\left(-i\pi \frac{r^2}{\mu^2 N} \right) \exp\left(i2\pi \frac{kr}{N} \right) \int_{-\infty}^{\infty} \Phi_f^{(s)}(\xi) \exp\left[-i\pi \frac{(x - \lambda Z \xi)^2}{\lambda Z} \right]
$$

$$
\times \exp(-i2\pi r \Delta f \xi) d\xi. \tag{5.62}
$$

Assume now that the frequency response of the hologram sampling device $\Phi_f^{(s)}(\xi)$ is a constant, which is equivalent to the assumption that its PSF is a delta-function. Practically, this means that the hologram recording photographic camera is assumed to have a very small fill-factor. In this simplifying assumption, obtain (see Appendix)

$$
OPSF(x,k) = \exp\left[i\pi \left(\frac{x^2}{\lambda Z} - \frac{k^2 \mu^2}{N} \right) \right] \text{frincd}\left(N; \frac{1}{\mu^2} - \frac{N \Delta f^2}{\lambda Z}; \frac{\Delta f x}{\lambda Z} - \frac{k}{N} \right). \tag{5.63}
$$

As one can see from this equation, OPSF of numerical reconstruction of Fresnel holograms recorded using cameras with very small fill-factor is just proportional to the frincd-function defined in the section "Invertibility of Discrete Fresnel Transforms and frincd-Function" (Equations 4.154 and Figures 4.14 and 4.15). Obviously, the major interest represents a special case of "in focus" reconstruction, when

$$
\mu^2 = \frac{\lambda Z}{N \Delta f^2}. \tag{5.64}
$$

In this case, numerical reconstruction PSF is the discrete sinc-function

$$OPSF(x,k) = \exp\left[i\frac{\lambda Z}{\Delta f^2}\left(\frac{\Delta f^2 x^2}{\lambda^2 Z^2} - \frac{k^2}{N^2}\right)\right]\text{sincd}\left[N;\pi\left(\frac{\Delta f\, x}{\lambda Z} - \frac{k}{N}\right)N\right]$$

$$= \exp\left[i\frac{\lambda Z}{\Delta f^2}\left(\frac{\Delta f^2 x^2}{\lambda^2 Z^2} - \frac{k^2}{N^2}\right)\right]\text{sincd}\left[N;\pi(x - k\Delta x)/\Delta x\right]. \tag{5.65}$$

where $\Delta x = \lambda Z/N\Delta f = \lambda Z/S_H$. As one can see, "in focus" reconstruction OPSF is essentially the same as that for numerical reconstruction of Fourier holograms (Equation 5.51) for the same assumption regarding the hologram sampling device. It has the same resolving power and provides aliasing free object reconstruction within the interval $S_o = \lambda Z/\Delta f$.

One can establish a link between the size of this interval and the value $\mu^2 = \lambda Z/N\Delta f^2 = N\lambda Z/S_H^2$ of the focusing parameter required for the reconstruction:

$$S_o = \lambda Z/\Delta f^{(s)} = \lambda Z N/S_H = \mu^2 S_H. \tag{5.66}$$

From this relationship, it follows that aliasing free reconstruction of the object from a hologram recorded on a distance defined by the focusing parameter μ^2 is possible if the object size does not exceed the value $\mu^2\, S_H$. Therefore, for $\mu^2 < 1$, allowed object size should be less than the hologram size, otherwise the aliasing caused by the periodicity of the discrete sinc-function will appear. This is, in particular, what Figures 5.17 and 5.19 illustrate.

Convolution Reconstruction Algorithm

In the *convolutional reconstruction algorithm*, convolutional discrete Fresnel transform is used for hologram reconstruction, and discrete reconstruction kernel is, with zero shifts of sampling grids in hologram and object wavefront domains,

$$DRK(r,k) = \text{frincd}\left(N;\mu^2;k - r\right) = \frac{1}{N}\sum_{s=0}^{N-1}\exp\left(i\pi\frac{\mu^2 s^2}{N}\right)\exp\left[-i2\pi\frac{(k-r)s}{N}\right].$$
$$\tag{5.67}$$

The Fourier series expansion defined by Equation 5.42 for this kernel is

$$\overline{DRK}(\xi,k) = \frac{1}{N}\sum_{r=0}^{N-1}\sum_{s=0}^{N-1}\exp\left(i\pi\frac{\mu^2 s^2}{N}\right)\exp\left[-i2\pi\frac{(k-r)s}{N}\right]\exp(-i2\pi r\Delta f\xi)$$

$$= \frac{1}{N}\sum_{s=0}^{N-1}\exp\left(i\pi\frac{\mu^2 s^2}{N}\right)\exp\left(-i2\pi\frac{ks}{N}\right)\sum_{r=0}^{N-1}\exp\left[i2\pi r\left(\frac{s}{N} - \Delta f\xi\right)\right].$$

$$\tag{5.68}$$

Then, obtain that the PSF of sampled hologram reconstruction is (see Appendix)

$$\overline{PSF}(x,\xi;k) = \overline{WPK}(x,\xi) \cdot \overline{DRK}(\xi,k) \propto \frac{\exp(i2\pi x\xi)\exp(-i\pi\lambda Z\xi^2)}{N}$$

$$\times \sum_{s=0}^{N-1} \exp\left(i\pi \frac{\mu^2 s^2}{N}\right)\exp\left(-i2\pi \frac{ks}{N}\right)\sum_{r=0}^{N-1}\exp\left[i2\pi r\left(\frac{s}{N} - \Delta f\xi\right)\right] \quad (5.69)$$

and that OPSF of numerical reconstruction of Fresnel holograms with Fourier reconstruction algorithm is

$$OPSF(x,k) \propto \int_{-\infty}^{\infty} \Phi_f^{(s)}(\xi)\exp(-i\pi\lambda Z\xi^2)\exp(-i2\pi x\xi)d\xi$$

$$\times \sum_{s=0}^{N-1} \exp\left(i\pi \frac{\mu^2 s^2}{N}\right)\exp\left(-i2\pi \frac{ks}{N}\right)\text{sincd}\left[N;\pi\left(s - N\Delta f\xi\right)\right]. \quad (5.70)$$

At least one important property can be immediately seen from Equation 5.70, that of periodicity of the OPSF over object sample index k with a period N. As, by the definition of the convolutional Fresnel transform, sampling interval Δx in the object plane is identical to the hologram sampling interval Δf, this periodicity of the PSF implies that object wavefront is reconstructed within the physical interval $N\Delta x = N\Delta f = S_H$, where S_H is the physical size of the hologram. Further detailed analysis is not feasible without bringing in numerical methods. Some results of such a numerical analysis are illustrated in Figure 5.21.

The top and middle plots in the figure reveal, in particular, that, though object sampling interval in the convolution method is set to be equal to the hologram sampling interval, the resolving power of the method is still defined by the same fundamental value $\lambda Z/S_H$ as that of the Fourier reconstruction algorithm and of the Fourier reconstruction algorithm for Fourier holograms. One can clearly see this when one compares the width of the main lobe of the PSF in Figure 5.21, the top plot, with the distance between vertical ticks, which indicate object sampling positions, and from observing in the middle plot in Figure 5.21, three times widening of the width of the main lobe of PSF that corresponds to the object-to-hologram distance parameter $\mu^2 = 0.45$ (PSF15) with respect to that for $\mu^2 = 0.15$ (PSF5). The bottom plot in Figure 5.21 shows a result of reconstruction of nine point sources placed uniformly within the object size. The plot vividly demonstrates that the hologram sampling device point function acts very similar to its action in the case of the Fourier reconstruction algorithm and of reconstruction of Fourier holograms: it modulates the reconstruction result with a function

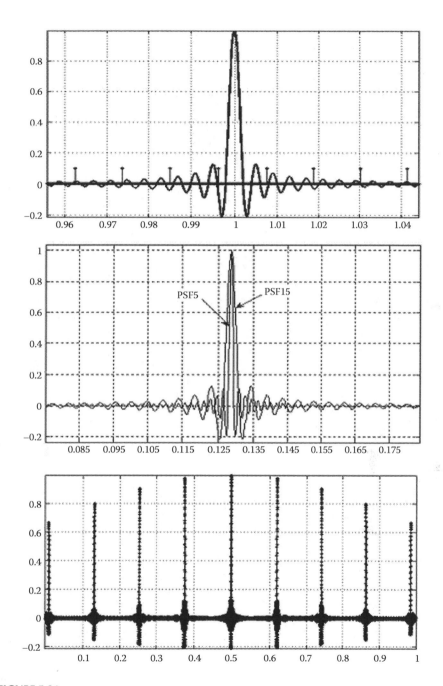

FIGURE 5.21
Point spread functions of the convolution algorithm: top—central lobe of OPSF shown along with boundaries of object sampling interval indicated by vertical ticks; middle—central lobes of OPSF of reconstructions for two distances between object and hologram (PSF5 and PSF15); bottom—reconstruction result for 9-point sources placed uniformly within object area.

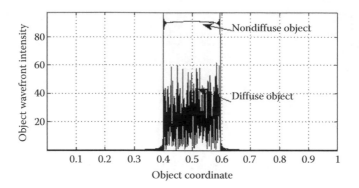

FIGURE 5.22
Results of computer simulation of reconstruction, by the convolution algorithm, of a Fresnel hologram of nondiffuse and diffuse objects with intensity profiles in form of a rectangular impulse that demonstrate appearance of heavy speckle noise for the diffuse object.

close to its Fourier transform, the frequency response of the hologram sampling device.

The finite width of the reconstruction algorithm PSFs not only limits the reconstruction resolving power, but also causes speckle noise in the reconstruction of diffuse objects for the same reason that was mentioned above in the discussion of OPSF of reconstruction of Fourier holograms. This phenomenon is illustrated in Figure 5.22, which compares reconstructions of intensity of nondiffuse and diffuse objects with intensity profiles in the form of a rectangular impulse.

Computer-Generated Display Holography

3D Imaging and Computer-Generated Holography

The inventions in holography by E. Leith and Yu. Denisyuk [10,11] were primarily motivated by the desire to create an efficient means for visualizing 3D images. Indeed, holographic imaging is an ultimate solution for 3D visualization. This is the only method that is capable of reproducing, in the most natural viewing conditions, 3D images that have all the visual properties of the original objects.

3D visual communication and display can be achieved through generating, at the viewer site, of holograms out of data that contain all relevant information regarding the scene to be viewed. Digital computers equipped with dedicated devices for fabricating holograms are ideal means for converting data on 3D scenes into optical holograms for visual perception. The core of the 3D digital holographic visual communication paradigm is the understanding that, for generating synthetic holograms at the viewer site, one does not need to produce, at the scene site, the hologram of the scene and to transmit it to the viewer's site. Neither does one need to necessarily imitate, at the viewer site, the full optical holograms of the scene. What one does need is to collect, at the scene site, a set of data that will be sufficient to generate, at the viewer site, a synthetic hologram of the scene that fits visual mechanisms for 3D perception.

The crucial issues in collecting, storing, and transmitting the data needed for the synthesis, at the viewer site, of display holograms are the volume of the required data and the computational complexity of the hologram synthesis. The upper bound of the amount of the data needed for the synthesis of display hologram's viewer site is the full volumetric description of the scene geometry and optical properties. However, a realistic estimation of the amount of data needed for generating a display hologram of the scene is by orders of magnitude lower than this upper bound due to the limitations of the human visual system. This also has a direct impact on the computational complexity of the hologram synthesis.

One of the most promising solutions for the synthesis of computer-generated display holograms, which are computationally inexpensive and at the same time are sufficient for granting 3D visual sensation, is generating *compound multiple-view holograms*. In this method, the scene to be viewed is represented by means of multiple-view images taken from different directions in the required view angle, and, for each image, its corresponding hologram is synthesized with an account of its position in the viewing angle (see Figure 5.23).

The mosaic of appropriately arranged elementary holograms forms a compound multiple-view hologram. This hologram, when viewed with two eyes, will reconstruct different aspects of scenes from different directions deter-

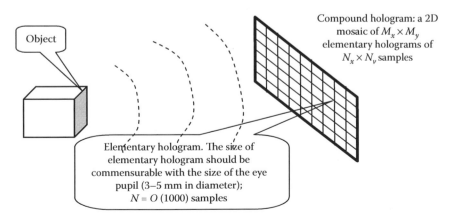

FIGURE 5.23
Principle of the synthesis of compound multiple-view holograms.

FIGURE 5.24
Viewing compound computer-generated hologram (a) and one of the views reconstructed from the hologram (b).

mined by the position of the viewer. The physical size of each hologram has to be, approximately, of the size of the viewer's eye pupil.

Figure 5.24 illustrates viewing a computer-generated multiple-view compound hologram. The hologram is composed of 900 elementary holograms of 256×256 pixels and contains 30×30 views, in spatial angle –90° ÷ 90°, of an object in the form of a cube. Being illuminated by a small LED lamp, the hologram can be used for viewing the reconstructed scene from different angles. Looking through the hologram with two eyes, viewers are able to see 3D image of a cube (Figure 5.24b) floating in the air.

For computing elementary holograms, discrete Fourier and discrete Fresnel transforms, which represent wave propagation integral transforms, can be used. The computational complexity of the synthesis of composite holograms can be estimated as follows. If individual images have $N_x \times N_y$-pixels, the complexity of the synthesis of elementary holograms with the use of FFTs is of the order of $N_x N_y \log N_x N_y$ operation. Therefore, the computational complexity of the synthesis of the composite hologram composed of $M_x \times M_y$ elementary holograms is of the order $M_x M_y N_x N_y \log N_x N_y$ operations. Note that the computational complexity of generating a single hologram of the same size is $O(M_x M_y N_x N_y \log M_x M_y N_x N_y)$, which is $1 + \log M_x M_y / \log N_x N_y$ times higher.

Recording Computer-Generated Holograms on Optical Media

Apparently, the major problem in computer-generated holography is not the computational complexity but rather the problem of converting digital data into a physical hologram. To perform this task, computer-controlled *spatial light modulators* (SLM) are needed.

Figure 5.25 illustrates computer-generated hologram recorded on an SLM and related definitions. The most important numerical characteristics of SLMs are the total number of cells, which can be exposed, sampling intervals, that is,

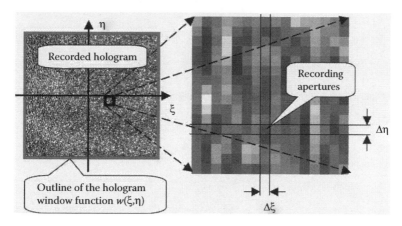

FIGURE 5.25
Computer-generated hologram recorded on an SLM and related definitions.

the distances $\Delta\xi$ and $\Delta\eta$ between neighboring separately and independently exposed resolution cells, and recording aperture, that is, PSF of the recorder.

Computer-generated holograms are arrays of complex number $\{\Gamma_{r,s}\}$ that represent hologram samples. For recording holograms, these samples should be appropriately encoded for recording according to the type of SLM. SLMs for recording computer-generated holograms can be classified into three categories: *amplitude-only SLM, phase-only SLM,* and *combined amplitude/phase SLM.*

Obviously, the most suitable for recording computer-generated holograms would be combined SLM that permit independent control of both the intensity and the phase transmittance or reflectance of the resolution cells by, correspondingly, amplitudes $\{|\Gamma_{r,s}|\}$ and phases $\{\theta_{r,s}\}$ of hologram samples represented as $\{\Gamma_{r,s} = |\Gamma_{r,s}|\exp(i\theta_{r,s})\}$.

In amplitude-only SLMs, the controlled optical parameter is their light intensity transmission or reflection factor in each resolution cell. Typical examples of amplitude SLMs are the photographic films used in photography and optical holography. Photographic media have relatively high resolution and dynamic range. However, they have substantial drawbacks: they require wet chemical development, and are not reversible. Recently, microlens array and micromirror array technology emerged and reports were published on using them as amplitude SLM for recording computer-generated holograms.

For recording holograms on amplitude SLMs, samples $\{\Gamma_{r,s}\}$ can be presented as an additive combination of numbers with a certain fixed phase. Specifically, the following three forms of such representation can be used:

- Orthogonal representation

$$\Gamma_{r,s} = \Gamma_{r,s}^{re} + i\Gamma_{r,s}^{im} = \Gamma_{r,s}^{re} + \Gamma_{r,s}^{im}\exp(i\pi/2), \tag{5.71}$$

where $\Gamma_{r,s}^{re}$ and $\Gamma_{r,s}^{im}$ are real and imaginary parts of $\Gamma_{r,s}$.

- Biorthogonal representation

$$\Gamma_{r,s} = \frac{1 + \text{sign}\left(\Gamma_{r,s}^{\text{re}}\right)}{2}\left|\Gamma_{r,s}^{\text{re}}\right| + \frac{1 + \text{sign}\left(\Gamma_{r,s}^{\text{im}}\right)}{2}\left|\Gamma_{r,s}^{\text{im}}\right|\exp(i\pi/2)$$

$$+ \frac{\left[1 - \text{sign}\left(\Gamma_{r,s}^{\text{re}}\right)\right]}{2}\left|\Gamma_{r,s}^{\text{re}}\right|\exp(i\pi) + \frac{\left[1 - \text{sign}\left(\Gamma_{r,s}^{\text{im}}\right)\right]}{2}\left|\Gamma_{r,s}^{\text{re}}\right|\exp(i3\pi/2), \quad (5.72)$$

where

$$\text{sign}(x) = \begin{cases} 1, & x > 0 \\ 0, & x = 0 \\ -1, & x < 0. \end{cases}$$

- Simplex representation

$$\Gamma_{r,s} = \Gamma_{r,s}^{(0)} + \Gamma_{r,s}^{(120)}\exp\left(i2\pi/3\right) + \Gamma_{r,s}^{(-120)}\exp(-i2\pi/3), \quad (5.73)$$

where $\Gamma_{r,s}^{(0)}$, $\Gamma_{r,s}^{(120)}$, and $\Gamma_{r,s}^{(-120)}$ are projections of $\Gamma_{r,s}$ treated as a vector on a complex plane to vectors $\exp(i0)$, $\exp(i2\pi/3)$, and $\exp(-i2\pi/3)$, respectively. Note that $\Gamma_{r,s}^{(0)}$, $\Gamma_{r,s}^{(120)}$, and $\Gamma_{r,s}^{(-120)}$ are nonnegative numbers and, for each $\Gamma_{r,s}$, one of it is equal to zero depending on which part of the complex plane bounded by vectors $\exp(i0)$, $\exp(i2\pi/3)$, and $\exp(-i2\pi/3)$ it belongs.

With these representations through real (Equation 5.71) or three (Equation 5.73) or four (Equation 5.72) nonnegative numbers, samples $\Gamma_{r,s}$ of holograms can be recorded on amplitude SLM using two, three, or four SLM resolution cells per each hologram sample.

Phase-only SLMs modulate the phase of optical beams. The most suitable for computer-controlled recording holograms are recently emerged *liquid crystal SLMs*. For recording holograms on phase SLM, samples $\left\{\Gamma_{r,s} = \left|\Gamma_{r,s}\right|\exp\left(i\theta_{r,s}\right)\right\}$ can be presented as an additive combination of numbers with fixed amplitude and appropriately selected phases:

$$\Gamma_{r,s} = \sum_{q=1}^{Q} A_0 \exp\left(i\phi_{r,s}^{(q)}\right). \quad (5.74)$$

Only cases $Q = 2$ and $Q = 3$ represent a practical interest. For $Q = 2$, one can obtain

$$\phi_{r,s}^{(1)} = \theta_{r,s} + \arccos\left(|\Gamma_{r,s}|/2A_0\right); \quad \phi_{r,s}^{(2)} = \theta_{r,s} - \arccos\left(|\Gamma_{r,s}|/2A_0\right). \quad (5.75)$$

For $Q = 3$

$$\phi_{r,s}^{(1)} = \theta_{r,s} - 2\arcsin\left(\frac{\sqrt{3}}{2}\sqrt{1 - |\Gamma_{r,s}|/3}\right); \quad \phi_{r,s}^{(2)} = \theta_{r,s}$$

$$(5.76)$$

$$\phi_{r,s}^{(3)} = \theta_{r,s} - 2\arcsin\left(\frac{\sqrt{3}}{2}\sqrt{1 - |\Gamma_{r,s}|/3}\right).$$

With these representations, hologram samples $\{\Gamma_{r,s}\}$ can be recorded on phase SLMs using two or three of the SLM resolution cells.

An important special case of recording display holograms on phase SLMs is the *kinoform* method. With this method, only phases $\{\theta_{r,s}\}$ of hologram samples are recorded and amplitudes are ignored. The advantage of this method is the high diffraction efficiency of the kinoform hologram: they do not absorb light used for hologram reconstruction. However, ignoring of amplitudes of hologram samples results in substantial distortions of reconstructed image. As we already mentioned in the section "Quantization in Digital Holography," these distortions can be minimized by means of assigning to object wavefront a specially optimized pseudorandom phase. Holograms shown in Figure 5.24a were recorded by the kinoform method.

A more detailed review of methods for hologram encoding for recording can be found in [12].

Optical Reconstruction of Computer-Generated Holograms

At the reconstruction stage, computer-generated holograms synthesized with the use of discrete representations of wave propagation transformations described in Chapter 4 are subjected to analog optical transformations in reconstruction optical set-ups such as those shown in Figure 5.26.

The effects of discretization of optical transforms, methods of encoding mathematical holograms into physical holograms and parameters of recorded holograms such as the number of hologram samples, hologram sampling intervals, and PSF of the hologram recording device influence the hologram reconstruction result. In this section, we will demonstrate this influence on an example of analysis of optical reconstruction of holograms recorded by the kinoform method.

Let

$$\Gamma_{r,s} = \sum_{k=0}^{N_1-1}\sum_{l=0}^{N_2-1} \tilde{A}_{k,l} \exp\left\{i2\pi\left[\frac{(k+u)(r+p)}{N_1} + \frac{(l+v)(s+q)}{N_2}\right]\right\} \quad (5.77)$$

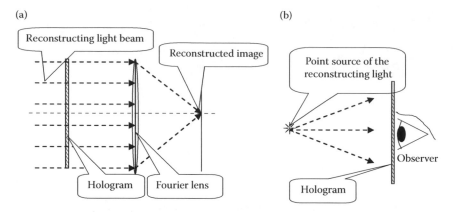

FIGURE 5.26
Set-ups for optical (a) and visual (b) reconstruction of computer-generated Fourier holograms.

be samples of the mathematical Fourier phase only hologram generated through SDFT($u,v;p,q$) from $N_1 \times N_2$ of object wavefront samples $\{\tilde{A}_{k,l}\}$ modified, using optimized pseudorandom diffuser described in the section "Quantization in Digital Holography," so as to produce a phase-only hologram. The resulting kinoform hologram recorded, with sampling intervals ($\Delta\xi$, $\Delta\eta$), by a hologram recorder with PSF h_{rec} (ξ, η) on a rectangular sampling grid along coordinates (ξ, η) can be, in denotations of Figure 5.25, written as

$$\tilde{\Gamma}(\xi,\eta) = w(\xi,\eta)\sum_r\sum_s \Gamma_{r,s}h_{\mathrm{rec}}\left(\xi - \xi_0 - r\Delta\xi, \eta - \eta_0 - s\Delta\eta\right), \qquad (5.78)$$

where (ξ_0, η_0) are shift parameters that depend on the geometry of positioning the hologram in the reconstruction set-up, $w(\xi, \eta)$ is a window function that defines the physical size of the recorded hologram: $0 < w (\xi, \eta) \le 1$, when (ξ, η) belong to the hologram area and $w(\xi, \eta) = 0$, otherwise.

At the reconstruction stage, the kinoform is to be subjected to optical Fourier transform in a reconstruction optical set-up to reconstruct an image

$$A_{\mathrm{rcnstr}}(x,y) = \int\limits_{-\infty}^{\infty}\int\limits_{-\infty}^{\infty} \tilde{\Gamma}(\xi,\eta)\exp\left(-i2\pi\frac{x\xi + y\eta}{\lambda Z}\right)d\xi\,d\eta$$

$$= \int\limits_{-\infty}^{\infty}\int\limits_{-\infty}^{\infty}\left[w(\xi,\eta)\sum_r\sum_s\Gamma_{r,s}h_{\mathrm{rec}}(\xi - \xi_0 - r\Delta\xi, \eta - \eta_0 - s\Delta\eta)\right]$$

$$\times \exp\left(-i2\pi\frac{x\xi + y\eta}{\lambda Z}\right)d\xi\,d\eta. \qquad (5.79)$$

Replace in Equation 5.79 the hologram window function $w(\xi, \eta)$ by its expression

$$w(\xi, \eta) = \int\limits_{-\infty}^{\infty} \int\limits_{-\infty}^{\infty} W\left(\bar{\xi}, \bar{\eta}\right) \exp\left(i2\pi \frac{\xi\bar{\xi} + \eta\bar{\eta}}{\lambda Z}\right) d\bar{\xi}\, d\bar{\eta} \qquad (5.80)$$

through its Fourier spectrum $W\left(\bar{\xi}, \bar{\eta}\right)$ and obtain (see Appendix)

$$A_{\mathrm{rcstr}}(x, y) = \int\limits_{-\infty}^{\infty} \int\limits_{-\infty}^{\infty} \int\limits_{-\infty}^{\infty} \int\limits_{-\infty}^{\infty} W\left(\bar{\xi}, \bar{\eta}\right) \exp\left(i2\pi \frac{\xi\bar{\xi} + \eta\bar{\eta}}{\lambda Z}\right) d\bar{\xi}\, d\bar{\eta}$$

$$\times \sum_{r}\sum_{s} \Gamma_{r,s} h_{\mathrm{rec}}\left(\xi - \xi_0 - r\Delta\xi, \eta - \eta_0 - s\Delta\eta\right)$$

$$\times \exp\left(-i2\pi \frac{x\xi + y\eta}{\lambda Z}\right) d\xi\, d\eta v$$

$$= \sum_{k=0}^{N_1-1}\sum_{l=0}^{N_2-1} A_{k,l} \exp\left[i2\pi\left(\frac{k+u}{2N_1}p + \frac{l+v}{N_2}q\right)\right]$$

$$\times \sum_{do_x=-\infty}^{\infty}\sum_{do_y=-\infty}^{\infty} \tilde{H}_{\mathrm{rec}}\left[\left(\frac{k+u}{N_1} + do_x\right)\frac{\lambda Z}{\Delta\xi}, \left(\frac{l+v}{N_2} + do_y\right)\frac{\lambda Z}{\Delta\eta}\right]$$

$$\times W\left(x - \frac{k+u}{N_1}\frac{\lambda Z}{\Delta\xi} - do_x\frac{\lambda Z}{\Delta\xi}; y - \frac{l+v}{N_2}\frac{\lambda Z}{\Delta\eta} - do_y\frac{\lambda Z}{\Delta\eta}\right). \qquad (5.81)$$

Selecting SDFT shift parameters $p = 0$ and $q = 0$, obtain finally

$$A_{\mathrm{rcstr}}(x, y) = \sum_{k=0}^{N_1-1}\sum_{l=0}^{N_2-1}\sum_{do_x=-\infty}^{\infty}\sum_{do_y=-\infty}^{\infty} \tilde{A}_{k,l}\tilde{H}_{\mathrm{rec}}\left[\left(\frac{k+u}{N_1} + do_x\right)\frac{\lambda Z}{\Delta\xi}, \left(\frac{l+v}{N_2} + do_y\right)\frac{\lambda Z}{\Delta\eta}\right]$$

$$\times W\left[x - \left(\frac{k+u}{N_1} + do_x\right)\frac{\lambda Z}{\Delta\xi}; y - \left(\frac{l+v}{N_2} + do_y\right)\frac{\lambda Z}{\Delta\eta}\right]$$

$$= \sum_{do_x=-\infty}^{\infty}\sum_{do_y=-\infty}^{\infty} \tilde{A}_{k,l}(x, do_x; y, do_y), \qquad (5.82)$$

where it is denoted

$$
\tilde{A}_{k,l}\left(x,do_x;y,do_y\right) = \sum_{k=0}^{N_1-1}\sum_{l=0}^{N_2-1} \tilde{H}_{rec}\left[\left(\frac{k+u}{N_1}-do_x\right)\frac{\lambda Z}{\Delta\xi},\left(\frac{l+v}{N_2}-do_y\right)\frac{\lambda Z}{\Delta\eta}\right]\tilde{A}_{k,l}
$$

$$
\times W\left(x-\frac{k+u}{N_1}\frac{\lambda Z}{\Delta\xi}-do_x\frac{\lambda Z}{\Delta\xi};y-\frac{l+v}{N_2}\frac{\lambda Z}{\Delta\eta}-do_y\frac{\lambda Z}{\Delta\eta}\right).
$$

$$(5.83)$$

Equations 5.82 and 5.83 have a clear physical interpretation:

- Object wavefront is reconstructed in a number of diffraction orders numbered by indices $\{do_x, do_y\}$.
- In the diffraction order $\{do_x, do_y\}$, the reconstructed wavefront is a result of interpolation of samples $\{\tilde{A}_{k,l}\}$; of the object wavefront modified for kinoform encoding, with an interpolation kernel $W(x; y)$, which is the Fourier transform of the recorded hologram window function $w(\xi, \eta)$, those samples being weighted by samples of the frequency response $\tilde{H}_{rec}(x,y)$ of the hologram recording device taken at coordinates

$$
\left\{x = \left(\frac{k+u}{N_1}+do_x\right)\frac{\lambda Z}{\Delta\xi},y=\left(\frac{l+v}{N_2}+do_y\right)\frac{\lambda Z}{\Delta\eta}\right\}.
$$

This interpretation is illustrated in Figure 5.27.

Computational Imaging Using Optics-Less Lambertian Sensors

Optics-Less Passive Sensors: Motivation

Conventional optical imaging systems use photosensitive planar arrays of detectors coupled with focused optics that form a map of the environment on the image plane. The optics carries this out at the speed of light, but comes with some disadvantages. Because of the law of diffraction, accurate mapping requires large lens sizes and complex optical systems. Also, lenses limit the field of view and are only available within a limited range of the electromagnetic spectrum. The ever-decreasing cost of computing and ever-increasing computer power makes it possible to make imaging devices smaller by replacing optical and mechanical components with computation. This motivates a search for optics-less computational imaging devices.

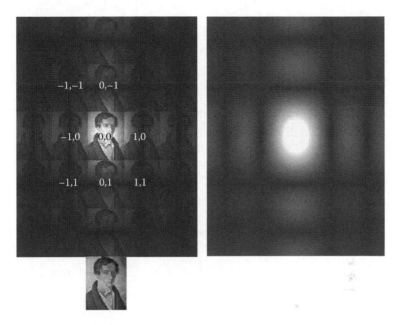

FIGURE 5.27
Left side: computer simulation of optical reconstruction of computer-generated kinoform of the image shown at the bottom; numbers indicate diffraction order indices. Right side: reconstruction masking function for a rectangular hologram recording aperture of size $\Delta\xi \times \Delta\eta$.

A good example of optics-less computational imaging is electronic recording and numerical reconstruction of optical holograms described in the section "Digital Image Formation by Means of Numerical Reconstruction of Holograms." This is a very attractive solution of the problem, but it requires active illumination of the imaged object by a coherent radiation and illumination of sensors by a reference beam coherent with the object beam in order to enable recording both amplitude and phase information of the diffracted object beam. One can avoid the need to record the phase information if object has certain *a priori* known redundancy, for instance, if it is known to be limited in space, or by means of recording intensity of diffracted object beam at two or more different distances. In these cases, different methods for *phase retrieval* can be used, which, in principle, are similar to iterative methods for image reconstruction from sparse nonuniform samples discussed in the section "Image Recovery from Sparse or Nonuniformly Sampled Data." We refer the readers to numerous publications on phase retrieval methods. In this section, we will show that optics-less incoherent imaging is also possible.

Imaging as a Parameter Estimation Task

One can treat images as sets of data that indicate locations in space and intensities of sources of radiation. These data can be regarded as parameters to be

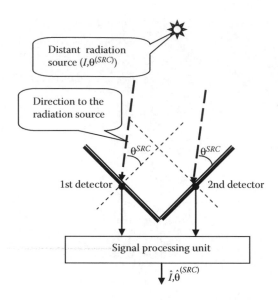

FIGURE 5.28
A two-detector sensor for the determination of directional angle and intensity of a distant radiation source.

estimated from sensor's signal and the imaging problem can be solved using optimal statistical parameter estimation approach outlined in the section "Quantifying Signal Processing Quality." To begin with, consider a simplistic task of determination of intensity and angular direction to a single distant radiation source. For determination of these two parameters, two detectors illuminated by the source under different angles would be sufficient, provided the detectors have a certain angular selectivity.

Consider a schematic diagram in Figure 5.28, which depicts a 2D model of a sensor consisting of two mutually perpendicular radiation detectors and a signal processing unit that computes from its input data estimates $\hat{I}$ and $\hat{\theta}^{(SRC)}$ of the source intensity I and directional angle $\theta^{(SRC)}$.

Assume that outputs s_1 and s_2 of radiation detectors are proportional to the source intensity I and detector's angular sensitivity function $\mathrm{AngSens}(\theta^{(SNS)})$, where $\theta^{(SNS)}$ is the angle between the direction to the source of radiation and the normal to the detector surface, and that they contain mutually independent Gaussian random noise components n_1 and n_2:

$$s_1 = I\,\mathrm{AngSens}\!\left(\theta_1^{(SNS)}\right) + n_1; \quad s_2 = I\,\mathrm{AngSens}\!\left(\theta_2^{(SNS)}\right) + n_2. \qquad (5.84)$$

Concerning the angular sensitivity of detectors, we opt the simplest assumption that the detector's output is proportional to the radiation energy per unit of the detector's surface projection to the plane perpendicular to the direction to the radiation source. This assumption leads to the Lambertian cosine law angular sensitivity function of the detectors:

$$\text{AngSens}\left(\theta^{(SNS)}\right) = \begin{cases} \cos\theta^{(SNS)}, & \left|\theta^{(SNS)}\right| < \pi/2 \\ 0, & \left|\theta^{(SNS)}\right| \geq \pi/2. \end{cases} \tag{5.85}$$

With this assumption, the output detectors depicted in Figure 5.28 will be

$$s_1 = I\cos\left(\theta^{(SRC)}\right) + n_1; \quad s_2 = I\sin\left(\theta^{(SRC)}\right) + n_2. \tag{5.86}$$

As it was described in the section "Statistical Models of Signals and Transformations," Equation 2.158, for the regarded Gaussian additive signal-independent noise model, maximum likelihood estimates of parameters $\hat{I}$ and $\hat{\theta}^{(SRC)}$ will be solutions of equation:

$$\left\{\hat{I}_{ML}, \hat{\theta}_{ML}^{(SRC)}\right\} = \underset{\left(I, \theta^{(SRC)}\right)}{\arg\min}\left[\left(s_1 - I\cos\left(\theta^{(SRC)}\right)\right)^2 + \left(s_1 - I\sin\left(\theta^{(SRC)}\right)\right)^2\right]. \tag{5.87}$$

They can be found by equating partial derivatives of the right part of Equation 5.87 over the sought variables, which gives the following quite obvious solutions (see Appendix):

$$\hat{\theta}_{ML}^{(SRC)} = \arctan\frac{s_2}{s_1}; \quad \hat{I}_{ML}^{(SRC)} = \sqrt{s_1^2 + s_2^2}. \tag{5.88}$$

In order to evaluate the estimation errors of these optimal ML, assume that these errors are sufficiently small. Then, estimation errors can be found (see Appendix) as increments/decrements $d\hat{\theta}_{ML}^{(SRC)}$ and $d\hat{I}_{ML}^{(SRC)}$ of estimates of source directional angle and intensity caused by increments/decrements $ds_1 = n_1$ and $ds_2 = n_2$ of observed signals s_1 and s_2 due to the noise components:

$$\varepsilon_\theta = \frac{n_2\cos\theta^{(SRC)} - n_1\sin\theta^{(SRC)}}{I}; \quad \varepsilon_I = n_1\cos\theta^{(SRC)} + n_2\sin s\theta^{(SRC)}. \tag{5.89}$$

As one can see, for sufficiently small noise, level estimation errors are proportional to noise in detector signals and, therefore they have normal distribution density. Their mean values are zero:

$$\begin{cases} AV_{\Omega_n}(\varepsilon_\theta) = \dfrac{AV_{\Omega_n}(n_2)\cos\theta - AV_{\Omega_n}(n_1)\sin\theta}{I} = 0 \\ AV_{\Omega_n}\varepsilon_I = AV_{\Omega_n}(n_1)\cos\theta + AV_{\Omega_n}(n_2)\sin\theta = 0 \end{cases} \tag{5.90}$$

and variances are proportional to the variance σ_n^2 of detector noise

$$AV_{\Omega_n}\left(\varepsilon_\theta^2\right) = AV_{\Omega_n}\left(\frac{n_2\cos\theta - n_1\sin\theta}{I}\right)^2$$

$$= \frac{AV_{\Omega_n}\left(n_2^2\right)\cos^2\theta + AV_{\Omega_n}\left(n_1^2\right)\sin^2\theta - 2AV_{\Omega_n}\left(n_1 n_2\right)\cos\theta\sin\theta}{I} = \frac{\sigma_n^2}{I_0^2}$$

$$= AV_{\Omega_n}\left(\varepsilon_I^2\right) = AV_{\Omega_n}\left(n_1\cos\theta + n_2\sin\theta\right)^2$$

$$= AV_{\Omega_n}\left(n_1^2\right)\cos^2\theta + AV_{\Omega_n}\left(n_2^2\right)\sin^2\theta + 2AV_{\Omega_n}\left(n_1 n_2\right)\cos\theta\sin\theta = \sigma_n^2, \quad (5.91)$$

where AV_{Ω_n} is the operator of averaging over ensemble Ω_n of realizations of noise.

One can extend the above idea of determination of parameters of a single radiation source to imaging of a given number of multiple radiation sources by means of a radiation sensor that consist of a set of bare radiation detectors with their natural angular selectivity to radiation arranged on a flat or curved surface and supplemented with signal processing unit that collects detector outputs and use them to compute optimal statistical estimations of sources' intensities and coordinates. We will refer to this sensor as *optics-less passive sensor* (OLP-sensor). The schematic diagram of OLP-sensors is depicted in Figure 5.29.

For the case of estimating, by an array of K elementary radiation detectors, of locations and intensities of a known number L of distant point radiation sources that can be specified by their directional angle and radiation intensity, a mathematical model of OLP-sensors can be formulated as follows.

Let I_k be the intensity of the k-th radiation source, X_k be a vector of spatial coordinates of the k th source, $\theta_l^{(SNS)}$ be the angle of the surface normal of the l-th detector with respect to a sensor's "optical axis," and s_l be its response to the radiation. Also let, for definiteness, assume that detectors' responses are contaminated by additive signal-independent and mutually independent noise components $\{n_l\}$ with normal distribution density. Then the equation

$$s_l = \sum_{k=1}^{K} I_k\, \text{AngSens}\left(\vartheta\left(X_k^{(SRC)}, \theta_l^{(SNS)}\right)\right) + n_l \qquad (5.92)$$

models the response of the l-th detector with angular sensitivity AngSens(ϑ) as a function of the angle ϑ between the normal to the detector surface and the direction to the radiation source.

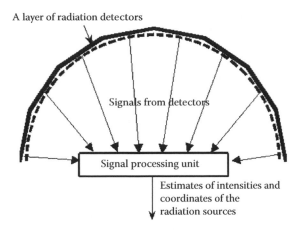

A layer of radiation detectors

Signals from detectors

Signal processing unit

Estimates of intensities and
coordinates of the
radiation sources

FIGURE 5.29
Schematic diagram of OLP-sensors.

The sensor's signal processor unit operates on output signals $\{S_n\}$ of all N detectors and generates statistically optimal estimate $\{\hat{I}_k\}$ and $\{\hat{\mathbf{X}}_k^{(SRC)}\}$ of intensities and coordinates of the radiation sources. These might be ML estimates, which require *a priori* knowledge of only the number of parameters to be estimated, or MAP estimates, for which one needs to know *a priori* probability densities of the sought parameters, or any other statistical estimates. Obviously, changing the type of the estimator requires only the corresponding reprogramming of the sensor signal processing unit and no hardware changes.

In particular, the ML estimator defined by Equation 2.171 takes the form

$$\left\{\hat{I}_k, \hat{\mathbf{X}}_k^{(SRC)}\right\} = \underset{\left\{I_k, \hat{\mathbf{X}}_k^{(SRC)}\right\}}{\arg\min} \left\{\sum_{l=1}^{L} \left[s_n - \sum_{k=1}^{K} I_k \, \mathrm{AngSens}\left(\vartheta\left(\mathbf{X}_k^{(SRC)}, \theta_l^{(SNS)}\right)\right)\right]^2\right\}. \quad (5.93)$$

Of a special importance is the case of estimating intensities of a given number of radiation sources in known locations, as, for instance, in nodes of a regular spatial grid. In this case, the ML-estimator is defined by the equation

$$\left\{\hat{I}_k\right\} = \underset{\{I_k\}}{\arg\min} \left\{\sum_{l=1}^{L} \left[s_n - \sum_{k=1}^{K} I_k \, \mathrm{AngSens}\left(\vartheta\left(\mathbf{X}_k^{(SRC)}, \theta_l^{(SNS)}\right)\right)\right]^2\right\}. \quad (5.94)$$

We call this mode of operation the *"imaging mode."*

Note that Equation 5.94 can be regarded as a special case of a general imaging equation:

$$\left\{\hat{I}_k\right\} = \underset{\{I_k\}}{\arg\min}\left\{\sum_{l=1}^{L}\left(s_l - \sum_{k=1}^{K} I_k h(k,l)\right)^2\right\}, \tag{5.95}$$

in which $h(k, l)$ is a PSF of a linear system that defines contribution of radiation from the k-th radiation source to the signal from the l-th detector.

In implementing the outlined parameter estimation approach, various *a priori* limitation regarding sought variables can be used. Some of them, such as limitations of spatial positions of the sources and nonnegativity of intensity estimates, are almost obvious. There might also be useful some additional *a priori* limitations associated with properties of imaged objects. Here are some of the most frequently used priors for image intensity signals of real objects:

- Minimum variance $VAR(I) = \sum_{k=1}^{K} \hat{I}_k^2 - \left(\sum_{k=1}^{K} \hat{I}_k^2\right)^2$
- Minimum *total variation* (average module of gradient) of intensity estimates $TVAR(I) = \sum_{k=1}^{K} Grad(\hat{I}_k)$
- Minimum bandwidth in a certain selected transform domain

The latter is the base of the compressive sensing approach already mentioned in the section "Application Examples."

OLP-sensors, being decision-making devices, are essentially nonlinear devices that cannot be described in terms of PSFs, which is customary for traditional optics-based imaging devices. The performance of OLP-sensors, and in particular, their resolving power is characterized, in the first-order approximation, by estimation error variances. The statistical theory of parameter estimation shows that, for parameter estimation from data corrupted by sufficiently small independent Gaussian additive noise, estimation errors have a normal distribution. We illustrated this conclusion above on the example of OLP-sensor for locating and determination of intensity of a single radiation source.

In general, finding solutions of Equations 5.93 through 5.95 is a very heavy computational task: it grows exponentially with the number of sources. Until recently, it made the practical use of this approach not feasible. With ever-growing computer power, this approach becomes more and more promising.

Optics-Less Passive Imaging Sensors: Possible Designs, Expected Performance, Advantages, and Disadvantages

There might be several possible designs of OLP-sensors depending on the shape of surface on which the sensor's detectors are installed. In the sensor arrays shown in Figure 5.30a, radiation detectors are placed on the outer

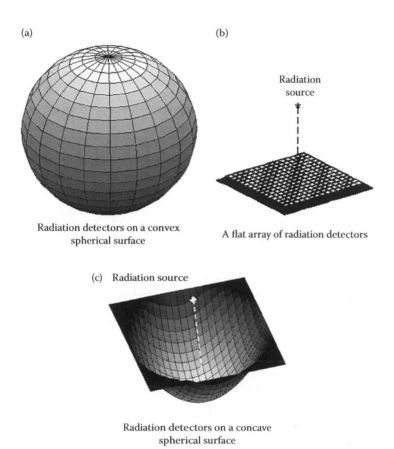

FIGURE 5.30
Examples of possible designs of OLP-sensors: convex spherical (a), flat (b), and concave spherical sensors (c).

surface of a sphere. Such sensors are potentially capable of localizing and imaging radiation sources throughout the solid angle.

Figure 5.30b shows an alternative design, an array of detectors on a flat surface. Because rays from distant sources are practically parallel and arrive at different detectors of the flat array from the same angles, they cannot be distinguished as coming from different sources. Therefore, such sensors are potentially capable of measuring coordinates and intensities of only sources, which are sufficiently close to them so that their rays arrive at different detectors from sufficiently different angles. Any intermediate designs consisting of detectors on curved convex or concave surfaces, such as the one shown in Figure 5.30c, are also possible.

Figures 5.31 through 5.33 enable obtaining a certain insight into potential imaging capabilities of OLP-sensors. Figure 5.31 shows simulation results for a model of the OLP convex spherical sensor consisting of 300 elementary

FIGURE 5.31

OLS spherical sensor in the "imaging" mode: original image is an array of sources that form characters "OLSS" (left), pattern of individual detectors' outputs on the surface of the sphere shown in gray scale (center, dark—low intensity, bright—high intensity), and estimates of source intensities (right). The sensor consisted of $15 \times 20 = 300$ detectors arranged within spatial angles $\pm\pi$ longitude and $\pm\pi/2.05$ latitude. The array of simulated radiation sources consisted of $19 \times 16 = 304$ sources with known directional angles within spatial angles $\pm\pi/2$ longitude and $\pm\pi/3$ latitude. Each detector had a noise standard deviation of 0.01, and source intensities were 0 (dark) or 1 (bright). Standard deviation of estimation errors of source intensities was found to be 0.0640.

detectors set to estimate intensities of 304 sources arranged, in known locations, to form the abbreviation "OLSS."

Figure 5.32 illustrates the performance of the flat sensor in the "imaging" mode and shows the results of reconstruction of the intensities of 8×16 sources arranged in a flat array as it is estimated by the flat sensor with 8×16 detectors at different distances from the sources. Detector noise standard deviation was set to 0.01 in the units of maximal radiation intensity at the sensor. The source intensities form an image of characters "SV." One can conclude from the figure that the flat OLS-sensor with a realistic signal-to-noise ratio (SNR) of 100 is capable of reasonably good imaging for sources situated in its close proximity, though the quality of the estimates rapidly decays with its distance to sources.

Figure 5.33a–d enables quantitative evaluation of the size and shape of the flat sensor's field of view in terms of the standard deviation of estimates of *X–Y* coordinates and intensity of a single source in different positions in front of the sensor.

In particular, one can see that within small distances from the sensor, accurate estimation of source coordinates and intensity is possible, but at distances greater than about half of the sensor length, the estimation accuracy drops by the order of magnitude. Experiments also revealed that the sensor's capability to resolve extremely close sources is limited by a value on the order of interdetector angular, in the case of the spherical sensor, or linear, in the case of the flat sensor, distance, which is intuitively well understood.

Estimation error standard deviation maps illustrated in Figure 5.34a–d are similar to those in Figure 5.33 with the difference that the sensor is bent to make it convex (a) and (b) or concave (c) and (d).

One can see from the figures that the bending surface of the sensor to a convex form widens the field of view of the sensor correspondingly, while bending to a concave form narrows it and concentrates it within a sector that encompasses the sensor.

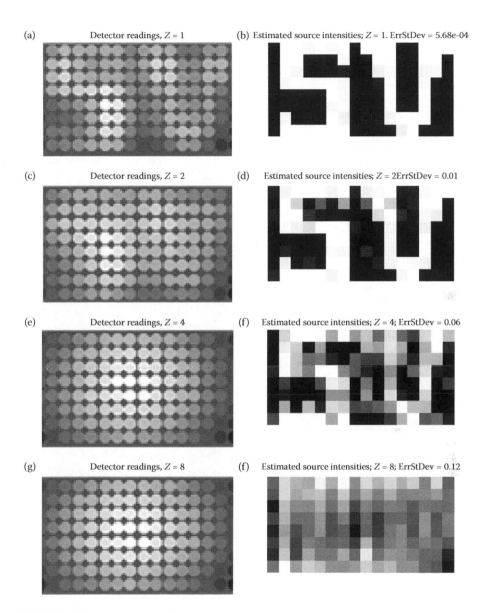

FIGURE 5.32
An illustration of sensing of 8×16 radiation sources arranged on a plane in the form of characters "SV" by a 3D model of a flat OLP sensor with 8×16 elementary detectors in the "imaging" mode for distances of sources from the sensor $Z = 1$–8 (in units of interdetector distance). SNR at detectors was kept, for all distances from sources, constant at 100 by making the source amplitude inversely proportional to the distance between the source and sensor planes. Detector noise was 0.01.

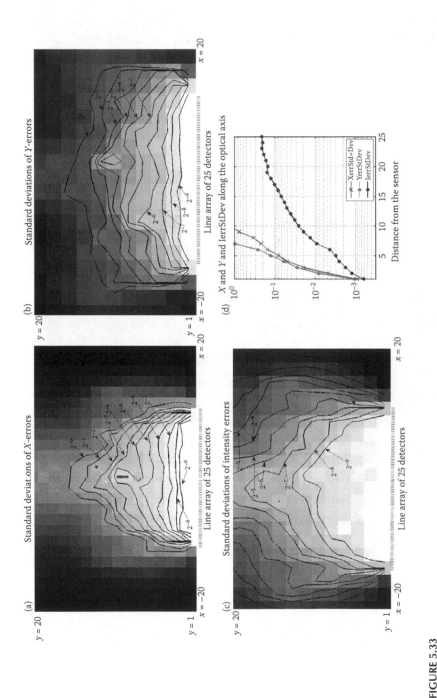

FIGURE 5.33

Maps of standard deviations of estimation errors of *X–Y* coordinates (a, b) and of intensity (c) of a radiation source as a function of the source position with respect to the surface of the flat line array of 25 detectors. Darker areas correspond to larger errors. Plot (d) shows standard deviations of *X*, *Y*, and intensity estimation errors as function of the distance from the sensor along the sensor "optical axis" (central sections of (a) through (c)).

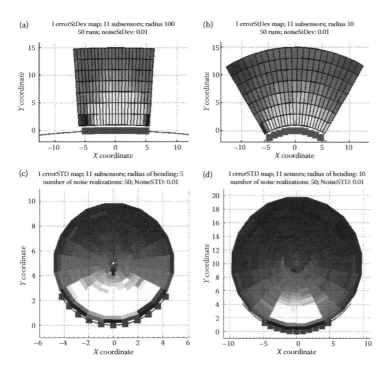

FIGURE 5.34

(a) and (b) Sensors on convex bent surfaces (1D model, 11 detectors, noise standard deviation 0.01): maps of standard deviations of estimation errors of source intensity as functions of the source position with respect to the sensor's surface for bending radius, in units of interdetector distance, 100 (a) and 10 (b). (c) and (d) Sensors on concave bent surfaces (1D model, 11 detectors, noise standard deviation 0.01): maps of standard deviations of estimation errors of source intensity as functions of the source position with respect to the sensor's surface for bending radius, in units of inter detector distance, 10 (c) and 5 (d). Darker areas correspond to larger errors.

The presented data confirm that OLP-sensors with a realistic SNR in its detectors are, in principle, capable of reasonably good localization and imaging of radiation sources. This implies that good directional vision without optics is possible even using the simplest possible detectors whose angular sensitivity is defined solely by the natural surface absorptivity such as the Lambertian one. Convex spherical sensors have an unlimited field of view, while flat sensors can localize and image radiation sources in their close proximity. Sensors on curved surfaces can work in intermediate zones.

As always, there is a trade-off between good and bad features of the OLP-sensors. The advantages include the following:

- No optics is needed, making this type of sensor applicable to virtually any type of radiation and to any wavelength.

- The field of view (of convex spherical sensors) is unlimited.
- The design is flexible: the shape of the sensor's surface can be easily changed to fit the required sensor's field of view.
- Diffraction-related limits are irrelevant to OLS-sensors whose resolving power is determined, given the detector's SNR, solely by the detector's size.

The cost for these advantages is high computational complexity, especially when good imaging properties for multiple sources are required.

Appendix

Derivation of Equation 5.47

$$\overline{DRK}(\xi, k) = \sum_{r=0}^{N-1} DRK(r, k) \exp\left(-i2\pi \tilde{r} \Delta f \xi\right)$$

$$\times \sum_{r=0}^{N-1} \exp\left[i2\pi \frac{(k+u)(r+v)}{N}\right] \exp\left[-i2\pi\left(r+v^{(s)}\right)\Delta f^{(s)} \xi\right]$$

$$\times \exp\left(-i2\pi v^{(s)} \Delta f^{(s)} \xi\right) \exp\left[i2\pi \frac{(k+u)v}{N}\right] \sum_{r=0}^{N-1} \exp\left[i2\pi\left(\frac{(k+u)}{N} - \Delta f^{(s)}\xi\right)r\right]$$

$$= \exp\left(-i2\pi v^{(s)} \Delta f^{(s)} \xi\right) \exp\left[i2\pi \frac{(k+u)v}{N}\right] \frac{\exp\left[i2\pi N\left(\frac{(k+u)}{N} - \Delta f^{(s)}\xi\right)\right] - 1}{\exp\left[i2\pi\left(\frac{(k+u)}{N} - \Delta f^{(s)}\xi\right)\right] - 1}$$

$$= \exp\left[-i2\pi\left(v^{(s)} + \frac{N-1}{2}\right)\Delta f \xi\right] \exp\left\{i2\pi \frac{(k+u)}{N}\left(v + \frac{N-1}{2}\right)\right\}$$

$$\times \frac{\sin\left[\pi N\left(\Delta f \xi - \frac{k+u}{N}\right)\right]}{\sin\left[\pi\left(\Delta f \xi - \frac{k+u}{N}\right)\right]}$$

Derivation of Equation 5.63

$$OPSF(x,k) = \exp\left[i\pi\left(\frac{x^2}{\lambda Z} - \frac{k^2\mu^2}{N}\right)\right] \times \sum_{r=0}^{N-1} \exp\left(-i\pi\frac{r^2}{\mu^2 N}\right)\exp\left(i2\pi\frac{kr}{N}\right)$$

$$\times \int_{-\infty}^{\infty} \exp\left[-i\pi\frac{(x-\lambda Z\xi)^2}{\lambda Z}\right]\exp(-i2\pi r\Delta f\xi)d\xi \times \exp\left[i\pi\left(\frac{x^2}{\lambda Z} - \frac{k^2\mu^2}{N}\right)\right]$$

$$\times \sum_{r=0}^{N-1} \exp\left(-i\pi\frac{r^2}{\mu^2 N}\right)\exp\left(i2\pi\frac{kr}{N}\right)\int_{-\infty}^{\infty} \exp\left[-i\pi\frac{(x-\lambda Z\xi)^2}{\lambda Z}\right]\exp(-i2\pi r\Delta f\xi)d\xi$$

$$= \exp\left[i\pi\left(\frac{x^2}{\lambda Z} - \frac{k^2\mu^2}{N}\right)\right]\sum_{r=0}^{N-1} \exp\left(-i\pi\frac{r^2}{\mu^2 N}\right)\exp\left(i2\pi\frac{kr}{N}\right)$$

$$\times \int_{-\infty}^{\infty} \exp\left[-i\pi\frac{\tilde{\xi}^2}{\lambda Z}\right]\exp\left(-i2\pi r\Delta f\frac{x-\tilde{\xi}}{\lambda Z}\right)d\xi$$

$$\times \exp\left[i\pi\left(\frac{x^2}{\lambda Z} - \frac{k^2\mu^2}{N}\right)\right]\sum_{r=0}^{N-1} \exp\left(-i\pi\frac{r^2}{\mu^2 N}\right)\exp\left(i2\pi\frac{kr}{N}\right)\exp\left(-i2\pi r\Delta f\frac{x}{\lambda Z}\right)$$

$$\times \int_{-\infty}^{\infty} \exp\left(-i\pi\frac{\tilde{\xi}^2}{\lambda Z}\right)\exp\left(i2\pi\frac{r\Delta f}{\lambda Z}\tilde{\xi}\right)d\xi$$

$$\propto \exp\left[i\pi\left(\frac{x^2}{\lambda Z} - \frac{k^2\mu^2}{N}\right)\right]\sum_{r=0}^{N-1} \exp\left(-i\pi\frac{r^2}{\mu^2 N}\right)$$

$$\times \exp\left[i2\pi\left(\frac{k}{N} - \frac{\Delta fx}{\lambda Z}\right)r\right]\exp\left(i\pi\frac{r^2\Delta f^2}{\lambda Z}\right)$$

$$\times \exp\left[i\pi\left(\frac{x^2}{\lambda Z} - \frac{k^2\mu^2}{N}\right)\right]\sum_{r=0}^{N-1} \exp\left[-i\pi\left(\frac{N\Delta f^2}{\lambda Z} - \frac{1}{\mu^2}\right)\frac{r^2}{N}\right]\exp\left[i2\pi\left(\frac{k}{N} - \frac{\Delta fx}{\lambda Z}\right)r\right]$$

$$= \exp\left[i\pi\left(\frac{x^2}{\lambda Z} - \frac{k^2\mu^2}{N}\right)\right]\mathrm{frincd}\left(N; \frac{1}{\mu^2} - \frac{N\Delta f^2}{\lambda Z}; \frac{\Delta fx}{\lambda Z} - \frac{k}{N}\right).$$

Derivation of Equation 5.69

$$OPSF(x,k) = \int_{-\infty}^{\infty} \Phi_f^{(s)}(\xi) \exp(-i\pi\lambda Z\xi^2) \exp(i2\pi x\xi) d\xi$$

$$\times \frac{1}{N} \sum_{s=0}^{N-1} \exp\left(i\pi \frac{\mu^2 s^2}{N}\right) \exp\left(-i2\pi \frac{ks}{N}\right) \sum_{r=0}^{N-1} \exp\left[i2\pi r\left(\frac{s}{N} - \Delta f\xi\right)\right]$$

$$= \int_{-\infty}^{\infty} \Phi_f^{(s)}(\xi) \exp(-i\pi\lambda Z\xi^2) \exp(i2\pi x\xi) d\xi$$

$$\times \sum_{s=0}^{N-1} \exp\left(i\pi \frac{\mu^2 s^2}{N}\right) \exp\left(-i2\pi \frac{ks}{N}\right)$$

$$\times \frac{\sin\left[\pi(s - N\Delta f\xi)\right]}{N \sin\left[\frac{\pi}{N}(s - N\Delta f\xi)\right]} \exp\left[i\pi \frac{N-1}{N}(s - N\Delta f\xi)\right]$$

$$\propto \int_{-\infty}^{\infty} \Phi_f^{(s)}(\xi) \exp(-i\pi\lambda Z\xi^2) \exp(i2\pi x\xi) d\xi \sum_{s=0}^{N-1} \exp\left(i\pi \frac{\mu^2 s^2}{N}\right) \exp\left(-i2\pi \frac{ks}{N}\right)$$

$$\times \operatorname{sincd}\left[N; \pi(s - N\Delta f\xi)\right].$$

Derivation of Equation 5.81

$$A_{\text{rcstr}}(x,y) = \int_{-\infty}^{\infty}\int_{-\infty}^{\infty}\int_{-\infty}^{\infty}\int_{-\infty}^{\infty} W(\bar{\xi}, \bar{\eta}) \exp\left(i2\pi \frac{\xi\bar{\xi} + \eta\bar{\eta}}{\lambda Z}\right) d\bar{\xi}\, d\bar{\eta}$$

$$\times \sum_r \sum_s \Gamma_{r,s} h_{\text{rec}}(\xi - \xi_0 - r\Delta\xi, \eta - \eta_0 - s\Delta\eta) \exp\left(-i2\pi \frac{x\xi + y\eta}{\lambda Z}\right) d\xi\, d\eta$$

$$= \int_{-\infty}^{\infty}\int_{-\infty}^{\infty} W(\bar{\xi}, \bar{\eta}) d\bar{\xi}\, d\bar{\eta} \sum_{r=-\infty}^{\infty} \sum_{s=-\infty}^{\infty} \Gamma_{r,s}$$

$$\times \int_{-\infty}^{\infty}\int_{-\infty}^{\infty} h_{\text{rec}}(\xi - \xi_0 - r\Delta\xi, \eta - \eta_0 - s\Delta\eta) \exp\left[-i2\pi \frac{(x - \bar{\xi})\xi + (y - \bar{\eta})\eta}{\lambda Z}\right] d\xi\, d\eta$$

$$\times \int\limits_{-\infty}^{\infty}\int\limits_{-\infty}^{\infty} W(\bar{\xi},\bar{\eta})d\bar{\xi}\,d\bar{\eta}\sum_{r=-\infty}^{\infty}\sum_{s=-\infty}^{\infty}\Gamma_{r,s}\exp\left[-i2\pi\frac{(x-\bar{\xi})(\xi_0+r\Delta\xi)+(y-\bar{\eta})(\eta_0+s\Delta\eta)}{\lambda Z}\right]$$

$$\times \int\limits_{-\infty}^{\infty}\int\limits_{-\infty}^{\infty} h_{\mathrm{rec}}(\xi,\eta)\exp\left[-i2\pi\frac{(x-\bar{\xi})\xi+(y-\bar{\eta})\eta}{\lambda Z}\right]d\xi\,d\eta.$$

Replacing the last multiplicand in this equation by the frequency response, in coordinates $\left(x-\bar{\xi},y-\bar{\eta}\right)$, of the hologram recording device:

$$H_{\mathrm{rec}}\left(x-\bar{\xi},y-\bar{\eta}\right) = \int\limits_{-\infty}^{\infty}\int\limits_{-\infty}^{\infty} h_{\mathrm{rec}}(\xi,\eta)\exp\left[-i2\pi\frac{(x-\bar{\xi})\xi+(y-\bar{\eta})\eta}{\lambda Z}\right]d\xi\,d\eta$$

obtain:

$$A_{\mathrm{rcstr}}(x,y) = \int\limits_{-\infty}^{\infty}\int\limits_{-\infty}^{\infty} W(\bar{\xi},\bar{\eta})H_{\mathrm{rec}}\left(x-\bar{\xi},y-\bar{\eta}\right)d\bar{\xi}\,d\bar{\eta}$$

$$\times\sum_{r=-\infty}^{\infty}\sum_{s=-\infty}^{\infty}\Gamma_{r,s}\exp\left[-i2\pi\frac{(x-\bar{\xi})(\xi_0+r\Delta\xi)+(y-\bar{\eta})(\eta_0+s\Delta\eta)}{\lambda Z}\right]$$

$$\times\int\limits_{-\infty}^{\infty}\int\limits_{-\infty}^{\infty} W(\bar{\xi},\bar{\eta})H_{\mathrm{rec}}\left(x-\bar{\xi},y-\bar{\eta}\right)\exp\left[-i2\pi\frac{(x-\bar{\xi})\xi_0+(y-\bar{\eta})\eta_0}{\lambda Z}\right]$$

$$\times\sum_{r=-\infty}^{\infty}\sum_{s=-\infty}^{\infty}\Gamma_{r,s}\exp\left[-i2\pi\frac{(x-\bar{\xi})r\Delta\xi+(y-\bar{\eta})s\Delta\eta}{\lambda Z}\right]d\bar{\xi}\,d\bar{\eta}$$

$$= \int\limits_{-\infty}^{\infty}\int\limits_{-\infty}^{\infty} W\left(x-\bar{\xi},y-\bar{\eta}\right)H_{\mathrm{rec}}\left(\bar{\xi},\bar{\eta}\right)\exp\left(-i2\pi\frac{\bar{\xi}\xi_0+\bar{\eta}\eta_0}{\lambda Z}\right)$$

$$\times\sum_{r=-\infty}^{\infty}\sum_{s=-\infty}^{\infty}\Gamma_{r,s}\exp\left(-i2\pi\frac{\bar{\xi}\Delta\xi r+\bar{\eta}\Delta\eta s}{\lambda Z}\right)d\bar{\xi}\,d\bar{\eta}.$$

Denoting

$$\tilde{H}_{\mathrm{rec}}^{(\xi_0,\eta_0)}\left(\bar{\xi},\bar{\eta}\right) = H_{\mathrm{rec}}\left(\bar{\xi},\bar{\eta}\right)\exp\left(-i2\pi\frac{\bar{\xi}\xi_0+\bar{\eta}\eta_0}{\lambda Z}\right)$$

and replacing $\Gamma_{r,s}$ by its expression through object wave front samples $\tilde{A}_{k,l}$ (Equation 5.77) results in:

$$A_{\text{rcstr}}(x,y) = \int\limits_{-\infty}^{\infty}\int\limits_{-\infty}^{\infty} W\left(x-\xi, y-\overline{\eta}\right)\tilde{H}_{\text{rec}}\left(\xi,\overline{\eta}\right)$$

$$\times \sum_{r=-\infty}^{\infty}\sum_{s=-\infty}^{\infty}\sum_{k=0}^{N_1-1}\sum_{l=0}^{N_2-1} \tilde{A}_{k,l}\exp\left\{i2\pi\left[\frac{(k+u)(r+p)}{N_1}+\frac{(l+v)(s+q)}{N_2}\right]\right\}$$

$$\times \exp\left(-i2\pi\frac{\overline{\xi}\Delta\xi r + \overline{\eta}\Delta\eta s}{\lambda Z}\right)d\overline{\xi}\,d\overline{\eta}$$

$$\times \int\limits_{-\infty}^{\infty}\int\limits_{-\infty}^{\infty} W\left(x-\xi, y-\overline{\eta}\right)\tilde{H}_{\text{rec}}\left(\xi,\overline{\eta}\right)\sum_{k=0}^{N_1-1}\sum_{l=0}^{N_2-1}\tilde{A}_{k,l}\exp\left[i2\pi\left(\frac{k+u}{N_1}p+\frac{l+v}{N_2}q\right)\right]$$

$$\times \sum_{r=-\infty}^{\infty}\sum_{s=-\infty}^{\infty}\exp\left[i2\pi\left(\frac{k+u}{N_1}r+\frac{l+v}{N_2}s\right)\right]\exp\left(-i2\pi\frac{\overline{\xi}\Delta\xi r + \overline{\eta}\Delta\eta s}{\lambda Z}\right)d\overline{\xi}\,d\overline{\eta}$$

$$= \int\limits_{-\infty}^{\infty}\int\limits_{-\infty}^{\infty} W\left(x-\xi, y-\overline{\eta}\right)\tilde{H}_{\text{rec}}\left(\xi,\overline{\eta}\right)\sum_{k=0}^{N_1-1}\sum_{l=0}^{N_2-1}\tilde{A}_{k,l}\exp\left[i2\pi\left(\frac{k+u}{N_1}p+\frac{l+v}{N_2}q\right)\right]$$

$$\times \sum_{r=-\infty}^{\infty}\exp\left[-i2\pi\left(\frac{\overline{\xi}\Delta\xi}{\lambda Z}-\frac{k+u}{N_2}\right)r\right]\sum_{s=-\infty}^{\infty}\exp\left[-i2\pi\left(\frac{\overline{\eta}\Delta\eta}{\lambda Z}-\frac{l+v}{N_2}\right)s\right]d\overline{\xi}\,d\overline{\eta}.$$

The last two multiplicands in this equation can be modified using the Poisson summation formula (Equation 3.82). Then obtain:

$$A_{\text{rcstr}}(x,y) = \int\limits_{-\infty}^{\infty}\int\limits_{-\infty}^{\infty} W\left(x-\xi, y-\overline{\eta}\right)\tilde{H}_{\text{rec}}\left(\xi,\overline{\eta}\right)\sum_{k=0}^{N_1-1}\sum_{l=0}^{N_2-1}\tilde{A}_{k,l}\exp\left[i2\pi\left(\frac{k+u}{2N_1}p+\frac{l+v}{N_2}q\right)\right]$$

$$\times \sum_{do_x=-\infty}^{\infty}\sum_{do_y=-\infty}^{\infty}\delta\left(\frac{\overline{\xi}\Delta\xi}{\lambda Z}-\frac{k+u}{N_2}-do_x\right)\delta\left(\frac{\overline{\eta}\Delta\eta}{\lambda Z}-\frac{l+v}{N_2}-do_y\right)d\overline{\xi}\,d\overline{\eta}$$

$$= \int\limits_{-\infty}^{\infty}\int\limits_{-\infty}^{\infty} W\left(x-\xi, y-\overline{\eta}\right)\tilde{H}_{\text{rec}}\left(\xi,\overline{\eta}\right)\sum_{k=0}^{N_1-1}\sum_{l=0}^{N_2-1}\tilde{A}_{k,l}\exp\left[i2\pi\left(\frac{k+u}{2N_1}p+\frac{l+v}{N_2}q\right)\right]$$

$$\times \sum_{r=-\infty}^{\infty}\sum_{s=-\infty}^{\infty}\delta\left(\frac{\overline{\xi}\Delta\xi}{\lambda Z}-\frac{k+u}{N_2}-do_x\right)\delta\left(\frac{\overline{\eta}\Delta\eta}{\lambda Z}-\frac{l+v}{N_2}-do_y\right)d\overline{\xi}\,d\overline{\eta}$$

$$= \sum_{k=0}^{N_1-1} \sum_{l=0}^{N_2-1} \tilde{A}_{k,l} \exp\left[i2\pi\left(\frac{k+u}{2N_1} p + \frac{l+v}{N_2} q \right)\right]$$

$$\times \sum_{do_x=-\infty}^{\infty} \sum_{do_y=-\infty}^{\infty} \tilde{H}_{\text{rec}}\left[\left(\frac{k+u}{N_1} + do_x \right)\frac{\lambda Z}{\Delta\xi}, \left(\frac{l+v}{N_2} + do_y \right)\frac{\lambda Z}{\Delta\eta} \right]$$

$$\times W\left(x - \frac{k+u}{N_1}\frac{\lambda Z}{\Delta\xi} - do_x \frac{\lambda Z}{\Delta\xi}; y - \frac{l+v}{N_2}\frac{\lambda Z}{\Delta\eta} - do_y \frac{\lambda Z}{\Delta\eta} \right).$$

Derivation of Equation 5.88

$$\frac{\partial}{\partial I}\left[\left(s_1 - I\cos\left(\theta^{(SRC)}\right)\right)^2 + \left(s_1 - I\sin\left(\theta^{(SRC)}\right)\right)^2 \right]$$

$$= -2\left(s_1 - I\cos\left(\theta^{(SRC)}\right)\right)\cos\left(\theta^{(SRC)}\right) - 2\left(s_2 - I\sin\left(\theta^{(SRC)}\right)\right)\sin\left(\theta^{(SRC)}\right)$$

$$= -2\left(\left[s_1 \cos\left(\theta^{(SRC)}\right) + s_2 \sin\left(\theta^{(SRC)}\right)\right] + 2I\left(\cos^2\left(\theta^{(SRC)}\right) + \sin^2\left(\theta^{(SRC)}\right)\right)\right) = 0,$$

$$\frac{\partial}{\partial\theta^{(SRC)}}\left[\left(s_1 - I\cos\left(\theta^{(SRC)}\right)\right)^2 + \left(s_1 - I\sin\left(\theta^{(SRC)}\right)\right)^2 \right]$$

$$= 2I\left(s_1 - I\cos\left(\theta^{(SRC)}\right)\right)\sin\left(\theta^{(SRC)}\right) - 2I\left(s_2 - I\sin\left(\theta^{(SRC)}\right)\right)\cos\left(\theta^{(SRC)}\right)$$

$$= 2I\left(s_1 \sin\left(\theta^{(SRC)}\right) - s_2 \cos\left(\theta^{(SRC)}\right)\right) = 0,$$

which gives

$$\begin{cases} I = s_1 \cos\theta^{(SRC)} + s_2 \sin\theta^{(SRC)} \\ s_1 \sin\theta^{(SRC)} = s_2 \cos\theta^{(SRC)}, \end{cases}$$

from which it follows:

$$\theta^{(SRC)} = \arctan\frac{s_2}{s_1}$$

$$I = \left(\frac{s_1^2}{s_2} + s_2 \right)\sin\theta = \frac{s_1^2 + s_2^2}{s_2}\sin\theta = \frac{s_1^2 + s_2^2}{s_2}\frac{\tan\theta}{\sqrt{1+\tan^2\theta}}$$

$$= \frac{s_1^2 + s_2^2}{s_2}\frac{s_2}{s_1\sqrt{1 + s_2^2/s_1^2}} = \sqrt{s_1^2 + s_2^2}.$$

Derivation of Equation 5.89

$$\varepsilon_\theta = d\hat{\theta}_{ML}^{(SRC)} = d\left(\arctan\frac{s_2}{s_1}\right) = ds_2\frac{\partial}{\partial s_2}\arctan\frac{s_2}{s_1} + ds_1\frac{\partial}{\partial s_1}\arctan\frac{s_2}{s_1}$$

$$= \frac{n_2}{s_1}\frac{1}{1+\left(\dfrac{s_2}{s_1}\right)^2} - \frac{n_1 s_2}{s_1^2}\frac{1}{1+\left(\dfrac{s_2}{s_1}\right)^2} = \frac{n_2 s_1 - n_1 s_2}{s_1^2 + s_1^2}$$

$$= \frac{n_2 I\cos\theta^{(SRC)} - n_1 I\sin\theta^{(SRC)}}{I^2} = \frac{n_2\cos\theta^{(SRC)} - n_1\sin\theta^{(SRC)}}{I}$$

$$\varepsilon_I = d\hat{I}_{ML}^{(SRC)} = ds_1\frac{\partial}{\partial s_1}\sqrt{s_1^2 + s_2^2} + ds_2\frac{\partial}{\partial s_2}\sqrt{s_1^2 + s_2^2} = n_1\frac{s_1}{\sqrt{s_1^2 + s_2^2}} + n_2\frac{s_2}{\sqrt{s_1^2 + s_2^2}}$$

$$= \frac{n_1 I\cos\theta^{(SRC)} + n_2 I\sin s\theta^{(SRC)}}{I} = n_1\cos\theta^{(SRC)} + n_2\sin s\theta^{(SRC)}.$$

Exercises

FocPlaneVar_Invar_reconstr_illustr.m

References

1. R. W. Gerchberg, W. O. Saxton, A practical algorithm for the determination of phase from image and diffraction plane pictures, *Optik*, 35, 237–246, 1972.
2. A. Papoulis, A new algorithm in spectral analysis and band-limited extrapolation, *IEEE Transactions on Circuits and Systems*, 22(9), 735–742, 1975.
3. R. A. Horn, C. R. Johnson, *Topics in Matrix Analysis*, Cambridge University Press, Cambridge, UK, 1991.
4. D. Donoho, Compressed sensing, *IEEE Transactions on Information Theory*, 52(4), 1289–1306, 2006.
5. J. W. Goodman, R. W. Lawrence, Digital image formation from electronically detected holograms, *Applied Physics Letters*, 11(3), 77–79, 1967.
6. M. A. Kronrod, N. S. Merzlyakov, L. P. Yaroslavsky, Reconstruction of a hologram with a computer, *Soviet Physics-Technical Physics*, 17(2), 419–420, 1972.
7. L. P. Yaroslavskii, N. S. Merzlyakov, *Methods of Digital Holography*, Consultance Bureau, N.Y., 1980. (English translation from Russian, In: *Methods of Digital Holography*, Editors L. P. Yaroslavskii, N.S. Merzlyakov, Moscow, Izdatel'stvo Nauka, 1977. 192 p.)

8. U. Schnars, W. Jüptner, Direct recording of holograms by a CCD target and numerical reconstruction, *Applied Optics*, 33(2), 179–181, 1994.
9. I. Yamaguchi, T. Zhang, Phase-shifting digital holography, *Optics Letters*, 22(16), 1268–1270, 1997.
10. E. N. Leith, J. Upatnieks, New techniques in wave front reconstruction, *JOSA*, 51, 1469–1473, 1961.
11. Y. N. Denisyuk, Photographic reconstruction of the optical properties of an object in its own scattered radiation field, *Dokl. Akad. Nauk SSSR*, 144, 1275–1279, 1962.
12. L. Yaroslavsky, Introduction to digital holography, Bentham E-book Series, In: *Digital Signal Processing in Experimental Research*, vol. 1, Editors L. Yaroslavsky and J. Astola, 2009, ISSN: 1879-4432, eISBN: 978-1-60805-079-6.

6

Image Resampling and Building Continuous Image Models

Accurate and fast image resampling is a key operation in many digital image processing applications such as multimodality data fusion, image mosaicking, image reconstruction from projections, image superresolution from image sequences, stabilization of video images distorted by atmosphere turbulence, target location and tracking with subpixel accuracy, and so on to name a few. Image resampling assumes reconstruction of a continuous approximation of the original nonsampled image by means of interpolation of available image samples to obtain samples "in-between" the available ones. Since image samples are obtained using shift (convolutional) discretization functions (Equation 3.4), continuous image approximation should be performed in computer through digital convolution. A number of convolutional interpolation methods are known, beginning from the simplest and the least accurate nearest-neighbor and linear (bilinear, for 2D case) interpolations to more accurate cubic (bicubic, for 2D case) and higher-order spline methods. How can one evaluate the interpolation accuracy of these methods? Is perfect interpolation of sampled data possible, which does not introduce to signals any distortions additional to those caused by signal sampling? This chapter answers these questions. In the section "Perfect Resampling Filter," we introduce the notion of the perfect resampling filter and show that discrete sinc interpolation as a discrete implementation of the ideal low-pass filtering dictated by the sampling theory is the gold standard for resampling sampled data. In the section "Fast Algorithms for Discrete Sinc Interpolation and Their Applications," we describe methods for efficient algorithmic implementation of discrete sinc interpolation for image subsampling, fractional shift, and rotation. In the section "Discrete Sinc Interpolation versus Other Interpolation Methods: Performance Comparison," we provide experimental evidence of the superiority of the discrete sinc interpolation compared to other convolutional interpolation methods. In the sections "Numerical Differentiation and Integration" and "Local ("Elastic") Image Resampling: Sliding Window Discrete Sinc Interpolation Algorithms," we illustrate applications of discrete sinc interpolation principles to accurate signal differentiation and integration and to image reconstruction from projections.

Perfect Resampling Filter

Consider optimization of signal interpolation by means of linear filtering implemented as digital convolution:

$$\tilde{a}_k = \sum_{n=0}^{N-1} h_n^{(intp)} a_{k-n},$$

(6.1)

where $\{\tilde{a}_k\}$ are samples of a signal obtained as a result of interpolation of initial signal samples $\{a_k\}$, $\{h_n^{(intp)}\}$ are samples of the interpolation filter PSF, and N is the number of available signal samples.

For the purposes of the design of the *perfect resampling filter*, one can regard signal coordinate shift as a general resampling operation. This is justified by the fact that samples of the resampled signal for any arbitrary signal resampling grid can be obtained one by one through the corresponding shifts of the original signal to the given sample position.

The *perfect shifting filter* is the filter that generates a shifted copy of the input signal with preservation of the original analog signal spectrum in its base band defined by the signal sampling rate. In order to further proceed with derivation of the perfect shifting filter, we will need the following characteristics of digital filters and their properties formulated in the sections "Digital Convolution" and "DFTs and Discrete Frequency Response of Digital Filter":

- The overall frequency response $OFR(f)$ of digital filter is its frequency response with respect to continuous signals that correspond to samples of the filter input signals.

- For ideal signal sampling and reconstruction antialiasing low-pass filters, overall frequency response $OFR(f)$ of digital filter coincides with its continuous frequency response $CFR(f)$ within the signal base band $[-1/2\Delta x, 1/2\Delta x]$ defined by the signal sampling interval Δx.

- Discrete frequency response $\{\eta_r\}$ of a digital filter with PSF $\{h_n\}$ applied to sampled signals of N samples is DFT of its PSF:

$$\eta_r = \frac{1}{\sqrt{N}} \sum_{n=0}^{N-1} h_n \exp\left(i2\pi \frac{nr}{N}\right).$$

(6.2)

- Continuous frequency response $CFR(f)$ of digital filter with discrete frequency response $\{\eta_r\}$ is a function interpolated from $\{\eta_r\}$:

$$CFR(f) = \sum_{r=0}^{N-1} \eta_r \, \text{sincd}[N; \pi(f - r\Delta f)/\Delta f],$$

(6.3)

with the discrete sinc-function

$$\text{sincd}(N;x) = \frac{\sin(x)}{N\sin(x/N)} \tag{6.4}$$

as an interpolation kernel.

In these formulas, f is the frequency parameter of the integral Fourier transform, Δf is spectrum sampling interval; for signal sampling interval Δx and cardinal sampling, $\Delta f = 1/N\Delta x$.

Coefficients $\{\eta_r\}$ of the discrete frequency response of digital filter are samples of its continuous frequency response $CFR(f)$ taken, with sampling interval Δf, at sampling points $\{r\Delta f\}$: $\{\eta_r = H_{CFR}(f = r\Delta f)\}$.

According to these definitions, overall continuous frequency response $OFR_{\delta x}^{(Shift)}(f)$ of the perfect shifting filter for the x-coordinate shift δx must be equal, in the signal base band $[-1/2\Delta x, 1/2\Delta x]$ to the frequency response $H_{\delta x}^{(Shift)}(f)$ of the continuous δx-shifting filter. The latter, by virtue of the Fourier transform shift theorem, is equal to $\exp(i2\pi f\delta x)$. Therefore

$$OFR_{\delta x}^{(Shift)}(f) = H_{\delta x}^{(Shift)}(f) = \exp(i2\pi f\delta x). \tag{6.5}$$

From the above-formulated properties of the overall, continuous, and discrete frequency responses of digital filters, it follows that discrete frequency response coefficients $\left\{\eta_r^{(Shift)}(\delta x)\right\}$ of the perfect δx-shifting filter, for indices from $r = 0$ to $r = (N-1)/2$ for odd N and to $r = N/2$ for even N, which correspond to the lowest and highest frequencies in the signal base band, must be samples, in sampling points $\{r/N\Delta x\}$, of its continuous frequency response. The rest of the coefficients should be set according to the symmetry property of DFT for real valued data (Equation 4.70). Thus, for odd number of signal samples N, coefficients $\left\{\eta_r^{(Shift)}(\delta x)\right\}$ must be set to

$$\begin{aligned} \eta_r^{(Shift)}(\delta x) &= \exp\left(i2\pi\frac{r\delta x}{N\Delta x}\right), & r &= 0,1,...,(N-1)/2 \\ \eta_r^{(Shift)}(\delta x) &= \eta_{N-r}^{*(Shift)}(\delta x), & r &= (N+1)/2,...,N-1 \end{aligned} \tag{6.6}$$

For even number of signal samples N, from the same requirement $\eta_r^{(Shift)}(\delta x) = \eta_{N-r}^{*(Shift)}(\delta x)$, it follows that the coefficient $\eta_{N/2}^{(Shift)}(\delta x)$, which corresponds to the signal highest frequency in its base band, must be a real number. Because of that, for even N this coefficient cannot be taken just as a sample of $\exp(i2\pi(r\delta x/N\Delta x))$ for $r = N/2$ and requires a special treatment. We select settings:

$$\eta_{r,opt}^{(intp)}(\delta\tilde{x}) = \begin{cases} \exp\left(i2\pi\dfrac{r\delta x}{N\Delta x}\right), & r = 0,1,...,N/2-1 \\[3mm] C\cos\left(\pi\dfrac{\delta x}{\Delta x}\right), & r = N/2 \end{cases} \tag{6.7}$$

$$\eta_{r,opt}^{(intp)}(\delta\tilde{x}) = \left(\eta_{N-r,opt}^{(intp)}(\delta x)\right)^{*}, \qquad r = N/2+1,...,N-1,$$

where C is a weight coefficient that defines signal spectrum shaping at its highest frequency component. In what follows we will consider, for even N, the following three options for C: Case_0: $C = 0$; Case_1: $C = 1$; Case_2, $C = 2$.

Applying to Equations 6.6 and 6.7 inverse discrete Fourier transform, one can obtain (see Appendix) that, for odd N, PSF of the perfect shifting filter defined by Equation 6.6 is

$$h_n^{(Shift)}(\delta x) = sincd[N, \pi(n - \delta\tilde{x}/\Delta x)] \tag{6.8}$$

and for even N, Case 0 and Case 2, perfect shifting filter PSFs are

$$h_n^{(Shift,0)}(\delta x) = \overline{sincd}[N-1; N; \pi(n - \delta x/\Delta x)] \tag{6.9}$$

and

$$h_n^{(Shift,2)}(\delta x) = \overline{sincd}[N+1; N; \pi(n - \delta x/\Delta x)], \tag{6.10}$$

respectively, where $\overline{sincd}(\cdot;\cdot;\cdot)$ is the general discrete sinc-function (Equation 4.54). Case_1 is obviously a combination of Case_0 and Case_2:

$$h_n^{(Shift,1)}(\delta x) = \left[h_n^{(Shift,0)}(\delta x) + h_n^{(Shift,2)}(\delta x)\right]/2 = \overline{sincd}[\pm1; N; \pi(n - \delta x/\Delta x)], \tag{6.11}$$

where

$$\overline{sincd}(\pm1; N; x) = [\overline{sincd}(N; N-1; x) + \overline{sincd}(N; N+1; x)]/2$$

$$= \left[\frac{\sin[(N-1)x/N] + \sin[(N+1)x/N]}{N\sin(x/N)}\right]\Big/2$$

$$= \frac{\sin(x)}{N\sin(x/N)}\cos(x/N)$$

$$= \cos(x/N)sincd(N; x). \tag{6.12}$$

Numerical interpolation using discrete sinc-functions (*sincd-functions*) given by Equations 6.8 through 6.11 as the interpolation kernel is called *discrete sinc interpolation*.

Discrete sinc-function is the point spread function of the ideal digital low-pass filter, whose discrete frequency response—DFT of its PSF—is a rectangular function and which is a discrete representation of the ideal continuous low-pass filter required, according to the sampling theory, for signal restoration from its samples. Figure 6.1 shows absolute value of the continuous frequency response of the discrete sinc interpolator (solid line) along with absolute values of samples of the discrete frequency response of the ideal low-pass digital filter (stems).

Three above versions of sincd-functions for Cases_0–2 are illustrated for comparison in Figure 6.2. As one can see from this figure, sincd-function for Case 1 converges to zero substantially faster than that for cases 0 and 2, thanks to its spectrum shaping: the spectral coefficient $\eta_{N/2}^{(int1)}$ corresponding to the highest frequency is halved. Due to this property, the Case 1 discrete sinc interpolation produces less ringing oscillations in the interpolated signal, which makes it preferable in practical applications.

From the above reasoning, it follows that discrete sinc interpolation secures perfect, for a given finite number of signal samples, resampling of discrete signals with preservation of the corresponding continuous signal spectra in their sampling points. All resampling filters with PSFs other than discrete sinc-function will distort samples of signal spectrum in the signal base band and, therefore, introduce interpolation error additional to signal distortion due to sampling. It is in this sense that discrete sinc interpolation can be regarded the "gold standard" of discrete signal interpolation. In the section "Discrete Sinc Interpolation versus Other Interpolation Methods: Performance Comparison," we will compare discrete sinc interpolation with

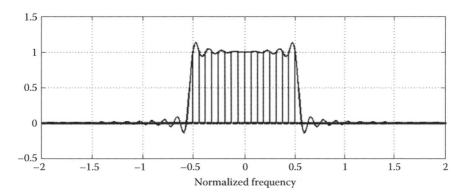

FIGURE 6.1

Continuous (solid line) and discrete (stems) frequency responses of the discrete sinc interpolator. Frequency indices are normalized so that interval [−0.5 ÷ 0.5] represents signal base band.

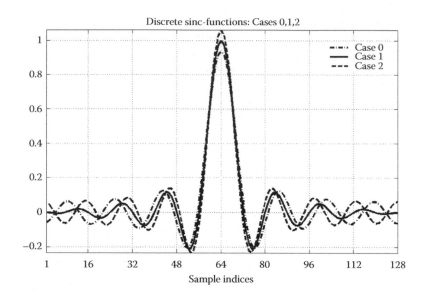

FIGURE 6.2
Comparison of three versions of sincd-functions.

other discrete interpolation methods and provide experimental evidence of its perfect performance and superiority.

Fast Algorithms for Discrete Sinc Interpolation and Their Applications

Signal Subsampling (Zooming-In) by Means of DFT or DCT Spectra Zero Padding

One of the basic image resampling tasks is image subsampling (zooming-in), that is, computing, from the given set of samples, a set of in-between samples. From properties of signal DFT spectra of sparse signals, discussed in the section "Convolutional Discrete Fresnel and Angular Spectrum Propagation Transforms" (Equations 4.85, 4.90, and 4.91), it straightforwardly follows that discrete sinc interpolated signal subsampling can be achieved by means of zero padding its DFT spectrum. Given subsampling (zoom) factor L, this algorithm is described by the equation

$$\tilde{a}_{\tilde{k}} = \text{IFFT}_{LN}\{\text{DFT\_ZP}_L[\text{FFT}_N(a_k)]\}, \tag{6.13}$$

where $\{a_k\}, k = 0,1,\ldots,N-1$ are input signal samples, $\{\tilde{a}_{\tilde{k}}\}$ are output signal samples, $\tilde{k} = 0, 1, \ldots, NL - 1$, $\text{FFT}_N(\cdot)$ and $\text{IFFT}_{LN}(\cdot)$ are N-point direct and LN-point inverse fast Fourier transform operators, respectively, and DFT_ZP$_L$[·] is a zero-padding operator, which forms, from N-points sequence of samples, an LN-points sequence by padding the former with $(L-1)N$ zeros. When N is an odd number, zeros are placed between $(N-1)/2$-th and $(N+1)/2$-th samples of the N-points sequence. When N is an even number, then

i. $(L-1)N+1$ zeros are placed between $N/2-1$-th and $N/2+1$-th samples of the N-points sequence and $N/2$-th sample is discarded, or

ii. $(L-1)N$ zeros are placed after $N/2$-th sample and then the sequence repeated beginning of its $N/2$-th sample, or

iii. $N/2$-th sample of the sequence is halved, $(L-1)N$ zeros are placed after it and then $N/2$-th through $(N-1)$-th samples of the sequence are placed at the end, $N/2$-th sample being also halved.

Cases (i)–(ii) implemented above-described Case_0, Case_2, and Case_1 discrete sinc interpolation, respectively. 2D image subsampling is implemented as separable in two consecutive steps over each of the two coordinates.

Although discrete sinc interpolation is perfect in preserving signal spectrum, it has one major drawback. As interpolation kernel discrete sinc-function decays to zero relatively slow, discrete sinc interpolation, being implemented through FFT, tends to produce heavy boundary effects that may propagate quite far from signal boundaries. It is especially an important issue in image processing. With the use of FFT in the resampling algorithm (Equation 6.13), images are treated as being periodic in both coordinates. Therefore, samples at their left and right and, respectively, upper and bottom borders are, virtually, immediate neighbors in the interpolation process. Therefore, any discontinuity between opposite border samples will cause heavy oscillations due to tales of the discrete sinc-function that propagate far away from the borders.

A simple and very efficient solution of this problem is zero padding in the domain of discrete cosine transform (DCT):

$$\tilde{a}_{\tilde{k}} = \text{IDCT}_{LN}\{\text{DCT\_ZP}_L[\text{DCT}_N(a_k)]\}, \tag{6.14}$$

where $\text{DCT}_N(\cdot)$ and $\text{IDCT}_{LN}(\cdot)$ are N-points fast direct and LN-points inverse discrete cosine transforms and $\text{DCT\_ZP}_L[\cdot]$ is a DCT spectrum zero-padding operator that places $(L-1)N$ zeros after the last $(N-1)$-th DCT spectrum sample. Similarly to the Case_1 DFT zero padding, for faster decay of the interpolation kernel, it is also advisable to halve the last two spectral samples that, for DCT, represent the signal highest frequency component.

Figure 6.3 demonstrates improvement, in terms of boundary effect oscillations, of image zooming in by means of DCT spectrum zero-padding algorithm compared to DFT spectrum zero padding.

FIGURE 6.3
Zooming of an image fragment outlined by white box (a) by means of zero padding its DFT spectrum (b) and zero-padding DCT spectrum (c).

As it is shown in Appendix, PSF of the DCT spectrum zero padding with halving its last two components is defined by the equation

$$
\tilde{\tilde{a}}_k = \frac{\alpha_0^{DCT}}{\sqrt{2LN}} + \frac{1}{N\sqrt{L}} \sum_{n=0}^{N-1} a_n \left\{ \frac{\sin\left[\pi(N+1/2NL)\left(\tilde{n}L - \tilde{k}\right)\right]}{\sin\left[\pi\left(\tilde{n}L - \tilde{k}\right)/2NL\right]} \cos\left[\pi\left(\tilde{n}L - \tilde{k}\right)/2L\right] \right.
$$

$$
+ \frac{\sin\left[\pi(N+1/2NL)\left(\tilde{n}L + \tilde{k}\right)\right]}{\sin\left[\pi\left(\tilde{n}L + \tilde{k}\right)/2NL\right]} \cos\left[\pi\left(\tilde{n}L + \tilde{k}\right)/2L\right]
$$

$$
+ \frac{\sin\left[\pi(N-1/2NL)\left(\tilde{n}L - \tilde{k}\right)\right]}{\sin\left[\pi\left(\tilde{n}L - \tilde{k}\right)/2NL\right]} \cos\left[\pi \frac{N-2}{2NL}\left(\tilde{n}L - \tilde{k}\right)\right]
$$

$$
+ \frac{\sin\left[\pi(N-1/2NL)\left(\tilde{n}L + \tilde{k}\right)\right]}{\sin\left[\pi\left(\tilde{n}L + \tilde{k}\right)/2NL\right]} \cos\left[\pi \frac{N-2}{2NL}\left(\tilde{n}L + \tilde{k}\right)\right] \right\}. \tag{6.15}
$$

Frequency responses and PSFs of DFT and DCT zero-padding signal 4×-subsampling are shown for comparison in Figure 6.4. One can see from these figures that they are quite close to one another. One can also see from Equation 6.15 that DCT zero-padding interpolation is not cyclic shift invariant and is implemented by two pairs of sincd-function interpolation kernels 1/2-shifted and mirror reflected with respect to each other. Its behavior is illustrated in Figure 6.5 for sampling positions at signal left and right boundaries and at the middle of the signal extent.

Computation-wise, signal L-times zooming by mean of zero padding its DFT or DCT spectra requires, with the use of fast Fourier transform or fast DCT algorithms, $O(\log N)$ operations per each of signal N samples for direct transform and $O(L \log NL)$ operation for inverse transform of zero-padded

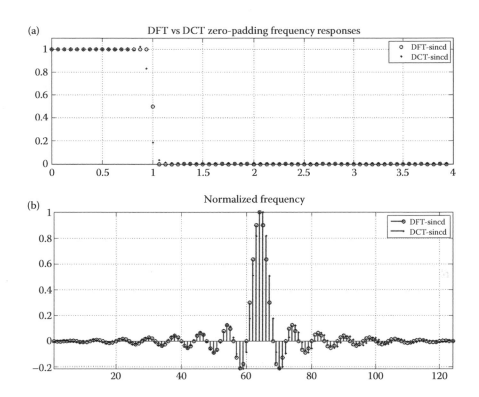

FIGURE 6.4
Frequency responses (a) and point spread functions (b) of DFT (DFT-sincd) and DCT (DCT-sincd) zero-padding signal 4×-subsampling ($N = 32$, $L = 4$).

spectra, which is quite inefficient for large L. Moreover, when conventional radix 2 FFT and fast DCT algorithms are used, zooming factor L should be selected to be an integer power of 2. These two facts limit to a certain degree applicability of the spectra zero-padding algorithms. However, their use might be a good practical solution when one works in an appropriate software environment, such as, for instance, MATLAB. Image zooming by means of DCT spectra zero padding can also be naturally used when images are represented in a compressed form such as in JPEG compression. In this case, zooming can be carried out without the need to decompress images.

DFT- and DCT-Based Signal Fractional Shift Algorithms and Their Basic Applications

In this section, we describe signal resampling algorithms that are based on the perfect $\delta\tilde{x}$-shifting filter introduced in the section "Perfect Resampling Filter." The filter is designed in DFT domain and, therefore, it can be implemented using fast Fourier transform with computational complexity of

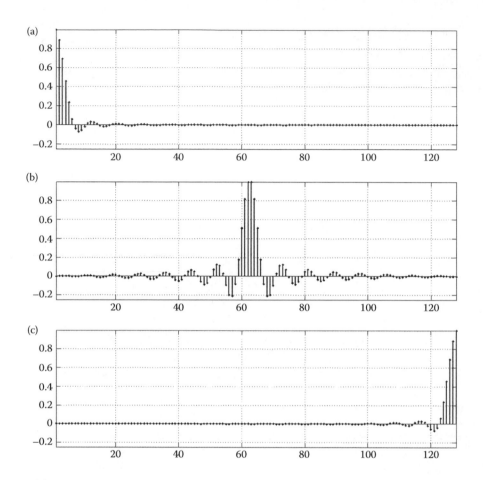

FIGURE 6.5
Point spread function of signal 4×-subsampling by means of DCT zero padding for three different sampling positions: at the beginning (a), in the middle (b), and at the end (c) of the signal interval.

$O(\log N)$ operations per output signal sample. The algorithm is described by the equation

$$a_k^{(\delta x)} = \text{IFFT}_N \left\{ \left\{ \eta_r^{(Shift)}(\delta x) \right\} \cdot [\text{FFT}_N(a_k)] \right\}, \quad k = 0, 1, ..., N - 1, \qquad (6.16)$$

where $\text{FFT}_N(\cdot)$ and $\text{IFFT}_N(\cdot)$ are direct and inverse N-point fast Fourier transforms, · symbolizes component-wise product elements of two arrays and $\left\{ \eta_r^{(Shift)}(u\Delta x) \right\}$ are defined, for odd and even N by the above Equations 6.6 and 6.7.

This algorithm is suitable for numerous applications. First of all, it allows generating arbitrarily shifted discrete sinc interpolated copies of signals.

Another application is signal/image zooming-in with an arbitrary integer zoom factor. For zoom factor L, signal/image zooming can be implemented through generating subsequently computed $L-1$ signal/image copies shifted by corresponding multiple of $1/L$ shifts:

$$a_k^{(l\delta x)} = \text{IFFT}_N \left\{ \left[\eta_r^{(Shift)} \left(\frac{l\Delta x}{L} \right) \right] \cdot [\text{FFT}_N(a_k)] \right\}, \quad l = 1,...,L-1. \quad (6.17)$$

The work of this algorithm is illustrated by a plot in Figure 6.6 obtained using program sincd_interpol_demo.m provided in Exercises.

Mathematically, such image zooming-in method is completely equivalent to the above-described DFT spectrum zero padding, but is much more computationally efficient. Its computational complexity for L-factor zooming is $O(\log N)$ per output sample rather than $O(\log LN)$ for the zero-padding method, and this complexity does not depend on whether L is power of two or not.

Obtained in this way, zoomed-in images represent the most perfect "continuous" models of images that can be generated for a prescribed subsampling interval. Such models can be used, in particular, for performing image resampling over an arbitrary sampling grid. In this process, required image samples, whose positions do not coincide with one of sampling nodes of the denser sampling grid of the zoomed-in image, can be approximated by the nearest available sample of the zoomed-in image as it is illustrated in Figure 6.7a. If the selected zoom factor is provided appropriately, the nearest-neighbor interpolation in combination with discrete sinc interpolated zooming-in will not compromise interpolation accuracy substantially. For instance, as one can see from the plot of the frequency response of the nearest-neighbor interpolator shown in Figure 6.7b, for

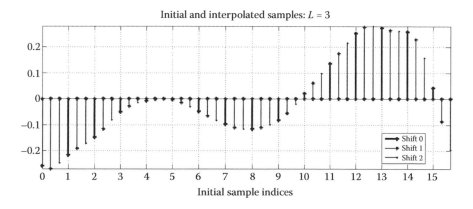

FIGURE 6.6
Initial signal samples (bold) and two sets of its subsamples shifted by 1/3 (shift 1) and 2/3 (shift 2) of the sampling interval by means of the perfect shifting filter.

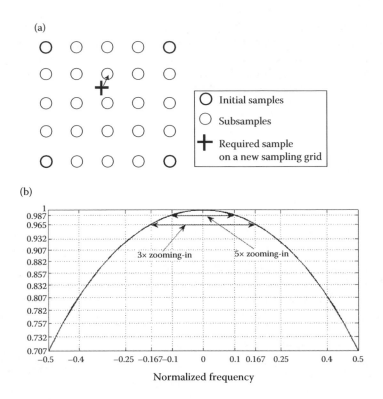

FIGURE 6.7
Nearest-neighbor interpolation in resampling of zoomed-in images (a) and its frequency response (b) within zoomed-in image baseband ([−0.5 ÷ 0.5]). Double arrows show base bands of initial images before their 5× and 3× zooming-in.

zoom factor 5, decay of the frequency response on the highest spatial frequency of the initial image is only 1.3% and for zoom factor 3, it is 3.5%, which can also be frequently tolerated.

One of the important applications, in which "continuous" image models are required, is fast location and tracking of moving targets in video sequences. In this case, template images of the target with arbitrary orientation and scale can be very rapidly computed by means of corresponding resampling of the magnified template image of the target obtained with a sufficiently large zoom factor. Other application examples are image reconstruction from projections using the direct Fourier reconstruction algorithm and image reconstruction from fan-beam projections, which will be discussed below in the section "Image Data Resampling for Image Reconstruction from Projections."

Two more applications worth of special mention are *"quasicontinuous"* *Fourier spectrum analysis* for detection and localization, with subpixel accuracy, of periodical components and image correlation analysis for detection and localization, with subpixel accuracy, of position of signal correlation peaks.

The "quasicontinuous" spectrum analysis with subpixel resolution in frequency domain $l\Delta f /L = 1/L\,N\Delta x$ can be performed by means of applying to the signal shifted DFT$(0,l/L)$ subsequently L times:

$$\alpha_r^{(l)} = \mathrm{FFT}_N\left[a_k \exp\left(i2\pi \frac{kl}{LN} \right)\right], \quad l = 1,...,L-1. \qquad (6.18)$$

As it is shown in Appendix, intermediate spectrum samples obtained in this way are discrete-sinc interpolated from samples $\{\alpha_r\}$ of the basic DFT spectrum

$$\alpha_r^{(l)} = \sum_{s=0}^{N-1} \alpha_r \, \mathrm{sincd}[N;\pi(s-r-l/N)]\exp\left[i\pi\frac{(N-1)(s-r-l/N)}{N} \right]. \qquad (6.19)$$

Correlation analysis with subpixel resolution of signals $\{a_k\}$ and $\{b_k\}$ can be performed using the algorithm that follows from the convolution theorem for DFTs (Equation 4.63):

$$c_k^{(l\delta x)} = \mathrm{IFFT}_N\left\{\left[\eta_r^{(Shift)}\left(\frac{l\Delta x}{L}\right)\right]\cdot[\mathrm{FFT}_N(a_k)]\cdot[\mathrm{FFT}_N(b_k)]\right\}, \quad l = 1,...,L-1. \qquad (6.20)$$

Being a cyclic convolution, the "DFT-based" signal fractional shift algorithm suffers from the same boundary effects as described above in DFT zero-padding algorithm. The efficient practical solution of the problem is computing convolution in DCT domain instead of DFT domain as it is described in the section "Signal Convolution in the DCT Domain." The DCT-based signal δx-shifting algorithm is defined by the equation

$$\tilde{a}_k = \frac{1}{\sqrt{2}} \mathrm{IDCT}\left\{\alpha_r^{(DCT)} \mathrm{Re}\,\tilde{\eta}_r^{\{Shift\}}(\delta x)\right\} - \left\{(-1)^k\right\}\cdot \mathrm{IDCT}\left\{\alpha_{N-r}^{(DCT)} \mathrm{Im}\,\tilde{\eta}_{N-r}^{\{Shift\}}(\delta x)\right\}$$

$$= \frac{1}{\sqrt{2N}}\left\{\alpha_0^{(DCT)}\tilde{\eta}_0^{\{Shift\}}(\delta x) + 2\sum_{r=1}^{N-1}\alpha_r^{(DCT)}\mathrm{Re}\left(\tilde{\eta}_r^{Shift}(\delta x)\right)\cos\left(\pi\frac{k+1/2}{N}r\right)\right.$$

$$\left. -2(-1)^k \sum_{r=1}^{N-1}\alpha_{N-r}^{(DCT)}\mathrm{Im}\left(\tilde{\eta}_{N-r}^{\{Shift\}}(\delta x)\right)\cos\left(\pi\frac{k+1/2}{N}r\right)\right\} \qquad (6.21)$$

that follows from Equation 4.119. The filter coefficients $\left\{\tilde{\eta}_r^{(Shift)}(\delta x)\right\}$ in this equation are defined by Equations 4.113 and 4.116, in which discrete sinc-functions defined by Equations 6.8 through 6.11 should be used as the convolution kernel $\{h_n\}$ of the filter in Equation 4.113. Because of this, the DCT-based algorithm is equivalent, in terms of the interpolation accuracy, to the above DFT-based perfect fractional shift algorithm.

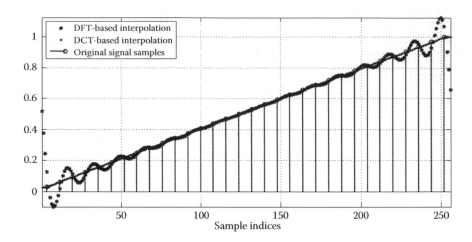

FIGURE 6.8
DFT-based versus DCT-based subsampling of a signal shown by dotted stems.

In application to signal L times zooming-in, the algorithm can be applied repeatedly $L - 1$ times similar to the DFT-based algorithm:

$$
\tilde{a}_k^{l\delta x} = \frac{1}{\sqrt{2}} \text{IDCT}\left\{ \alpha_r^{(DCT)} \text{Re}\left(\tilde{\eta}_r^{\{Shift\}}(\delta x) \right)^l \right\} - \left\{ (-1)^k \right\} \cdot \text{IDCT}\left\{ \alpha_{N-r}^{(DCT)} \text{Im}\left(\tilde{\eta}_{N-r}^{\{Shift\}}(\delta x) \right)^l \right\}
$$

$$
= \frac{1}{\sqrt{2N}} \left\{ \alpha_0^{(DCT)} \left(\tilde{\eta}_0^{(Shift)}(\delta x) \right)^l + 2 \sum_{r=1}^{N-1} \alpha_r^{(DCT)} \text{Re}\left(\left(\tilde{\eta}_r^{(Shift)}(\delta x) \right)^l \right) \cos\left(\pi \frac{k+1/2}{N} r \right) \right.
$$

$$
\left. - 2(-1)^k \sum_{r=1}^{N-1} \alpha_{N-r}^{(DCT)} \text{Im}\left(\left(\tilde{\eta}_r^{(Shift)}(\delta x) \right)^l \right) \cos\left(\pi \frac{k+1/2}{N} r \right) \right\} \qquad l = 1, \dots, L-1
$$

$$(6.22)$$

using, at l-th step, $\left(\tilde{\eta}_r^{(Shift)}(\delta x) \right)^l$ as the filter coefficients.

With respect to boundary effects, DCT-based fractional shift algorithm is as efficient as the above-described DCT spectrum zero-padding algorithm. It is illustrated in Figure 6.8 on an example of subsampling (magnification) of a saw-tooth signal.

As one can see on the figure, oscillations that propagate from signal borders for DFT-based discrete sinc interpolation almost completely disappear when DCT-based discrete sinc interpolation algorithm is used.

Fast Image Rotation Using the Fractional Shift Algorithms

Rotation of a 2D coordinate system by an angle θ as a geometrical transformation of signal coordinates can be described as a multiplication of signal coordinate vector (x,y) by a rotation matrix $\mathbf{ROT}_\theta$:

$$\mathbf{ROT}_\theta \begin{bmatrix} \tilde{x} \\ \tilde{y} \end{bmatrix} = \begin{bmatrix} \cos\theta & -\sin\theta \\ \sin\theta & \cos\theta \end{bmatrix} \begin{bmatrix} x \\ y \end{bmatrix}. \tag{6.23}$$

In computers, physical coordinates (x,y) are represented, given sampling intervals $(\Delta x, \Delta y)$, by integer indices of pixels $\{k,l\}$:

$$\begin{bmatrix} x \\ y \end{bmatrix} = \begin{bmatrix} k\Delta x \\ l\Delta y \end{bmatrix} \tag{6.24}$$

and the rotation matrix is applied to the vector of indices:

$$\mathbf{ROT}_\theta \begin{bmatrix} k \\ l \end{bmatrix} = \begin{bmatrix} \cos\theta & -\sin\theta \\ \sin\theta & \cos\theta \end{bmatrix} \begin{bmatrix} k \\ l \end{bmatrix}. \tag{6.25}$$

Equation 6.25 describes the resampling rule that should be applied to input image pixel indices to generate indices of its rotated copy. In order to reduce the computational complexity of this transformation, it is advisable to factorize the rotation matrix into a product of three matrices each of which modifies only one coordinate:

$$\mathbf{ROT}_\theta = \begin{bmatrix} \cos\theta & -\sin\theta \\ \sin\theta & \cos\theta \end{bmatrix} = \begin{bmatrix} 1 & -\tan(\theta/2) \\ 0 & 1 \end{bmatrix} \begin{bmatrix} 1 & 0 \\ \sin\theta & 1 \end{bmatrix} \begin{bmatrix} 1 & -\tan(\theta/2) \\ 0 & 1 \end{bmatrix}. \tag{6.26}$$

This implementation of image rotation is known as the *three-pass rotation algorithm*. It carries out, on each of three passes, only shifts along one of the coordinates: along rows on the first pass, along columns on the second pass, and again along rows on the third pass as it is illustrated in Figure 6.9. Specifically, in rotation of an image of $N_x \times N_y$ pixels ($k = 0,1,\ldots,N_x - 1$, $l = 0,1,\ldots,N_y - 1$) around point ($0 \le k_0 \le N_x - 1$; $0 \le l_0 \le N_y - 1$) on the first pass k-th row is shifted by $\delta x_k/\Delta x = -\tan(\theta/2)(k - k_0)$, on the second pass l-th column is shifted by $\delta x_k/\Delta x = \sin\theta(l - l_0)$, and on the third pass again k-th row is shifted by $\delta x_k/\Delta x = -\tan(\theta/2)(k - k_0)$. The above-described DFT- or DCT-based signal fractional shift algorithms are ideally suited for performing these shifts.

As far as these algorithms implement a cyclic convolution, image rotation by this method entails characteristic aliasing artifacts at image borders. They are illustrated in Figure 6.9b. One can avoid them by inscribing the image into an array of correspondingly larger size, as it is shown in Figure 6.9a or by using only the aliasing free image part inside the circle of the diameter equal to the image linear size (see Figure 6.9d).

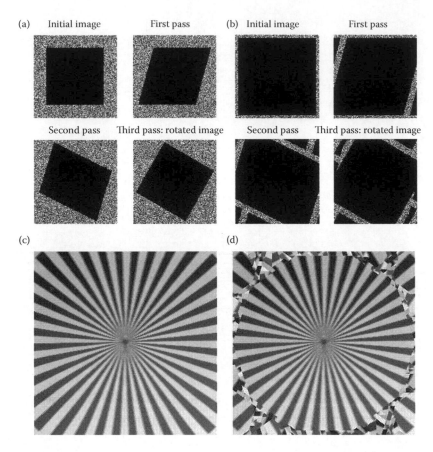

FIGURE 6.9
The principle of the three-pass image rotation algorithm and aliasing artifacts associated with implementation of interpolation as cyclic convolution: (a) rotation without aliasing; (b) rotation aliasing artifacts owing to the cyclicity of the convolution; (c) and (d) initial image and $10 \times 36°$-rotated image, which shows aliasing artifacts outside the circle of the diameter equal to the image linear size.

Image Zooming and Rotation Using "Scaled" and Rotated DFTs

DFT- and DCT-based fractional shift algorithms enable efficient signal/image discrete sinc interpolated subsampling and zooming-in with an integer zoom factor. Direct- and inverse-scaled discrete Fourier transform (ScDFT($u_x, v_f; \sigma$)):

$$
\alpha_r^{u,v;\sigma} = \frac{1}{\sqrt{\sigma N}} \sum_{k=0}^{N-1} a_k \exp\left[i2\pi \frac{(k+u_x)(r+v_f)}{\sigma N} \right],
$$

$$
a_k = \frac{1}{\sqrt{\sigma N}} \sum_{r=0}^{N-1} \alpha_r^{u,v;\sigma} \exp\left[-i2\pi \frac{(k+u_x)(r+v_f)}{\sigma N} \right]
$$

$$(6.27)$$

introduced in Chapter 4 (Equations 4.36 and 4.37), when applied with different scale parameters enable signal/image discrete sinc interpolated rescaling with an arbitrary scale factor. In order to prove this, let scale parameters for direct and inverse transforms be σ_d and σ_i, and shift parameters be $u_x^{(d)}, v_f^{(d)}$, and $u_x^{(i)}, v_f^{(i)}$, respectively. Then from Equations 6.27, obtain, with selection $v_f^{(i)} = v_f^{(d)} = v_f = -(N-1)/2$ (see Appendix):

$$a_k = \frac{1}{\sqrt{\sigma_i N}} \sum_{r=0}^{N-1} \alpha_r^{u,v;\sigma_d} \exp\left[-i2\pi \frac{\left(k + u_x^{(i)}\right)\left(r + v_f^{(i)}\right)}{\sigma_i N}\right]$$

$$= \frac{1}{\sqrt{\sigma_i \sigma_d}} \sum_{n=0}^{N-1} a_n \operatorname{sincd}\left[N; \pi\left(\frac{n + u_x^{(d)}}{\sigma_d} - \frac{k + u_x^{(i)}}{\sigma_i}\right)\right], \qquad (6.28)$$

which means discrete sinc interpolated signal resampling in scaled coordinates. In applications, one can set either of the scale parameters σ_d or σ_i to one.

As it is shown in Chapter 4, scaled DFT can be represented as a digital convolution, and therefore, by the convolution theorem, can be computed using FFT. Comparing with the above-described zooming-in algorithm, this method is not as fast, but it allows performing completely arbitrary image magnifying–demagnifying. It is especially important for the magnification factors in the range 0.5–2, which otherwise can be achieved with a comparable accuracy, using the above-described zooming-in algorithm, only through subsampling highly oversampled image.

One can supplement image rescaling with rotation using rotated scaled DFT introduced in Chapter 4 (Equation 4.52) with appropriately selected scale and rotation angle parameters. Similarly to the above-described signal shift or signal rescaling using shifted DFT or scaled DFT, one can use rotated DFT for image rescaling and rotation by applying RotDFT with appropriate scale and rotation angle parameters to signal SDFT spectrum. It is shown in Appendix that image $\tilde{a}_{k,l}$ obtained through inverse shifted scaled DFT of the image spectrum computed using shifted DFT $(u_x, u_y; v_f, v_p)$ is a discrete sinc interpolated rotated and scaled copy of initial image $\{a_{m,n}\}$:

$$\tilde{a}_{k,l} = \frac{1}{\sigma N} \sum_{r=0}^{N-1}\sum_{s=0}^{N-1} \left\{ \frac{1}{N} \sum_{m=0}^{N-1}\sum_{n=0}^{N-1} a_{m,n} \exp\left[i2\pi\left(\frac{\tilde{m}\tilde{r}}{N} + \frac{\tilde{n}\tilde{s}}{N}\right)\right] \right\}$$

$$\times \exp\left[i2\pi\left(\frac{\tilde{k}\cos\theta + \tilde{l}\sin\theta}{\sigma N}\tilde{r}^{(\sigma)} - \frac{\tilde{k}\sin\theta - \tilde{l}\cos\theta}{\sigma N}\tilde{s}^{(\sigma)}\right)\right]$$

$$= \frac{1}{\sigma} \sum_{m=0}^{N-1}\sum_{n=0}^{N-1} a_{m,n} \operatorname{sincd}\left\{N; \pi\left[\tilde{m} + \frac{\tilde{k}\cos\theta + \tilde{l}\sin\theta}{\sigma}\right]\right\}$$

$$\times \operatorname{sincd}\left\{N; \pi\left[\tilde{n} + \frac{\tilde{l}\cos\theta - \tilde{k}\sin\theta}{\sigma}\right]\right\}, \qquad (6.29)$$

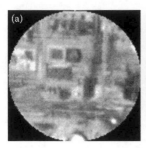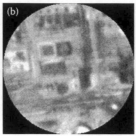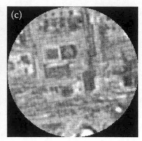

FIGURE 6.10
Simultaneous image rescaling, rotation, denoising, sharpening, and enhancement using RotScDFT: (a) initial image; (b) 10°-rotated and 1.7 times magnified image; (c) 10°-rotated, 1.7 times magnified, denoised, sharpened, and enhanced image.

where $\tilde{m} = m + u_1$, $\tilde{n} = n + u_2$, $\tilde{r} = r + v_{f_1}$; $s = \tilde{s} + v_{f_2}$; $\tilde{k} = k + u_1^{(\sigma)}$, $\tilde{l} = l + u_2^{(\sigma)}$, $\tilde{r} = r + v_{f_1}$; $\tilde{s}^{(\sigma)} = s + v_{f_2}^{(\sigma)}$, and $v_{f_1} = v_{f_2}^{(\sigma)} = v_{f_2} = v_{f_2}^{(\sigma)} = -(N - 1/2)$.

To conclude this discussion of discrete sinc-interpolation algorithms mention that, as these algorithms are implemented through processing in DFT or DCT domain, one can combine rotation and scaling with adaptive image restoration and enhancement through nonlinear modification of its spectrum such as soft/hard thresholding, P-th low dynamic range compression and alike described in Chapter 8. Figure 6.10 illustrates this option. It shows a result of a simultaneous image rotation and scaling along with a result of rotation, scaling, denoising, and enhancement by means of thresholding of low-energy spectral coefficients combined with rising of absolute values of remaining (not zeroed by thresholding) image spectral coefficients to a power $P < 1$ (in this particular case, $P = 0.5$).

Discrete Sinc Interpolation versus Other Interpolation Methods: Performance Comparison

In this section, we provide experimental data of comparison of discrete sinc interpolation with other more traditional numerical interpolation methods in terms of the interpolation accuracy and signal preservation. Compared with discrete sinc interpolation are methods offered by the MATLAB image processing toolbox: nearest-neighbor interpolation, linear (bilinear) interpolation and cubic (bicubic) spline interpolation. PSFs and frequency responses of the compared interpolation methods are shown in Figure 6.11.

Plots of frequency responses of the methods clearly reveal major drawback of the traditional interpolation methods: they tend to substantially attenuate high-frequency signal components within the signal base band (frequency interval $[-0.5 \div 0.5]$ in Figure 6.11c) and introduce substantial aliasing

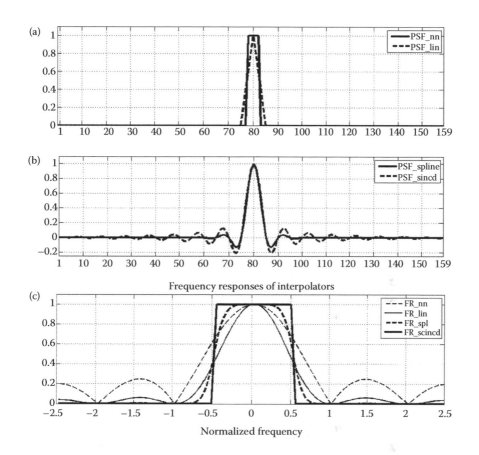

FIGURE 6.11
Point spread functions (a, b) and frequency responses (c) of nearest-neighbor (nn), linear (lin), bicubic spline (spl), and discrete sinc (sincd) interpolators.

frequency components outside this interval (in the case of signal subsampling). In imaging, this tendency results in image blurring that can heavily worsen visual image quality and its applicability for further analysis, for instance, object recognition, target location, and alike.

In Figures 6.12 through 6.14, results of comparison experiments carried out using program RotateComparis_demo_CRC.m (see Exercises) are presented that vividly illustrate this phenomenon on examples of multiple rotation of test images of a piece of printed text (Figure 6.12a) and of a pseudorandom image with uniform spectrum that was low-pass prefiltered to 0.7 of the base band in both directions ("Prus" image, Figure 6.12b) to exclude aliasing artifacts outside the circle of the radius equal to the highest horizontal and vertical spatial frequency.

With this program, for nearest-neighbor, bilinear, and spline interpolation methods, rotations were carried out using MATLAB program imrotate.m

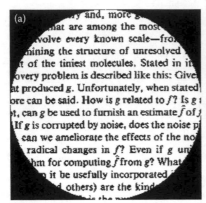

FIGURE 6.12
Test images for comparison of interpolation methods: "Text" image (a) and a pseudorandom image (b) with uniform spectrum within 0.7 of the base band ("Prus" image).

from the image processing toolbox. For discrete sinc interpolated rotation, the program contains a code that implements the three-step rotation algorithm through DFT-based fractional shift algorithm described in the previous section.

Images shown in Figure 6.13 clearly show that, after 60 rotations through 18° each, standard interpolation methods completely destroy the readability of the text, while discrete sinc interpolated rotated image is virtually not distinguishable from the original one.

In the analysis of interpolation errors, it is very instructive to compare their power spectra to see which spectral components suffered more. For this purpose, test image "Prus" is very appropriate. Figure 6.14 presents spectra of rotation errors computed as spectra of differences between the initial test image "Prus" and the results of its rotation by 1080° carried out in 60 steps using bilinear, bicubic, and discrete sinc interpolations. One can see in the figure that in the case of discrete sinc interpolation, error spectrum is practically zero within the base band circle, while for bicubic and for bilinear interpolation, error spectrum intensity is low only for low spatial frequencies and grows quite substantially to high frequencies.

DFT- and DCT-based image resampling algorithms are advantageous over more traditional methods not only in terms of the interpolation accuracy, but also in terms of the computational efficiency, which is evidenced, in particular, by the above-described experiments in image rotation using MATLAB implementations of the algorithms: numerical data on computation time data presented in the caption to Figure 6.13 show that discrete sinc interpolation works substantially faster than bicubic interpolation.

The computational efficiency of the interpolation error-free discrete sinc interpolation algorithms is rooted in the use of fast Fourier and fast DCT transforms. Perhaps, the best last concluding remark of this discussion of the

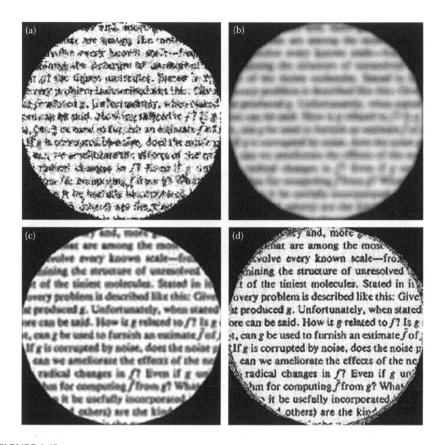

FIGURE 6.13

Discrete sinc interpolation versus conventional numerical interpolation methods used for $60 \times 18°$ rotations of the test image "Text": (a) nearest-neighbor interpolation, computation time (CT) 7.27 time units; (b) bilinear interpolation, C = 11.1 time units; (c) bicubic interpolation, CT = 17.7 time units; (d) discrete sinc interpolation, CT = 14.16 time units.

discrete sinc interpolation methods and their applications would be mentioning that fast Fourier transform algorithm was invented about 200 years ago by Carl Friedrich Gauss for the purpose of facilitating numerical interpolation, in fact by the method which we now call discrete sinc interpolation, of sampled data of astronomical observation [1].

Numerical Differentiation and Integration

Perfect Digital Differentiation and Integration

Signal numerical differentiation and integration are operations that require measuring infinitesimal increments of signals and their arguments. Therefore,

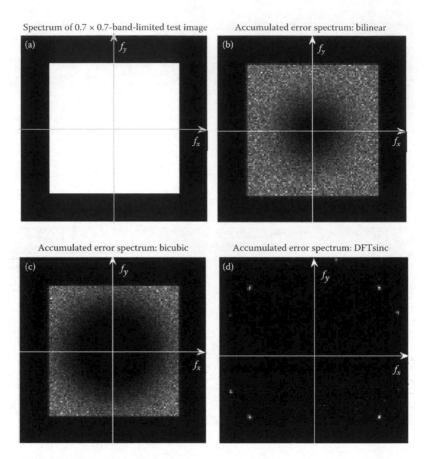

FIGURE 6.14
Spectrum of test image "Prus" (a) and spectra of rotation error after $60 \times 18°$ rotations of the image using bilinear (b), bicubic (c), and discrete sinc interpolation (c). All spectra are shown in frequency coordinates (white arrows) centered at spectrum zero frequency (dc-component), image lightness being proportional to error spectra intensity; bright points in (d) are spectral aliasing components intentionally left at the borders of the image base band to secure, for display purposes, the same image dynamic range as that of (b) and (c).

numerical computing signal derivatives and integrals assume one of another method of building "continuous" models of signals specified by their samples through explicit or implicit interpolation between available signal samples. Because differentiation and integration are shift invariant linear operations, methods of computing signal derivatives and integrals from their samples can be conveniently designed and compared in the Fourier transform domain.

Let the Fourier transform spectrum of a continuous signal $a(x)$ be $\alpha(f)$:

$$a(x) = \int_{-\infty}^{\infty} \alpha(f)\exp(-i2\pi fx)\,df. \qquad (6.30)$$

Then the Fourier spectrum of its derivative

$$\dot{a}(x) = \frac{d}{dx} a(x) = \int_{-\infty}^{\infty} [(-i2\pi f)\alpha(f)] \exp(-i2\pi fx) \, df \qquad (6.31)$$

will be $(-i2\pi f)\alpha(f)$ and the Fourier spectrum of its integral

$$\bar{a}(x) = \int a(x) \, dx = \int_{-\infty}^{\infty} \left[\left(-\frac{1}{i2\pi f} \right) \alpha(f) \right] \exp(-i2\pi fx) \, df \qquad (6.32)$$

will be $\alpha(f)/(-i2\pi f)$. Therefore, signal differentiation and integration can be regarded as signal linear filtering with filter frequency responses, respectively

$$H^{(diff)}(f) = -i2\pi f \qquad (6.33)$$

and

$$H^{(intg)}(f) = i/2\pi f. \qquad (6.34)$$

Now, let signal $a(x)$ be represented by its samples $\{a_k\}$, $k = 0,1,\ldots,N-1$ and let $\{\alpha_r\}$ be a set of DFT coefficients of the discrete signal $\{a_k\}$:

$$a_k = \frac{1}{\sqrt{N}} \sum_{r=0}^{N-1} \alpha_r \exp\left(-i2\pi \frac{kr}{N} \right). \qquad (6.35)$$

Then, following the argumentation of the section "Fast Algorithms for Discrete Sinc Interpolation and Their Applications" for the optimal resampling filter and using the relationship of Equation 4.79 that establishes mutual correspondence between continuous signal frequency and frequency index of its DFT, one can conclude that samples $\left\{ \eta_{r,opt}^{(diff)} \right\}$ and $\left\{ \eta_{r,opt}^{(intg)} \right\}$ of continuous frequency response of perfect numerical differentiation and integration filters are defined for even N as

$$\eta_r^{(diff)} = \begin{cases} -i2\pi r/N, & r = 0,1,\ldots,N/2-1 \\ -\pi/2, & r = N/2 \\ i2\pi(N-r)/N, & r = N/2+1,\ldots,N-1 \end{cases}, \qquad (6.36)$$

$$\eta_{r,opt}^{(intg)} = \begin{cases} 0, & r = 0 \\ iN/2\pi r, & r = 1,\ldots,N/2-1 \\ -\pi/2, & r = N/2 \\ iN/2\pi(N-r), & r = N/2+1,\ldots,N-1 \end{cases}, \qquad (6.37)$$

and for odd N as

$$
\eta_r^{(diff)} = \begin{cases} -i2\pi r/N, & r = 0,1,...,(N-1)/2 - 1 \\ i2\pi(N-r)/N, & r = (N+1)/2,...,N-1 \end{cases},
\tag{6.38}
$$

$$
\eta_r^{(intg)} = \begin{cases} iN/2\pi r, & r = 0,1,...,(N-1)/2 - 1 \\ iN/2\pi(N-r), & r = (N+1)/2,...,N-1 \end{cases}.
\tag{6.39}
$$

One can show that numerical differentiation and integration according to Equations 6.36 through 6.39 imply discrete sinc interpolation of signals. Note that the coefficients $\eta_{N/2}^{(diff)}$ and $\eta_{N/2}^{(intg)}$ in Equations 6.36 and 6.37 are halved that correspond to the above-described Case_1 of discrete sinc interpolation (Equation 6.12).

Equations 6.36 through 6.39 imply the following algorithmic implementation for computing derivatives and integrals of signals specified by their samples:

$$
\{\dot{a}_k\} = \mathrm{IFFT}_N\left(\{\eta_r^{(diff)}\} \cdot \mathrm{FFT}(\{a_k\})\right),
\tag{6.40}
$$

$$
\{\bar{a}_k\} = \mathrm{IFFT}_N\left(\{\eta_r^{(intg)}\} \cdot \mathrm{FFT}(\{a_k\})\right),
\tag{6.41}
$$

where $\mathrm{FFT}(\cdot)$ and $\mathrm{IFFT}(\cdot)$ are direct and inverse fast Fourier transforms and $\cdot$ symbolizes element-wise multiplication of vectors. Thanks to the use of fast Fourier transform, the computational complexity of the algorithms is $O(\log N)$ operations per signal sample. Digital filter described by Equation 6.40 is called the *discrete ramp-filter*.

Like all DFT-based discrete sinc interpolation algorithms, DFT-based differentiation and integration algorithms, being the most accurate in term of preserving signal spectral components within the base band, suffer from boundary effects. Especially vulnerable in this respect is DFT-based differentiation because of discontinuities at signal borders due to their periodical replication in processing in the DFT domain. This drawback can be sufficiently alleviated by means of extension of the signals to double length through of mirror reflection at their boundaries before applying the above-described DFT-based algorithms. For such extended signals, DFT-based differentiation and integration are reduced to using fast DCT algorithms instead of FFT:

$$
\{\dot{a}_k\} = -\frac{2\pi}{N\sqrt{2}} \mathrm{IDCT}\left\{(N-r)\alpha_{N-r}^{(DCT)}\right\}
$$

$$
= \frac{2\pi}{N\sqrt{2N}}(-1)^k \sum_{r=1}^{N-1}(N-r)\alpha_{N-r}^{(DCT)}\cos\left(\pi\frac{k+1/2}{N}r\right)
\tag{6.42}
$$

$$\{\bar{a}_k\} = \frac{N}{2\pi\sqrt{2}}\left\{\text{IDCT}\left(\frac{\alpha_{N-r}^{(\text{DCT})}}{N-r}\right)\right\} = \frac{\sqrt{N}}{2\pi\sqrt{2}}(-1)^k \sum_{r=1}^{N-1} \frac{\alpha_{N-r}^{(\text{DCT})}}{N-r}\cos\left(\pi\frac{k+1/2}{N}r\right) \quad (6.43)$$

where $\{\alpha_r^{(\text{DCT})}\}$ are DCT transform coefficients of the signal. These formulas can be obtained as special cases of fast digital convolution algorithms described in the section "Signal Convolution in the DCT Domain" if one substitutes into Equation 4.121 frequency responses of differentiation and integration filters given by Equations 6.36 through 6.39. We will refer to the filters defined by Equations 6.42 and 6.43 as *DCT-based differentiation ramp-filter* and *DCT-based integration filter*, respectively.

Traditional Numerical Differentiation and Integration Algorithms versus DFT/DCT-Based Ones: Performance Comparison

In numerical mathematics, alternative methods of numerical computing signal derivatives and integrals are commonly used that are implemented through signal discrete convolution in the signal domain:

$$\dot{a}_k = \sum_{n=0}^{N_h-1} h_n^{(\text{diff})} a_{k-n}, \quad (6.44)$$

$$\bar{a}_k = \sum_{n=0}^{N_h-1} h_n^{\text{intg}} a_{k-n} \quad (6.45)$$

The following simplest differentiating kernels of two and five samples are recommended in manuals on numerical methods [2]:

$$h_n^{\text{diff}(1)} = [-0.5, \ 0, \ 0.5] \quad (6.46)$$

and

$$h_n^{\text{diff}(2)} = [-1/12, 8/12, 0, -8/12, 1/12]. \quad (6.47)$$

Both are based on an assumption that, on intersample distances, signals can be expanded into Taylor series. We will refer to them as D1 and D2 differentiation methods.

Most known numerical integration methods are the *Newton–Cotes quadrature rules* [2]. The first three rules are the trapezoidal, the Simpson, and the 3/8-Simpson rules. In all the methods, the value of the integral in the first point is not defined because it affects only the result's constant bias and should be arbitrarily chosen. When it is chosen to be equal to zero, the trapezoidal, Simpson, and 3/8-Simpson numerical integration methods are defined, for k as a running sample index, by equations, respectively:

$$\bar{a}_1^{(T)} = 0, \quad \bar{a}_k^{(T)} = \bar{a}_{k-1}^{(T)} + \frac{1}{2}(a_{k-1} + a_k), \tag{6.48}$$

$$\bar{a}_1^{(S)} = 0, \quad \bar{a}_k^{(S)} = \bar{a}_{k-2}^{(S)} + \frac{1}{3}(a_{k-2} + 4a_{k-1} + a_k), \tag{6.49}$$

$$\bar{a}_0^{(3/8S)} = 0, \quad \bar{a}_k^{(3/8S)} = \bar{a}_{k-3}^{(3/8S)} + \frac{3}{8}(a_{k-3} + 3a_{k-2} + 3a_{k-1} + a_k). \tag{6.50}$$

As it was mentioned in the section "DFTs and Discrete Frequency Response of Digital Filter," continuous and overall frequency responses of digital filters are determined by their discrete frequency responses (DFT of their PSF). Applying N-point discrete Fourier transform to Equations 6.48 through 6.50, obtain for discrete frequency responses $\eta_r^{diff(1)}$, $\eta_r^{diff(2)}$, $\eta_r^{int,T}$, $\eta_r^{int,S}$, $\eta_r^{(int,3/8S)}$ of the above-described numerical differentiation and integration methods, respectively:

$$\eta_r^{diff(1)} \propto \sin(2\pi r/N); \quad r = 0,1,...,N-1 \tag{6.51}$$

$$\eta_r^{diff(2)} \propto \frac{8\sin(2\pi r/N) - \sin(4\pi r/N)}{12}; \quad r = 0,1,...,N-1 \tag{6.52}$$

$$\eta_r^{(int,T)} = \frac{\bar{\alpha}_r^{(Tr)}}{\alpha_r} = \begin{cases} 0, & r - 0, \\ -\dfrac{\cos(\pi r/N)}{2i\sin(\pi r/N)}, & r = 1,...,N-1 \end{cases} \tag{6.53}$$

$$\eta_r^{(int,S)} = \frac{\bar{\alpha}_r^{(S)}}{\alpha_r} = \begin{cases} 0, & r = 0 \\ -\dfrac{\cos(2\pi r/N) + 2}{3i\sin(2\pi r/N)}, & r = 1,...,N-1 \end{cases} \tag{6.54}$$

$$\eta_r^{(int,3/8S)} = \frac{\bar{\alpha}_r^{(3S)}}{\alpha_r} = \begin{cases} 0, & r = 0 \\ -\dfrac{\cos(3\pi r/N) + 3\cos(\pi r/N)}{i\sin(3\pi r/N)}, & r = 1,...,N-1 \end{cases} \tag{6.55}$$

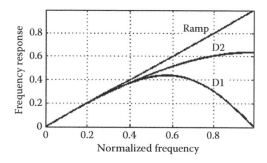

FIGURE 6.15

Absolute values of frequency responses of differentiation filters described by Equation 6.51 (curve D1), Equation 6.52 (curve D2), and Equations 6.40 and 6.42 ("Ramp"-filter).

These frequency responses along with frequency responses of DFT-based differentiation and integration filters (Equations 6.51 through 6.55) are shown, for comparison, in Figures 6.15 and 6.16, respectively.

One can see from these figures that standard numerical differentiation and integration methods entail certain and sometimes very substantial distortions of signal spectral contents on high frequencies. All these methods attenuate signal high frequencies, and Simpson and 3/8-Simpson integration methods, being slightly more accurate than the trapezoidal method in the middle of the signal base band, even tend to generate substantial artifacts if signals contain higher frequencies. Frequency response of the 3/8-Simpson rule tends to infinity for 2/3 of the maximum frequency, and the frequency response of the Simpson rule has almost the same tendency for the maximal

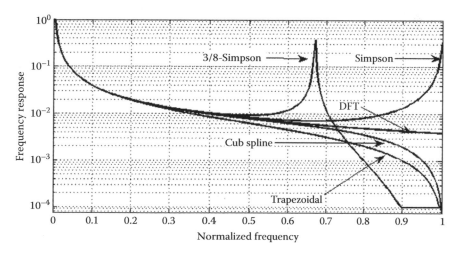

FIGURE 6.16

Absolute values of frequency responses of numerical integration filters described by Equations 6.53 through 6.55 and that of the DFT-based method (Equations 6.41 and 6.43).

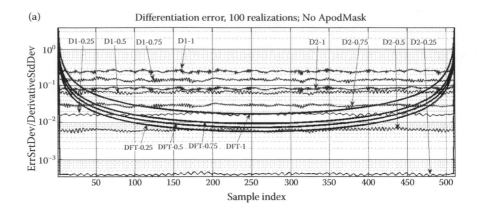

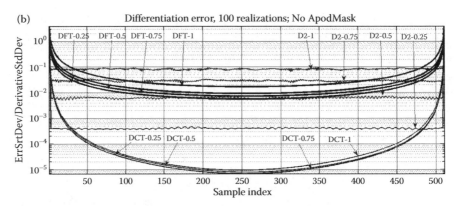

FIGURE 6.17

Experimental data on signal sample-wise normalized standard deviation of the differentiation error for D1, D2, and DFT-based differentiation methods (a) and for D2, DFT-, and DCT-based methods. (b) Numbers at curves indicate fraction (from one quarter to one) of test signal bandwidth with respect to the base bandwidth as defined by the sampling rate.

frequency in the base band. This means, in particular, that noise that might be present in input data and round off computation errors will be overamplified by Simpson and 3/8-Simpson in these frequencies.

Quantitative evaluation of performance of the considered differentiation methods can be carried out using a simulation program differentiator_comparison_CRC.m provided in Exercises. The program implements statistical simulation of differentiation by the considered methods of realizations of pseudorandom signals with uniform spectrum in the range from 1/16 of the base band to the entire base band. Figures 6.17 and 6.18 present results of such an experimental evaluation. In the simulation, 16 series of statistical experiments with 100 experiments in each run were carried out. In the runs, realizations of pseudorandom signals of 32,704 samples with uniform Fourier spectrum were generated in order to imitate, by means of 32-fold

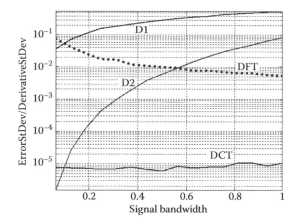

FIGURE 6.18
Normalized standard deviation of the differentiation error averaged over 100 samples in the middle of the signal (samples from 200-th to 300-th) for D1, D2, DFT, and DCT differentiation methods as a function of test signal bandwidth (in fractions of the base band defined by the signal sampling rate).

oversampling, continuous signals. In each run, generated pseudorandom signals were low-pass filtered to 1/32 of its base-band using the ideal low-pass filter implemented in DFT domain. The filtered signal was then used as a model of a continuous signal and its derivative was computed using DFT domain ramp-filter and used as an estimate of the signal ideal derivative. Then the central half of this signal, which encompasses 16,352 samples taken 8196 samples apart from the signal borders, was subsampled with rate 32 to generate 511 signal samples that were used in the differentiation by the four methods: D1 method (Equation 6.46), D2 method (Equation 6.47), DFT-based method (Equation 6.40), and DCT-based method (Equation 6.42). The corresponding central part of the ideal derivative signal was also subsampled with rate 32 and was used as a reference to evaluate the differentiation error for tested methods. The differentiation error was computed as a difference between the "ideal" derivative and the results of applying tested differentiation methods divided by the standard deviation of the "ideal" derivative over all samples, thus producing estimations normalized to the derivative signal energy. Finally, the standard deviation of the normalized error over 100 realizations was found for each signal sample. The obtained results are plotted sample-wise in Figure 6.17a for methods D1, D2, and DFT and in Figure 6.17b for methods D2, DFT, and DCT.

In Figure 6.17a, one can see that, indeed, the simplest method D1 performs very poorly, while method D2 outperforms the DFT method for signals with bandwidth <0.5 of the base band because of the boundary effects for the latter. It can also be seen that the accuracy of the DFT differentiation method substantially improves with the distance from signal boundaries. However, even for samples that are far away from signal boundaries, boundary effects badly deteriorate its differentiation accuracy. Data presented in Figure 6.17b evidence that the DCT-based differentiation method does successfully over-come the boundary effect problem and substantially outperforms both D1 and D2 methods even for narrow band signals.

In order to exclude boundary effects in evaluating and comparing dif-ferentiation errors, Figure 6.18 presents plots of normalized error standard deviation computed on average over the central 100 samples (from 200th to 300th) of the test signals with different bandwidth.

These plots convincingly evidence that method D2 provides better accu-racy only for signals with bandwidth <0.05 of the base band, and even for such signals, normalized error standard deviation for the DCT method is anyway <10^{-5}. For signals with broader bandwidth, the accuracy of the DCT differentiation method is better than for other methods by at least two orders of magnitude. One can also find, using this simulation program, that mask-ing signal with a window ("apodization") function, which gradually nulls signal samples in the vicinity of its border, substantially improves the dif-ferentiation accuracy of both DFT and DCT methods even further.

Also note that, as it follows from the above results, traditional numerical methods maintain a good differentiation accuracy if signals are very sub-stantially oversampled, which actually undermines their only advantage, their low computational complexity.

One more way to evaluating accuracy of numerical differentiation and integration is iterative application to a test signal successive differentiation and integration and comparison of restored signals with the initial signal. Plots in Figure 6.19 obtained using program diffrentiat_integrat_error_ CRC.m (Exercises) illustrate results performed in this way comparison of DCT-based differentiation and integration and of conventional D2 method of differentiation and trapezoidal method of integration. One can clearly see on these plots that conventional methods of differentiation and integration tend to very substantially blur signals. Using this program, one can also see that DCT-based differentiation/integration procedures have two to three orders of magnitude lower standard deviation of the signal restoration error.

Local ("Elastic") Image Resampling: Sliding Window Discrete Sinc Interpolation Algorithms

The above-described perfect DFT- and DCT-based fractional shift algorithms are computationally very efficient for performing regular shifts of all image pixels. For image resampling in arbitrary sampling grids, they can be used as it was described above for generating highly oversampled "quasicontin-uous" image models, which require additional large memory buffers. An

(a)

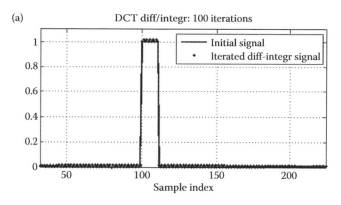

(b)

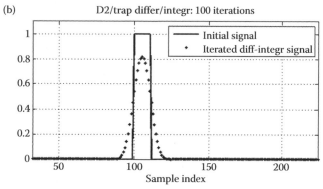

FIGURE 6.19
Comparison signal restoration after iterative successive 100 differentiations and integrations applied to a rectangular test signal for DCT-based differentiation and integration methods (a) and for D2 differentiator and trapezoidal rule integrator (b).

alternative solution is implementation of discrete sinc interpolation in sliding window processing.

In signal interpolation in sliding window, the perfect shifting filter is generated and applied, in each window position, only for pixels within the window, and only interpolated signal samples that correspond to the window central sample have to be computed in each window position from signal samples within the window. The interpolation function in this case is a discrete sinc-function, whose extent is equal to the window size rather than to the whole image size required for the perfect discrete sinc interpolation. Therefore, sliding window discrete sinc interpolation cannot provide the perfect interpolation, which the above-described global discrete sinc interpolation does. Figure 6.20 illustrates 1D frequency responses of sliding window, for window size of 15 pixels, and of global (full image size) discrete sinc interpolations for image 3× zooming.

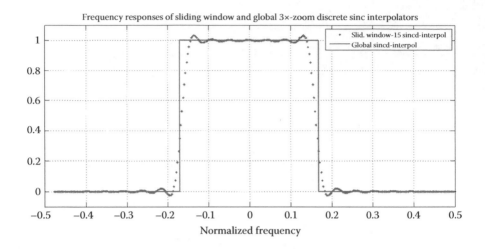

FIGURE 6.20

Frequency responses of sliding window of 15 samples (dotted line) and of the perfect (global) discrete sinc interpolations (solid line) for signal 3×-zooming.

Such an implementation of the discrete sinc interpolation can be regarded as a special case of signal domain convolution interpolation methods. As it follows from the above theory, it, in principle, has the highest interpolation accuracy among all convolution interpolation methods with the same interpolation kernel support. Additionally, being implemented in the DFT or DCT domains, it offers an option of image resampling with simultaneous restoration and enhancement by means of methods of local adaptive filtering described in Chapter 8.

In local adaptive filtering carried out in sliding window, in each position of the window transform, coefficients of window samples are computed and then nonlinearly modified to obtain transform coefficients of the output signal samples in the window. These coefficients are then used to generate an estimate of the window central pixel by inverse transform computed for the window central pixel. Such a filtering can be implemented in the domain of any transform. Therefore, one can, in a straightforward way, combine the sliding window DFT or DCT domain discrete sinc interpolation signal resampling and filtering for signal restoration and enhancement.

Figure 6.21 illustrates the application of such a combined filtering/interpolation for image irregular-to-regular resampling combined with denoising.

In this example, the left image is distorted by known displacements of pixels with respect to regular equidistant positions and by an additive noise. In the right image, these displacements are compensated and noise is substantially reduced with the above-described sliding window resampling/denoising algorithm.

FIGURE 6.21
An example of image resampling from irregular to regular sampling grid and denoising: (a) noisy and irregularly sampled image; (b) resampled (rectified) and simultaneously denoised image.

Image Data Resampling for Image Reconstruction from Projections

Discrete Radon Transform: An Algorithmic Definition and Filtered Back Projection Method for Image Reconstruction

Precise data resampling is a crucial issue in image reconstruction from projections. As it was shown in the section "Imaging from Projections and Radon Transform" of Chapter 2, image reconstruction from projections is based on properties of integral Radon transform. The main problem in discrete representation of the integral Radon transform is definition, for computing image projections, of line integral under arbitrary angle over an image sampling grid.

Any definition of the discrete line integral should assume one or another method of image interpolation for finding image values along the projection line in its points that do not coincide with nodes of image sampling grid. For a rectangular sampling grid, only column-wise, row-wise, and 45° diagonal-wise integrations do not require any interpolation. One possible solution of this problem is line integration over a "continuous" image model, obtained by means of the above-described method of image zooming-in with discrete sinc interpolation. Another, and a more computationally efficient, solution is the following algorithmic definition of the *discrete Radon transform*:

$$Pr(\theta_r, k) = \text{SUM}_l\left[\text{ROT}_{\theta_r}(\{a_{k,l}\})\right], \tag{6.56}$$

where $\{Pr(\theta_r, k)\}$ are samples of the r-th projection, taken under angle θ_r, of the image defined by its samples $\{a_{k,l}\}$ over a square sampling grid $\{k,l\}$; $\text{ROT}_{\theta_r}(\cdot)$ is an operator of image rotation by the angle θ_r around the center of the sampling grid and $\text{SUM}_l[\cdot]$ is the operator of summation of samples of the rotated image over index l.

With this definition, the required image interpolation is carried out in the process of image rotation. In order to secure the least possible interpolation error, image rotation should be performed with discrete sinc interpolation. We will assume that for the implementation of the rotation operator, the above-described fast three-step rotation algorithm is used, which preserves the number of image samples, although the above-described image rotation algorithm in scaled coordinates can, in principle, be used as well.

The filtered back projection algorithm for image reconstruction from projections is defined by Equation 2.107. According to the above definition of the discrete Radon transform, this algorithm can be implemented as the following:

$$a_{k,l} = \text{SUM}_r\left\{\text{ROT}_{-\theta_r}\{\text{BckP}_l\{\text{RAMPF}\{Pr(\theta_r,k)\}\}\}\right\}, \tag{6.57}$$

where $\{Pr(\theta_r,k)\}$ are samples of the image r-th projection, taken under angle θ_r, $\text{RAMPF}\{\cdot\}$ is an operator of ramp-filtering for differentiation described in the section "Numerical Differentiation and Integration," $\text{BckP}_l\{\cdot\}$ is a "back projection" operator implemented as replication of the operand over index l and $\text{SUM}_r\{\cdot\}$ is a summation operator that sums up replicated filtered projections over the entire set of projection angles $\{\theta_r\}$.

Figure 6.22 generated using program radon_invradon_demo_CRC.m provided for exercises (Exercises) illustrates discrete Radon transform and image reconstruction using the filtered back projection method.

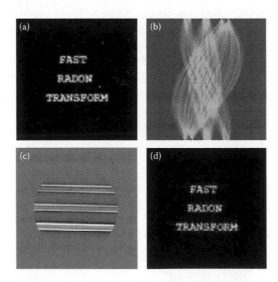

FIGURE 6.22
Test image (a), a set of its projections with projection angle as the vertical coordinate (b), an example of filtered projection projected backward along the horizontal projected line (c), and reconstructed image (d).

Direct Fourier Method of Image Reconstruction

According to the direct Fourier method of image reconstruction from their parallel projections (see the section "Imaging from Projections and Radon Transform"), 1D Fourier spectra of projections should be placed under their corresponding angles in a polar coordinate system in frequency domain to form a 2D image spectrum, which then can be used for image reconstruction by inverse Fourier transform. As one can see from Figure 6.23 that shows a polar coordinate system of spectral samples in a Cartesian coordinate system, spectral samples are nonuniformly spaced in Cartesian coordinates and are very sparse, especially high-frequency ones. In principle, one can attempt to apply, for reconstruction of spectral samples on a uniform dense grid, iterative algorithms for image recovery from sparse nonuniform samples considered in Chapter 5 using *a priori* redundancy of images associated with empty area around body slice, which is always present. However, this method also does not seem feasible because of very high sparsity of spectral samples on high frequencies, which is not compensated by the image redundancy.

Another option is computing 2D inverse Fourier transform using shifted DFTs with shift parameters in frequency domain specific for each particular spectrum sample according to its position in 2D Cartesian coordinate system, in which SDFT is defined. However, in this case, fast Fourier transform,

Polar (o) and Cartesian (+) sampling grids

FIGURE 6.23
Spectral samples (small circles) in a Cartesian sampling grid (crosses).

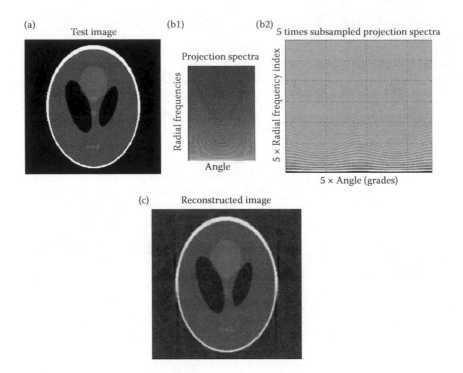

(a) Test image   (b1) Projection spectra   (b2) 5 times subsampled projection spectra

(c) Reconstructed image

FIGURE 6.24
An illustrative example of image restoration from projections using the direct Fourier reconstruction method: (a) test image; (b1) projection spectra; (b2) projection spectra five times subsampled in both coordinates for polar-to-Cartesian coordinate conversion (only half of the spectral coefficients that correspond to frequencies from zero to the highest one in the base band are displayed; the others that are complex conjugate to them are not shown); (c) image reconstructed by means of inverse DFT applied to 2D spectrum obtained by resampling zoomed-in projection spectra.

which exists only for uniform sampling grids, is not applicable. That makes such solution impractical by virtue of its high computational complexity.

A feasible practical option for solving this problem is polar-to-Cartesian coordinate conversion of spectra of projections to 2D image spectrum by means of resampling, in Cartesian coordinates, of a "continuous" model of the projection spectra in polar coordinate system formed through separable (radial frequencies-wise and angle index-wise) zooming-in projection spectra using discrete sinc interpolation. Figure 6.24 presents an illustrative example of image reconstruction from projections achieved through inverse DFT of image 2D spectrum obtained by resampling of image spectrum in polar coordinate system.

Image Reconstruction from Fan-Beam Projections

Described above methods of image reconstruction from projections assume image projection in parallel x-ray beams. This is the original classical image

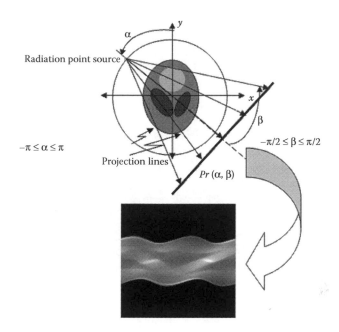

FIGURE 6.25
Geometry of image fan projections.

reconstruction method, for which well-developed reconstruction algorithms are available. In practice, in commercial CT scanners fan-beam projection rather than parallel-beam one is used because fan-beam projections can be obtained with a point source of x-ray, which is much easier for fabrication than collimated parallel beam sources.

The geometry of fan projection is sketched in Figure 6.25. Point source of radiation makes full 360° revolution around the object and integrals of absorption of radiation by the object over projection lines form, for each position angle α of the point source, projections $Pr(\alpha, \beta)$ as a function of the ray angles β. All sets of projections for $-\pi \leq \alpha \leq \pi$ is then used for image reconstruction.

In principle, for inverting Radon transform in fan-beam projection geometry, dedicated reconstruction algorithms are required. There is however an alternative and attractive option of converting, by an appropriate resampling, the set of fan projections to a set of parallel projections and to enable in this way image reconstruction using algorithms for image reconstruction from parallel projections. For the resampling, one can use the above-described algorithms for resampling by means of image zooming-in using global or local discrete sinc interpolation. The latter can be, if required, combined with denoising as it was discussed in the section "Local ("Elastic") Image Resampling: Sliding Window Discrete Sinc Interpolation Algorithms." This process of converting one type of projections into another type is called *data rebinning*. Figure 6.26 illustrates this method of image reconstruction.

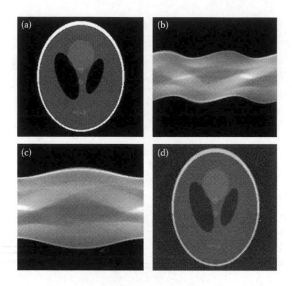

FIGURE 6.26

Image reconstruction from fan projections: (a) initial test image; (b) its fan-beam projections; (c) parallel projections converted from the fan-beam projections; (d) image reconstructed from converted parallel projections.

Appendix

Derivation of Equations 6.6 and 6.7

From Equation 6.2, for odd number of signal samples N, we have

$$h_n^{(Shift)}(\delta x) = \frac{1}{\sqrt{N}} \sum_{r=0}^{N-1} \eta_r^{(Shift)}(\delta x) \exp\left(-i2\pi \frac{nr}{N}\right)$$

$$= \frac{1}{\sqrt{N}} \left\{ \sum_{r=0}^{(N-1)/2} \eta_r^{(Shift)}(\delta x) \exp\left(-i2\pi \frac{nr}{N}\right) \right.$$

$$\left. + \sum_{r=(N+1)/2}^{N-1} \eta_r^{(Shift)}(\delta x) \exp\left(-i2\pi \frac{nr}{N}\right) \right\}$$

$$= \frac{1}{\sqrt{N}} \left\{ \sum_{r=0}^{(N-1)/2} \eta_r^{(Shift)}(\delta x) \exp\left(-i2\pi \frac{nr}{N}\right) \right.$$

$$\left. + \sum_{r=1}^{(N-1)/2} \eta_{N-r}^{(Shift)}(\delta x) \exp\left[-i2\pi \frac{n}{N}(N-r)\right] \right\}$$

$$= \frac{1}{\sqrt{N}} \left\{ \sum_{r=0}^{(N-1)/2} \eta_r^{(Shift)} \exp\left(-i2\pi \frac{nr}{N} r\right) + \sum_{r=1}^{(N-1)/2} \eta_{N-r}^{(Shift)} \exp\left(i2\pi \frac{nr}{N}\right) \right\}$$

$$= \frac{1}{\sqrt{N}} \left\{ \sum_{r=0}^{(N-1)/2} \eta_r^{(Shift)}(\delta x) \exp\left(-i2\pi \frac{nr}{N}\right) + \sum_{r=1}^{(N-1)/2} \eta_r^{*(Shift)}(\delta x) \exp\left(i2\pi \frac{nr}{N}\right) \right\}$$

$$= \frac{1}{N} \left\{ \sum_{r=0}^{(N-1)/2} \exp\left(i2\pi \frac{\delta x}{N\Delta x} r\right) \exp\left(-i2\pi \frac{nr}{N}\right) \right.$$

$$\left. + \sum_{r=1}^{(N-1)/2} \exp\left(-i2\pi \frac{\delta x}{N\Delta x}\right) \exp\left(i2\pi \frac{nr}{N}\right) \right\}$$

$$= \frac{1}{N} \left\{ \sum_{r=0}^{(N-1)/2} \exp\left(-i2\pi \frac{n - \delta x/\Delta x}{N} r\right) + \sum_{r=1}^{(N-1)/2} \exp\left(i2\pi \frac{n - \delta x/\Delta x}{N} r\right) \right\}.$$

Denote temporarily

$$\tilde{n} = n - \delta x/\Delta x.$$

Then obtain

$$h_n^{(Shift)}(\delta x) = \frac{1}{N} \left\{ \frac{\exp\left[-i\pi \frac{\tilde{n}}{N}(N+1)\right] - 1}{\exp\left(-i2\pi \frac{\tilde{n}}{N}\right) - 1} + \frac{\exp\left[i\pi \frac{\tilde{n}}{N}(N+1)\right] - \exp\left(i2\pi \frac{\tilde{n}}{N}\right)}{\exp\left(i2\pi \frac{\tilde{n}}{N}\right) - 1} \right\}$$

$$= \frac{1}{N} \left\{ \frac{\exp(-i\pi\tilde{n}) - \exp\left(i\pi \frac{\tilde{n}}{N}\right)}{\exp\left(-i\pi \frac{\tilde{n}}{N}\right) - \exp\left(i\pi \frac{\tilde{n}}{N}\right)} + \frac{\exp(i\pi\tilde{n}) - \exp\left(i\pi \frac{\tilde{n}}{N}\right)}{\exp\left(i\pi \frac{\tilde{n}}{N}\right) - \exp\left(-i\pi \frac{\tilde{n}}{N}\right)} \right\}$$

$$= \frac{1}{N} \frac{-\exp(-i\pi\tilde{n}) + \exp\left(i\pi \frac{\tilde{n}}{N}\right) + \exp(i\pi\tilde{n}) - \exp\left(i\pi \frac{\tilde{n}}{N}\right)}{\exp\left(i\pi \frac{\tilde{n}}{N}\right) - \exp\left(-i\pi \frac{\tilde{n}}{N}\right)}$$

$$= \frac{1}{N} \frac{\sin(\pi\tilde{n})}{\sin\left(\pi \frac{\tilde{n}}{N}\right)} = \text{sincd}[N, \pi(n - \delta x/\Delta x)].$$

For even number of signal samples *N*, we have from Equation 6.2

$$h_n^{(intp)}(\delta x) = \frac{1}{\sqrt{N}} \sum_{r=0}^{N-1} \eta_r^{(Shift)}(\delta x) \exp\left(-i2\pi \frac{nr}{N}\right)$$

$$= \frac{1}{\sqrt{N}} \left\{ \sum_{r=0}^{N/2-1} \eta_r^{(Shift)}(\delta x) \exp\left(-i2\pi \frac{nr}{N}\right) + \sum_{r=N/2}^{N-1} \eta_r^{(Shift)}(\delta x) \exp\left(-i2\pi \frac{nr}{N}\right) \right\}$$

$$= \frac{1}{\sqrt{N}} \left\{ \sum_{r=0}^{N/2-1} \eta_r^{(Shift)}(\delta x) \exp\left(-i2\pi \frac{nr}{N}\right) + \sum_{r=1}^{N/2} \eta_{N-r}^{(Shift)} \exp\left[-i2\pi \frac{n}{N}(N-r)\right] \right\}$$

$$= \frac{1}{\sqrt{N}} \left\{ \sum_{r=0}^{N/2-1} \eta_r^{(Shift)} \exp\left(-i2\pi \frac{nr}{N}\right) + \sum_{r=1}^{N/2} \eta_{N-r}^{(Shift)}(\delta x) \exp\left(i2\pi \frac{nr}{N}\right) \right\}$$

$$= \frac{1}{\sqrt{N}} \left\{ \sum_{r=0}^{(N-1)/2} \eta_r^{(Shift)}(\delta x) \exp\left(-i2\pi \frac{nr}{N}\right) + \sum_{r=1}^{N/2} \eta_r^{*(Shift)}(\delta x) \exp\left(i2\pi \frac{nr}{N}r\right) \right\}$$

$$= \frac{1}{N} \left\{ \sum_{r=0}^{N/2-1} \exp\left(i2\pi \frac{\delta \tilde{x}}{N\Delta x}r\right) \exp\left(-i2\pi \frac{nr}{N}\right) \right.$$

$$+ \sum_{r=1}^{N/2-1} \exp\left(-i2\pi \frac{\delta \tilde{x}}{N\Delta x}r\right) \exp\left(i2\pi \frac{nr}{N}\right) + \eta_{N/2}^{(Shift)}(\delta x) \exp\left(i\pi N \frac{n}{N}\right) \right\}$$

$$= \frac{1}{N} \left\{ \sum_{r=0}^{N/2-1} \exp\left(-i2\pi \frac{n - \delta \tilde{x}/\Delta x}{N}r\right) + \sum_{r=1}^{N/2-1} \exp\left(i2\pi \frac{n - \delta \tilde{x}/\Delta x}{N}r\right) \right.$$

$$+ \eta_{N/2}^{(Shift)}(\delta x) \exp(i\pi n) \right\}$$

$$= \frac{1}{N} \left\{ \sum_{r=0}^{N/2-1} \exp\left(-i2\pi \frac{\tilde{n}r}{N}\right) + \sum_{r=1}^{(N-1)/2} \exp\left(i2\pi \frac{\tilde{n}}{N}r\right) \right.$$

$$+ \eta_{N/2}^{(Shift)}(\delta x) \exp\left(i\pi \frac{\delta \tilde{x}}{\Delta x}\right) \exp(i\pi \tilde{n}) \right\}$$

$$= \frac{1}{N} \left\{ \frac{\exp(-i\pi \tilde{n}) - 1}{\exp(-i2\pi(\tilde{n}/N)) - 1} + \frac{\exp(-i\pi \tilde{n}) - \exp(i2\pi(\tilde{n}/N)r)}{\exp(i2\pi(\tilde{n}/N)) - 1} \right.$$

$$+ \eta_{N/2}^{(Shift)}(\delta x) \exp\left(i\pi \frac{\delta \tilde{x}}{\Delta x}\right) \exp(i\pi \tilde{n}) \right\}$$

$$= \frac{1}{N} \left\{ \frac{\exp(-i\pi(N-1/N)\tilde{n}) - \exp(i\pi(\tilde{n}/N))}{\exp(-i\pi(\tilde{n}/N)) - \exp(i\pi(\tilde{n}/N))} + \frac{\exp(i\pi(N-1/N)\tilde{n}) - \exp(i\pi(\tilde{n}/N)r)}{\exp(i\pi(\tilde{n}/N)) - \exp(-i\pi(\tilde{n}/N))} \right.$$

$$\left. + \eta_{N/2}^{(Shift)} \exp\left(i\pi \frac{\delta\tilde{x}}{\Delta x} \right) \exp(i\pi\tilde{n}) \right\}$$

$$= \frac{1}{N} \left\{ \frac{\exp(i\pi(N-1/N)\tilde{n}) - \exp(i\pi(\tilde{n}/N)r) - \exp(-i\pi(N-1/N)\tilde{n}) + \exp(i\pi(\tilde{n}/N))}{\exp(i\pi(\tilde{n}/N)) - \exp(-i\pi(\tilde{n}/N))} \right.$$

$$\left. + \eta_{N/2}^{(Shift)}(\delta x) \exp\left(i\pi \frac{\delta\tilde{x}}{\Delta x} \right) \exp(i\pi\tilde{n}) \right\}$$

$$= \frac{1}{N} \left\{ \frac{\exp(i\pi(N-1/N)\tilde{n}) - \exp(-i\pi(N-1/N)\tilde{n})}{\exp(i\pi(\tilde{n}/N)) - \exp(-i\pi(\tilde{n}/N))} + \eta_{N/2}^{(Shift)} \exp\left(i\pi \frac{\delta\tilde{x}}{\Delta x} \right) \exp(i\pi\tilde{n}) \right\}$$

$$= \frac{1}{N} \left\{ \frac{\sin(\pi(N-1/N)\tilde{n})}{\sin(\pi(\tilde{n}/N))} + \eta_{N/2}^{(Shift)}(\delta x) \exp(i\pi(\delta\tilde{x}/\Delta x)) \exp(i\pi\tilde{n}) \right\}.$$

Case_0: $\eta_{N/2,opt}^{(intp)} = 0$;

$$h_n^{(intp0)}(\delta\tilde{x}) = \overline{\mathrm{sincd}}[N; N-1; \pi(n - \delta\tilde{x}/\Delta x)].$$

Case_2: $\eta_{N/2}^{(Shift)} = 2\cos(\pi(\delta\tilde{x}/\Delta x))$;

$$h_n^{(intp0)}(\delta\tilde{x}) = \overline{\mathrm{sincd}}[N; N+1; \pi(n - \delta\tilde{x}/\Delta x)],$$

where

$$\overline{\mathrm{sincd}}\{N; M; x\} = \frac{1}{N} \frac{\sin(Mx/N)}{\sin(x/N)}.$$

Proof for Case_2. Find $\eta_{N/2}^{(Shift2)}$ that satisfies the condition:

$$\frac{\sin(\pi(N-1/N)\tilde{n})}{\sin(\pi(\tilde{n}/N))} + \eta_{N/2}^{(Shift2)} \exp\left(i\pi \frac{\delta\tilde{x}}{\Delta x} \right) \exp(i\pi\tilde{n}) = \frac{\sin(\pi(N+1/N)\tilde{n})}{\sin(\pi(\tilde{n}/N))}.$$

From the above equation,

$$\eta_{N/2}^{(Shift)} = \frac{\sin(\pi(N+1/N)\tilde{n}) - \sin(\pi(N-1/N)\tilde{n})}{\sin(\pi(\tilde{n}/N))} \exp(-i\pi\tilde{n}) \exp\left(-i\pi \frac{\delta\tilde{x}}{\Delta x} \right)$$

$$= 2\cos(\pi\tilde{n})\exp(-i\pi\tilde{n})\exp\left(-i\pi\frac{\delta\tilde{x}}{\Delta x}\right) = 2\cos(\pi\tilde{n})\exp(-i\pi n)$$

$$= 2\cos\left[\pi\left(n - \frac{\delta\tilde{x}}{\Delta x}\right)\right](-1)^n = 2\cos(\pi n)(-1)^n\cos\left(\pi\frac{\delta\tilde{x}}{\Delta x}\right) = 2\cos\left(\pi\frac{\delta\tilde{x}}{\Delta x}\right).$$

PSF of Signal Zooming by Means of Zero Padding of Its DCT Spectrum

Consider analytical expressions that describe signal zooming by means of zero padding its DCT spectrum. Let α_r^{DCT} be DCT spectrum of signal $\{a_k\}$,

$$\alpha_r^{\text{DCT}} = \frac{2}{\sqrt{N}}\sum_{k=0}^{N-1} a_k\cos\left[\pi\frac{(k+1/2)r}{N}\right] = \frac{1}{\sqrt{2N}}\sum_{k=0}^{2N-1}\tilde{a}_k\exp\left[i2\pi\frac{(k+1/2)r}{2N}\right],$$

where $k = 0,\ldots,N-1$ and

$$\tilde{a}_k = \begin{cases} a_k, & k = 0,\ldots,N-1 \\ a_{2N-1-k}, & k = N,\ldots,LN-1 \end{cases}.$$

Form a zero pad spectrum:

$$\tilde{\alpha}_r^{\text{DCT},L} = \begin{cases} \alpha_r^{\text{DCT}}, & r = 0,\ldots,N-1 \\ 0, & r = N,\ldots,LN-1 \end{cases}.$$

Being a DCT spectrum, this spectrum has an odd-symmetry property (Equation 4.95):

$$\tilde{\alpha}_r^{\text{DCT},L} = -\tilde{\alpha}_{2LN-r}^{\text{DCT},L}; \quad \tilde{\alpha}_{LN}^{\text{DCT},L} = 0.$$

Compute inverse DCT of the zero pad spectrum, using representation of DCT through SDFT(1/2,0) and this symmetry property (4.95):

$$\tilde{\tilde{a}}_k = \frac{1}{\sqrt{2LN}}\sum_{r=0}^{2N-1}\tilde{\alpha}_r^{\text{DCT},L}\exp\left[-i2\pi\frac{(k+1/2)r}{2N}\right]$$

$$= \frac{1}{\sqrt{2LN}}\left\{\sum_{r=0}^{N-1}\alpha_r^{\text{DCT}}\exp\left[-i2\pi\frac{(k+1/2)}{2LN}r\right]\right.$$

$$\left. + \sum_{r=2LN-N+1}^{2LN-1}\tilde{\alpha}_r^{\text{DCT}}\exp\left[-i2\pi\frac{(k+1/2)}{2LN}r\right]\right\}$$

$$= \frac{1}{\sqrt{2LN}} \left\{ \sum_{r=0}^{N-1} \alpha_r^{DCT} \exp\left[-i2\pi \frac{(k+1/2)}{2LN} r \right] \right.$$

$$\left. + \sum_{r=1}^{N-1} \tilde{\alpha}_{2LN-r}^{DCT} \exp\left[-i2\pi \frac{(k+1/2)}{2LN} (2LN-r) \right] \right\}$$

$$= \frac{1}{\sqrt{2LN}} \left\{ \alpha_0^{DCT} + \sum_{r=1}^{N-1} \alpha_r^{DCT} \exp\left[-i2\pi \frac{(k+1/2)}{2LN} r \right] \right.$$

$$\left. - \sum_{r=1}^{N-1} \alpha_r^{DCT} \exp\left[-i2\pi \frac{(k+1/2)}{2LN} (2LN-r) \right] \right\}$$

$$= \frac{1}{\sqrt{2LN}} \left\{ \alpha_0^{DCT} + \sum_{r=0}^{N-1} \alpha_r^{DCT} \exp\left[-i2\pi \frac{(k+1/2)}{2LN} r \right] \right.$$

$$\left. + \sum_{r=0}^{N-1} \alpha_r^{DCT} \exp\left[i2\pi \frac{(k+1/2)}{2LN} r \right] \right\}.$$

Replace in this formula α_r^{DCT} with its expressions through SDFT(1/2,0) in the right part of Equation 4.93 and obtain

$$\tilde{\tilde{a}}_k = \frac{1}{\sqrt{2LN}} \left\{ \alpha_0^{DCT} + \sum_{r=1}^{N-1} \frac{1}{\sqrt{2N}} \sum_{n=0}^{2N-1} \tilde{a}_n \exp\left[i2\pi \frac{(n+1/2)r}{2N} \right] \exp\left[-i2\pi \frac{(k+1/2)}{2LN} r \right] \right.$$

$$\left. + \sum_{r=1}^{N-1} \frac{1}{\sqrt{2N}} \sum_{n=0}^{2N-1} \tilde{a}_n \exp\left[i2\pi \frac{(n+1/2)r}{2N} \right] \exp\left[i2\pi \frac{(k+1/2)}{2LN} r \right] \right\}$$

$$= \frac{\alpha_r^{DCT}}{\sqrt{2LN}} + \frac{1}{\sqrt{2LN}} \sum_{r=1}^{N-1} \frac{1}{\sqrt{2N}} \sum_{n=0}^{2N-1} \tilde{a}_n \left\{ \exp\left[i\pi \left(n+1/2 - \frac{k+1/2}{L} \right) r \Big/ N \right] \right.$$

$$\left. + \exp\left[i\pi \left(n+1/2 + \frac{k+1/2}{L} \right) r \Big/ N \right] \right\}.$$

Denote $\tilde{n} = n + 1/2$ and $\tilde{k} = k + 1/2$. Then obtain

$$\tilde{\tilde{a}}_k = \frac{\alpha_r^{DCT}}{\sqrt{2LN}} + \frac{1}{\sqrt{2LN}} \sum_{r=1}^{N-1} \frac{1}{\sqrt{2N}} \sum_{n=0}^{2N-1} \tilde{a}_n \left\{ \exp\left[i\pi \left(\tilde{n} - \frac{\tilde{k}}{L} \right) r \Big/ N \right] + \exp\left[i\pi \left(\tilde{n} + \frac{\tilde{k}}{L} \right) r \Big/ N \right] \right\}$$

$$
= \frac{\alpha_0^{DCT}}{\sqrt{2LN}} + \frac{1}{\sqrt{2LN}} \sum_{r=1}^{N-1} \frac{1}{\sqrt{2N}} \left[\sum_{n=0}^{N-1} \tilde{a}_n \left\{ \exp\left[i\pi\left(\tilde{n} - \frac{\tilde{k}}{L} \right) r \Big/ N \right] + \exp\left[i\pi\left(\tilde{n} + \frac{\tilde{k}}{L} \right) r \Big/ N \right] \right\} \right.
$$

$$
\left. + \sum_{n=N}^{2N-1} \tilde{a}_n \left\{ \exp\left[i\pi\left(\tilde{n} - \frac{\tilde{k}}{L} \right) r \Big/ N \right] + \exp\left[i\pi\left(\tilde{n} + \frac{\tilde{k}}{L} \right) r \Big/ N \right] \right\} \right]
$$

$$
= \frac{\alpha_0^{DCT}}{\sqrt{2LN}} + \frac{1}{\sqrt{2LN}} \sum_{r=1}^{N-1} \frac{1}{\sqrt{2N}} \left[\sum_{n=0}^{N-1} \tilde{a}_n \left\{ \exp\left[i\pi\left(\tilde{n} - \frac{\tilde{k}}{L} \right) r \Big/ N \right] + \exp\left[i\pi\left(\tilde{n} + \frac{\tilde{k}}{L} \right) r \Big/ N \right] \right\} \right.
$$

$$
\left. + \sum_{n=0}^{N-1} \tilde{a}_{2N-1-n} \left\{ \exp\left[i\pi\left(2N - \tilde{n} - \frac{\tilde{k}}{L} \right) r \Big/ N \right] + \exp\left[i\pi\left(2N - \tilde{n} + \frac{\tilde{k}}{L} \right) r \Big/ N \right] \right\} \right]
$$

$$
= \frac{\alpha_0^{DCT}}{\sqrt{2LN}} + \frac{1}{\sqrt{2LN}} \sum_{r=1}^{N-1} \frac{1}{\sqrt{2N}} \left[\sum_{n=0}^{N-1} a_n \left\{ \exp\left[i\pi\left(\tilde{n} - \frac{\tilde{k}}{L} \right) r \Big/ N \right] + \exp\left[i\pi\left(\tilde{n} + \frac{\tilde{k}}{L} \right) r \Big/ N \right] \right\} \right.
$$

$$
\left. + \sum_{n=0}^{N-1} \tilde{a}_{2N-1-n} \left\{ \exp\left[-i\pi\left(\tilde{n} + \frac{\tilde{k}}{L} \right) r \Big/ N \right] + \exp\left[-i\pi\left(\tilde{n} - \frac{\tilde{k}}{L} \right) r \Big/ N \right] \right\} \right]
$$

$$
= \frac{\alpha_0^{DCT}}{\sqrt{2LN}} + \frac{1}{2N\sqrt{L}} \sum_{n=0}^{N-1} a_n \left\{ \sum_{r=1}^{N-1} \exp\left[i\pi\left(\tilde{n} - \frac{\tilde{k}}{L} \right) r \Big/ N \right] + \sum_{r=1}^{N-1} \exp\left[i\pi\left(\tilde{n} + \frac{\tilde{k}}{L} \right) r \Big/ N \right] \right.
$$

$$
\left. + \sum_{r=1}^{N-1} \exp\left[-i\pi\left(\tilde{n} + \frac{\tilde{k}}{L} \right) r \Big/ N \right] + \sum_{r=1}^{N-1} \exp\left[-i\pi\left(\tilde{n} - \frac{\tilde{k}}{L} \right) r \Big/ N \right] \right\}
$$

$$
= \frac{\alpha_0^{DCT}}{\sqrt{2LN}} + \frac{1}{2N\sqrt{L}} \sum_{n=0}^{N-1} a_n \left\{ \frac{\exp\left[i\pi\left(\tilde{n}L - \tilde{k} \right)/L \right] - \exp\left[i\pi\left(\tilde{n}L - \tilde{k} \right)/LN \right]}{\exp\left[i\pi\left(\tilde{n}L - \tilde{k} \right)/LN \right] - 1} \right.
$$

$$
+ \frac{\exp\left[\left(\tilde{n}L + \tilde{k} \right)/L \right] - \exp\left[i\pi\left(\tilde{n}L + \tilde{k} \right)/LN \right]}{\exp\left[\pi\left(\tilde{n}L + \tilde{k} \right)/LN \right] - 1}
$$

$$
+ \frac{\exp\left[-\pi\left(\tilde{n}L + \tilde{k} \right)/L \right] - \exp\left[-\pi\left(\tilde{n}L + \tilde{k} \right)/LN \right]}{\exp\left[-\pi\left(\tilde{n}L + \tilde{k} \right)/LN \right] - 1}
$$

$$
\left. + \frac{\exp\left[-i\pi\left(\tilde{n}L - \tilde{k} \right)/L \right] - \exp\left[-i\pi\left(\tilde{n}L - \tilde{k} \right)/LN \right]}{\exp\left[-i\pi\left(\tilde{n}L - \tilde{k} \right)/LN \right] - 1} \right\}
$$

$$
= \frac{\alpha_0^{DCT}}{\sqrt{2LN}} + \frac{1}{2N\sqrt{L}} \sum_{n=0}^{N-1} a_n \left\{ \frac{\sin\left[\pi \dfrac{N+1}{2NL}\left(\tilde{n}L - \tilde{k} \right) \right]}{\sin\left[\pi\left(\tilde{n}L - \tilde{k} \right)/2NL \right]} \exp\left[i\pi\left(\tilde{n}L - \tilde{k} \right)/2L \right] \right.
$$

$$+ \frac{\sin\left[\pi \dfrac{N+1}{2NL}(\tilde{n}L + \tilde{k})\right]}{\sin[\pi(\tilde{n}L + \tilde{k})/2NL]} \exp[i\pi(\tilde{n}L + \tilde{k})/2L]$$

$$+ \frac{\sin\left[\pi \dfrac{N+1}{2NL}(\tilde{n}L + \tilde{k})\right]}{\sin[\pi(\tilde{n}L + \tilde{k})/2NL]} \exp[-i\pi(\tilde{n}L + \tilde{k})/2L]$$

$$\left. + \frac{\sin\left[\pi \dfrac{N+1}{2NL}(\tilde{n}L - \tilde{k})\right]}{\sin[\pi(\tilde{n}L - \tilde{k})/2NL]} \exp[-i\pi(\tilde{n}L - \tilde{k})/2L] \right\}.$$

From this, we finally obtain:

$$\tilde{\tilde{a}}_k = \frac{\alpha_0^{DCT}}{\sqrt{2LN}} + \frac{1}{N\sqrt{L}} \sum_{n=0}^{N-1} a_n \left\{ \frac{\sin\left[\pi \dfrac{N+1}{2NL}(\tilde{n}L - \tilde{k})\right]}{\sin[\pi(\tilde{n}L - \tilde{k})/2NL]} \cos[\pi(\tilde{n}L - \tilde{k})/2L] \right.$$

$$\left. + \frac{\sin\left[\pi \dfrac{N+1}{2NL}(\tilde{n}L + \tilde{k})\right]}{\sin[\pi(\tilde{n}L + \tilde{k})/2NL]} \cos[\pi(\tilde{n}L + \tilde{k})/2L] \right\}.$$

By analogy with DFT zero padding, one can improve the speed of convergence of DCT zero-padding PSF by halving DCT spectral coefficients that correspond to the highest frequency. For signals of N samples, these are coefficients with indices $N - 2$ and $N - 1$. To find an analytical expression for this case, we first compute, using above two equations, L-zoomed signal for the case when those coefficients are zeroed:

$$\tilde{\tilde{a}}_k = \frac{\alpha_r^{DCT}}{\sqrt{2LN}} + \frac{1}{2N\sqrt{L}} \sum_{n=0}^{N-1} a_n \left\{ \sum_{r=1}^{N-3} \exp\left[i\pi\left(\tilde{n} - \frac{\tilde{k}}{L}\right)r \middle/ N\right] \right.$$

$$+ \sum_{r=1}^{N-3} \exp\left[i\pi\left(\tilde{n} + \frac{\tilde{k}}{L}\right)r \middle/ N\right]$$

$$\left. + \sum_{r=1}^{N-3} \exp\left[-i\pi\left(\tilde{n} + \frac{\tilde{k}}{L}\right)r \middle/ N\right] + \sum_{r=1}^{N-3} \exp\left[-i\pi\left(\tilde{n} - \frac{\tilde{k}}{L}\right)r \middle/ N\right] \right\}$$

$$
= \frac{\alpha_0^{DCT}}{\sqrt{2LN}} + \frac{1}{N\sqrt{L}} \sum_{n=0}^{N-1} a_n \left\{ \frac{\sin\left[\pi\dfrac{N-1}{2N}\left(\tilde{n} - \dfrac{\tilde{k}}{L}\right)\right]}{\sin\left[\pi\left(\tilde{n} - \dfrac{\tilde{k}}{L}\right)\middle/2N\right]} \cos\left[\pi\dfrac{N-2}{N}\left(\tilde{n} - \dfrac{\tilde{k}}{L}\right)\middle/2\right] \right.
$$

$$
\left. + \frac{\sin\left[\pi\dfrac{N-1}{2N}\left(\tilde{n} + \dfrac{\tilde{k}}{L}\right)\right]}{\sin\left[\pi\left(\tilde{n} + \dfrac{\tilde{k}}{L}\right)\middle/2N\right]} \cos\left[\pi\dfrac{N-2}{N}\left(\tilde{n} + \dfrac{\tilde{k}}{L}\right)\middle/2\right] \right\}.
$$

Then *L*-zoomed signal obtained by zero-padding signal DCT spectrum and halving its highest frequency components is defined by the equation

$$
\tilde{a}_k = \frac{\alpha_0^{DCT}}{\sqrt{2LN}} + \frac{1}{N\sqrt{L}} \sum_{n=0}^{N-1} a_n \left\{ \frac{\sin[\pi(N+1/2NL)(\tilde{n}L - \tilde{k})]}{\sin[\pi(\tilde{n}L - \tilde{k})/2NL]} \cos\left(\pi\frac{\tilde{n}L - \tilde{k}}{2L}\right) \right.
$$

$$
+ \frac{\sin[\pi(N+1/2NL)(\tilde{n}L + \tilde{k})]}{\sin[\pi(\tilde{n}L + \tilde{k})/2NL]} \cos\left(\pi\frac{\tilde{n}L + \tilde{k}}{2L}\right)
$$

$$
+ \frac{\sin[\pi(N-1/2NL)(\tilde{n}L - \tilde{k})]}{\sin[\pi(\tilde{n}L - \tilde{k})/2NL]} \cos\left[\pi\frac{N-2}{2NL}(\tilde{n}L - \tilde{k})\right]
$$

$$
\left. + \frac{\sin[\pi(N-1/2NL)(\tilde{n}L + \tilde{k})]}{\sin[\pi(\tilde{n}L + \tilde{k})/2NL]} \cos\left[\pi\frac{N-2}{2NL}(\tilde{n}L + \tilde{k})\right] \right\}.
$$

Derivation of Equation 6.18

$$
\alpha_r^{(l)} = \mathrm{FFT}_N\left[a_k \exp\left(i2\pi\frac{kl}{LN}\right)\right] = \frac{1}{\sqrt{N}} \sum_{k=0}^{N-1} a_k \exp\left(i2\pi\frac{kl}{LN}\right) \exp\left(i2\pi\frac{kr}{N}\right).
$$

Replace in this equation $\{a_k\}$ with its expression through its DFT spectrum, $\{\alpha_r\}$ and obtain

$$
\alpha_r^{(l)} = \frac{1}{N} \sum_{k=0}^{N-1} \sum_{s=0}^{N-1} \alpha_r \exp\left(-i2\pi\frac{ks}{N}\right) \exp\left(i2\pi\frac{kl}{LN}\right) \exp\left(i2\pi\frac{kr}{N}\right)
$$

$$
= \frac{1}{N} \sum_{s=0}^{N-1} \alpha_r \sum_{k=0}^{N-1} \exp\left(-i2\pi\frac{s-r-l/N}{N}k\right) = \frac{1}{N} \sum_{s=0}^{N-1} \alpha_r \frac{\exp[-i2\pi(s-r-l/N)]-1}{\exp[-i2\pi(s-r-l/N)/N]-1}
$$

$$= \frac{1}{N} \sum_{s=0}^{N-1} \alpha_r \frac{\exp[i\pi(s - r - l/N)] - \exp[-i\pi(s - r - l/N)]}{\exp[i\pi((s - r - l/N)/N)] - \exp[-i\pi((s - r - l/N)/N)]}$$

$$\times \exp\left[i\pi \frac{(N - 1)(s - r - l/N)}{N}\right]$$

$$= \sum_{s=0}^{N-1} \alpha_r \frac{\sin[\pi(s - r - l/N)]}{N \sin[\pi((s - r - l/N)/N)]} \exp\left[i\pi \frac{(N - 1)(s - r - l/N)}{N}\right]$$

$$= \sum_{s=0}^{N-1} \alpha_r \, \mathrm{sincd}[N; \pi(s - r - l/N)] \exp\left[i\pi \frac{(N - 1)(s - r - l/N)}{N}\right].$$

Derivation of Equation 6.28

$$a_k = \frac{1}{\sqrt{\sigma_i} N} \sum_{r=0}^{N-1} \alpha_r^{u,v;\sigma_d} \exp\left[-i2\pi \frac{\left(k + u_x^{(i)}\right)\left(r + v_f^{(i)}\right)}{\sigma_i N}\right]$$

$$= \frac{1}{\sqrt{\sigma_i} N} \sum_{k=0}^{N-1} \left\{ \frac{1}{\sqrt{\sigma_d} N} \sum_{n=0}^{N-1} a_n \exp\left[i2\pi \frac{\left(n + u_x^{(d)}\right)\left(r + v_f^{(d)}\right)}{\sigma_d N}\right] \right\}$$

$$\times \exp\left[-i2\pi \frac{\left(k + u_x^{(i)}\right)\left(r + v_f^{(i)}\right)}{\sigma_i N}\right]$$

$$= \frac{1}{\sqrt{\sigma_i} N} \sum_{k=0}^{N-1} \left\{ \frac{1}{\sqrt{\sigma_d} N} \sum_{n=0}^{N-1} a_n \exp\left(i2\pi \frac{\tilde{n}^{(d)} \tilde{r}^{(d)}}{\sigma_d N}\right) \right\} \exp\left(-i2\pi \frac{\tilde{k}^{(i)} \tilde{r}^{(i)}}{\sigma_i N}\right)$$

$$= \frac{1}{\sqrt{\sigma_i \sigma_d} N} \sum_{n=0}^{N-1} a_n \left\{ \sum_{r=0}^{N-1} \exp\left[i \frac{2\pi}{N} \left(\frac{\tilde{n}^{(d)}}{\sigma_d} \tilde{r}^{(d)} - \frac{\tilde{k}^{(i)}}{\sigma_i} \tilde{r}^{(i)}\right)\right] \right\}$$

$$= \frac{1}{\sqrt{\sigma_i \sigma_d} N} \sum_{n=0}^{N-1} a_n \left\{ \sum_{r=0}^{N-1} \exp\left[i \frac{2\pi}{N} \left(\frac{\tilde{n}^{(d)}}{\sigma_d} r + \frac{\tilde{n}^{(d)}}{\sigma_d} v_f^{(d)} - \frac{\tilde{k}^{(i)}}{\sigma_i} r - \frac{\tilde{k}^{(i)}}{\sigma_i} v_f^{(i)}\right)\right] \right\}$$

$$= \frac{1}{\sqrt{\sigma_i \sigma_d} N} \sum_{n=0}^{N-1} a_n \exp\left[i \frac{2\pi}{N} \left(\frac{\tilde{n}^{(d)}}{\sigma_d} v_f^{(d)} - \frac{\tilde{k}^{(i)}}{\sigma_i} v_f^{(i)}\right)\right]$$

$$\times \left\{ \sum_{r=0}^{N-1} \exp\left[i \frac{2\pi}{N} \left(\frac{\tilde{n}^{(d)}}{\sigma_d} - \frac{\tilde{k}^{(i)}}{\sigma_i}\right) r\right] \right\}$$

$$
= \frac{1}{\sqrt{\sigma_i \sigma_d}\, N} \sum_{n=0}^{N-1} a_n \exp\left[i \frac{2\pi}{N} \left(\frac{\tilde{n}^{(d)}}{\sigma_d} v_f^{(d)} - \frac{\tilde{k}^{(i)}}{\sigma_i} v_f^{(i)} \right) \right] \frac{\exp[i2\pi(\tilde{n}^{(d)}/\sigma_d - \tilde{k}^{(i)}/\sigma_i)] - 1}{\exp\left[i \frac{2\pi}{N}(\tilde{n}^{(d)}/\sigma_d - \tilde{k}^{(i)}/\sigma_i) \right] - 1}
$$

$$
= \frac{1}{\sqrt{\sigma_i \sigma_d}\, N} \sum_{n=0}^{N-1} a_n \exp\left[i \frac{2\pi}{N} \left(\frac{\tilde{n}^{(d)}}{\sigma_d} v_f^{(d)} - \frac{\tilde{k}^{(i)}}{\sigma_i} v_f^{(i)} \right) \right]
$$

$$
\times \frac{\sin[\pi(\tilde{n}^{(d)}/\sigma_d - \tilde{k}^{(i)}/\sigma_i)]}{\sin[\pi(\tilde{n}^{(d)}/\sigma_d - \tilde{k}^{(i)}/\sigma_i)/N]} \exp\left[i\pi \frac{N-1}{N} \left(\frac{\tilde{n}^{(d)}}{\sigma_d} - \frac{\tilde{k}^{(i)}}{\sigma_i} \right) \right].
$$

With selection $v_f^{(d)} = v_f^{(d)} = -(N-1/2)$ obtain

$$
a_k = \frac{1}{\sqrt{\sigma_i \sigma_d}\, N} \sum_{n=0}^{N-1} a_n \exp\left[i \frac{2\pi}{N} \left(\frac{\tilde{n}^{(d)}}{\sigma_d} - \frac{\tilde{k}^{(i)}}{\sigma_i} \right) v_f \right] \times \frac{\sin[\pi(\tilde{n}^{(d)}/\sigma_d - \tilde{k}^{(i)}/\sigma_1)]}{\sin[\pi(\tilde{n}^{(d)}/\sigma_d - \tilde{k}^{(i)}/\sigma_1)/N]}
$$

$$
\times \exp\left[i\pi \frac{N-1}{N} \left(\frac{\tilde{n}^{(d)}}{\sigma_d} - \frac{\tilde{k}^{(i)}}{\sigma_i} \right) \right]
$$

$$
= \frac{1}{\sqrt{\sigma_i \sigma_d}\, N} \sum_{n=0}^{N-1} a_n \exp\left[i \frac{2\pi}{N} \left(\frac{\tilde{n}^{(d)}}{\sigma_d} - \frac{\tilde{k}^{(i)}}{\sigma_i} \right) \left(v_f + \frac{N-1}{2} \right) \right] \frac{\sin[\pi(\tilde{n}^{(d)}/\sigma_d - \tilde{k}^{(i)}/\sigma_i)]}{\sin[\pi(\tilde{n}^{(d)}/\sigma_d - \tilde{k}^{(i)}/\sigma_i)/N]}
$$

$$
= \frac{1}{\sqrt{\sigma_i \sigma_d}} \sum_{n=0}^{N-1} a_n \frac{\sin[\pi(\tilde{n}^{(d)}/\sigma_d - \tilde{k}^{(i)}/\sigma_i)]}{N \sin[\pi(\tilde{n}^{(d)}/\sigma_d - \tilde{k}^{(i)}/\sigma_i)/N]}
$$

$$
= \frac{1}{\sqrt{\sigma_i \sigma_d}} \sum_{n=0}^{N-1} a_n \operatorname{sincd}[N; \pi(\tilde{n}^{(d)}/\sigma_d - \tilde{k}^{(i)}/\sigma_i)]
$$

$$
= \frac{1}{\sqrt{\sigma_i \sigma_d}} \sum_{n=0}^{N-1} a_n \operatorname{sincd}\left[N; \pi \left(\frac{n + u_x^{(d)}}{\sigma_d} - \frac{k + u_x^{(i)}}{\sigma_i} \right) \right].
$$

Derivation of Equation 6.29

Let

$$
\alpha_{r,s} = \frac{1}{N} \sum_{m=0}^{N-1} \sum_{n=0}^{N-1} a_{m,n} \exp\left[i2\pi \left(\frac{\tilde{\tilde{m}}\tilde{\tilde{r}}}{N} + \frac{\tilde{\tilde{n}}\tilde{\tilde{s}}}{N} \right) \right],
$$

where $\tilde{m} = m + u_x$, $\tilde{n} = n + u_y$, $\tilde{r} = r + v_f$, $\tilde{s} = s + v_p$, be SDFT of an image represented by its samples $\{a_{m,n}\}$. Apply to this spectrum-rotated DFT defined by Equation 4.91 and obtain

$$
\tilde{a}_{k,l} = \frac{1}{\sigma N} \sum_{r=0}^{N-1} \sum_{s=0}^{N-1} \left\{ \frac{1}{N} \sum_{m=0}^{N-1} \sum_{n=0}^{N-1} a_{m,n} \exp\left[i2\pi\left(\frac{\tilde{m}\tilde{r}}{N} + \frac{\tilde{n}\tilde{s}}{N} \right) \right] \right\}
$$

$$
\times \exp\left[i2\pi\left(\frac{\tilde{k}\cos\theta + \tilde{l}\sin\theta}{\sigma N}\tilde{r} - \frac{\tilde{k}\sin\theta - \tilde{l}\cos\theta}{\sigma N}\tilde{s} \right) \right]
$$

$$
= \frac{1}{\sigma N^2} \sum_{m=0}^{N-1} \sum_{n=0}^{N-1} a_{m,n} \times \left\{ \sum_{r=0}^{N-1} \sum_{s=0}^{N-1} \exp\left[i2\pi\left(\frac{\tilde{m} + (\tilde{k}\cos\theta + \tilde{l}\sin\theta)/\sigma}{N}r \right. \right. \right.
$$

$$
\left. \left. \left. + \frac{\tilde{n} - (\tilde{k}\sin\theta - \tilde{l}\cos\theta)/\sigma}{N}s \right) \right] \right\}
$$

$$
\times \exp\left[i2\pi\left(\frac{\tilde{m}v_0^{(r)} + (\tilde{k}\cos\theta + \tilde{l}\sin\theta)v_\sigma^{(r)}/\sigma}{N} + \frac{\tilde{n}v_0^{(s)} + (\tilde{k}\sin\theta - \tilde{l}\cos\theta)v_\sigma^{(s)}/\sigma}{N} \right) \right]
$$

$$
= \frac{1}{\sigma N^2} \sum_{m=0}^{N-1} \sum_{n=0}^{N-1} a_{m,n} \frac{\sin\{\pi[\tilde{m} + (\tilde{k}\cos\theta + \tilde{l}\sin\theta)/\sigma]\}}{\sin\left[\pi\frac{\tilde{m} + (\tilde{k}\cos\theta + \tilde{l}\sin\theta)/\sigma}{N} \right]} \frac{\sin\{\pi[\tilde{n} + (\tilde{l}\cos\theta - \tilde{k}\sin\theta)/\sigma]\}}{\sin\left[\pi\frac{\tilde{n} + (\tilde{l}\cos\theta - \tilde{k}\sin\theta)/\sigma}{N} \right]}
$$

$$
\times \exp\left[i2\pi \frac{\tilde{m}\left(v_0^{(r)} + \frac{N-1}{2} \right) + (\tilde{k}\cos\theta + \tilde{l}\sin\theta)\left(v_\sigma^{(r)} + \frac{N-1}{2} \right)/\sigma}{N} \right]
$$

$$
\times \exp\left[i2\pi \frac{\tilde{n}\left(v_0^{(s)} + \frac{N-1}{2} \right) + (\tilde{k}\sin\theta - \tilde{l}\cos\theta)\left(v_\sigma^{(s)} + \frac{N-1}{2} \right)/\sigma}{N} \right].
$$

Natural settings of shift parameters that cancel phase shift factors will be

$$
v_0^{(r)} = v_\sigma^{(r)} = v_0^{(s)} = v_\sigma^{(s)} = -\frac{N-1}{2}.
$$

Then, obtain finally

$$\tilde{a}_{k,l} = \frac{1}{\sigma} \sum_{m=0}^{N-1} \sum_{n=0}^{N-1} a_{m,n} \operatorname{sincd}\left\{ N; \pi \left[\tilde{\tilde{m}} + \frac{\tilde{k}\cos\theta + \tilde{l}\sin\theta}{\sigma} \right] \right\}$$

$$\times \operatorname{sincd}\left\{ N; \pi \left[\tilde{\tilde{n}} + \frac{\tilde{l}\cos\theta - \tilde{k}\sin\theta}{\sigma} \right] \right\}.$$

Exercises

sincd_interpol_demo_CRC.m
RotateComparis_demo_CRC.m
differentiator_comparison_CRC.m
diffrentiat_integrat_error_CRC.m
radon_invradon_demo_CRC.m

References

1. C. F. Gauss, Nachclass: Theoria interpolationis methodo nova tractata, *Werke*, Band 3, 265–327, Königlishe Gesellshaft der Wissenshaften, Göttingen, 1866 (cited after M.T. Heideman, D.H. Johnson, and C.S. Burrus, Gauss and the history of the fast Fourier transform, *IEEE ASSP Magazine*, 1(4), 14–81, 1984).
2. W. H. Press, B. P. Flannery, S. A. Teukolsky, W. T. Vetterling. *Numerical Recipes. The Art of Scientific Computing.* Cambridge University Press, Cambridge, 1987.

7

Image Parameter Estimation: Case Study— Localization of Objects in Images

The evaluation of numerical parameters of objects from image data is required in many applications. Typical examples are measuring the number of objects, object orientations, dimensions and coordinates, and object tracking in video sequences. As special cases of this problem, one can also regard object recognition, when it is required to determine object index in the list of possible objects, image segmentation, when one needs to determine pixel belonging to one of several classes, fitting parameterized 2D curves (so-called "snakes" and "active contours" models) and 3D geometrical models.

As a rule, algorithms for parameter measurement are to be designed for the use for an arbitrary image from multitude of images generated by an imaging system. Therefore, the most appropriate approach to solve this problem is a statistical optimization one.

In this chapter, we address the problem of estimation of image numerical parameter using as a typical task, the task of target localization, or measurement of coordinates of a target object in images. Being of great practical importance by itself, this task has a fundamental value for developing and understanding methods for solving other tasks of image parameter estimation and can be regarded as a guiding example both in terms of the design of parameter estimation algorithms and in terms of evaluating their potential performance. We consider the task of target localization in two formulations: target localization on empty background in the presence of additive Gaussian noise (the section "Localization of Target Objects in the Presence of Additive Gaussian Noise") and target localization in cluttered images that contain many nontarget objects that camouflage the target one (the section "Target Localization in Cluttered Images").

Localization of Target Objects in the Presence of Additive Gaussian Noise

Optimal Localization Device for Target Localization in Noncorrelated Gaussian Noise

Let the target object located at coordinates $\{x_0, y_0\}$ of an image plane $\{x, y\}$ be specified by its $N_x \times N_y$ samples $\{a_{k,l}(x_0, y_0)\}$, $k = 0,1,\ldots,N_x - 1$, $l = 0,1,\ldots,N_y - 1$.

Consider a discrete model, in which samples $\{b_{k,l}\}$ of an observed input image that contains a target object can be regarded as a sum of samples of the target object signal and samples $\{n_{k,l}\}$ of a random process that represents imaging system sensor noise:

$$b_{k,l} = a_{k,l}(x_0, y_0) + n_{k,l} \qquad (7.1)$$

and assume that noise samples $\{n_k\}$ are statistically independent on the signal $\{a_{k,l}(x_0,y_0)\}$, are uncorrelated, and have a Gaussian probability distribution density with zero mean and variance σ_n^2. We discussed this classical ASIN-model in the sections "Models of Random Interferences" in Chapter 2. Although the model is quite simplistic, it is a good start model because, first, it enables solving the problem of statistically optimal localization analytically and, second, reveals some fundamental properties of statistically optimal localization. Moreover, there are a number of practical tasks to which this model is well adequate, such as the tasks of localization of constellations in stellar navigation and tracing target objects observed on a uniform background.

For the model of Equation 7.1, we obtained in the section "Basics of Optimal Statistical Parameter Estimation" in Chapter 2 (Equation 2.175) that the optimal MAP-estimates of object coordinates are solutions of equation

$$\{\hat{x}_0, \hat{y}_0\}_{\text{MAP}} = \underset{(x_0, y_0)}{\arg\min} \left\{ \sum_{k=0}^{N_x-1} \sum_{l=0}^{N_y-1} \left| b_{k,k} - a_{k,l}(x_0, y_0) \right|^2 - 2\sigma_n^2 \ln P(x_0, y_0) \right\}, \qquad (7.2)$$

where $P(x_0, y_0)$ is *a priori* probability distribution function of the object coordinates. The term $\sum_{k=0}^{N_x-1} \sum_{l=0}^{N_y-1} |b_{k,l}|^2$ of this equation does not depend on the coordinates (x_0, y_0). We can assume that the term $\sum_{k=0}^{N_x-1} \sum_{l=0}^{N_y-1} |a_k(x_0, y_0)|^2$ does not depend on the coordinates (x_0, y_0) either, because it refers to the same object in different coordinates. Therefore, Equation 7.2 is reduced to

$$\{\hat{x}_0, \hat{y}_0\}_{MAP} = \underset{(x_0, y_0)}{\arg\max} \left\{ \sum_{k=0}^{N_x-1} \sum_{l=0}^{N_y-1} b_{k,l} a_{k,l}(x_0, y_0) + \sigma_n^2 \ln P(x_0, y_0) \right\} \qquad (7.3)$$

Correspondingly, the ML-estimate of target coordinates can be obtained as

$$\{\hat{x}_0, \hat{y}_0\}_{ML} = \underset{(x_0, y_0)}{\arg\max} \left\{ \sum_{k=0}^{N_x-1} \sum_{l=0}^{N_y-1} b_{k,l} a_{k,l}(x_0, y_0) \right\} \qquad (7.4)$$

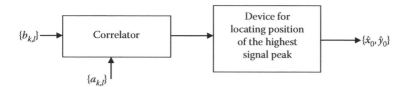

FIGURE 7.1
Flow diagram of the ML-optimal localization device.

Equation 7.4 implies that the device for obtaining optimal ML-estimates of the target object should contain two units: correlator unit, which computes mutual correlation function $\sum_{k=0}^{N_x-1}\sum_{l=0}^{N_y-1} b_{k,l}a_{k,l}(x_0, y_0)$ of the observed signal $\{b_{k,l}\}$ and the target object signal $\{a_{k,l}(x_0, y_0)\}$, or *template*, in all possible ranges of its coordinates, and a unit for locating the position of the highest signal peak at the correlator output (Figure 7.1).

Optimal MAP-estimator also consists of a correlator and a decision-making device locating the maximum in the correlation pattern. The only difference between ML- and MAP-estimators is that, in the MAP-estimator, the correlation pattern is biased by the appropriately normalized logarithm of the object coordinates' *a priori* probability distribution.

As it was described in the section "2D Discrete Fourier Transforms" in Chapter 4, digital correlation can be in a computationally efficient way implemented in the DFT domain by means of multiplication of input signal DFT spectrum by the complex conjugate of the target object DFT spectrum. Such an implementation is called *matched filtering*. Correspondingly, the filter that implements this operation is called *matched filter*.

Performance of ML-Optimal Estimators: Normal and Anomalous Localization Errors

Due to the random noise present in the signal, any localization device will always estimate object coordinates with a certain error. The estimation accuracy can be generally characterized by the probability density $p(\varepsilon_x, \varepsilon_y)$ of the estimation errors:

$$\varepsilon_x = x_0 - \hat{x}_0, \quad \varepsilon_y = y_0 - \hat{y}_0. \tag{7.5}$$

For evaluation of $p(\varepsilon_x, \varepsilon_y)$, one should analyze the distribution density of the coordinates of the highest peak of the output signal of the correlator within the area of all possible positions of the target object. This signal is a sum of the target object signal autocorrelation function and of a correlated Gaussian noise resulted from filtering the input white noise by the matched

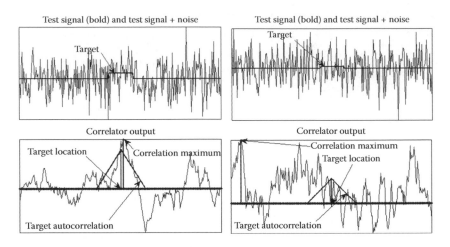

FIGURE 7.2
Normal (left column) and anomalous (right column) localization errors. Upper plots show the target signal with and without noise; bottom plots show the corresponding correlator outputs.

filter. This mixture is a nonstationary random process. This fact complicates the analysis very substantially. An approximate analytical solution of the problem can be obtained through the following reasoning.

Consider Figure 7.2. The two upper plots in Figure 7.2 represent a target signal with realizations of noise of two different levels, and the bottom plots show the corresponding correlator outputs. One can see that for the lower noise level (bottom left plot), the maximum of the correlator output signal is located in a close vicinity of the target location, while for the higher noise level, the maximum of the correlation is located far away from the target location.

Therefore, two essentially different types of possible localization errors have to be distinguished. One type of the errors occurs when noise intensity is relatively low with respect to that of the target object signal. In such cases, correlator output maxima are located in a close vicinity of the target location, and, therefore, localization errors are relatively small. The second type of the errors is represented by large errors, which occur when correlator output maxima are found very far away from the actual position of the target object. Small localization errors are caused by distortion of the target autocorrelation peak shape by the noise. Large errors are a result of prominent large noise outbursts occurring outside the area occupied by the object. We call the first type of error *normal errors* because, as we shall see, their probability distribution density can be approximated by a Gaussian (normal) distribution. The errors of the second type called *anomalous errors*. One can say that normal errors characterize the accuracy of coordinate measurements and anomalous errors characterize the measurement reliability. The program localization_demo_CRC.m provided in Exercises illustrates the phenomenon of anomalous errors in localization of objects in images.

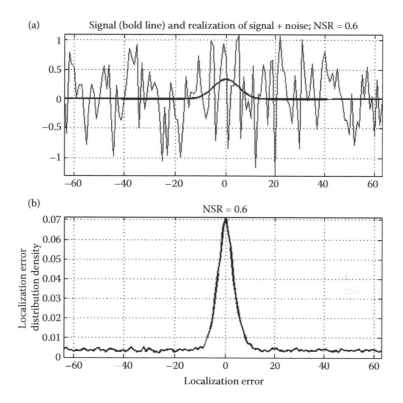

FIGURE 7.3
(a) A test target impulse (bold line) and a realization of uncorrelated Gaussian noise.
(b) Empirical probability density of the localization error obtained through statistical simulation of localization of this impulse in uncorrelated Gaussian noise.

Figure 7.3 shows an example of the empirical distribution density of
errors of localization of an impulse in its mixture with uncorrelated additive Gaussian noise, obtained using the demo program loclzerr_CRC.m (see
Exercises). As one can see, the localization error distribution density contains
a peak around zero, which can be associated with normal errors, and almost
uniform tales, which can be associated with anomalous errors. The uniform
distribution density of anomalous errors has quite an obvious explanation.
Outside the area occupied by the object, only the noise component is present. Because noise is a spatially homogeneous random process, extremely
arge noise outbursts, which cause anomalous localization errors, are equally
probable everywhere in this area.

The detailed treatment of numerical characteristics of normal and anomalous errors is brought out in Appendix. In "Distribution Density and Variances
of Normal Localization Errors" in Appendix, it is shown that normal errors
are zero mean random variables with Gaussian distribution density and variances defined, in the units of sampling intervals, by the relationships

$$\overline{\varepsilon_{\Delta x}^2} = \frac{1}{4\pi^2} \frac{\overline{f_{\Delta y}^2}}{\left(\overline{f_{\Delta x}^2}\right)\left(\overline{f_{\Delta y}^2}\right) - \left(\overline{f_{\Delta xy}^2}\right)^2} \frac{\sigma_n^2}{E_a} \, ;$$

$$\overline{\varepsilon_{\Delta y}^2} = \frac{1}{4\pi^2} \frac{\overline{f_{\Delta x}^2}}{\left(\overline{f_{\Delta x}^2}\right)\left(\overline{f_{\Delta y}^2}\right) - \left(\overline{f_{\Delta xy}^2}\right)^2} \frac{\sigma_n^2}{E_a} \, ; \qquad (7.6)$$

$$\overline{\varepsilon_{\Delta xy}^2} = \frac{1}{4\pi^2} \frac{\overline{f_{\Delta xy}^2}}{\overline{f_{\Delta x}^2}\,\overline{f_{\Delta y}^2} - \left(\overline{f_{\Delta xy}^2}\right)^2} \frac{\sigma_n^2}{E_a} \, .$$

The parameters of these equations are defined through DFT spectrum $\{\alpha_{r,s}\}$:

$$\alpha_{r,s} = \frac{1}{\sqrt{N_x N_y}} \sum_{k=0}^{N_x-1} \sum_{l=0}^{N_y-1} a_{k,l} \exp\left[i2\pi\left(\frac{kr}{N_x} + \frac{ls}{N_y} \right) \right] \qquad (7.7)$$

of the target object signal $\{a_{k,l}\}$ as follows:

$$E_a = \sum_{k=0}^{N_x-1} \sum_{l=0}^{N_y-1} |a_k|^2 = \sum_{r=0}^{N_x-1} \sum_{s=0}^{N_y-1} |\alpha_r|^2 \, ;$$

$$\overline{f_{\Delta x}^2} = \frac{1}{N_x^2} \frac{\sum_r r^2 \sum_s |\alpha_{r,s}|^2}{\sum_r \sum_s |\alpha_{r,s}|^2} \, ; \quad \overline{f_{\Delta y}^2} = \frac{1}{N_y^2} \frac{\sum_s s^2 \sum_r |\alpha_{r,s}|^2}{\sum_r \sum_s |\alpha_{r,s}|^2} \, ; \qquad (7.8)$$

$$\overline{f_{\Delta xy}^2} = \frac{1}{N_x N_y} \frac{\sum_r \sum_s rs |\alpha_{r,s}|^2}{\sum_r \sum_s |\alpha_{r,s}|^2} \, ,$$

where summation is carried out over indices r,s of spatial frequencies from the lowest frequencies ($r = 1$, $\varepsilon = 1$) to the highest ones ($N_x/2$, $N_y/2$) or ($(N_x - 1)/2$, $(N_y - 1)/2$) depending on whether N_x and N_y are even or odd numbers.

On plots in Figure 7.4, one can compare these theoretical estimates of normal noise standard deviation of localization errors with experimental data obtained using MATLAB program loclzerr_CRC.m (see Exercises) for the test signal shown in the upper plot of Figure 7.3. The plots clearly show

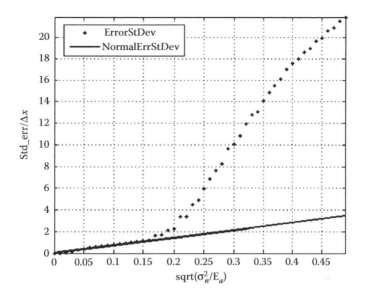

FIGURE 7.4
Theoretical (solid line) and experimental (dot line) curves of standard deviation of localization error as functions of noise-to-signal ratio.

that for sufficiently low noise-to-signal ratios theoretical estimates fit experimental data very well. However, with the growth of the noise-to-signal ratio, the error standard deviation starts growing faster and faster, which evidences the appearance of anomalous errors.

As it was already mentioned, anomalous errors have uniform distribution density over the area of search. Therefore, their sufficient numerical statistical characteristics are their probability P_{Ae}, that is, the area of the uniform tails of the error probability density. In "Evaluation of the Probability of Anomalous Localization Errors" in Appendix, it is shown that the probability of anomalous errors can be evaluated by the integral

$$P_{Ae} = \frac{1}{\sqrt{2\pi}} \int_{-\infty}^{\infty} \exp\left(-\frac{n^2}{2}\right) \left\{ 1 - \left[\Phi\left(\sqrt{\frac{E_a}{\sigma_n^2}} + n \right) \right]^{Q-1} \right\} dn, \qquad (7.9)$$

where Q is the number of positions the target object can occupy, without overlapping, in the area of search and

$$\Phi(x) = \frac{1}{\sqrt{2\pi}} \int_{-\infty}^{x} \exp\left(-\frac{n^2}{2}\right) dn \qquad (7.10)$$

is the "error integral."

In "Evaluation of the Probability of Anomalous Localization Errors" in Appendix, it is also shown that the probability of anomalous localization errors P_{Ae} as a function of SNR E_a/σ_n^2 for large Q features a threshold behavior:

$$\lim_{Q\to\infty} P_{Ae} = \frac{1}{\sqrt{2\pi}} \int_{-\infty}^{\infty} \exp\left(-\frac{n^2}{2}\right) \left\{1 - \left[\Phi\left(\sqrt{\frac{E_a}{\sigma_n^2}} + n\right)\right]^{Q-1}\right\} dn$$

$$= \begin{cases} 0, & \text{if } E_a/\sigma_n^2 > 2\ln Q \\ 1, & \text{if } E_a/\sigma_n^2 \le 2\ln Q \end{cases}. \tag{7.11}$$

This feature implies that if the area of search is large enough with respect to the size of the object, the probability of anomalous errors may become enormously high when input SNR E_a/σ_n^2 is lower than the fundamental threshold defined by Equation 7.11. This also means that, when the area of search of a given target increases, the SNR must also be increased in order to keep the probability of anomalous errors low. Moreover, for any given intensity of noise, there exists a trade-off between localization accuracy defined by the variance of normal errors and localization reliability described by the probability of anomalous errors. Increasing the accuracy achieved by means of increasing parameters $\overline{f_{\Delta x}^2}$, $\overline{f_{\Delta y}^2}$, $\overline{f_{\Delta x \Delta y}^2}$ in Equation 7.8 through widening the object signal spectrum results, with signal energy fixed, in increasing the probability of anomalous errors, because widening signal spectrum width is equivalent to narrowing the target object, and, consequently, to increasing of the ratio Q of the area of search to the object signal area.

Figure 7.5 shows results of numerical verification of the above relationships by computer simulation of the optimal localization device. The data plotted in the figure were obtained by computer simulation of localization of uniformly painted square of 8×8 pixels in noisy images of different size (program localization_demo_CRC.m).

In conclusion, it should be noted that the relationships of Equation 7.6 for variances of normal errors and the threshold relationship of Equation 7.11 for probability of anomalous errors have a very general fundamental value. Variances of normal errors determine, in a special case of a point source target object, the variance of error in measuring the coordinates of the point source. That has a direct relation to such an important characteristics of imaging system as their resolving power, that is, their capability to resolve two closely located point sources. In the case of point sources, target object signal is the system PSF and imaging system resolving power is conventionally evaluated using the *Rayleigh's criterion*: two point sources are considered resolved if the minimum between two corresponding PSF peaks does not exceed 80% of the peak maxima. The above analysis of the accuracy of target localization reveals that the Rayleigh's criterion does not take into account

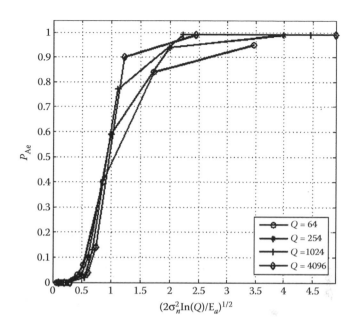

FIGURE 7.5

Experimental data for the probability of anomalous errors as a function of noise-to-signal ratio $2\sqrt{\sigma_n^2 \ln Q / E_a}$ for different values of Q. The theoretical threshold value $2\sqrt{\sigma_n^2 \ln Q / E_a} = 1$.

imaging system noise, which is yet another factor that, along with the system PSF, determines system resolving power. It shows, in particular, that system resolving power might be arbitrarily high if noise level is sufficiently low and, vice versa, it can be very low if noise level is high even if system PSF is sharp enough to satisfy Rayleigh's criterion.

As for the probability of anomalous errors, which characterizes localization reliability, note that $\ln Q = \ln 2 \times \log_2 Q$, and $\log_2 Q$ can be regarded as the entropy, in bits, of the results of determining in which position of Q different object positions the target object is located, or, generally, which of Q possible values the parameter under measurement has. Therefore, Equation 7.11 gives the absolute lower bound for the object signal energy per bit of measurement information required for reliable parameter estimation in the presence of white Gaussian noise with variance σ_n^2:

$$E_a / \log_2 Q > 2\sigma_n^2 \ln 2. \tag{7.12}$$

Target Object Localization in the Presence of Nonwhite (Correlated) Additive Gaussian Noise

In this section, we extend the above results obtained for the additive uncorrelated noise model to the case of nonwhite, or correlated, noise. Let the noise

DFT power spectrum be $\sigma_n^2 |H_{r,s}^{(n)}|^2$, where $|H_{r,s}^{(n)}|^2$ is a normalized spectrum shaping function such that

$$\sum_{r=0}^{N_x-1}\sum_{s=0}^{N_y-1}\left|H_{r,s}^{(n)}\right|^2 = 1. \tag{7.13}$$

Pass the observed signal plus noise mixture through a filter with frequency response

$$\eta_{r,s}^{(wht)} = \frac{1}{|H_{r,s}^{(n)}|}. \tag{7.14}$$

We will refer to this filter as to the *whitening filter*. At the output of the whitening filter, noise power spectrum becomes uniform with spectral density σ_n^2, while the target object signal spectrum is modified to $\alpha_{r,s} / |H_{r,s}^{(n)}|$, that is, the above-discussed model of additive uncorrelated noise for the modified target signal is applicable to the signal at the output of the whitening filter. Therefore, one can conclude that for the case of additive correlated noise, an ML-optimal localization device should consist of the whitening filter followed by a filter matched to the modified target object signal and by a device for localizing the signal maximum. In this way, we arrive at the optimal localization device shown in a flow diagram of Figure 7.6 that provides ML-estimation of the target object coordinates.

The whitening filters and the filter matched to the modified target object can be combined into one filter with discrete frequency response

$$\eta_{r,s} = \frac{\alpha_{r,s}^*}{|H_{r,s}^{(n)}|^2}. \tag{7.15}$$

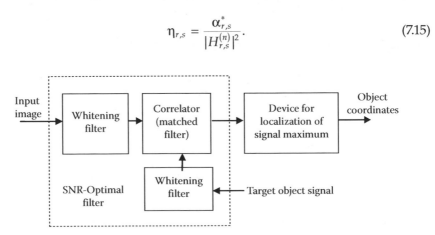

FIGURE 7.6
Flow diagram of the ML-optimal localization device for the case of correlated Gaussian noise.

The above reasoning also implies that, for a target signal observed in a mixture with additive correlated Gaussian noise, the filter defined by Equation 7.15 generates a signal that is monotonically related with *a posteriori* probability of target coordinates.

The filter defined by Equation 7.15 has yet another important feature. It provides the highest possible, for all linear filters, ratio of its response to the target object signal to standard deviation of noise at its output (SNR). In order to prove this, consider the response of an arbitrary filter with a discrete frequency response $\eta_{r,s}$ to the target signal $a_{k,l}$ located at coordinates $(\tilde{x}_0 = x_0/\Delta x; \tilde{y}_0 = y_0/\Delta y)$ and noise $n_{k,l}$ with spectral density $\sigma_n^2 |H_{r,s}^{(n)}|^2$. The filter output to the target signal at the point $(\tilde{x}_0, \tilde{y}_0)$ is equal to

$$
b_{k,l} = \frac{1}{\sqrt{N_x, N_y}} \sum_{r=0}^{N_x-1} \sum_{s=0}^{N_y-1} \alpha_{r,s} \eta_{r,s} \exp\left\{ -i2\pi\left(\frac{\tilde{k} - \tilde{x}_0}{N_x} r + \frac{\tilde{l} - \tilde{y}_0}{N_y} l \right) \right\} \Bigg|_{\substack{\tilde{k}=\tilde{x}_0 \\ \tilde{l}=\tilde{y}_0}}
$$

$$
= \frac{1}{\sqrt{N_x, N_y}} \sum_{r=0}^{N_x-1} \sum_{s=0}^{N_y-1} \alpha_{r,s} \eta_{r,s}. \tag{7.16}
$$

The standard deviation of noise at the filter output can be found as

$$
\tilde{\sigma}_n = \sigma_n \left(\sum_{r=0}^{N_x-1} \sum_{s=0}^{N_y-1} |\eta_{r,s}|^2 |H_{r,s}^{(n)}|^2 \right)^{1/2}. \tag{7.17}
$$

Their ratio, or the SNR, is then equal to

$$
SNR = \frac{b_{\tilde{x}_0, \tilde{y}_0}}{\tilde{\sigma}_n} = \frac{\dfrac{1}{\sqrt{N_x, N_y}} \sum\limits_{r=0}^{N_x-1} \sum\limits_{s=0}^{N_y-1} \alpha_{r,s} \eta_{r,s}}{\sigma_n \left(\sum\limits_{r=0}^{N_x-1} \sum\limits_{s=0}^{N_y-1} |\eta_{r,s}|^2 |H_{r,s}^{(n)}|^2 \right)^{1/2}}. \tag{7.18}
$$

By virtue of the Cauchy–Bunyakovsky–Schwarz inequality

$$
\sum_n a_n b_n^* \leq \left(\sum_n a_n^2 \right)^{1/2} \left(\sum_n b_n^2 \right)^{1/2}, \tag{7.19}
$$

the following inequality holds for this ratio:

$$SNR = \frac{\frac{1}{\sqrt{N_x, N_y}} \sum_{r=0}^{N_x-1} \sum_{s=0}^{N_y-1} \left(\frac{\alpha_{r,s}}{|H_{r,s}^{(n)}|}\right) \left(\eta_{r,s}|H_{r,s}^{(n)}|\right)}{\sigma_n \left(\sum_{r=0}^{N_x-1} \sum_{s=0}^{N_y-1} |\eta_{r,s}|^2 |H_{r,s}^{(n)}|^2\right)^{1/2}} \leq \frac{1}{\sigma_n \sqrt{N_x N_y}} \left(\sum_{r=0}^{N_x-1} \sum_{s=0}^{N_y-1} \frac{|\alpha_{r,s}|^2}{|H_{r,s}^{(n)}|}\right)^{1/2},$$

(7.20)

which reaches its upper bound for $\eta_{r,s}$ that satisfies Equation 7.15. We will refer to this filter as to the *SNR-optimal filter*.

$$\eta_{r,s}^{(opt)} = \frac{\alpha_{r,s}^*}{|H_{r,s}^{(n)}|^2} = \arg\max_{\eta_{r,s}}(SNR).$$

(7.21)

Localization Accuracy for the SNR-Optimal Filter

Because the SNR-optimal filter is the matched filter for the target object signal modified by the whitening operation, the potential localization accuracy of the optimal filter can be derived from Equations 7.6 and 7.8 for localization accuracy in the presence of white noise in which target object DFT spectrum $\{\alpha_{r,s}\}$ is replaced by its whitened spectrum $\alpha_{r,s}/|H_{r,s}^{(n)}|$.

It is very instructive to compare the potential localization accuracy in the cases of white and nonwhite noise for the same input noise variance. For the sake of simplicity, we consider a 1D case. In this case, we have from Equation 7.6 for nonwhite noise that

$$\overline{\left(\varepsilon_{\Delta x}^{(nwn)}\right)^2} = \frac{1}{4\pi^2 (f_{\Delta x}^{(nwn)})^2} \frac{\sigma_n^2}{E_a^{(nwn)}} = \frac{1}{4\pi^2 f_{\Delta x}^2} \frac{\sigma_n^2}{E_a} \frac{f_{\Delta x}^2 E_a}{(f_{\Delta x}^{(nwn)})^2 E_a^{(nwn)}} = \overline{\varepsilon_{\Delta x}^2} G,$$

(7.22)

where

$$E_a^{(nwn)} = \sum_{r=0}^{N_x-1} \frac{|\alpha_r|^2}{|H_r^{(2)}|};$$

$$\overline{\left(f_{\Delta x}^{(nwn)}\right)^2} = \frac{1}{N_x^2} \frac{\sum_r r^2 \frac{|\alpha_r|^2}{|H_r^{(n)}|^2}}{\sum_r \frac{|\alpha_r|^2}{|H_r^{(n)}|^2}}$$

(7.23)

and

$$G = \frac{\overline{\left(\varepsilon_{\Delta x}^{(nwn)}\right)^2}}{\varepsilon_{\Delta x}^2} \frac{\overline{f_{\Delta x}^2 E_a}}{\left(f_{\Delta x}^{(nwn)}\right)^2 E_a^{(nwn)}} = \frac{\sum_r r^2 |\alpha_r|^2}{\sum_r r^2 \frac{|\alpha_r|^2}{|H_r^{(n)}|^2}} \tag{7.24}$$

is the ratio of localization error variances for nonwhite and white noise of the same intensity. As, by the definition (Equation 7.13), $\sum_{r=0}^{N_x-1} |H_r^{(n)}|^2 = 1$, all $|H_r^{(n)}|^2$ do not exceed one. This implies that $r^2 |\alpha_r|^2 / |H_r^{(n)}|^2 \geq r^2 |\alpha_r|^2$ and, therefore $G \leq 1$, which means that the potential localization accuracy in the presence of nonwhite noise is always better than that for white noise with the same variance.

This conclusion is well intuitively understood. Consider, for instance, a trivial special case, when noise is band-limited and its bandwidth is less than that of the signal. SNR-optimal filter is in this case a band pass filter, which lets through all input signal frequencies, where noise spectrum vanishes to zero and blocks all other frequencies, where noise spectrum is nonzero. The resulting signal is therefore noise-free and the potential localization error variance is equal to zero.

Optimal Localization in Color and Multicomponent Images

The above-presented results for optimal localization in monochrome images can be straightforwardly extended to localization in color, or, generally, multicomponent images, observed, in each component, in the presence of additive Gaussian noise.

Let $\{a_{k,l}^{(m)}(x_0, y_0)\}$, $k = 0,1,\ldots,N_x - 1$, $l = 0,1,\ldots,N_y - 1$ be the m-th component of the target object image $m = 1,2,\ldots,M$ (for color images $M = 3$) located in coordinates (x_0, y_0) with coordinate probability density $P(x_0, y_0)$. Also let $\{b_{k,l}^{(m)}\}$ and $\{n_{k,l}^{(m)}\}$ be samples of the corresponding components of the observed image, in which the target is to be localized, and of the additive Gaussian noise such that

$$b_{k,l}^{(m)} = a_{k,l}^{(m)}(x_0, y_0) + n_{k,l}^{(m)}. \tag{7.25}$$

In the assumption that components of the additive noise as well as samples of noise within each component are all mutually uncorrelated, obtain that *a posteriori* probability of observing signal $\{b_{k,l}^{(m)}\}$ provided the target object is located at coordinates (x_0, y_0) is

$$P\{n_{k,l}^{(m)} = b_{k,l}^{(m)} - a_{k,l}^{(m)}(x_0, y_0)\} \propto \prod_{m=1}^{M} \prod_{k=0}^{N-1} \exp\left\{-\frac{1}{2\sigma_n^{(m)2}}\left[b_{k,l}^{(m)} - a_{k,l}^{(m)}(x_0, y_0)\right]^2\right\}, \tag{7.26}$$

where $\sigma_n^{(m)2}$ denotes the variance of the m-th noise component. Therefore, optimal MAP- and ML-localization devices are defined, respectively, by equations

$$\{\hat{x}_0, \hat{y}_0\} = \underset{(x_0, y_0)}{\arg\max} \left\{ \sum_{m=1}^{M} \frac{1}{\sigma_n^{(m)2}} \left[\sum_{k=0}^{N_x-1} \sum_{l=0}^{N_y-1} b_{k,l}^{(m)} a_{k,l}^{(m)}(x_0, y_0) \right] - \sum_{m=1}^{M} \ln P(x_0, y_0) \right\} \quad (7.27)$$

and

$$\{\hat{x}_0, \hat{y}_0\} = \underset{(x_0, y_0)}{\arg\max} \left\{ \sum_{m=1}^{M} \frac{1}{\sigma_n^{(m)2}} \sum_{k=0}^{N_x-1} \sum_{l=0}^{N_y-1} b_{k,l}^{(m)} a_{k,l}^{(m)}(\hat{x}_0, \hat{y}_0) \right\}, \quad (7.28)$$

which implies that multicomponent optimal localization device should consist of M parallel component-wise correlators, or matched filters, of an adder for weighted summation of correlators' outputs and of a unit for determining coordinates of the signal maximum at the adder's output. Its schematic diagram is shown in Figure 7.7.

Variances of normal localization errors, which characterize the accuracy of optimal localization, and probability of anomalous errors in multicomponent images can be found with just the same technique as it was done for single-component images. For details, readers may refer to Ref. [1].

As in the case of single-component images, for localization in multicomponent images with correlated noise components, input image and target object image prewhitening can be used in order to reduce the problem to the case of

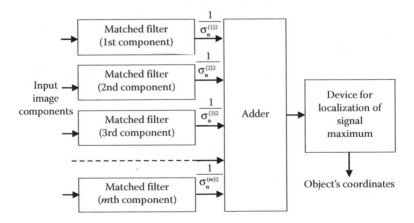

FIGURE 7.7
Schematic flow diagram of the optimal device for localization of a target object in multicomponent images.

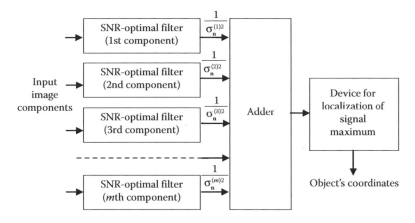

FIGURE 7.8
Schematic flow diagram of ML-optimal device for localization of target objects in multicomponent images with correlated component noise.

uncorrelated noise. The optimal localization device of Figure 7.7 should then be modified as it is shown in Figure 7.8.

SNR-optimal filters in this device are filters with discrete frequency responses $\{\eta_{r,s}^{(m,\text{opt})}\}$ defined by the equation

$$\eta_{r,s}^{(m,\text{opt})} = \frac{\alpha_{r,s}^{(m)*}}{\left|H_{r,s}^{(m)}\right|^2},$$ (7.29)

in which $\alpha_{r,s}^{(m)*}$ is a complex conjugate DFT spectrum of the target object m-th component image and $|H_{r,s}^{(m)}|^2$ is the DFT spectral density of the m-th noise component normalized according to Equation 7.13.

Object Localization in the Presence of Multiple Nonoverlapping Nontarget Objects

In this section, we discuss an extended image model, in which observed image signal contains, along with the target object signal and additive noncorrelated Gaussian sensor noise, some number Q of nontarget objects that do not overlap one another and the target object. A good example of the situation, in which such a model is appropriate, would be the task of locating a specific character in a printed text.

Inasmuch as nontarget objects do not overlap the target object, the only obstacle for locating target object in the close vicinity of its actual location is additive sensor noise. Therefore, the design of the filter for the highest localization accuracy, that is, the lowest variance of normal errors, is governed by the same reasoning as that for the above-described additive noise model with no nontarget objects. The optimal filter in this case will also be the matched filter and the variance of normal errors will be defined by the same formulas as those in the cases of the absence of nontarget objects (Equation 7.6). The presence of nontarget objects affects only the probability of anomalous errors.

In order to estimate the probability of anomalous errors of target localization in the presence of multiple nontarget objects, consider first a simple special case, when the area of search contains certain number Q of identical nontarget objects. Let the maximal value of their cross-correlation function with the target object be $R_{a,q}$. Then, by analogy with Equation 7.9, the probability of anomalous localization errors (probability of false detection) for the matched filter localization device can be evaluated as follows:

$$P_{Ae} = \frac{1}{\sqrt{2\pi}} \int_{-\infty}^{\infty} \exp\left(-\frac{n^2}{2}\right) \left\{ 1 - \left[\Phi\left(\frac{E_a - R_{a,q}}{\sqrt{E_a \sigma_n^2}} + n \right) \right]^Q \right\} dn. \tag{7.30}$$

As one can see, the presence of nontarget objects very substantially increases the probability of anomalous errors compared to that for the case of target localization on empty background, because $E_a - R_{a,q} < E_a$ and this difference can even be negative, if cross-correlation of target object with nontarget object is higher than target object autocorrelation.

In a more general case, when in the area of search there might be some random number Q of nontarget objects, which belong to one of C classes according to the maximal values $R_{a,q}^{(c)}$ of their cross-correlation function with the target object, it is shown in Ref. [1] that the probability of anomalous localization (false detection) errors can be found as

$$P_a = \sum_Q P(Q) \left\{ \frac{1}{\sqrt{2\pi}} \int_{-\infty}^{\infty} \exp\left(-\frac{n^2}{2}\right) \left\{ 1 - \left[\sum_{c=1}^{C} P(c)\Phi\left(\frac{E_a - R_{a,q}^{(c)}}{\sqrt{E_a \sigma_n^2}} + n \right) \right]^Q \right\} dn \right\}, \tag{7.31}$$

where $P(Q)$ is the probability of total number Q of nontarget objects and $P(c)$ is the probability of nontarget objects of c-th class ($c = 1,2,...,C$).

Figure 7.9 illustrates the phenomenon of false identification of a target object with nontarget objects on an example of character recognition. It can also be observed using MATLAB program localization_demo_CRC.m provided in Exercises.

In order to decrease the probability of anomalous errors in the presence of nontarget objects, it is necessary to suppress cross-correlation peaks for nontarget objects with respect to the autocorrelation peak of the target object. This requires an appropriate modification of the matched filter. Therefore, for localization of a target in the presence of nontarget objects,

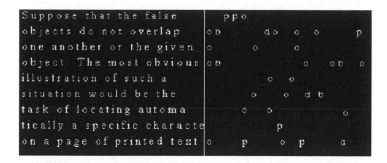

FIGURE 7.9
Detection, using matched filtering, of a character "o" in a printer text. Left image: noisy image of a printed text with standard deviation of additive noise 15 (within signal range 0–255). Right image: results of detection of character "o"; one can see quite a number of false detections.

it is not possible to simultaneously secure the minimum of normal error variance and the minimum of the probability of anomalous errors with the same estimator, and optimal localization must be carried out in two steps. The first step is target detection with minimal probability of anomalous errors. A coarse but reliable estimate of the object coordinates is obtained at this stage. For the accurate estimation of target coordinates, the second step is needed, in which localization with minimal variance of normal errors is obtained in a reduced area of search in the vicinity of the location found at the first step. The optimal localization device for the second step is the matched filter. The problem of the design of optimal device for reliable detection of targets on the background of a clutter of nontarget objects is addressed in the next section.

Target Localization in Cluttered Images

Formulation of the Approach

Consider now the most general problem of locating targets in images that contain a target object and a clutter of nontarget objects that obscure the target object. There are many image processing tasks, in which the localization of a target object in a clutter of nontarget objects is required. Here are some of them, to name a few:

- Navigation using terrain maps
- Detection, localization and tracking of various formations in medical
- Detection and localization of defects in nondestructive testing images

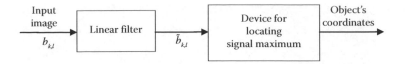

FIGURE 7.10
Schematic diagram of the localization device.

- Detection and localization of specific patterns in fingerprints
- Detection and localization of specific objects in surveillance images, for instance, face detection

As it follows from the discussion in the section "Localization of Target Objects in the Presence of Additive Gaussian Noise," those background non-target objects represent the main obstacle for reliable localization of target objects. Our purpose therefore is to find out how can one minimize the danger of false identification of the target object with one of many nontarget objects.

In practical tasks of target localization, it is required, as a rule, to secure the most reliable target localization in a particular observed image rather than on average over a hypothetic ensemble of images, which rarely can be specified. Therefore, optimization and adaptation of the localization algorithm for individual images are desired.

This requirement of adaptivity will be imperative in our approach. The second imperative requirement will be that of low computational complexity of localization algorithms. Bearing this in mind, we shall restrict the discussion with the same type of localization devices as those described in the section "Localization of Target Objects in the Presence of Additive Gaussian Noise," that is, the devices that consist of a linear filter followed by a unit for locating signal maximum at the filter output (Figure 7.10).

Owing to the existence of fast and recursive algorithms of digital linear filtering, such devices have low computational complexity in their computer implementation. They also have promising optical and electro-optical implementations [2]. In such type of devices, it is the linear filter that has to be optimized and is the subject for adaptation.

SCR-Optimal Adaptive Correlator

Consider first the task, illustrated in Figure 7.11, of locating a precisely defined target object on a given image.

Let $\{b_{k,l}\}$, $k = 0,\ldots,N_x - 1$; $l = 0,1,\ldots,N_y - 1$, be samples of an input image that contains, in coordinates (x_0, y_0), a target image defined by its samples $\{a_{k,l}(x_0, y_0)\}$. The position of the target object has to be found as a position of the highest signal peak at the output of the linear filter of the localization device defined in Figure 7.10. Also let $\{\eta_{r,s}\}$ be the discrete frequency response of the

FIGURE 7.11
Input image (left) and target object (right image, highlighted).

linear filter, and $\{\alpha_{r,s}\}$ and $\{\beta_{r,s}\}$ be the DFT spectral coefficients of the target object and of the input image, respectively ($r = 0,\ldots,N_x - 1$; $s = 0,1,\ldots,N_y - 1$).

Then filter output at the location of the target object is

$$\tilde{a}_{x_0,y_0} = \frac{1}{\sqrt{N_xN_y}} \sum_{r=0}^{N_x-1}\sum_{s=0}^{N_y-1} \alpha_{r,s}\exp\left[i2\pi\left(\frac{x_0r}{N_x} + \frac{y_0s}{N_y}\right)\right]\exp\left[-i2\pi\left(\frac{kr}{N_x} + \frac{ls}{N_y}\right)\right]\Bigg|_{\substack{k=x_0\\l=y_0}}$$

$$= \frac{1}{\sqrt{N_xN_y}} \sum_{r=0}^{N_x-1}\sum_{s=0}^{N_y-1} \alpha_{r,s}\eta_{r,s}, \tag{7.32}$$

and filter response to the entire input image is

$$\tilde{b}_{k,l} = \frac{1}{\sqrt{N_xN_y}} \sum_{r=0}^{N_x-1}\sum_{s=0}^{N_y-1} \beta_{r,s}\eta_{r,s}\exp\left[-i2\pi\left(\frac{kr}{N_x} + \frac{ls}{N_y}\right)\right]. \tag{7.33}$$

The rate of target object false detections is determined by the number of possible target positions within filter output image $\tilde{b}_{k,l}$, in which $\tilde{b}_{k,l} \geq \tilde{a}_{x_0,y_0}$. One can try to minimize this number by an appropriate selection of the filter frequency response $\{\eta_{r,s}\}$. No analytical solution of this minimization problem is feasible, because there is no analytical relationship between the frequency response of the filter and the probability density of its output signal. The only statistical parameters of the filter output signal that can be determined given the filter frequency response are signal mean value

$$\bar{b}_{k,l} = \frac{1}{\sqrt{N_xN_y}} \sum_{r=0}^{N_x-1}\sum_{s=0}^{N_y-1} \beta_{0,0}\eta_{0,0} \tag{7.34}$$

and variance

$$\bar{\tilde{b}}^2 = \frac{1}{N_x N_y} \sum_{r=0}^{N_x-1} \sum_{s=0}^{N_y-1} |\beta_{r,s}|^2 |\eta_{r,s}|^2. \tag{7.35}$$

The latter follows from Parseval's relationship. Therefore, for optimization of the localization device filter, we will rely upon the Tchebyshev's inequality, which establishes a relationship between probability that a random variable x exceeds some threshold x_{thr} and the variable's mean value $\bar{x}$ and standard deviation σ.

$$\text{Probability}\left(|x - \bar{x}| \geq x_{\text{thr}}\right) \leq x_{\text{thr}}^2 / \sigma^2. \tag{7.36}$$

Filter output signal mean value does not affect localization of the output signal global maximum. Therefore, in order to simplify further analysis, we will assume it set to be zero by selecting $\eta_{0,0} = 0$. Then, using the Tchebyshev's inequality (Equation 7.36), we obtain for the probability of false identification of the target object located in coordinates (x_0, y_0) with one of nontarget (background) objects:

$$P_{\text{Ae}}(x_0, y_0) = \text{Probability}\left(b_{k,l}^{(\text{bg},x_0,y_0)} - \overline{b^{(\text{bg},x_0,y_0)}} \geq \tilde{a}_{x_0,y_0} \right) \leq \frac{\overline{\tilde{b}^{(\text{bg},x_0,y_0)^2}}}{\tilde{a}_{x_0,y_0}^2}, \tag{7.37}$$

where $\{b_{k,l}^{(\text{bg},x_0,y_0)}\}$ are filter output values at points outside the target location, $\overline{\tilde{b}_{x_0,y_0}^{(\text{bg},x_0,y_0)^2}}$ is their variance, and $\tilde{a}_{x_0,y_0}$ is the filter output value, defined by Equation 7.32, at the target location. The optimal design of the filter requires minimization of $P_{\text{Ae}}(x_0, y_0)$ on average over all ranges of possible target object coordinates:

$$\eta_{r,s}^{(\text{opt})} = \underset{\eta_{r,s}}{\arg\min} \, AV_{x_0,y_0}\left[P_{\text{Ae}}(x_0, y_0) \right]. \tag{7.38}$$

According to Equation 7.37, this is equivalent to

$$\eta_{r,s}^{(\text{opt})} = \underset{\eta_{r,s}}{\arg\min} \, AV_{x_0,y_0}\left[\frac{\overline{\tilde{b}^{(\text{bg},x_0,y_0)^2}}}{\tilde{a}_{x_0,y_0}^2} \right] = \underset{\eta_{r,s}}{\arg\max} \, \frac{\tilde{a}_{x_0,y_0}^2}{AV_{x_0,y_0}\left(\overline{\tilde{b}^{(\text{bg},x_0,y_0)^2}} \right)}. \tag{7.39}$$

assuming that the filter response to the target object in its location does not depend on the location. We call the ratio

$$SCR = \frac{\tilde{a}_{x_0,y_0}^2}{AV_{x_0,y_0}\left(\overline{\tilde{b}^{(\text{bg},x_0,y_0)^2}} \right)} \tag{7.40}$$

signal-to-clutter ratio (SCR).

Equation 7.39 implies that, for minimizing P_{Ae}, one should maximize SCR. If the DFT power spectrum $|\beta_{r,s}^{(bg,x_0,y_0)}|^2$ of the input image background component is known, $\bar{b}^{(bg,x_0,y_0)2}$ can be found using Parseval's relationship:

$$\bar{b}^{(bg,x_0,y_0)2} = \frac{1}{N_x N_y} \sum_{r=0}^{N_x-1} \sum_{s=0}^{N_y-1} \left|\beta_{r,s}^{(bg,x_0,y_0)}\right|^2 |\eta_{r,s}|^2 \tag{7.41}$$

and therefore

$$AV_{x_0,y_0}\left(\bar{b}^{(bg,x_0,y_0)2}\right) = \frac{1}{N_x N_y} \sum_{r=0}^{N_x-1} \sum_{s=0}^{N_y-1} AV_{x_0,y_0}\left(\left|\beta_{r,s}^{(bg,x_0,y_0)}\right|^2\right) |\eta_{r,s}|^2. \tag{7.42}$$

Substitute this equation and Equation 7.32 in Equation 7.40 and obtain:

$$SCR = \frac{\left(\sum_{r=0}^{N_x-1} \sum_{s=0}^{N_y-1} \alpha_{r,s}\eta_{r,s}\right)^2}{\sum_{r=0}^{N_x-1} \sum_{s=0}^{N_y-1} AV_{x_0,y_0}\left(\left|\beta_{r,s}^{(bg,x_0,y_0)}\right|^2\right) |\eta_{r,s}|^2}. \tag{7.43}$$

By virtue of Cauchy–Bunyakovsky–Schwarz inequality (Equation 7.19), the SCR defined by this equation has an upper bound:

$$SCR = \frac{\left(\sum_{r=0}^{N_x-1} \sum_{s=0}^{N_y-1} \alpha_{r,s}\eta_{r,s}\right)^2}{\sum_{r=0}^{N_x-1} \sum_{s=0}^{N_y-1} AV_{x_0,y_0}\left(\left|\beta_{r,s}^{(bg,x_0,y_0)}\right|^2\right) |\eta_{r,s}|^2} = \frac{\left(\sum_{r=0}^{N_x-1} \sum_{s=0}^{N_y-1} \left(\frac{\alpha_{r,s}}{|AV\beta|}\right)\left(\eta_{r,s}|AV\beta|\right)\right)^2}{\sum_{r=0}^{N_x-1} \sum_{s=0}^{N_y-1} |AV\beta|^2 |\eta_{r,s}|^2}$$

$$\leq \frac{\sum_{r=0}^{N_x-1} \sum_{s=0}^{N_y-1} \frac{|\alpha_{r,s}|^2}{|AV\beta|^2} \sum_{r=0}^{N_x-1} \sum_{s=0}^{N_y-1} |\eta_{r,s}|^2 |AV\beta|^2}{\sum_{r=0}^{N_x-1} \sum_{s=0}^{N_y-1} |AV\beta|^2 |\eta_{r,s}|^2} = \sum_{r=0}^{N_x-1} \sum_{s=0}^{N_y-1} \frac{|\alpha_{r,s}|^2}{|AV\beta|^2}$$

$$= \sum_{r=0}^{N_x-1} \sum_{s=0}^{N_y-1} \frac{|\alpha_{r,s}|^2}{AV_{x_0,y_0}\left(\left|\beta_{r,s}^{(bg,x_0,y_0)}\right|^2\right)} \tag{7.44}$$

that is reached when

$$\eta_{r,s} = \eta_{r,s}^{(\text{opt})} = \underset{\eta_{r,s}}{\arg\max}(SCR) = \frac{\alpha_{r,s}^*}{AV_{x_0,y_0}\left(\left|\beta_{r,s}^{(\text{bg},x_0,y_0)}\right|^2\right)}, \tag{7.45}$$

where * denotes complex conjugation.

One can see that the filter defined by Equation 7.45 is analogous to the SNR-optimal filter (Equation 7.21) introduced in the section "Localization of Target Objects in the Presence of Additive Gaussian Noise" for object localization in the presence of correlated Gaussian noise. The numerator of its frequency response is the frequency response of the filter matched to the target object, just as in the SNR-optimal filter for target locating on the background of additive white Gaussian noise. The denominator of its frequency response is image background component power spectrum $AV_{x_0,y_0}(|\beta_{r,s}^{(\text{bg},x_0,y_0)}|^2)$ averaged over all possible positions of the target object. It replaces the additive Gaussian noise power spectrum in the filter of Equation 7.21. This makes optimal filter defined by Equation 7.45 to be adaptive to the input image. We will call this filter *SCR-optimal adaptive correlator*.

In order to implement the SCR-optimal adaptive correlator, one needs knowledge of the averaged power spectrum $AV_{x_0,y_0}(|\beta_{r,s}^{(\text{bg},x_0,y_0)}|^2)$ of the background component of the image. It has to be estimated from the power spectrum of the input image. For this, one has to specify in which way target and background components are combined in the input image.

Consider the following two models for this relationship: additive model and implant model. For the additive model, target image and background components are summed up to form the input image:

$$b_{k,l} = a_{k,l}(x_0, y_0) + b_{k,l}^{(\text{bg},x_0,y_0)}. \tag{7.46}$$

The additive model seems to be adequate to the cases of "transmissive" imaging, such as x-ray imaging.

For the implant model, target object and background image component complement each other and the latter is a part of the input image

$$b_{k,l}^{(\text{bg},x_0,y_0)} = \left(1 - w_{k,l}^{(\text{tg},x_0,y_0)}\right)b_{k,l} \tag{7.47}$$

selected by a certain target window function $0 \le w_{k,l}^{(\text{tg},x_0,y_0)} \le 1$, which is non-zero in points that belong to the target object and to zero in nontarget object points. The implant model is more adequate to "reflective" imaging, such as, for instance, conventional photography.

For the additive model, the spectrum of the background image component can be found as

$$\beta_{r,s}^{(bg,x_0,y_0)} = \beta_{r,s} - \alpha_{r,s} \exp\left[i2\pi\left(\frac{rx_0}{N_x} + \frac{sy_0}{N_y} \right) \right]. \tag{7.48}$$

It is shown in Appendix (Equation 7A.47) that in this case its averaged power spectrum can be evaluated as

$$AV_{x_0,y_0}\left(\left| \beta_{r,s}^{(bg,x_0,y_0)} \right|^2 \right) = |\beta_{r,s}|^2 + |\alpha_{r,s}|^2. \tag{7.49}$$

For the implant model (Equation 7.47), it is shown in Appendix that

$$AV_{x_0,y_0}\left(\left| \beta_{r,s}^{(bg,x_0,y_0)} \right|^2 \right) = \left(1 - \frac{2\operatorname{Re}\omega_{r,s}^{(tgt,0,0)}}{\sqrt{N_x N_y}} \right)|\beta_{r,s}|^2$$

$$+ \frac{1}{N_x} \sum_{\tilde{r}=0}^{N_x-1} \sum_{\tilde{s}=0}^{N_y-1} |\beta_{\tilde{r},\tilde{s}}|^2 \left| \omega_{r-\tilde{r},s-\tilde{s}}^{(tgt,0,0)} \right|^2, \tag{7.50}$$

where $\{\omega_{r,s}^{(tgt,0,0)}\}$ are DFT spectral coefficients of the target object window function $w_{k,l}^{tgt,0,0}$ for target located in the origin of coordinates:

$$\omega_{r,s}^{(tgt,0,0)} \frac{1}{\sqrt{N_x N_y}} \sum_{k=0}^{N_x-1} \sum_{s=0}^{N_y-1} w_{k,l}^{(tgt,0,0)} \exp\left[i2\pi\left(\frac{kr}{N_x} + \frac{ls}{N_y} \right) \right]. \tag{7.51}$$

In a special case when the target window function is a rectangle of $(2K + 1)$ $(2L + 1)$ pixels averaged over unknown target object coordinates, the power spectrum of the background image component can be, for the implant model, evaluated as (see Appendix, Equation 7A.60)

$$AV_{x_0,y_0}\left(\left| \beta_{r,s}^{(bg,x_0,y_0)} \right|^2 \right)$$

$$= \left(1 - 2\frac{(2K+1)(2L+1)}{N_x N_y} \overline{\operatorname{sincd}}(2K+1; N_x; r) \, \overline{\operatorname{sincd}}(2L+1; N_y; s) \right)|\beta_{r,s}|^2$$

$$+ \left(\frac{2K+1}{N_x} \right)^2 \left(\frac{2L+1}{N_y} \right)^2 \sum_{\tilde{r}=0}^{N_x-1} \sum_{\tilde{s}=0}^{N_y-1} |\beta_{\tilde{r},\tilde{s}}|^2 \left| \overline{\operatorname{sincd}}(2K+1; N_x; r - \tilde{r}) \right|^2$$

$$\times \left| \overline{\operatorname{sincd}}(2L+1; N_y; s - \tilde{s}) \right|^2. \tag{7.52}$$

As target objects usually occupy only a relatively small part of the input image area (the largest relative area of the target is about 1/9 for the case, when there are 3×3 possible positions of the target), the contribution of the target object into power spectrum of the input image is relatively small. In view of this, both additive and implant models imply that one can use, as a zero-order approximation to the averaged power spectrum of the image background component, either power spectrum of the input image:

$$AV_{x_0,y_0}\left(\left|\beta_{r,s}^{(bg,x_0,y_0)}\right|^2\right) \approx \left|\beta_{r,s}\right|^2 \tag{7.53}$$

or input image power spectrum $\left|\beta_{r,s}\right|^2$ smoothed by a certain smoothing window function $\omega_{r,s}^{smth}$:

$$AV_{x_0,y_0}\left(\left|\beta_{r,s}^{(bg,x_0,y_0)}\right|^2\right) \approx \sum_{\tilde{r}=0}^{N_x-1}\sum_{\tilde{s}=0}^{N_y-1}\omega_{\tilde{r},\tilde{s}}^{smth}\left|\beta_{r-\tilde{r},s-\tilde{s}}\right|^2 ; \quad \sum_{\tilde{r}=0}^{N_x-1}\sum_{\tilde{s}=0}^{N_y-1}\omega_{\tilde{r},\tilde{s}}^{smth} = 1. \tag{7.54}$$

This smoothing corresponds to windowing input image by an *"apodization" function*, which is equal to unity in the center of the image and gradually decays to zero to the image borders. Such a windowing, known in optics as *apodization*, is a useful method for reducing border effects in evaluating image DFT spectra, caused by cyclicity of DFT (see the section "2D Discrete Fourier Transforms" in Chapter 4).

In the conclusion of this section, we will illustrate, using MATLAB program corr_comparison_CRC.m provided in Exercises, advantages, in terms of the localization reliability, of the described SCR-optimal adaptive correlator over the matched filter correlator, which is optimal for target location in Gaussian noise in the absence of nontarget objects and clutter.

Figure 7.12 enables comparison of the performance of the matched filter and SCR-optimal adaptive correlators in the localization of corresponding image fragments in stereoscopic images. This comparison shows that the matched filter correlator is practically incapable of reliable localization of small image fragments of one of the stereoscopic images in the second image, while the SCR-optimal adaptive correlator successfully does the job. It suppresses the background image component substantially and therefore secures much better discrimination capability of the localization device.

Local Adaptive SCR-Optimal Correlators

The described SCR-optimal adaptive correlator minimizes the probability of anomalous errors of target localization in images by means of adaptation of its filter frequency response to the power spectrum of the background image component. This approach is justified if images are spatially homogeneous in terms of their spectra, that is, power spectra of arbitrary image fragments do not differ much from the power spectrum of the entire image. While this is

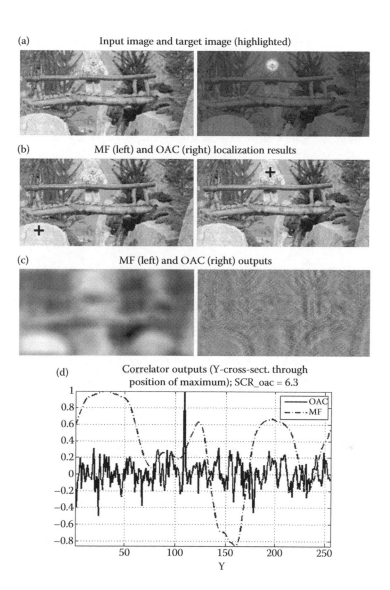

(a) Input image and target image (highlighted)

(b) MF (left) and OAC (right) localization results

(c) MF (left) and OAC (right) outputs

(d) Correlator outputs (Y-cross-sect. through position of maximum); SCR_oac = 6.3

FIGURE 7.12
Comparison of discrimination capability of matched filter and SCR-optimal adaptive correlators in localization of a fragment of one of two stereoscopic images on the second image. (a) Left and right stereoscopic images; target fragment is highlighted by a target window function (circle of 31 pixels in diameter). (b) Results (marked by a cross) of localization of the target fragment by the matched filter correlator (left image: false detection) and by SCR-optimal adaptive correlator (right image: correct detection). (c) Output images of the matched filter (left) and of the SCR-optimal adaptive correlator (right): note a bright spot in the location of the target fragment. (d) Rows of outputs of the matched filter (dash-dotted line) and of SCR-optimal adaptive correlator drawn through the corresponding highest peak; note a target peak at location of the target fragment, which is substantially higher than responses of the filter to the background clutter (signal-to-clutter ratio is 6.3^2).

true for many images, which belong to the class of so-called texture images, such as those shown in Figure 7.13a–c, this spectral homogeneity is an exemption rather than a rule. Generally, images are nonhomogeneous and their local power spectra may vary substantially (see, e.g., image in Figure 7.13d).

A natural extension of the above-developed SCR-optimal adaptive correlators to nonhomogeneous images is designing and applying the correlator filter locally in sliding window of the size commensurable with size of image fragments that can be regarded as being homogeneous in terms of their power spectra. In such an implementation, in each position $\{k,l\}$ $(k = 1,...,N_x; l = 1,...,N_y)$ of the window SCR-adaptive filter is applied to the image within the window and SCR is computed as ratio of the squared filter output signal $\hat{b}_{k,l}$ to the variance $\widehat{b^2_{k,l}}$ of the filter output signal within the window (local variance):

$$SCR_{k,l} = \frac{\hat{b}^2_{k,l}}{\widehat{b^2_{k,l}}}. \tag{7.55}$$

Estimate of the target coordinates is then found as a position of the global maximum over the $SCR_{k,l}$ map. We will refer to sliding window SCR-optimal adaptive correlators with frequency response

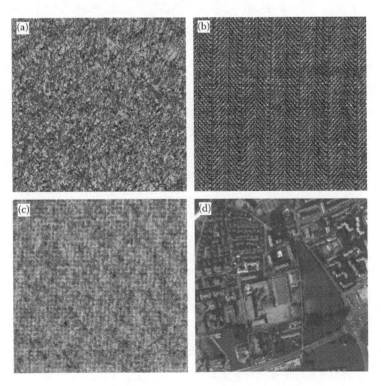

FIGURE 7.13
Examples of texture (a,b,c) and nontexture (d) images.

$$\eta_{r,s}^{(k,l,\text{opt})} = \frac{\alpha_{r,s}^*}{\left|\beta_{r,s}^{(\text{bg},k,l)}\right|^2},\qquad(7.56)$$

where $\left|\beta_{r,s}^{(\text{bg},k,l)}\right|^2$ is an estimate (obtained as it is described in the previous section) of the image background component within the window in its (k,l)-th position, as to *SCR-optimal local adaptive correlators*.

SCR-optimal local adaptive correlators do substantially outperform global optimal adaptive correlators in terms of SCR they provide in real-life images. Figure 7.14 generated using MATLAB program lcoptcorr_CRC.m provided in Exercises illustrates comparison of the SCR-optimal global and

(a) Local SCR-opt. adapt. correlator: localization result (marked by cross); SCR = 7.3

(b) Global SCR-opt. adapt. correlator: localization result (marked by cross)

(c)   Global (left) and local (right; WSz = 29×29; Rtrgt = 7) correlations

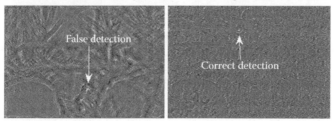

FIGURE 7.14

Comparison of SCR-optimal local and global adaptive correlators. (a) Result (marked by cross) of correct localization by the local adaptive correlator in sliding window of 29×29 pixels, on left image of a fragment of right image (highlighted; diameter of the target object circular window function 15 pixels). (b) Images: result of false detection (marked by a bold cross on the left image) of the same target by the global correlator. (c) Corresponding output images of global (left) and local (right) correlators.

local adaptive correlators in the localization of a small fragment of one of the stereoscopic images in the second image. As one can see, the same filter, which in the example shown in Figure 7.12, properly detected a target of 31 pixels in diameter, fails, when applied globally rather than locally, to properly localize a smaller target of 15 pixels in diameter, while this small target is successfully localized by the SCR-optimal local adaptive correlator.

Object Localization in Blurred Images

Images produced by imaging systems are frequently not sharp enough due to the weakening of their high-frequency components by optics or due to other technical reasons of low resolution of the imaging system. They also are distorted by the presence of a certain level of random noise, produced by image sensors. These image distortions and their correction are treated in the next chapter. In this section, we will discuss how they affect the performance of SCR-optimal adaptive correlators in target localization in images.

Let the discrete frequency response of the imaging system be $\{\eta_{r,s}^{(ims)}\}$ and let the noise be an additive random process with uniform power spectrum v_n^2. Then, the original target object spectrum $\alpha_{r,s}$ will be modified at the output of the imaging system to $\alpha_{r,s}\eta_{r,s}^{(ims)}$. The image background component power spectrum $|\beta_{r,s}^{(bg)}|^2$ will also be modified by the imaging system to $|\beta_{r,s}^{(bg)}|^2|\eta_{r,s}^{(ims)}|^2 + v_n^2$. With an account of these factors, SCR-optimal adaptive correlators should then have the following frequency response:

$$H_{opt}^{(bl)}(f_x, f_y) = \frac{\alpha_{r,s}^* \eta_{r,s}^{*(ims)}}{\overline{|\beta_{r,s}^{(bg)}|^2}\,|\eta_{r,s}^{(ims)}|^2 + v_n^2}, \qquad (7.57)$$

where $\overline{|\beta_{r,s}^{(bg)}|^2}$ is an estimate of the undistorted image background component power spectrum averaged over all possible positions of the target object.

Modify Equation 7.57 in the following way:

$$\eta_{r,s}^{(blr,opt)} = \frac{\alpha_{r,s}^*}{\overline{|\beta_{r,s}^{(bg)}|^2}}\, \frac{\eta_{r,s}^{*(ims)}\,\overline{|\beta_{r,s}^{(bg)}|^2}}{|\beta_{r,s}^{(bg)}|^2\,|\eta_{r,s}^{(ims)}|^2 + v_n^2} = \frac{\alpha_{r,s}^*\eta_{r,s}^{*(ims)}}{\overline{|\beta_{r,s}^{(bg)}|^2}}\left\{\frac{1}{\eta_{r,s}^{(ims)}}\frac{\overline{|\beta_{r,s}^{(bg)}|^2}\,|\eta_{r,s}^{(ims)}|^2}{|\beta_{r,s}^{(bg)}|^2\,|\eta_{r,s}^{(ims)}|^2 + v_n^2}\right\}. \qquad (7.58)$$

Equation 7.58 can be treated as a frequency response of two filters in cascade: SCR-optimal adaptive filter with frequency response designed for undistorted image

$$\eta_{r,s}^{(opt)} = \frac{\alpha_{r,s}^*}{\overline{|\beta_{r,s}^{(bg)}|^2}}, \qquad (7.59)$$

which is preceded by the filter with frequency response

$$\eta_{r,s}^{(\text{dblr})} = \frac{1}{\eta_{r,s}^{(\text{ims})}} \frac{\left|\beta_{r,s}^{(\text{bg})}\right|^2 \left|\eta_{r,s}^{(\text{ims})}\right|^2}{\left|\beta_{r,s}^{(\text{bg})}\right|^2 \left|\eta_{r,s}^{(\text{ims})}\right|^2 + v_n^2} = \frac{1}{\eta_{r,s}^{(\text{ims})}} \frac{SNR_{r,s}}{SNR_{r,s} + 1,} \tag{7.60}$$

where

$$SNR_{r,s} = \frac{\left|\beta_{r,s}^{(\text{bg})}\right|^2 \left|\eta_{r,s}^{(\text{ims})}\right|^2}{v_n^2} \tag{7.61}$$

is SNR in the distorted image on spatial frequency with indices (r,s). This filter has an *inverse filter* component (term $1/\eta_{r,s}^{(\text{ims})}$), which, when applied to the input image distorted by the imaging system frequency response $\eta_{r,s}^{(\text{ims})}$, will reverse this distortion, and a component with frequency response $SNR_{r,s}/(SNR_{r,s} + 1)$, which depends on the component-wise SNR in the system. This second component "regularizes" the inverse filtering component by means of suppressing spectral components with low SNR. In Chapter 8, we will show that this filter component is an *empirical Wiener filter* for image deblurring. Thus, SCR-adaptive filtering by the filter of Equation 7.57 can be interpreted as a two-stage procedure: image deblurring and subsequent SCR-optimal adaptive filtering of the deblurred image carried out in one stage. This means that target localization on blurred images does not, in principle, require preliminary image deblurring correction, because this job is automatically performed by the SCR-optimal adaptive correlator designed with an account for image blur.

Evaluate how much image blur deteriorates performance of SCR-optimal adaptive correlators in target localization. From Equation 7.44, it follows that in the presence of image blur and noise, SCR attained by the SCR-optimal correlator is

$$SCR^{(\text{blr})} = \sum_{r=0}^{N_x-1} \sum_{s=0}^{N_y-1} \frac{\left|\alpha_{r,s}\right|^2}{\left|\beta_{r,s}^{(\text{bg})}\right|^2 + v_{r,s}^2} = \sum_{r=0}^{N_x-1} \sum_{s=0}^{N_y-1} \frac{\left|\alpha_{r,s}\right|^2}{\left|\beta_{r,s}^{(\text{bg})}\right|^2} \frac{\left|\beta_{r,s}^{(\text{bg})}\right|^2}{\left|\beta_{r,s}^{(\text{bg})}\right|^2 + v_{r,s}^2}$$

$$= \sum_{r=0}^{N_x-1} \sum_{s=0}^{N_y-1} \frac{\left|\alpha_{r,s}\right|^2}{\left|\beta_{r,s}^{(\text{bg})}\right|^2} \frac{SNR_{r,s}}{SNR_{r,s} + 1} \leq \sum_{r=0}^{N_x-1} \sum_{s=0}^{N_y-1} \frac{\left|\alpha_{r,s}\right|^2}{\left|\beta_{r,s}^{(\text{bg})}\right|^2}. \tag{7.62}$$

The right part of this inequality is SCR that can be attained if there were no signal blur and noise.

The ratio $SNR_{r,s}/(SNR_{r,s}+1)$ is a measure of deterioration of SCR on each particular signal frequency component. In particular, it tells that while $\{SNR_{r,s}\}$ are sufficiently high, image blur may have a marginal effect on the localization reliability of the SCR-optimal adaptive correlator.

Object Localization and Edge Detection: Selection of Reference Objects for Target Tracking

It is widely believed in the image processing community that the information conveyed by images is contained mostly in image high-frequency components and in edges and image analysis should start from edge detection or enhancement. The theory of the SCR-optimal adaptive correlator provides a rational explanation for this belief. Represent Equation 7.44 for the optimal adaptive correlator in the following way:

$$\eta_{r,s}^{(opt)} = \frac{\alpha_{r,s}^{*}}{\left|\beta_{r,s}^{(bg)}\right|^{2}} = \frac{\alpha_{r,s}^{*}}{\left(\left|\beta_{r,s}^{(bg)}\right|^{2}\right)^{1/2}} \frac{1}{\left(\left|\beta_{r,s}^{(bg)}\right|^{2}\right)^{1/2}}. \tag{7.63}$$

In this representation, the SCR-optimal adaptive correlator is regarded as consisting of two filters in cascade. The filter represented by the right-hand factor in Equation 7.63

$$\eta_{r,s}^{(whtng)} = \frac{1}{\left(\left|\beta_{r,s}^{(bg)}\right|^{2}\right)^{1/2}} \tag{7.64}$$

is an analog of the whitening filter introduced in the section "Target Localization in Cluttered Images" (Equation 7.14) for SNR-optimal target localization in correlated Gaussian noise. This filter, being applied to the input image, makes the image background component spectrum almost uniform. The filter represented by the left-hand factor in Equation 7.63 is a matched filter for the target object, modified by the same whitening operator.

This representation implies two conclusions. First of all, as power spectra of images usually (though not always) tend to decay on high frequencies, spectrum whitening usually results in emphasizing high-frequency image components with respect to its low-frequency components. It is in this sense that one can say that high-frequency image components are more important for image object recognition than low-frequency components. Second, if one visually compares image before and after whitening such as, for instance, images shown in Figure 7.15, one can see that image whitening results in what is visually interpreted as edge enhancement.

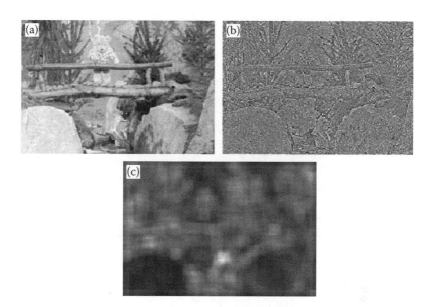

FIGURE 7.15
Test image (a), whitened test image (b), and its local variances in the window of 15×15 pixels (c).

In general, the whitening tends to suppress those image frequency components that have high energy and that are, therefore, responsible for features common to the majority of objects represented in image. Low-energy frequency components, which are responsible for "uncommon," or rare objects and object features, are, on the contrary, emphasized. In a sense, one can say that whitening is an operation that automatically enhances dissimilarities and suppresses similarities of objects in images. This is well illustrated by a result of whitening of a test image of geometrical figures and printed characters shown in Figure 7.16. One can see in the figure that the whitening does enhance edges (object borders), but the enhancement is quite selective. Vertical and horizontal edges that are common to all figures and characters are enhanced much less than circumferences, slanted edges, and corners in geometrical figures and characters and those are also enhanced very selectively according to their rate of occurrences in the image.

It is very instructive to also consider PSFs of whitening filters for different images. Table 7.1 shows the central 5×5 samples of PSFs of whitening operators (inverse 2D DFTs of frequency responses, Equation 7.64) for test images shown in Figures 7.14a and 7.16a.

One can see that those PSFs for considered two substantially different images are quite similar. Their most remarkable common feature is that the central peak of PSFs is surrounded by negative closest neighbors (shown nonbold in Table 7.1). In this respect, these two particular whitening filter

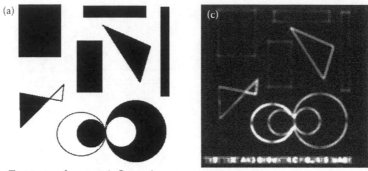

Test text and geometric figures images

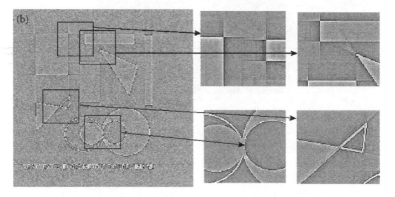

FIGURE 7.16
Test image of geometrical figures (a), whitened image and its magnified fragments (b), and its local variances in the window of 7×7 pixels (c).

TABLE 7.1

PSFs of the Whitening Operators

PSF of the Whitening Operator for Image of Figure 7.14a

0.002	0.023	0.043	0.037	0.029
0.003	−0.006	−0.294	−0.035	−0.005
0.047	−0.249	1.000	−0.249	0.047
−0.005	−0.035	−0.294	−0.006	0.003
0.029	0.037	0.043	0.023	0.002

PSF of the Whitening Operator for Image of Figure 7.16a

0.012	0.003	−0.024	0.0221	−0.03
0.014	−0.019	−0.2	−0.023	0.042
0.03	−0.243	1.000	−0.243	0.03
0.042	−0.023	−0.2	−0.019	0.014
−0.03	0.022	−0.024	0.003	0.012

PSFs resemble very much the PSF of filters for computing image spatial *Laplacian,* two most common versions of which are as follows:

$$\mathbf{L}_1 = \begin{bmatrix} 0 & -0.25 & 0 \\ -0.25 & 1 & -0.25 \\ 0 & -0.25 & 0 \end{bmatrix},$$

$$\mathbf{L}_2 = \begin{bmatrix} -0.125 & -0.125 & -0.125 \\ -0.125 & 1 & -0.125 \\ -0.125 & -0.125 & -0.125 \end{bmatrix}.$$

(7.65)

This observation suggests a rational explanation for the common belief in the importance of the Laplacian operator in image processing: the Laplacian operator can be considered as an empirical approximation to adaptive whitening operators.

The remarkable fact is also that visual systems of humans and vertebrates feature similar capacity of neurons to reduce the activity of its spatially nearest neighbors. This action known as *lateral inhibition* is actually what whitening and Laplacian operators perform, thanks to negative weights of the samples of the filter PSF that surround the central peak. This feature exhibits itself as the so-called *Mach effect*: visual impression of edge enhancement in images. Figure 7.17 illustrates this phenomenon.

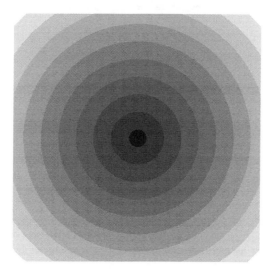

FIGURE 7.17
Mach effect: image visually appears darker at dark sides of edges and brighter on bright sides.

FIGURE 7.18
Illustration of importance of whitened spectra for object recognition: images in the bottom row are obtained by means of exchange power spectra of images of the upper row retaining image-whitened spectra. Note that images are still well recognizable, though Mona Lisa's smile charm evaporated.

The importance of whitened edge-contained image component for object recognition can also be illustrated by an experiment with exchanging power spectra of images (Figure 7.18), in which the power spectrum of one image is replaced by the power spectrum of another, while phase spectra retain the original. Both images after this exchange remain very well recognizable, though one can notice that Mona Lisa's smile charm evaporated.

This means that low-frequency image components also do carry important information. In Figure 7.19 shown are image of "Mona Lisa" low-pass filtered to quarter of the image base band (left image) and its complement high-frequency component (right image). As one can see, Mona Lisa's famous smile is retained in the low-frequency component, while in the high-frequency component, it volatilized.

Consider now the issue of selecting target objects in cases when target objects have to be selected from image fragments for, for example, object

FIGURE 7.19
An illustration of importance of low-frequency image components. The left image is a component of the "Mona Lisa" image low-pass filtered to quarter of the image base band. It perfectly reproduces Mona Lisa's smile charm. The right image is its high-frequency complement. The charm volatilized.

tracking in videos, in image registration and matching with a map, in stereo image analysis, in robot vision navigation, and in multimodal medical imaging. In such applications, the question is how to make the choice of the target objects to the best advantage.

A natural figure of merit of target objects that determines potential reliability of their localization is the SCR that can be attained for these objects in SCR-optimal adaptive correlators. One can obtain from Equation 7.62 that maximal attained SCR for the SCR-optimal adaptive correlator is equal to the energy of the whitened spectrum of the target object:

$$SCR = \sum_{r=0}^{N_x-1} \sum_{s=0}^{N_y-1} \frac{\left|\alpha_{r,s}\right|^2}{\left|\beta_{r,s}^{(bg)}\right|^2}. \tag{7.66}$$

Therefore, the higher the energy of the whitened spectrum, the higher is the SCR that one can expect at the SCR-optimal adaptive correlator output for this object. This implies that the best candidates for target objects will be image fragments that have maximal energy of their "whitened" power spectrum, that is, image fragments with intensive high-frequency components, which are visually interpreted as containing the most intensive edges or texture.

This conclusion is illustrated in Figures 7.15 and 7.16, where the "goodness" of the fragments of 15×15 pixels (Figure 7.15c) and 7×7 pixels (Figure 7.16c) in terms of the *SCR*-factor is represented by the pixel gray levels for test images of Figures 7.15a and 7.16a. One can see from these images that edge-rich areas are indeed the most appropriate potential candidates for reference object. One can also make a remarkable observation that, for rectangles in the image of geometrical figures, corners have much higher "goodness." This result is intuitively very well understandable: edges in all rectangles are similar and cannot be regarded as their specific features, while corners are more specific.

Appendix

Distribution Density and Variances of Normal Localization Errors

Although we consider working with images in their sampled representation, target localization can be performed, with appropriate signal subsampling, with subpixel accuracy and localization errors computed in units of sampling intervals may have arbitrary noninteger values. Therefore, it would be appropriate to use continuous image models for the analysis of statistical characteristics of localization errors.

Let $a(x, y)$ and $b(x, y)$ be, respectively, continuous models of the target object (template) image and of the observed image, in which it should be localized. For the considered case of additive signal-independent noise model, $b(x, y) = a(x, y) + n(x, y)$, where $n(x, y)$ stands for realizations of noise. Then, the optimal ML-estimate of target object coordinates is a solution of the following continuous analog of Equation 7.4:

$$\{\hat{x}_0, \hat{y}_0\}_{\text{ML}} = \underset{(x_0,y_0)}{\arg\min} \left\{ \int_{-\infty}^{\infty} \int_{-\infty}^{\infty} b(x,y)a(x - x_0, y - y_0)\mathrm{d}x\mathrm{d}y \right\}$$

$$= \underset{(x_0,y_0)}{\arg\min} \left\{ \int_{-\infty}^{\infty} \int_{-\infty}^{\infty} \left[a(x,y) + n(x,y) \right] a(x - x_0, y - y_0)\mathrm{d}x\mathrm{d}y \right\}, \quad (7A.1)$$

that is, it is a position of the maximum correlation between the observed image $b(x, y)$ and the template image $a(x, y)$ taken in all possible range of its coordinates $\{x_0, y_0\}$.

Consider the correlator output signal

$$R_{ab}(x_0, y_0) = \int_{-\infty}^{\infty} \int_{-\infty}^{\infty} \left[a(x,y) + n(x,y) \right] a(x - x_0, y - y_0)\mathrm{d}x\mathrm{d}y$$

$$= R_a(x - x_0, y - y_0) + \tilde{n}(x,y), \quad (7A.2)$$

where

$$R_a(x_0, y_0) = \int\limits_{-\infty}^{\infty} \int\limits_{-\infty}^{\infty} a(x,y)\, a(x - x_0, y - y_0)\, dx\, dy \qquad (7A.3)$$

is an autocorrelation function of the target object signal, and

$$\tilde{n}(x_0, y_0) = \int\limits_{-\infty}^{\infty} \int\limits_{-\infty}^{\infty} n(x,y)a(x - x_0, y - y_0)\, dx\, dy \qquad (7A.4)$$

is a correlated Gaussian random process resulting from filtering input Gaussian noise by the matched filter. Its correlation function can be found as

$$R_{\tilde{n}} = AV_{\Omega_N}\left[\tilde{n}(x,y) \cdot \tilde{n}(\bar{x}, \bar{y})\right]$$

$$= \int\limits_{-\infty}^{\infty} \int\limits_{-\infty}^{\infty} \int\limits_{-\infty}^{\infty} \int\limits_{-\infty}^{\infty} AV_{\Omega_N}\{n(x,y)n(\bar{x}, \bar{y})\}\, a(x - x_0, y - y_0)\, a(\bar{x} - \bar{x}_0, \bar{y} - \bar{y}_0)\, dx\, dy\, d\bar{x}\, d\bar{y}$$

$$= \int\limits_{-\infty}^{\infty} \int\limits_{-\infty}^{\infty} \int\limits_{-\infty}^{\infty} \int\limits_{-\infty}^{\infty} \sigma_n^2 \delta(x - \bar{x}, y - \bar{y})\, a(x - x_0, y - y_0)\, a(\bar{x} - \bar{x}_0, \bar{y} - \bar{y}_0)\, dx\, dy\, d\bar{x}\, d\bar{y}$$

$$= \sigma_n^2 \int\limits_{-\infty}^{\infty} \int\limits_{-\infty}^{\infty} a(x - x_0, y - y_0)\, a(x - \bar{x}_0, y - \bar{y}_0)\, dx\, dy \; = \sigma_n^2\, R_a\, (x_0 - \bar{x}_0, y_0 - \bar{y}_0),$$

$$(7A.5)$$

where $AV_{\Omega_N}(\cdot)$ denotes statistical averaging over the noise ensemble Ω_N, $AV_{\Omega_N}\{n(x,y)n(\bar{x}, \bar{y})\} = \sigma_n^2 \delta(x - \bar{x}, y - \bar{y})$ is the correlation function of uncorrelated noise and σ_n^2 is noise variance.

In the case of small (normal) errors, signal maxima at the output of the correlator are located in a close vicinity of the point (x_0, y_0), the position of maximum of $R_a(x, y)$. Then the following system of equations determines the coordinates $(\hat{x}_0, \hat{y}_0)$ of the maximum of $R(x, y)$:

$$\begin{cases} \dfrac{\partial}{\partial x} R(x,y) = \dfrac{\partial}{\partial x} R_a(x - x_0, y - y_0) + \dfrac{\partial}{\partial x} \tilde{n}(x,y) = 0 \\[2mm] \dfrac{\partial}{\partial y} R(x,y) = \dfrac{\partial}{\partial y} R_a(x - x_0, y - y_0) + \dfrac{\partial}{\partial y} \tilde{n}(x,y) = 0 \end{cases} . \qquad (7A.6)$$

Let the solution of this system be $\hat{x}_0 = x_0 + \varepsilon_x$; $\hat{y}_0 = y_0 + \varepsilon_y$ and localization errors ε_x and ε_y are small enough to permit the first-order Taylor expansion of $R(x, y)$ around point $\{x_0, y_0\}$. Then, we have

$$\frac{\partial}{\partial x} R_a(x - x_0, y - y_0)\bigg|_{\substack{x=x_0+\varepsilon_x \\ y=y_0+\varepsilon_y}}$$

$$= \frac{\partial}{\partial x} R_a(x - x_0, y - y_0)\bigg|_{\substack{x=x_0 \\ y=y_0}} + \varepsilon_x \frac{\partial^2}{\partial x^2} R_a(x - x_0, y - y_0)\bigg|_{\substack{x=x_0 \\ y=y_0}}$$

$$+ \varepsilon_y \frac{\partial^2}{\partial x \partial y} R_a(x - x_0, y - y_0)\bigg|_{\substack{x=x_0 \\ y=y_0}}$$

$$= \frac{\partial}{\partial x} R_a(x - x_0, y - y_0)\bigg|_{\substack{x=x_0 \\ y=y_0}} + D_x\varepsilon_x + D_{xy}\varepsilon_y = -\frac{\partial}{\partial x} \tilde{n}(x, y)\bigg|_{\substack{x=x_0 \\ y=y_0}}. \quad (7A.7)$$

Similarly, one can obtain

$$\frac{\partial}{\partial x} R_a(x - x_0, y - y_0)\bigg|_{\substack{x=x_0 \\ y=y_0}} + D_{xy}\varepsilon_x + D_y\varepsilon_y = -\frac{\partial}{\partial y} \tilde{n}(x, y)\bigg|_{\substack{x=x_0+n_x \\ y=y_0+n_y}}, \quad (7A.8)$$

where

$$D_x = \frac{\partial^2}{\partial x^2} R_a(x - x_0, y - y_0)\bigg|_{\substack{x=x_0 \\ y=y_0}}; \quad D_y = \frac{\partial^2}{\partial y^2} R_a(x - x_0, y - y_0)\bigg|_{\substack{x=x_0 \\ y=y_0}};$$

$$D_{xy} = \frac{\partial^2}{\partial x \partial y} R_a(x - x_0, y - y_0)\bigg|_{\substack{x=x_0 \\ y=y_0}}. \qquad (7A.9)$$

In the point (x_0, y_0) of the target location, the autocorrelation function $R_a(x - x_0, y - y_0)$ has a maximum. Therefore

$$\frac{\partial}{\partial x} R_a(x - x_0, y - y_0)\bigg|_{\substack{x=x_0 \\ y=y_0}} = 0 \quad \text{and} \quad \frac{\partial}{\partial y} R_a(x - x_0, y - y_0)\bigg|_{\substack{x=x_0 \\ y=y_0}} = 0. \quad (7A.10)$$

Substitute these equalities in Equations 7A.7 and 7A.8 and obtain the following system of equations:

$$\begin{cases} D_x \varepsilon_x + D_{xy} \varepsilon_y = -\dfrac{\partial}{\partial x} \tilde{n}(x,y) \Big|_{\substack{x=x_0 \\ y=y_0}} \\[3mm] D_{xy} \varepsilon_x + D_y \varepsilon_y = -\dfrac{\partial}{\partial y} \tilde{n}(x,y) \Big|_{\substack{x=x_0+n_x \\ y=y_0+n_y}} \end{cases} \tag{7A.11}$$

from which the following relationships for errors ε_x and ε_y in x and y directions follow:

$$\varepsilon_x = \frac{D_y}{D_x D_y - D_{xy}^2} v_x - \frac{D_{xy}}{D_x D_y - D_{xy}^2} v_y;$$

$$\varepsilon_y = \frac{D_x}{D_x D_y - D_{xy}^2} v_y - \frac{D_{xy}}{D_x D_y - D_{xy}^2} v_x, \tag{7A.12}$$

where

$$v_x = -\frac{\partial}{\partial x} \tilde{n}(x,y) \Big|_{\substack{x=x_0+\varepsilon_x \\ y=y_0+\varepsilon_y}} \; ; \quad v_y = -\frac{\partial}{\partial y} \tilde{n}(x,y) \Big|_{\substack{x=x_0+\varepsilon_x \\ y=y_0+\varepsilon_y.}} \tag{7A.13}$$

Derivatives v_x and v_y of the random Gaussian process $\tilde{n}(x,y)$ are Gaussian random variables. Therefore, from Equation 7A.12, it follows that small ML-coordinate estimation errors $\{\varepsilon_x, \varepsilon_y\}$ have a Gaussian distribution with zero mean and variances:

$$\overline{\varepsilon_x^2} = AV_{\Omega_N} \left(|\varepsilon_x|^2 \right) = \left(\frac{D_y}{D_x D_y - D_{xy}^2} \right)^2 AV_{\Omega_N} \left(|v_x|^2 \right) + \left(\frac{D_{xy}}{D_x D_y - D_{xy}^2} \right)^2 AV_{\Omega_N} \left(|v_y|^2 \right)$$

$$- 2 \frac{D_y D_{xy}}{(D_x D_y - D_{xy}^2)^2} AV_{\Omega_N} (v_x v_y); \tag{7A.14}$$

$$\overline{\varepsilon_y^2} = AV_{\Omega_N} (\varepsilon_y)^2 = \left(\frac{D_x}{D_x D_y - D_{xy}^2} \right)^2 AV_{\Omega_N} \left(|v_y|^2 \right) + \left(\frac{D_{xy}}{D_x D_y - D_{xy}^2} \right)^2 AV_{\Omega_N} \left(|v_x|^2 \right)$$

$$- 2 \frac{D_x D_{xy}}{(D_x D_y - D_{xy}^2)^2} AV_{\Omega_N} (v_x v_y) \tag{7A.15}$$

$$\overline{\varepsilon_{xy}^2} = AV_{\Omega_N} (\varepsilon_x \varepsilon_y) = \frac{D_x D_y - (D_{xy})^2}{(D_x D_y - D_{xy}^2)^2} AV_{\Omega_N} (v_x v_y)$$

$$- \frac{D_y D_{xy}}{(D_x D_y - D_{xy}^2)^2} AV_{\Omega_N} \left(|v_x|^2 \right) + \frac{D_x D_{xy}}{(D_x D_y - D_{xy}^2)^2} AV_{\Omega_N} \left(|v_x|^2 \right). \tag{7A.16}$$

The parameters involved in Equations 7.45 through 7.47 can be found using the relationship between signal correlation functions and power spectra (Equation 2.127) and properties of the Fourier transform:

$$D_x = \frac{\partial^2}{\partial x^2} R_a(x - x_0, y - y_0)\bigg|_{\substack{x=x_0 \\ y=y_0}}$$

$$= \frac{\partial^2}{\partial x^2} \int_{-\infty}^{\infty}\int_{-\infty}^{\infty} |a(f_x, f_y)|^2 \exp\{-i2p[f_x(x - x_0) + f_y(y - y_0)]\}df_x df_y\bigg|_{\substack{x=x_0 \\ y=y_0}}$$

$$= -4\pi^2 \int_{-\infty}^{\infty}\int_{-\infty}^{\infty} f_x^2 |\alpha(f_x, f_y)|^2 df_x df_y = -4\pi^2 \overline{f_x^2} E_a, \qquad (7A.17)$$

where $\alpha(f_x, f_y)$ is a Fourier spectrum of the target object image

$$\alpha(f_x, f_y) = \int_{-\infty}^{\infty}\int_{-\infty}^{\infty} a(x, y)\exp[i2\pi(f_y x + f_y y)]df_x df_y. \qquad (7A.18)$$

E_a is its energy

$$E_a = \int_{-\infty}^{\infty}\int_{-\infty}^{\infty} |\alpha(f_x, f_y)|^2 df_x df_y \qquad (7A.19)$$

and $\overline{f_x^2}$ is energy of template image derivative along the axis f_x relative to the signal energy:

$$\overline{f_x^2} = \frac{\displaystyle\int_{-\infty}^{\infty}\int_{-\infty}^{\infty} f_x^2 |\alpha(f_x, f_y)|^2 df_x df_y}{\displaystyle\int_{-\infty}^{\infty}\int_{-\infty}^{\infty} |\alpha(f_x, f_y)|^2 df_x df_y.} \qquad (7A.20)$$

Similarly

$$D_y = -4\pi^2 \overline{f_y^2} \int_{-\infty}^{\infty}\int_{-\infty}^{\infty} |\alpha(f_x, f_y)|^2 df_x df_y = -4\pi^2 \overline{f_y^2} E_a \qquad (7A.21)$$

and

$$D_{xy} = -4\pi^2 \int\limits_{-\infty}^{\infty} \int\limits_{-\infty}^{\infty} f_x f_y \left|\alpha(f_x, f_y)\right|^2 df_x df_y = -4\pi^2 \overline{f_{xy}^2} E_a, \qquad (7A.22)$$

where

$$\overline{f_y^2} = \frac{\displaystyle\int\limits_{-\infty}^{\infty}\int\limits_{-\infty}^{\infty} f_y^2 \left|\alpha(f_x, f_y)\right|^2 df_x df_y}{\displaystyle\int\limits_{-\infty}^{\infty}\int\limits_{-\infty}^{\infty} \left|\alpha(f_x, f_y)\right|^2 df_x df_y} \qquad (7A.23)$$

and

$$\overline{f_{xy}^2} = \frac{\displaystyle\int\limits_{-\infty}^{\infty}\int\limits_{-\infty}^{\infty} f_x f_y \left|\alpha(f_x, f_y)\right|^2 df_x df_y}{\displaystyle\int\limits_{-\infty}^{\infty}\int\limits_{-\infty}^{\infty} \left|\alpha(f_x, f_y)\right|^2 df_x df_y}. \qquad (7A.24)$$

The second moments $AV_{\Omega_N}(|v_x|^2)$, $AV_{\Omega_N}(|v_y|^2)$, and $AV_{\Omega_N}(v_x v_y)$ of the noise component in the formulas (7A.14 through 7A.16) can also be found in the spectral domain. For instance, the spectrum of v_x, being a spectrum of derivative, is equal to $4\pi^2 f_x^2$ times power spectrum of $R_n(x, y)$, which, by virtue of Equation 7A.5, is equal to $\sigma_n^2 \left|\alpha(f_x, f_y)\right|^2$. Therefore

$$AV_{\Omega_N}(v_x^2) = 4\pi^2 \sigma_n^2 \int\limits_{-\infty}^{\infty}\int\limits_{-\infty}^{\infty} f_x^2 \left|\alpha(f_x, f_y)\right|^2 df_x df_y = 4\pi^2 \overline{f_x^2} \sigma_n^2 E_a. \qquad (7A.25)$$

In a similar way, one can obtain

$$AV_{\Omega_N}(v_y^2) = 4\pi^2 \sigma_n^2 \int\limits_{-\infty}^{\infty}\int\limits_{-\infty}^{\infty} f_y^2 \left|\alpha(f_x, f_y)\right|^2 df_x df_y = 4\pi^2 \sigma_n^2 E_a \overline{f_y^2} \qquad (7A.26)$$

and

$$AV_{\Omega_N}(v_x v_y) = 4\pi^2 \sigma_n^2 \int\limits_{-\infty}^{\infty}\int\limits_{-\infty}^{\infty} f_x f_y \left|\alpha(f_x, f_y)\right|^2 df_x df_y = 4\pi^2 \sigma_n^2 E_a \overline{f_{xy}^2}. \qquad (7A.27)$$

After substituting these parameters into Equations 7A.14 through 7A.16, we finally arrive, for the optimal ML-estimator, at the following relationships for variances of normal errors in object coordinate estimation in the presence of additive white zero mean Gaussian noise:

$$\overline{\varepsilon_x^2} = \frac{1}{4\pi^2} \frac{\overline{f_y^2}}{\overline{f_x^2}\,\overline{f_y^2} - \left(\overline{f_{xy}^2}\right)^2} \frac{\sigma_n^2}{E_a}; \quad \overline{\varepsilon_y^2} = \frac{1}{4\pi^2} \frac{\overline{f_x^2}}{\overline{f_x^2}\,\overline{f_y^2} - \left(\overline{f_{x,y}^2}\right)^2} \frac{\sigma_n^2}{E_a};$$

$$\overline{\varepsilon_{xy}^2} = \frac{1}{4\pi^2} \frac{\overline{f_{xy}^2}}{\overline{f_x^2}\,\overline{f_y^2} - \left(\overline{f_{xy}^2}\right)^2} \frac{\sigma_n^2}{E_a}.$$

(7A.28)

The power spectra $|\alpha(f_x, f_y)|^2$ of real-valued signals feature the property of central symmetry: $|\alpha(f_x, f_y)|^2 = |\alpha(-f_x, -f_y)|^2$. If the object signal power spectrum is symmetrical with respect to the coordinate axes as well: $|\alpha(f_x, f_y)|^2 = |\alpha(f_x, -f_y)|^2 = |\alpha(-f_x, f_y)|^2$, Equation 7A.28 takes the following simpler form:

$$\overline{\varepsilon_x^2} = \frac{1}{\overline{f_x^2}} \frac{\sigma_n^2}{4\pi^2 E_a}; \quad \overline{\varepsilon_y^2} = \frac{1}{\overline{f_y^2}} \frac{\sigma_n^2}{4\pi^2 E_a}; \quad \overline{\varepsilon_{xy}^2} = 0. \tag{7A.29}$$

The last equation implies that, if the signal power spectrum is axes-symmetrical, normal localization errors along coordinates x and y are uncorrelated. This situation takes place if the object spectrum $\alpha(f_x, f_y)$, and, correspondingly, the target object signal $a(x,y)$ are separable functions of the coordinates. Equation 7A.29 is applicable to 1D signals.

Equation 7A.28 shows that variances of normal localization errors are fully determined by the SNR σ_n^2/E_a and the energy of derivatives of the target object signal. These are the only characteristics of the object shape that affect the potential accuracy of its localization.

In digital processing, it is natural to evaluate error variances in units of signal sampling intervals $\Delta x, \Delta y$. Denote

$$\overline{\varepsilon_{\Delta x}^2} = \overline{\varepsilon_x^2}/\Delta x^2; \quad \overline{\varepsilon_{\Delta y}^2} = \overline{\varepsilon_y^2}/\Delta y^2; \quad \overline{\varepsilon_{\Delta xy}^2} = \overline{\varepsilon_{xy}^2}/\Delta x \Delta y. \tag{7A.30}$$

Then, from Equation 7.59, obtain

$$\overline{\varepsilon_{\Delta x}^2} = \frac{1}{4\pi^2} \frac{\overline{f_y^2}}{\left(\overline{f_x^2}\Delta x^2\right)\left(\overline{f_y^2}\right) - \overline{f_{xy}^2}\Delta x^2} \frac{\sigma_n^2}{E_a} = \frac{1}{4\pi^2} \frac{\overline{f_y^2}\Delta y^2}{\left(\overline{f_x^2}\Delta x^2\right)\left(\overline{f_y^2}\Delta y^2\right) - \left(\overline{f_{xy}^2}\Delta x \Delta y\right)^2} \frac{\sigma_n^2}{E_a}$$

$$= \frac{1}{4\pi^2} \frac{\overline{f_{\Delta y}^2}}{\left(\overline{f_{\Delta x}^2}\right)\left(\overline{f_{\Delta y}^2}\right) - \left(\overline{f_{\Delta xy}^2}\right)^2} \frac{\sigma_n^2}{E_a};$$

$$\overline{\varepsilon_{\Delta y}^2} = \frac{1}{4\pi^2} \frac{\overline{f_{\Delta x}^2}}{\left(\overline{f_{\Delta x}^2}\right)\left(\overline{f_{\Delta y}^2}\right) - \left(\overline{f_{\Delta xy}^2}\right)^2} \frac{\sigma_n^2}{E_a};$$

$$\overline{\varepsilon_{\Delta xy}^2} = \frac{1}{4\pi^2} \frac{\overline{f_{\Delta xy}^2}}{\overline{f_{\Delta x}^2}\,\overline{f_{\Delta y}^2} - \left(\overline{f_{\Delta xy}^2}\right)^2} \frac{\sigma_n^2}{E_a}, \tag{7A.31}$$

where

$$\overline{f_{\Delta x}^2} = \overline{f_x^2}\Delta x^2 = \frac{\overline{f_x^2}}{1/\Delta x^2}; \quad \overline{f_{\Delta y}^2} = \overline{f_y^2}\Delta y^2 = \frac{\overline{f_y^2}}{1/\Delta y^2};$$

$$\overline{f_{\Delta xy}^2} = \overline{f_{xy}^2}\Delta x\Delta y = \frac{\overline{f_{xy}^2}}{(1/\Delta x)(1/\Delta y)}. \tag{7A.32}$$

For separable 2D template signals, obtain correspondingly

$$\overline{\varepsilon_{\Delta x}^2} = \frac{1}{4\pi^2 \overline{f_{\Delta x}^2}} \frac{\sigma_n^2}{E_a}; \quad \overline{\varepsilon_{\Delta y}^2} = \frac{1}{4\pi^2 \overline{f_{\Delta y}^2}} \frac{\sigma_n^2}{E_a}; \quad \sigma_{\Delta xy}^2 = 0. \tag{7A.33}$$

These equations are also valid for 1D signals.

The above formulas for variances of localization errors are expressed through parameters of continuous signals. In order to express them in terms of parameters of corresponding sampled signals, one has to apply relationships between sampled signals and reconstruct from them continuous signals given by the sampling theory (see Chapter 3), characterizations of digital filters in terms of equivalent continuous filters outlined in the section "DFTs and Discrete Frequency Response of Digital Filter" in Chapter 4 and Parseval's relationships (2.7 through 2.9). In this treatment, we will assume ideal sampling and reconstruction, which imply band-limited signals with X–Y base bands (−1/2 $\Delta x \div 1/2\Delta x$; $-1/2\Delta y \div 1/2\Delta y$).

According to the Parseval's relationship (Equation 2.8), the relationship of Equation 4.79 between continuous signal frequency and frequency index of its DFT and Equations 6.36 and 6.38 for frequency responses of signal differentiators, obtain for sampled images of $N_x \times N_y$ pixels:

$$E_a = \int_{-\infty}^{\infty}\int_{-\infty}^{\infty} |\alpha(f_x, f_y)|^2 df_x df_y = \sum_{k=0}^{N_x-1}\sum_{l=0}^{N_y-1} |a_k|^2 \tag{7A.34}$$

$$\overline{f_{\Delta x}^2} = \frac{\displaystyle\int\limits_{-1/2\Delta x}^{1/2\Delta x}\int\limits_{-1/2\Delta y}^{1/2\Delta y} f_x^2 \Delta x^2 \left|\alpha(f_x, f_y)\right|^2 df_x df_y}{\displaystyle\int\limits_{-1/2\Delta x}^{1/2\Delta x}\int\limits_{-1/2\Delta y}^{1/2\Delta y} \left|\alpha(f_x, f_y)\right|^2 df_x df_y} = \frac{2\displaystyle\sum_r \sum_s \left(\frac{r}{N_x}\right)^2 \left|\alpha_{r,s}\right|^2}{2\displaystyle\sum_r \sum_s \left|\alpha_{r,s}\right|^2}$$

$$= \frac{1}{N_x^2} \frac{\displaystyle\sum_r r^2 \sum_s \left|\alpha_{r,s}\right|^2}{\displaystyle\sum_r \sum_s \left|\alpha_{r,s}\right|^2} ; \tag{7A.35}$$

$$\overline{f_{\Delta y}^2} = \frac{\displaystyle\int\limits_{-1/2\Delta x}^{1/2\Delta x}\int\limits_{-1/2\Delta y}^{1/2\Delta y} f_y^2 \Delta y^2 \left|\alpha(f_x, f_y)\right|^2 df_x df_y}{\displaystyle\int\limits_{-1/2\Delta x}^{1/2\Delta x}\int\limits_{-1/2\Delta y}^{1/2\Delta y} \left|\alpha(f_x, f_y)\right|^2 df_x df_y} = \frac{1}{N_y^2} \frac{\displaystyle\sum_s s^2 \sum_r \left|\alpha_{r,s}\right|^2}{\displaystyle\sum_r \sum_s \left|\alpha_{r,s}\right|^2} ; \tag{7A.36}$$

$$\overline{f_{\Delta xy}^2} = \frac{\displaystyle\int\limits_{-1/2\Delta x}^{1/2\Delta x}\int\limits_{-1/2\Delta y}^{1/2\Delta y} f_x f_y^2 \Delta x \Delta y \left|\alpha(f_x, f_y)\right|^2 df_x df_y}{\displaystyle\int\limits_{-1/2\Delta x}^{1/2\Delta x}\int\limits_{-1/2\Delta y}^{1/2\Delta y} \left|\alpha(f_x, f_y)\right|^2 df_x df_y} = \frac{1}{N_x N_y} \frac{\displaystyle\sum_r \sum_s rs \left|\alpha_{r,s}\right|^2}{\displaystyle\sum_r \sum_s \left|\alpha_{r,s}\right|^2} , \tag{7A.37}$$

where

$$\alpha_{r,s} = \frac{1}{\sqrt{N_x N_y}} \sum_{k=0}^{N_x-1} \sum_{l=0}^{N_y-1} a_{k,l} \exp\left[i2\pi\left(\frac{kr}{N_x} + \frac{ls}{N_y}\right)\right] \tag{7A.38}$$

is the DFT of sampled version $\{a_{k,l}\}$ of the target object signal. Summations in Equations 7A.35 through 7A.37 are carried out over indices r,s of spatial frequencies from the lowest frequencies ($r = 1$, $s = 1$) to the highest ones (($N_x/2$, $N_y/2$) or ($N_x - 1$)/2, ($N_y - 1$)/2) depending on whether N_x and N_y are even or odd numbers.

Evaluation of the Probability of Anomalous Localization Errors

We defined anomalous errors as those that occur when the localization device wrongly locates the target object at output of the matched filter outside the area it occupies, that is, in the area that contains only the noise component of the image. Because noise at the matched filter output results from filtering the input white Gaussian noise, it is spatially homogeneous in this area. Therefore, large noise outbursts may be found with equal probability everywhere in an area outside the target object in its actual location. This implies that the tales of the localization error distribution density are uniform and one can characterize anomalous errors by simply their total probability determined by the area under the distribution density tales.

Analytical evaluation of the probability of anomalous errors by methods of the theory of random processes requires cumbersome computation and cannot be carried out without certain simplifying assumptions. Its reasonably good estimation can be obtained by the following simple reasoning.

The output of the matched filter outside the area occupied by the object is a correlated Gaussian random process $\tilde{n}(x,y)$ with correlation function defined by Equation 7A.5:

$$R_n(x,y) = \sigma_n^2 R_a(x,y). \tag{7A.39}$$

Let ΔS be its *correlation interval*, that is, a minimal distance at which correlation between process values can be regarded as negligibly small. Because the correlation function of noise at the matched filter output is, according to Equation 7A.39, proportional to that of the object signal, ΔS has an order of magnitude of the area occupied by the signal at the matched filter output. Therefore, in the area of the search S there are approximately $Q = S/\Delta S$ uncorrelated samples of Gaussian random process with variance $\tilde{\sigma}_n^2 = R_n(0,0) = \sigma_n^2 R_a(0,0) = \sigma_n^2 E_a$.

Let us take one of this samples, the one at the point of the actual location of the target object. According to Equation 7A.2, its value is equal to $R_a(0,0) + R_n = E_a + R_n$, where R_a is a Gaussian zero mean random value with variance $\tilde{\sigma}_n^2$. One can then find the probability of anomalous errors P_{Ae} as a value complementary to the probability that none of the rest of $(Q-1)$ uncorrelated samples of Gaussian noise outside the object location exceeds this value:

$$P_{Ae} = 1 - \frac{1}{\sqrt{2\pi\tilde{\sigma}_n^2}} \int_{-\infty}^{\infty} \exp\left(-\frac{\tilde{n}^2}{2\tilde{\sigma}_n^2}\right) d\tilde{n} \left[\frac{1}{\sqrt{2\pi\tilde{\sigma}_n^2}} \int_{-\infty}^{E_a+\tilde{n}} \exp\left(-\frac{t^2}{2\tilde{\sigma}_n^2}\right) dt\right]^{Q-1}$$

$$= \frac{1}{\sqrt{2\pi}} \int_{-\infty}^{\infty} \exp\left(-\frac{n^2}{2}\right) \left\{1 - \left[\Phi\left(\frac{E_a}{\tilde{\sigma}_n} + n\right)\right]^{Q-1}\right\} dn$$

$$= \frac{1}{\sqrt{2\pi}} \int_{-\infty}^{\infty} \exp\left(-\frac{n^2}{2}\right) \left\{1 - \left[\Phi\left(\frac{E_a}{\sqrt{E_a\sigma_n^2}} + n\right)\right]^{Q-1}\right\} dn$$

$$= \frac{1}{\sqrt{2\pi}} \int_{-\infty}^{\infty} \exp\left(-\frac{n^2}{2}\right) \left\{1 - \left[\Phi\left(\sqrt{\frac{E_a}{\sigma_n^2}} + n\right)\right]^{Q-1}\right\} dn, \tag{7A.40}$$

where

$$\Phi(x) = \frac{1}{\sqrt{2\pi}} \int_{-\infty}^{x} \exp\left(-\frac{n^2}{2}\right) dn \qquad (7A.41)$$

is the "error integral."

Formula 7A.40 is known in communication theory as the one that determines the probability of errors in communication channels with Q *orthogonal signals* and additive white Gaussian noise. The remarkable feature of this integral is its threshold behavior for large Q. Consider

$$\lim_{Q\to\infty}\left(\ln\left[\Phi\left(\sqrt{\frac{E_a}{\sigma_n^2}}+n\right)\right]^{Q-1}\right) = \lim_{Q\to\infty}\left(Q\ln\left[\Phi\left(\mu\sqrt{\ln Q}\right)\right]\right) = \lim_{Q\to\infty}\left(\frac{\ln\left[\Phi(\mu\sqrt{\ln Q})\right]}{1/Q}\right),$$

where $\mu = \sqrt{E_a/\sigma_n^2 \ln Q}$. Because $\lim_{Q\to\infty}\Phi(\mu\sqrt{\log Q}+n) = 1$, and, therefore, $\lim_{Q\to\infty}\ln[\Phi(\mu\sqrt{\log Q}+n)] = 0$, the above limit is an indeterminate form 0/0. Then by L'Hospital's rule, we have

$$\lim_{Q\to\infty}\left(\ln\left[\Phi\left(\sqrt{\frac{E_a}{\sigma_n^2}}+n\right)\right]^{Q-1}\right) = \lim_{Q\to\infty}\left(\frac{\frac{d}{dQ}\ln\left[\Phi\left(\mu\sqrt{\ln Q}\right)\right]}{\frac{d}{dQ}(1/Q)}\right)$$

$$= \lim_{Q\to\infty}\left(\frac{\frac{d}{dQ}\left[\Phi\left(\mu\sqrt{\ln Q}\right)\right]}{\Phi\left(\mu\sqrt{\ln Q}\right)(-1/Q^2)}\right) = \lim_{Q\to\infty}\left(-Q^2\exp\left(-\frac{\mu^2}{2}\ln Q\right)\frac{d}{dQ}\left(\mu\sqrt{\ln Q}\right)\right)$$

$$= \lim_{Q\to\infty}\left(-\frac{\mu}{2}\frac{Q^{2-\frac{\mu^2}{2}}}{Q\sqrt{\ln Q}}\right) = \lim_{Q\to\infty}\left(-\frac{\mu}{2}\frac{Q^{1-\frac{\mu^2}{2}}}{\ln Q}\right) = \begin{cases} -\infty, & \frac{\mu^2}{2} < 1 \\ 0, & \frac{\mu^2}{2} \geq 1 \end{cases}. \qquad (7A.42)$$

Therefore

$$\lim_{Q\to\infty}\left[\Phi\left(\sqrt{\frac{E_a}{\sigma_n^2}}+n\right)\right]^{Q-1} = \begin{cases} 0, & E_a/2\sigma_n^2 < \ln Q \\ 1, & E_a/2\sigma_n^2 \geq \ln Q \end{cases}$$

and

$$\lim_{Q \to \infty} P_{Ae} = \frac{1}{\sqrt{2\pi}} \int_{-\infty}^{\infty} \exp\left(-\frac{n^2}{2}\right)\left\{1 - \left[\Phi\left(\sqrt{\frac{E_a}{\sigma_n^2}} + n\right)\right]^{Q-1}\right\} dn$$

$$= \begin{cases} 0, & \text{if } E_a/\sigma_n^2 > 2\ln Q \\ 1, & \text{if } E_a/\sigma_n^2 \leq 2\ln Q \end{cases}. \tag{7A.43}$$

Derivation of Equations 7.49, 7.50, and 7.51

For the additive model:

$$AV_{x_0,y_0}\left(\left|\beta_{r,s}^{(bg,x_0,y_0)}\right|^2\right)$$

$$= AV_{x_0,y_0}\left\{\beta_{r,s} - \alpha_{r,s}\exp\left[i2\pi\left(\frac{rx_0}{N_x} + \frac{sy_0}{N_y}\right)\right]\right\}\left\{\beta_{r,s}^* - \alpha_{r,s}^*\exp\left[-i2\pi\left(\frac{rx_0}{N_x} + \frac{sy_0}{N_y}\right)\right]\right\}$$

$$= \left|\beta_{r,s}\right|^2 + \left|\alpha_{r,s}\right|^2 - \alpha_{r,s}\beta_{r,s}^* AV_{x_0,y_0}\left\{\exp\left[i2\pi\left(\frac{rx_0}{N_x} + \frac{sy_0}{N_y}\right)\right]\right\}$$

$$- \alpha_{r,s}^*\beta_{r,s} AV_{x_0,y_0}\left\{\exp\left[-i2\pi\left(\frac{rx_0}{N_x} + \frac{sy_0}{N_y}\right)\right]\right\}. \tag{7A.44}$$

In the assumption that target coordinates have uniform distribution density over the input image

$$AV_{x_0,y_0}\left\{\exp\left[\pm i2\pi\left(\frac{rx_0}{N_x} + \frac{sy_0}{N_y}\right)\right]\right\}$$

$$= \frac{1}{N_x N_y} \int_{-N_x/2}^{N_x/2} \exp\left(\pm i2\pi\frac{rx_0}{N_x}\right)dx_0 \int_{-N_y/2}^{N_y/2} \exp\left(\pm i2\pi\frac{sy_0}{N_x}\right)dy_0$$

$$= \frac{1}{N_x N_y} N_x \frac{\exp(\pm i\pi r) - \exp(\pm i\pi r)}{\pm i2\pi r} N_y \frac{\exp(\pm i\pi s) - \exp(\pm i\pi s)}{\pm i2\pi s}$$

$$= \frac{\sin(\pi r)}{\pi r}\frac{\sin(\pi s)}{\pi s} = \delta(r)\delta(s). \tag{7A.45}$$

Then

$$AV_{x_0,y_0}\left(\left|\beta_{r,s}^{(bg,x_0,y_0)}\right|^2\right) = \left|\beta_{r,s}\right|^2 + \left|\alpha_{r,s}\right|^2 - (\alpha_{r,s}\beta_{r,s}^* + \alpha_{r,s}^*\beta_{r,s})\delta(r)\delta(s)$$

$$= \left|\beta_{r,s}\right|^2 + \left|\alpha_{r,s}\right|^2 - 2\alpha_{0,0}\beta_{0,0}\delta(r)\delta(s)$$

$$= \begin{cases} \left|\beta_{0,0}\right|^2 + \left|\alpha_{0,0}\right|^2 - 2\alpha_{0,0}\beta_{0,0}, & r,s = 0 \\ \left|\beta_{r,s}\right|^2 + \left|\alpha_{r,s}\right|^2, r = 1,\dots,N_x - 1, s = 1,\dots,N_y - 1. \end{cases} \quad (7A.46)$$

Spectral coefficients with indices $r,s = 0$ are responsible for the signal dc-component, which is irrelevant for localization of signal maximum. Therefore, one can, for evaluation of $AV_{x_0,y_0}(\left|\beta_{r,s}^{(bg,x_0,y_0)}\right|^2)$, use the relationship

$$AV_{x_0,y_0}\left(\left|\beta_{r,s}^{(bg,x_0,y_0)}\right|^2\right) = \left|\beta_{r,s}\right|^2 + \left|\alpha_{r,s}\right|^2. \quad (7A.47)$$

For the implant model, consider, for the sake of simplicity, a 1D case:

$$b_k^{(bg)} = \left(1 - w_k^{(tgt,x_0)}\right)b_k = b_k - w_k^{(tgt,x_0)}b_k. \quad (7A.48)$$

Then, for the DFT spectrum of the background component, we, using the DFT convolution theorem (the section "Discrete Representation of Fourier Integral Transform" in Chapter 4), have

$$\beta_r^{(bg,x_0)} = \beta_r - \frac{1}{\sqrt{N_x}}\beta_r \circ \omega_r^{(bg,x_0)}$$

$$\left|\beta_r^{(bg,x_0)}\right|^2 = \left[\beta_r - \frac{1}{\sqrt{N_x}}\beta_r \circ \omega_r^{(tgt,x_0)}\right]\left[\beta_r^* - \frac{1}{\sqrt{N_x}}\beta_r^* \circ \omega_r^{*(tgt,x_0)}\right]$$

$$= \left|\beta_r\right|^2 + \frac{\left|\beta_r \circ \omega_r^{(tgt,x_0)}\right|^2}{N_x} - \frac{\beta_r\left(\beta_r^* \circ \omega_r^{*tgt,x_0}\right) + \beta_r^*\left(\beta_r \circ \omega_r^{(tgt,x_0)}\right)}{\sqrt{N_x}}, \quad (7A.49)$$

where $\{\beta_r\}$ is DFT spectrum of the input image and symbol $\circ$ denotes cyclic convolution. Find the averaged power spectrum of the background component in assumption of uniform distribution density of the target coordinate x_0:

$$AV_{x_0}\left(\left|\beta_r^{(bg,x_0)}\right|^2\right)$$

$$= AV_{x_0}\left[\left|\beta_r\right|^2 + \frac{\left|\beta_r \circ \omega_r^{(tgt,x_0)}\right|^2}{N_x} - \frac{\beta_r\left(\beta_r^* \circ \omega_r^{*tgt,x_0}\right) + \beta_r^*\left(\beta_r \circ \omega_r^{(tgt,x_0)}\right)}{\sqrt{N_x}}\right]$$

$$= \left| \beta_r \right|^2 + \frac{A V_{x_0} \left[\left| \beta_r \circ \omega_r^{(tgt, x_0)} \right|^2 \right]}{N_x}$$

$$- \frac{\beta_r \left[\beta_r^* \circ A V_{x_0} \left(\omega_r^{*(tgt, x_0)} \right) \right] + \beta_r^* \left[\beta_r \circ A V_{x_0} \left(\omega_r^{(tgt, x_0)} \right) \right]}{\sqrt{N_x}}. \qquad (7A.50)$$

For $A V_{x_0}(\omega_r^{(tgt, x_0)})$ and $A V_{x_0}(\omega_r^{*(tgt, x_0)})$, we have

$$A V_{x_0} \left(\omega_r^{(tgt, x_0)} \right) = \omega_r^{(tgt, 0)} A V_{x_0} \left[\exp \left(-i 2\pi \frac{r x_0}{N_x} \right) \right] = \omega_r^{(tgt, 0)} \frac{1}{N_x} \int_{-N_x/2}^{N_x/2} \exp \left(-i 2\pi \frac{r x_0}{N_x} \right) dx_0$$

$$= \omega_r^{(tgt, 0)} \frac{1}{N_x} \frac{\exp(-i\pi r) - \exp(i\pi r)}{(-i 2\pi r)/N_x} = \omega_r^{(tgt, 0)} \frac{\sin(\pi r)}{\pi r} = \omega_r^{(tgt, 0)} \delta(r) \tag{7A.51}$$

$$A V_{x_0} \left(\omega_r^{*(tgt, x_0)} \right) = \omega_r^{*(tgt, 0)} \delta(r), \tag{7A.52}$$

where $\omega_r^{(tgt, x_0)}$ and $\omega_r^{*(tgt, 0)}$ are the DFT spectra of the target window function and its complex conjugates in the target position in the origin of coordinate. Then

$$\beta_r \circ A V_{x_0} \left(\omega_r^{(tgt, x_0)} \right) = \sum_{\tilde{r}=0}^{N_x - 1} \beta_{\tilde{r}} \omega_{r - \tilde{r}}^{(tgt, 0)} \delta(r - \tilde{r}) = \beta_r \omega_r^{(tgt, o)}, \tag{7A.53}$$

$$\beta_r^* \circ A V_{x_0} \left(\omega_r^{*(tgt, x_0)} \right) = \sum_{\tilde{r}=0}^{N_x - 1} \beta_{\tilde{r}}^* \omega_{r - \tilde{r}}^{*(tgt, 0)} \delta(r - \tilde{r}) = \beta_r^* \omega_r^{*(tgt, o)}. \tag{7A.54}$$

For $A V_{x_0} \left[\left| \beta_r \circ \omega_r^{(tgt, x_0)} \right|^2 \right]$, we have

$$A V_{x_0} \left[\left| \beta_r \circ \omega_r^{(tgt, x_0)} \right|^2 \right] = A V_{x_0} \left[\sum_{\tilde{r}=0}^{N_x - 1} \beta_{\tilde{r}} \omega_{r - \tilde{r}}^{(tgt, x_0)} \sum_{\tilde{\tilde{r}}=0}^{N_x - 1} \beta_{\tilde{\tilde{r}}}^* \omega_{r - \tilde{\tilde{r}}}^{*(tgt, x_0)} \right]$$

$$= \sum_{\tilde{r}=0}^{N_x - 1} \sum_{\tilde{\tilde{r}}=0}^{N_x - 1} \beta_{\tilde{r}} \beta_{\tilde{\tilde{r}}}^* \omega_{r - \tilde{\tilde{r}}}^{*(tgt, 0)} \omega_{r - \tilde{r}}^{(tgt, 0)} A V_{x_0} \left[\exp \left(i 2\pi \frac{\tilde{\tilde{r}} - \tilde{r}}{N_x} x_0 \right) \right]$$

$$= \sum_{\tilde{r}=0}^{N_x - 1} \sum_{\tilde{\tilde{r}}=0}^{N_x - 1} \beta_{\tilde{r}} \beta_{\tilde{\tilde{r}}}^* \omega_{r - \tilde{\tilde{r}}}^{*(tgt, 0)} \omega_{r - \tilde{r}}^{(tgt, 0)} \delta(\tilde{\tilde{r}} - \tilde{r}) = \sum_{\tilde{r}=0}^{N_x - 1} \sum_{\tilde{\tilde{r}}=0}^{N_x - 1} \left| \beta_{\tilde{r}} \right|^2 \left| \omega_{r - \tilde{r}}^{(tgt, 0)} \right|^2.$$

$$(7A.55)$$

Substitute Equations 7A.51 through 7A.55 into Equation 7A.50 to finally obtain

$$AV_{x_0}\left(\left|\beta_r^{(bg,x_0)}\right|^2\right) = |\beta_r|^2 + \frac{\sum_{\tilde{r}=0}^{N_x-1}\left|\beta_{\tilde{r}}\right|^2\left|\omega_{r-\tilde{r}}^{(tgt,0)}\right|^2}{N_x} - \frac{\beta_r\beta_r^*\omega_r^{*(tgt,0)} + \beta_r^*\beta_r\omega_r^{(tgt,0)}}{\sqrt{N_x}}$$

$$= |\beta_r|^2 + \sum_{\tilde{r}=0}^{N_x-1}\left|\beta_{\tilde{r}}\right|^2\left|\omega_{r-\tilde{r}}^{(tgt,0)}\right|^2 - |\beta_r|^2\left(\omega_r^{(tgt,0)} + \omega_r^{*(tgt,0)}\right)$$

$$= \left(1 - \frac{2\,\text{Re}\,\omega_r^{(tgt,0)}}{\sqrt{N_x}}\right)|\beta_r|^2 + \frac{1}{N_x}\sum_{\tilde{r}=0}^{N_x-1}\left|\beta_{\tilde{r}}\right|^2\left|\omega_{r-\tilde{r}}^{(tgt,0)}\right|^2. \tag{7A.56}$$

Let, for instance, $w_k^{(tgt,0)}$ is a rectangular window function

$$w_k^{(tgt,0)} = \begin{cases} 1, & k = 0,1,\ldots,K \\ 0, & k = K+1, K+2,\ldots, N-K-1. \\ 1, & k = N-K, N-k+1,\ldots, N-1 \end{cases} \tag{7A.57}$$

Then

$$\omega_r^{(tgt,0)} = \frac{1}{\sqrt{N_x}}\sum_{k=0}^{N_x-1} w_k^{(tgt,0)}\exp\left(i2\pi\frac{kr}{N_x}\right)$$

$$= \frac{1}{\sqrt{N_x}}\left[\sum_{k=0}^{K}\exp\left(i2\pi\frac{kr}{N_x}\right) + \sum_{k=N-K}^{N_x-1}\exp\left(i2\pi\frac{kr}{N_x}\right)\right]$$

$$= \frac{1}{\sqrt{N_x}}\left[\frac{\exp\left(i2\pi\frac{K+1}{N_x}r\right) - 1}{\exp\left(i2\pi\frac{r}{N_x}\right) - 1} + \frac{\exp\left(i2\pi\frac{N_x r}{N_x}\right) - \exp\left(i2\pi\frac{N_x-K}{N_x}r\right)}{\exp\left(i2\pi\frac{r}{N_x}\right) - 1}\right]$$

$$= \frac{1}{\sqrt{N_x}}\left[\frac{\exp\left(i2\pi\frac{K+1}{N_x}r\right) - 1 + 1 - \exp\left(-i2\pi\frac{K}{N_x}r\right)}{\exp\left(i2\pi\frac{r}{N_x}\right) - 1}\right]$$

$$= \frac{1}{\sqrt{N_x}} \left[\frac{\exp\left(i2\pi \dfrac{K+1/2}{N_x}r\right) - 1 + 1 - \exp\left(-i2\pi \dfrac{K+1/2}{N_x}r\right)}{\exp\left(i\pi \dfrac{r}{N_x}\right) - \exp\left(-i\pi \dfrac{r}{N_x}\right)} \right]$$

$$= \frac{1}{\sqrt{N_x}} \frac{\sin\left(\pi \dfrac{2K+1}{N_x}r\right)}{\sin\left(\pi \dfrac{r}{N_x}\right)} = \frac{2K+1}{\sqrt{N_x}} \overline{\mathrm{sincd}}\left(2K+1; N_x; r\right), \qquad (7A.58)$$

where $\overline{\mathrm{sincd}}(.;.;.)$ is discrete sinc-function defined by Equation 4.54. Substitute Equation 7A.58 into Equation 7A.50 and obtain the following estimate of the averaged background image component spectrum:

$$AV_{x_0}\left(\left|\beta_r^{(bg,x_0)}\right|^2\right) = \left(1 - 2\frac{2K+1}{N_x}\overline{\mathrm{sincd}}(2K+1; N_x; r)\right)|\beta_r|^2$$

$$+ \left(\frac{2K+1}{N_x}\right)\sum_{\bar{r}=0}^{N_x-1}|\beta_{\bar{r}}|^2\left|\overline{\mathrm{sincd}}\right|(2K+1; N_x; r - \qquad (7A.59)$$

Equations 7A.56 and 7A.59 can be extended to the 2D case as follows:

$$AV_{x_0 y_0}\left(\left|\beta_{r,s}^{(bg,x_0,y_0)}\right|^2\right) = \left(1 - \frac{2\,\mathrm{Re}\,\omega_{r,s}^{(tgt,0,0)}}{\sqrt{N_x N_y}}\right)|\beta_{r,s}|^2 + \frac{1}{N_x}\sum_{\bar{r}=0}^{N_x-1}\sum_{\bar{s}=0}^{N_y-1}|\beta_{\bar{r},\bar{s}}|^2\left|\omega_{r-\bar{r},s-\bar{s}}^{(tgt,0,0)}\right|^2 \quad (7A.60)$$

$$AV_{x_0,y_0}\left(\left|\beta_{r,s}^{(bg,x_0,y_0)}\right|^2\right)$$

$$= \left(1 - 2\frac{(2K+1)(2L+1)}{N_x N_y}\overline{\mathrm{sincd}}(2K+1; N_x; r)\overline{\mathrm{sincd}}(2L+1; N_y; s)\right)|\beta_{r,s}|^2$$

$$+ \left(\frac{2K+1}{N_x}\right)^2\left(\frac{2L+1}{N_y}\right)^2$$

$$\times \sum_{\bar{r}=0}^{N_x-1}\sum_{\bar{s}=0}^{N_y-1}|\beta_{\bar{r},\bar{s}}|^2\left|\overline{\mathrm{sincd}}(2K+1; N_x; r - \tilde{r})\right|^2\left|\overline{\mathrm{sincd}}(2L+1; N_y; s - \tilde{s})\right|^2, \qquad (7A.61)$$

where $(2K+1)$ and $(2L+1)$ are dimensions of a rectangular target window function $w_{k,l}^{(tgt,0,0)}$.

Exercises

localization_demo_CRC.m
loclzerr_CRC.m
corr_comparison_CRC.m
lcoptcorr_CRC.m

References

1. L. P. Yaroslavsky, The theory of optimal methods for localization of objects in pictures, In: *Progress in Optics*, Editor, E. Wolf, Vol. XXXII, Elsevier Science Publishers, Amsterdam, 1993.
2. L. P. Yaroslavsky, *Digital Holography and Digital Image Processing: Principles, Methods, Algorithms*, Kluwer Academic Publishers, Boston, 2004.

8

Image Perfecting

This chapter is an introduction to methods for perfecting visual and/or metrological image quality. We begin, in the section "Image Perfecting as a Processing Task," with the formulation of the image perfecting task. Then, in the sections "Possible Approaches to Restoration of Image Distorted by Blur and Contaminated by Noise" and "MMSE-Optimal Linear Filters for Image Restoration," we present possible approaches to correction of image blur and cleaning images from additive noise, and introduce MMSE-optimal linear filters that perform this processing and are implemented, for the sake of simplification of their design and minimization of their computational complexity, in a domain of certain orthogonal transform, which features fast transform algorithm. In the section "Sliding Window Transform Domain Adaptive Image Restoration," we extend MMSE-optimal linear filters for working in sliding window, which makes them local adaptive. In the section "Multicomponent Image Restoration and Data Fusion," we further extend these filters to multicomponent image restoration and consider the issue of fusing image data from different sources. In the section "Filtering Impulse Noise," we discuss nonlinear filtering methods for cleaning impulse noise in images. In the section "Correcting Image Grayscale Nonlinear Distortions," methods for correcting image grayscale distortions are addressed, and finally, in the section "Nonlinear Filters for Image Perfecting," we provide a survey and classification of nonlinear filters for image denoising and enhancement.

Image Perfecting as a Processing Task

Image perfecting as an image processing task has two stages: *image restoration* and *image enhancement*. Image restoration is a processing aimed at correcting signal distortions that may occur in imaging systems due to technical limitations in their design, implementation, and working conditions. Typical examples of image distortions are image blur and noisiness. These distortions much affect the suitability of images for visual analysis and, as it was shown in Chapter 7, determine image potentials for localization and recognition of objects. Two more examples of image distortions are geometrical distortions and point-wise nonlinear grayscale distortions.

The methods for distortion correction are based on the canonical model of imaging systems shown in Figure 8.1.

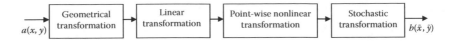

FIGURE 8.1
A canonical model of imaging systems.

The model assumes the existence of a "true" image $a(x,y)$, which the system would produce if there were no signal distortions in it, and represents the system output image $b(\tilde{x}, \tilde{y})$ as a result of transformations of the "true" image by a combination of geometrical transformation, linear transformations, point-wise nonlinear transformations, and stochastic transformations.

Geometrical transformations are transformations of image coordinates. They are specified by relationships $(x, y) \Rightarrow (\tilde{x}, \tilde{y})$ between the initial image coordinate system (x,y) and the transformed one $(\tilde{x}, \tilde{y})$.

Linear transformations are specified through the system PSF or frequency response (see the section "Signal Transformations" in Chapter 2). Frequency response of the ideal imaging system is assumed to be uniform for all frequencies in the image base band defined by the sampling rate. Frequency responses of real imaging systems are not uniform. Usually they decay, more or less rapidly, on high frequencies, which results in image blur.

Point-wise transformations are specified through the system transfer function (Equation 2.28). Ideally, the system transfer function is a linear function. Deviations of system transfer functions from linear ones cause distortions of the gray-level scale (nonlinear distortions).

Stochastic transformations model random interferences, or noise, in images. They are specified by statistical noise models such as those described in the section "Models of Signal Random Interferences" (Chapter 2).

The goal of image restoration is estimating, with a certain required accuracy, the "true" signal $a(x,y)$, given distorted signal $b(\tilde{x}, \tilde{y})$ produced by the imaging system. This problem is frequently referred to as the *inverse problem*.

If the system does not introduce any random distortions and system's parameters such as PSF and transfer function are known, the inverse problem has a trivial solution: in order to restore signal $a(x,y)$ from signal $b(\tilde{x}, \tilde{y})$, the latter should be subjected, in the order inverse to that distortion factors have in the model, to transformations inverse to those introduced by the system. This is what, in practice, people do for correcting geometrical distortions and nonlinear distortions.

However, this does not work well for correcting distortions caused by linear transformations. Applying inverse transformation for correcting linear distortions frequently results in overamplification of random disturbances that are unavoidably present in images on the output of imaging systems. This frequently makes images "corrected" by the inverse filtering even worse for visual perception than noncorrected images; hence, image restoration methods smarter than simple inverse filtering are required.

The goal of image enhancement is producing, out of raw or corrected ("restored") images, enhanced images, which better fit the needs and capabilities of human operators in image analysis and decision making. The human visual system has certain limitations in its capability to perceive information carried by images. For instance, in grayscale images, the human visual system cannot detect gray-level contrasts that are lower than a certain contrast sensitivity threshold. It is also incapable of apprehending image attributes other than variations of its intensity, which might be decisive for image analysis. From the other side, human vision perceives colors, has 3D capability through stereo vision, and is capable of very efficient detection of image variations in time. Therefore, converting images into a form that makes use of capabilities of human visual system to perceive information to the highest possible degree is required in addition to image restoration.

Possible Approaches to Restoration of Images Distorted by Blur and Contaminated by Noise

Let $\{b_{k,l}\}$, $(k = 0,1,\ldots,N_x - 1, l = 0,1,\ldots,N_y - 1)$, be samples of a distorted image, which are in a certain way, defined by the imaging system model, determined by samples $\{a_{k,l}\}$ of the "true" image to be restored from $\{b_{k,l}\}$, and let $\{\hat{a}_{k,l}\}$ be samples of the restored image. Consider an imaging system model that describes image blur and contamination by additive signal-independent zero mean Gaussian noise. For such a model

$$b_{k,l} = (h \circ a)_{k,l} + n_{k,l}, \tag{8.1}$$

where $\{n_{k,l}\}$ are samples of noise, $(h \circ a)_{k,l}$ are samples of digital convolution of samples $\{a_{k,l}\}$ of the "true" image with imaging system PSF $\{h_{k,l}\}$, which specifies the linear transformation responsible for image blur. Our goal is to find an optimal, in a certain sense that we are going to specify, image restoration algorithm, that is, mapping $\{b_{k,l}\} \Rightarrow \{\hat{a}_{k,l}\}$.

An immediate option for solving this problem is to consider the task of image restoration as a parameter estimation task, that is, a task of estimating the set of image samples $\{a_{k,l}\}$ from the observed set of samples $\{b_{k,l}\}$ of the input image as we did in the sections "Computational Imaging Using Optics-Less Lambertian Sensors" in Chapter 5 and "Target Localization in Cluttered Images" in Chapter 7. As shown in the section "Statistical Models of Signals and Transformations" in Chapter 2, the statistically optimal estimates $\{\hat{a}_{k,l}\}$ of $\{a_{k,l}\}$ are ML- and MAP-estimates:

$$\{\hat{a}_{k,l}\}^{(ML)} = \arg\min_{a_{k,l}} \sum_{k=0}^{N_x-1} \sum_{l=0}^{N_y-1} \left[b_{k,l} - (h \circ a)_{k,l} \right]^2; \tag{8.2}$$

$$\{\hat{a}_{k,l}\}^{(MAP)} = \arg\min_{a_{k,l}}\left\{\sum_{k=0}^{N_x-1}\sum_{l=0}^{N_y-1}\left[b_{k,l} - (h \circ a)_{k,l}\right]^2 - 2\sigma_n^2 \ln P(\{a_{k,l}\})\right\}, \quad (8.3)$$

where σ_n^2 is noise variance and $P\big(\{\hat{a}_{k,l}\}\big)$ is *a priori* probability of the set $\{\hat{a}_{k,l}\}$ of image samples.

We showed in the section "Computational Imaging Using Optics-Less Lambertian Sensors" (Chapter 5) that this approach, in principle, works. The main problem with it is the extremely high, even for moderate image sizes $N_x \times N_y$, dimensionality of the minimization task, which leads to prohibitively high computational expenses. It is especially true for ML-estimation. In MAP-estimation, the volume of the signal space, in which global minimum is searched, can, in principle, be narrowed by *a priori* probabilities. The problem is, however, how one can specify those probabilities. During several decades of trying to develop meaningful probabilistic mathematical models for images, a number of models have been suggested, beginning from the simplest Gaussian random process model, to more sophisticated Gibbs and 2D Markov random process models, to name a few. Being mathematically more elegant than the other, those models, however, failed to produce results of practical rather than only of pure academic value.

In order to ease the solution of the problem of narrowing the search space, one can replace *a priori* probability distribution of input images (term $[-2\sigma_n^2 \ln P(\{a_{k,l}\})]$) in Equation 8.3) by another functional $\Re(\{a_{k,l}\})$ of images that characterizes the "desirability" of the solution $\{\hat{a}_{k,l}\}$ (the more desirable the solution, the smaller $\Re(\{a_{k,l}\})$ must be). Considering that mean squared deviation of $(h \circ a)_{k,l}$ from $\{b_{k,l}\}$ in Equation 8.3 normalized by $N_x \times N_y$ is, for large $N_x \times N_y$, a statistically good estimate for variance σ_n^2 of additive noise, one can reformulate Equation 8.3 in terms of constrained optimization:

$$\{\hat{a}_{k,l}\}^{(Opt)} = \arg\min_{a_{k,l}} \Re(\{a_{k,l}\}) \text{ subject to } \frac{1}{N_x N_y}\sum_{k=0}^{N_x-1}\sum_{l=0}^{N_y-1}\left[b_{k,l} - (h \circ a)_{k,l}\right]^2 = \sigma_n^2$$

$$(8.4)$$

or, using Lagrange multiplier λ

$$\{\hat{a}_{k,l}\}^{(Opt)} = \arg\min_{a_{k,l}}\left\{\Re(\{a_{k,l}\}) + \lambda\sum_{k=0}^{N_x-1}\sum_{l=0}^{N_y-1}\left[b_{k,l} - (h \circ a)_{k,l}\right]^2\right\}. \quad (8.5)$$

This approach to solving inverse problem is called *regularization*.

A classic example of the regularization functional $\Re(\{a_{k,l}\})$ is L2 norm of the sought solution:

$$\Re(\{a_{k,l}\}) = \sum_{k=0}^{N_x-1}\sum_{l=0}^{N_y-1}\left|a_{k,l}\right|^2,\tag{8.6}$$

which links the desirability of the solution to its variance. This prior is known to be related to the so-called *Tikhonov regularization* and to *Wiener filtering*. We will detail the Wiener filtering approach in the next section.

L2 norm forces the solution to not deviate much from its mean value by means of punishing its deviations from the mean, large deviations being punished substantially stronger than small ones. This does enable noise reduction and produces "smooth" image estimates but at the expense of substantial loss of image sharpness, which, as we saw in Chapter 7, is of crucial importance for image analysis.

As a better measure of image "goodness," L2 norm the image Laplacian was suggested:

$$\Re(\{a_{k,l}\}) = \sum_{k=0}^{N_x-1}\sum_{l=0}^{N_y-1}\left|Laplacian\{a_{k,l}\}\right|^2\tag{8.7}$$

(for the definition of Laplacian operator and its usefulness, see the section "Object Localization and Edge Detection: Selection of Reference Objects for Target Tracking" (Chapter 7)). The rationale behind this measure is the fact that most of the image energy is contained in a low-frequency component of image spectra, whereas most of the white noise energy is contained in its high-frequency components, hence the higher differential measure, such as Laplacian, that is, image second spatial derivative, the more probable it is due to the noise. However, punishing large values of image Laplacian also contradicts with desirable properties of images: as we saw in the section "Object Localization and Edge Detection: Selection of Reference Objects for Target Tracking" (Chapter 7), the energy of object Laplacian is a good measure of their potential detectability in a clutter of other objects.

In order to soften the punishment, the L1 norm can be used instead of the L2 norm. An example of such a softened differential measure is using image *total variation* as a regularizing functional:

$$\Re(\{a_{k,l}\}) = \sum_{k=0}^{N_x-1}\sum_{l=0}^{N_y-1}\left|Gradient\{a_{k,l}\}\right|,\tag{8.8}$$

where the *Gradient* operator computes sums of absolute values of image spatial derivatives.

Considerable progress in the search of an appropriate regularizing functional was achieved recently, when it was suggested to formulate functional

$\Re(\{a_{k,l}\})$ in terms of sparsity of transform coefficients of the solution, or, in other words, to search solutions among band-limited, in a certain transform, approximations to $\{a_{k,l}\}$. This approach was inspired by advances in transform image compression methods, such as JPEG or wavelet image coding (see the section "Basics of Image Data Compression" in Chapter 3). The measure of transform coefficient sparsity is their L0 norm:

$$\Re(\{a_{k,l}\}) = \sum_{r=0}^{N_x-1}\sum_{s=0}^{N_y-1} \left| T\{a_{k,l}\} \right|_{r,s}^0, \tag{8.9}$$

which computes the number of transform $T\{.\}$ nonzero coefficients. It turned out, however, that such a norm is not well suited for numerical optimization algorithms and can be replaced by the L1 norm:

$$\Re(\{a_{k,l}\}) = \sum_{r=0}^{N_x-1}\sum_{s=0}^{N_y-1} \left| T\{a_{k,l}\} \right|_{r,s}. \tag{8.10}$$

This transform domain "sparsity" approach is being actively pursued at present, and interested readers are referred to numerous publications under key words "compressive sensing" and "sparse representations." Being, however, still based on numerical optimization, this approach does solve the problem of the computational complexity of the optimization. Aside from that, it shares the important drawback common to all the above-listed approaches: all of them disregard image spatial inhomogeneity. We already demonstrated the importance of taking image spatial inhomogeneity into account in the section "Local Adaptive SCR-Optimal Correlators" (Chapter 7), where we introduced local adaptive SCR-optimal correlators for target location. Later in this chapter, we introduce *local adaptive filtering* methods that explicitly account for image spatial inhomogeneity.

A constructive and computationally efficient alternative to the above-outlined parameter estimation approach is a "waveform estimation" approach that can be traced back to classical works by the fathers of information theory N. Wiener [1], A. Kolmogorov [2], C. E. Shannon [3], and V. A. Kotelnikov [4]. In this approach, the inaccuracy of restoration of signals is evaluated by L2 norm of the restoration error, that is, by the mean-squared difference between the "true" and restored signals, and restoration is supposed to be performed by linear filtering of the distorted signals, which is designed to minimize this mean-squared restoration error. We will refer to this approach as *minimum mean squared error (MMSE) approach*. Linear filtering that minimizes MSE restoration error is also called *Wiener filtering*. In terms of the general parameter estimation approach, the MMSE approach is aimed at minimization of the second moment of the *a posteriori* probability of the estimate.

MMSE approach-based linear filtering is motivated by the following reasoning. In the absence of noise, linear distortions introduced by linear filtering of image signal in imaging systems can, in principle, be corrected by the inverse linear filtering. One can expect that, in the presence of small noise, linear filtering will still be sufficient, except that a certain "small" modification of the inverse filter will be required to keep the restoration inaccuracy small.

In what follows, we outline the MMSE approach in detail, discuss its potentials and limitations, and demonstrate that its application in sliding window, similar to that described in the section "Local Adaptive SCR-Optimal Correlators" (Chapter 7), is a promising solution of the problem of restoration of noisy, and blurred images both in terms of restoration accuracy and in terms of computational efficiency.

MMSE-Optimal Linear Filters for Image Restoration

Transform Domain MSE-Optimal Scalar Filters

Consider designing a linear filter that generates, from a copy $\{b_{k,l}\}$ of a "true" signal $\{a_{k,l}\}$ distorted in the imaging system by linear filtering and by additive noise, its estimate $\{\hat{a}_{k,l}\}$ with minimal mean squared error:

$$\{\hat{a}_k\} = \underset{R\{b_k\}=\{\hat{a}_k\}}{\arg\min}\left\{AV_{\Omega_A}AV_{\Omega_N}\left(\sum_{k=0}^{N_x-1}\sum_{l=0}^{N_y-1}\left|a_{k,l}-\hat{a}_{k,l}\right|^2\right)\right\} \qquad (8.11)$$

evaluated on average (averaging operators AV_{Ω_A} and AV_{Ω_N}) over image Ω_A and noise Ω_N ensembles.

In order to simplify further formula derivations, we will use 1D notation unless otherwise indicated.

In general, linear filtering of a discrete signal can be conveniently described as multiplication of a vector of input signal samples $\mathbf{B} = \{b_k\}$ by a filter matrix $\mathbf{H}$ (see Chapter 4, Equation 4.8):

$$\hat{\mathbf{A}} = \mathbf{H}\cdot\mathbf{B}, \qquad (8.12)$$

where $\hat{\mathbf{A}} = \{\hat{a}_k\}$ is a vector of filter output signal samples. For signals of N samples, a general filter matrix $\mathbf{H}$ has dimensions $N\times N$. The specification of such a filter requires determining N^2 filter coefficients, and the filtering itself requires performing N operations per signal sample. In image processing, the computational complexity of both determination of filter coefficients and the filtering might become too high because of high dimensionality of image arrays. Fast transform algorithms, with which transform coefficients can be computed for $O(N \log N)$ operations, allow to dramatically ease the filter design, and decrease its computational complexity. This motivates considering only *"scalar" filtering*, specified by diagonal matrices, assuming that it is implemented in a domain of orthogonal transforms that can be computed with fast algorithms. Scalar filtering is described by the equation

$$\hat{\mathbf{A}} = \mathbf{T}^{-1} \cdot \mathbf{H_d} \cdot \mathbf{T} \cdot \mathbf{B}, \tag{8.13}$$

where $\mathbf{T}$ and $\mathbf{T}^{-1}$ are, respectively, direct and inverse orthogonal transforms, and $\mathbf{H} = \mathbf{diag}\{\eta_r\}$, $(r = 0,1,\ldots,N-1)$ is a diagonal filter matrix. Such a scalar filtering implies the following relationship between transform coefficients $\{\hat{\alpha}_r\} = \mathbf{T} \cdot \hat{\mathbf{A}}$ and $\{\beta_r\} = \mathbf{T} \cdot \mathbf{B}$ of filter output and input signal samples:

$$\hat{\alpha}_r = \eta_r \beta_r. \tag{8.14}$$

In the assumption of orthogonality of the transform $\mathbf{T}$, one can, by virtue of Parseval's relationship (Equation 2.20), reformulate the filter optimality condition defined by Equation 8.20 in terms of signal transform coefficients:

$$\{\hat{\alpha}_r\} = \underset{\{\eta_r\}}{\arg\min}\left\{ AV_{\Omega_A}AV_{\Omega_N}\left(\sum_{r=0}^{N-1}|\alpha_r - \hat{\alpha}_r|^2 \right) \right\}$$

$$= \underset{\{\eta_r\}}{\arg\min}\left\{ AV_{\Omega_A}AV_{\Omega_N}\left(\sum_{r=0}^{N-1}|\alpha_r - \eta_r\beta_r|^2 \right) \right\}. \tag{8.15}$$

Through computing derivatives over the sought variables and equaling them to zero, one can obtain from Equation 8.15 (see Appendix) that MSE-optimal scalar filter coefficients $\{\eta_r\}$ are normalized cross-correlation coefficients between spectral coefficients β_r and α_r of the filter input and "true" signals:

$$\eta_r^{(opt)} = \frac{AV_{\Omega_A}AV_{\Omega_N}\left(\alpha_r \beta_r^* \right)}{AV_{\Omega_A}AV_{\Omega_N}\left(|\beta_r|^2 \right)}. \tag{8.16}$$

Equation 8.16 implies that, in order to implement the optimal scalar Wiener filter, one should know cross-correlation $\left\{ AV_{\Omega_A}AV_{\Omega_N}\left(\alpha_r \beta_r^* \right) \right\}$ between spectral coefficients of filter input signal and "true" signal and power spectrum $\left\{ AV_{\Omega_A}AV_{\Omega_N}\left(|\beta_r|^2 \right) \right\}$ of the input signal in the selected transform domain. We will refer to filters defined by Equation 8.16 as *scalar Wiener filters*.

Empirical Wiener Filters for Image Denoising

Consider the additive signal-independent noise model (Section "Models of Signal Random Interferences" in Chapter 2), in which filter input signal samples $\{b_k\}$ are a sum of "true" signal samples $\{a_k\}$ and samples $\{n_k\}$ of signal-independent zero mean random noise:

$$b_k = a_k + n_k. \tag{8.17}$$

In the spectral domain of the transform T, the same relationship holds for signal and noise spectral coefficients $\{\beta_r\}$, $\{\alpha_r\}$, and $\{v_r\}$:

$$\beta_r = \alpha_r + v_r. \tag{8.18}$$

For this model, one can obtain that

$$AV_{\Omega_A} AV_{\Omega_N} \left(\alpha_r \beta_r^* \right) = AV_{\Omega_A} AV_{\Omega_N} \left[\alpha_r \left(\alpha_r^* + v_r^* \right) \right] = AV_{\Omega_A} \left(|\alpha_r|^2 \right) \tag{8.19}$$

and

$$AV_{\Omega_A} AV_{\Omega_N} \left(|\beta_r|^2 \right) = AV_{\Omega_A} AV_{\Omega_N} \left[\left(\alpha_r + v_r \right) \left(\alpha_r^* + v_r^* \right) \right]$$

$$= AV_{\Omega_A} \left(|\alpha_r|^2 \right) + A V_{\Omega_A} \left(|v_r|^2 \right) \tag{8.20}$$

because for zero mean noise $AV_{\Omega_N}(v_r^*) = AV_{\Omega_N}(v_r) = 0$. Therefore, coefficients $\left\{ \eta_r^{(WF)} \right\}$ of scalar Wiener filter for suppressing additive signal-independent noise are defined as

$$\eta_r^{(WF)} = \frac{AV_{\Omega_A} \left(|\alpha_r|^2 \right)}{AV_{\Omega_A} \left(|\alpha_r|^2 \right) + AV_{\Omega_N} \left(|v_r|^2 \right)} = \frac{SNR_r}{1 + SNR_r} \tag{8.21}$$

or, in 2D notations, as

$$\eta_r^{(WF)} = \frac{AV_{\Omega_A} \left(|\alpha_{r,s}|^2 \right)}{AV_{\Omega_A} \left(|\alpha_{r,s}|^2 \right) + AV_{\Omega_N} \left(|v_{r,s}|^2 \right)} = \frac{SNR_{r,s}}{1 + SNR_{r,s}}, \tag{8.22}$$

where

$$SNR_{r,s} = \frac{AV_{\Omega_A} \left(|\alpha_{r,s}|^2 \right)}{AV_{\Omega_N} \left(|v_{r,s}|^2 \right)} \tag{8.23}$$

is SNR on the (r,s)-th image spectral component. Thus, scalar Wiener filter weight coefficients are defined, for each image spectral coefficient, by the $SNR_{r,s}$ for this coefficient. The lower the SNR for a particular signal spectral component, the lower will be the contribution of this component to the filter output signal.

Equation 8.22 can be given yet another interpretation that links Wiener filtering with SCR-optimal filtering for target location that we discussed in Chapter 7. Suppose that the image contains a single object with spectrum $\{\alpha_r\}$. For such an object, Equation 8.22 for scalar Wiener filter takes the form

$$\eta_r^{(WF)} = \frac{\left(|\alpha_{r,s}|^2\right)}{\left(|\alpha_{r,s}|^2\right) + AV_{\Omega_N}\left(|\nu_{r,s}|^2\right)} = \alpha_{r,s}\frac{\alpha_{r,s}^*}{\left(|\alpha_{r,s}|^2\right) + AV_{\Omega_N}\left(|\nu_{r,s}|^2\right)}. \qquad (8.24)$$

Equation 8.24 represents this filter as consisting of two filters in a cascade. The first filter

$$\left\{\frac{\alpha_{r,s}^*}{\left(|\alpha_{r,s}|^2\right) + AV_{\Omega_N}\left(|\nu_{r,s}|^2\right)}\right\}$$

is similar to target location filters considered in Chapter 7. It produces at its output a peak at the location of the object. If this peak were a delta-function, it would draw, after applying to it the second filter $\{\alpha_{r,s}\}$, the object in the detected location, as if the filter recognizes the object. Because the peak is not a delta-function, Wiener filter draws at the object location the object in a certain way blurred by the peak.

In order to implement the scalar Wiener filter, one has to know power spectra $AV_{\Omega_A}\left(|\alpha_{r,s}|^2\right)$ and $AV_{\Omega_N}\left(|\nu_{r,s}|^2\right)$ of the "true" image and of noise in the selected transform domain. Noise power spectrum $AV_{\Omega_N}\left(|\nu_{r,s}|^2\right)$ might be known from the specification certificate of the imaging device. Otherwise, it can be measured empirically directly in noisy input images using methods described in Appendix. Denote noise power spectrum estimate by $\overline{|\nu_{r,s}|^2}$.

As for the "true" image power spectrum $AV_{\Omega_A}\left(|\alpha_{r,s}|^2\right)$, it is almost never known. However, one can attempt to estimate it using Equation 8.20 and an estimate $\overline{|\beta_{r,s}|^2}$ of the power spectrum $AV_{\Omega_A}AV_{\Omega_N}\left(|\beta_{r,s}|^2\right)$ of the input noisy image, which can be obtained by means of one or another method of averaging of the image spectrum $|\beta_{r,s}|^2$ (for instance, as we will see later in an illustrative example of Figure 8.2, 1D row-wise image power spectrum can be estimated by means of averaging of power spectra of image rows). With

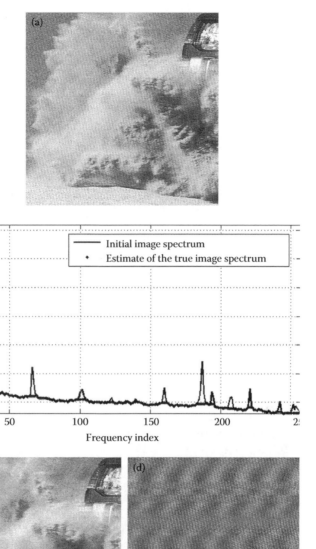

FIGURE 8.2
Filtering moiré interferences in an image: (a) input image; (b) plots of averaged DFT power spectra of the input image rows (solid line) and of an estimate of true image spectrum (dotted line) used for detecting moire noise spectral components; (c) output image cleaned from moire noise by means of rejecting filtering; (d) difference between input and output: eliminated noise pattern.

such empirical estimates $\overline{|\beta_{r,s}|^2}$ and $\overline{|v_{r,s}|^2}$, an estimate $\overline{|\alpha_{r,s}|^2}$ of the true image power spectrum $AV_{\Omega_A}\left(|\alpha_{r,s}|^2\right)$ can be obtained as

$$\overline{|\alpha_{r,s}|^2} = \overline{|\beta_{r,s}|^2} - \overline{|v_{r,s}|^2}. \tag{8.25}$$

Because empirical estimates of image and noise spectra may deviate from accurate statistical ones, for which AN infinitely large number of noise realizations are required, Equation 8.25 might give negative values for some estimates of $\overline{|\alpha_{r,s}|^2}$, which contradicts the property of power spectra to be nonnegative. In order to prevent this, the following modified spectrum estimation can be used:

$$\overline{|\alpha_{r,s}|^2} = \max\left[\overline{|\beta_{r,s}|^2} - \overline{|v_{r,s}|^2} \;;\quad 0\right]. \tag{8.26}$$

With these estimates for image and noise power spectra, we arrive at the following empirical substitute for the scalar Wiener filter:

$$\eta_{r,s}^{(EWF)} = \max\left[\frac{\overline{|\beta_{r,s}|^2} - \overline{|v_{r,s}|^2}}{\overline{|\beta_{r,s}|^2}} \;;\quad 0\right]. \tag{8.27}$$

We will refer to this filter as the *empirical Wiener filter*.

If the imaging system noise is known to be uncorrelated with variance σ_n^2, the empirical Wiener filter takes the form

$$\eta_{r,s}^{(EWF)} = \max\left[\frac{\overline{|\beta_{r,s}|^2} - \sigma_n^2}{\overline{|\beta_{r,s}|^2}} \;;\quad 0\right]. \tag{8.28}$$

As one can see from Equations 8.27 and 8.28, the weight coefficients of scalar Wiener filters assume values in the range between zero and one. A version of the empirical Wiener filter of Equation 8.28 with binary coefficients

$$\eta_{r,s}^{(rjct)} = \begin{cases} 1, & \text{if } \overline{|\beta_{r,s}|^2} \geq Thr \\ 0, & \text{otherwise} \end{cases}, \tag{8.29}$$

where *Thr* is a rejecting threshold called the *rejecting filter*. As it follows from Equation 8.28, the rejecting threshold *Thr* has a value of the order of magnitude of the noise variance σ_n^2. Rejecting filters eliminate from the input images spectra all components whose intensity is lower than the rejecting threshold $Thr \cong \sigma_n^2$; all other image spectral components are preserved by rejecting filters. Therefore, rejecting filters replace signals with their band-limited approximations and, by virtue of the Parseval's relationship for orthonormal transforms, they are exact solutions of the optimization equation (8.4), in which transform coefficient sparsity defined by Equation 8.9 is used as a regularization functional $\Re(\{a_{k,l}\})$. In image processing applications, this feature might be an advantage of rejecting filters before empirical Wiener filters that minimize restoration MSE by the expense of distorting all image components, distortions being heavier the lower the SNR for those components.

From Equation 8.27, it follows that, having used for the filter design estimates of true signal power spectrum given by Equation 8.26, empirical Wiener filters produce signals with power spectrum $[(|\alpha_r|^2)^2 |\beta_r|^2/(|\beta_r|^2)^2; \ 0]$, which deviates from those estimates. Using image power spectrum preservation as a criterion for signal restoration (instead of MMSE-criterion), one can arrive at scalar filters with coefficients

$$\eta_{r,s}^{(SPF)} = \max\left[\frac{|\beta_{r,s}|^2 - \sigma_n^2}{|\beta_{r,s}|^2}; \quad 0\right]^{1/2}. \tag{8.30}$$

We call them *spectrum preservation filters*. Modifications of input image spectra carried out by rejecting and spectrum preservation filters are sometimes called *hard thresholding* and *soft thresholding*, respectively.

The described filters proved to be very efficient in image denoising if signal and/or noise spectra are well separated in the transform domain. A typical example of such a situation is filtering of narrow band noise, whose spectrum has only a few components in the transform domain. Figures 8.2 through 8.4 provide examples of filtering such a narrow band noise.

Figure 8.2 demonstrates filtering "moiré" noise patterns in an image. Such interferences frequently appear, in particular, in images digitized by optical scanners. The image in Figure 8.2a is an example of an image contaminated with "moiré" noise, and the plot in Figure 8.2b (solid line) shows an averaged DFT power spectrum of input image rows. One can clearly see anomalous peaks of the noise spectrum in the input image power spectrum. These peaks can be quite easily detected in the averaged power spectrum, for example, by means of checking local violations of spectrum monotonic decay with higher frequencies, as it is implemented in the program demoire1_CRC.m (see Exercises) used to generate this figure. Obtained in this way, the locations of detected peaks are used for zeroing values of the spectral coefficients in those points. The obtained rejecting filter applied row-wise to spectra of noisy image rows generates output-filtered image shown in Figure 8.2c. The difference between input and output images shown in Figure 8.2d presents the noise component eliminated in the output image. Note that this type of interference can also be successfully filtered out in DCT and Walsh transform domains as well.

Figure 8.3 illustrates prefiltering of the off-axis hologram of Figure 5.18a. The DFT spectrum of the hologram (Figure 8.3b) indicates that the hologram contains a periodical noise component, which exhibits itself in horizontal and vertical stripes, and an intensive parasitic zero-order diffraction component (bright central spot in the spectrum). The periodical noise can be attributed to imperfect analog-to-digital conversion in the digital camera used for hologram recording; dc and parasitic low-frequency components result from recording intensities of object and reference beams (last two terms in Equation 5.1). Both periodical and zero-order diffraction terms, which impair image reconstructed from the hologram, as one can see in Figure 8.3e, can be substantially weakened by rejecting filtering the hologram by a filter with frequency response shown in Figure 8.3c. As a result, a noticeable improvement of quality of the reconstructed image shown in Figure 8.3f is achieved.

Figure 8.4 illustrates a cleaning image from *banding noise*. This type of noise is a frequent distortion in imaging systems that use mechanical scanning for generating images. Banding noise exhibits itself in chaotic behavior of mean values of image signal in the direction of scan, that is, in the direction of image rows. Figure 8.4a shows an example of an image distorted by banging noise. Row-wise image mean values shown as a function of the row number in Figure 8.4b are image Radon transform coefficients in this direction. Normally, row-wise mean values are quite smooth functions of the row index. One can estimate the smooth component of this function, that is, "true" values of Radon transform spectral coefficients, by means of one or another smoothing operation, for instance, by means of computing spectrum local mean or local median in sliding window (for details of local mean and local median, see the section "Filter Classification Tables and Particular Examples"). A result of such a smoothing is shown by a bold line in Figure 8.4b. The difference between the initial and smoothed curves shown in Figure 8.4d gives estimates of banding noise interferences for each image row. Subtracting these differences from all pixels in the corresponding row eliminates the noise as one can see in Figure 8.4c.

In terms of the discussed filters, this can be treated as image Radon spectrum preservation filtering. Program filtering_stripes_CRC.m provided in Exercises implements the described filtering.

The denoising capability of Wiener filtering of uncorrelated (white) noise is much lower. When filtering white noise, Wiener filter tends to weaken low-energy image spectral components. In the case of filtering in DFT or in DCT domains, these are usually image high-frequency coefficients, which are, as we saw in section "Object Localization and Edge Detection: Selection of Reference Objects for Target Tracking" (Chapter 7), very important for object detection and recognition. In addition, Wiener filtering converts input white noise into a residual output correlated noise although with a lower variance.

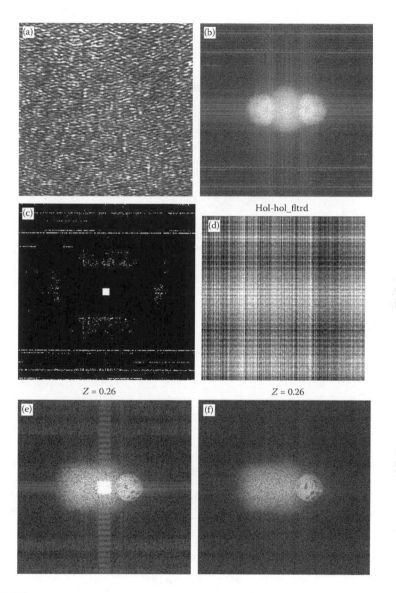

FIGURE 8.3

Cleaning periodic interferences and zero-order diffraction in Fresnel holograms before their numerical reconstruction: (a) a digitized hologram; (b) DFT spectrum of the hologram centered at x–y zero frequencies and enhanced for display purposes; spectrum reveals the presence of quasiperiodic interferences (horizontal and vertical stripes) and substantial zero-order diffraction term (bright spot at zero frequencies); (c) pattern of 2D rejecting filter discrete frequency response coefficients (black for 1 and white for 0) used for hologram prefiltering; (d) difference between initial hologram and filtered one demonstrating noise pattern removed by the filtering (image is enhanced for display purposes); (e) image reconstructed from the initial hologram; (f) image reconstructed from the filtered hologram, which shows removal of interferences and zero-order diffraction term seen in figure (e).

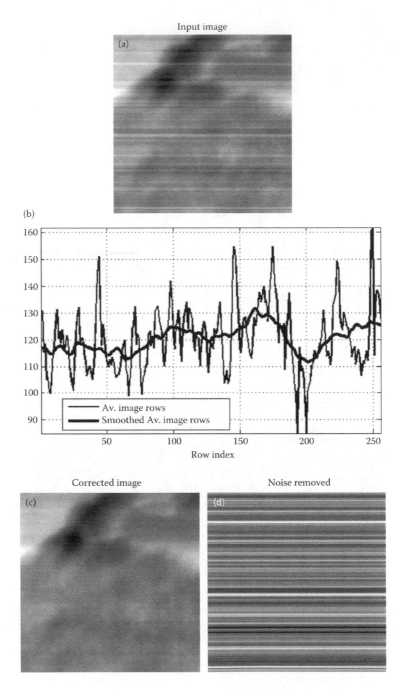

FIGURE 8.4
Filtering banding noise: (a) initial noisy image; (b) plots of row-wise image mean values (solid line) and of a result of their smoothing by local averaging in sliding window (bold line); (c) filtered image cleaned from noise; (d) removed noise component (enhanced for display purposes).

It is well known that human vision is much more sensitive to correlated noise than to white noise of the same intensity. Therefore, Wiener filtering for image denoising may even worsen image visual quality rather than improving it.

A good method for evaluating image restoration capability of filtering is its computer simulation. In the simulation, one can implement the ideal Wiener filter because both signal and noise spectra are known. In this way, one can determine potentials of the Wiener filtering and its real capability in its implementation as an empirical Wiener filter.

Figures 8.5 through 8.7 show the results of computer simulation of Wiener filtering of two test images with additive white noise using the program WienerFilter_CRC.m for two noise levels (PSNR = 3 and PSNR = 2), respectively. The empirical Wiener filter was implemented in this example according to Equation 8.27. As one can see from the figures, ideal Wiener filtering does improve image quality, when noise level is not too high. For the empirical Wiener filter, improvement is much less appreciable. In particular, the empirical Wiener filter is incapable of restoring the readability of text for PSNR = 2 (noise standard deviation is 128 in the range 0–255). The difference images between initial noisy images and the results of the Wiener filtering (restoration error) clearly reveals that the filtering tends to destroy image edges.

As we have seen, the noise-suppressing capability of scalar empirical Wiener filters depends on appropriate selection of transform domain, where image and noise spectra would be separated to the highest possible degree. For highly correlated noise, a transform in which noise spectrum is concentrated in a few components should be selected. For uncorrelated (white) noise, this separation is much less possible as the spectrum of uncorrelated noise in any transform remains to be uniform, and certain separation signal and noise are only possible due to the energy compaction property of the transform with respect to image signals. As we mentioned, DCT transform is one of the most promising in this respect. Another option is using wavelet transforms, some of which also exhibit good energy compaction property. A version of the rejecting and spectrum preservation filters that are implemented with wavelet transform image decomposition is known as *wavelet shrinkage* filtering by hard and soft thresholding, respectively. Figure 8.8 shows a flow diagram of signal denoising by means of the wavelet shrinkage.

Empirical Wiener Filters for Image Deblurring

Consider now an imaging system model in which the following relationship holds for transform domain spectra $\{\beta_{r,s}\}$, $\{\alpha_{r,s}\}$, and $\{v_{r,s}\}$ of input signal, ideal signal, and noise:

$$\beta_{r,s} = \lambda_{r,s}\alpha_{r,s} + v_{r,s}, \tag{8.31}$$

where $\{\lambda_r\}$ is a set of coefficients that specify the imaging systems in the transform domain. For DFT, coefficients $\{\lambda_r\}$ are samples of the imaging system

(a)   Noisy image; PSNR = 3

Image recovery and, more generally, sign
oblems that are among the most fundam
ey involve every known scale—from the
termining the structure of unresolved sta
ent of the tiniest molecules. Stated in its
covery problem is described like this. Given
at produced g. Unfortunately, when stated
ore can be said. How is g related to f? Is g i
t, can g be used to furnish an estimate f of f
If g is corrupted by noise, does the noise pl
s, can we ameliorate the effects of the nois
ice radical changes in f? Even if g uniqu
gorithm for computing f from g? What abo
n? Can it be usefully incorporated in our
These (and others) are the kinds of quest

(b)   Ideal Wiener filter denoising

Image recovery and, more generally, sign
oblems that are among the most fundam
ey involve every known scale—from the
termining the structure of unresolved star
ent of the tiniest molecules. Stated in its
covery problem is described like this: Given
at produced g. Unfortunately, when stated
ore can be said. How is g related to f? Is g i
t, can g be used to furnish an estimate f of f
If g is corrupted by noise, does the noise pl
s, can we ameliorate the effects of the nois
ice radical changes in f? Even if g uniqu
gorithm for computing f from g? What abo
n? Can it be usefully incorporated in our
These (and others) are the kinds of quest

(c)
Ideal Wiener filter restoration error, std = 45.3

(d)
Empirical Wiener filter denoising

Image recovery and, more generally, sign
oblems that are among the most fundam
ey involve every known scale—from the
ermining the structure of unresolved sta
ent of the tiniest molecules. Stated in its
covery problem is described like this: Given
at produced g. Unfortunately, when stated
ore can be said. How is g related to f? Is g i
t, can g be used to furnish an estimate f of f
If g is corrupted by noise, does the noise pl
s, can we ameliorate the effects of the nois
ice radical changes in f? Even if g uniqu
gorithm for computing f from g? What abo
n? Can it be usefully incorporated in our
These (and others) are the kinds of quest

(e) Empirical Wiener filter restoration error, std = 66.4

FIGURE 8.5
Denoising of a test image (a) with PSNR = 3 (noise standard deviation 84.7 in the range 0–255)
by the ideal (b) and empirical (d) Wiener filters; (c) and (e) are corresponding restoration errors.

(a)   Noisy image; PSNR = 2   (b)   Ideal Wiener filter

(c)   (d)
Ideal Wiener filter restoration error, std = 69.8   Empirical Wiener filter; Gamma = 1

(e) Empirical Wiener filter restoration error, std = 86

FIGURE 8.6
Denoising of a test image (a) with PSNR = 2 (noise standard deviation 128 in the range 0–255) by the ideal (b) and empirical (d) Wiener filters; (c) and (e) are corresponding restoration errors.

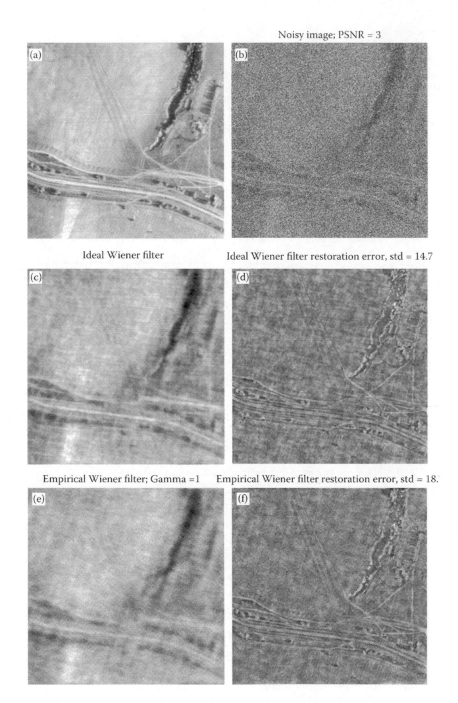

FIGURE 8.7
Denoising of a test noisy image (b) with PSNR = 3 (noise standard deviation 84.7 in the range 0–255) by the ideal (c) and empirical (e) Wiener filters. Noiseless image is shown for comparison in (a); (d) and (f) are corresponding restoration errors.

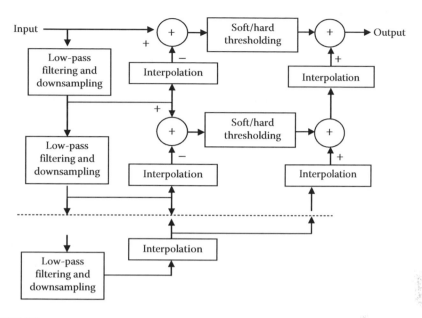

FIGURE 8.8
Wavelet shrinkage: signal denoising in wavelet transform domain. Units "soft/hard" thresholding implement signal spectrum preservation and rejecting filters of Equations 8.30 and 8.29, respectively.

frequency response. In ideal imaging systems, they should all be equal to unity. In reality, they usually decay with the frequency index r, which results, in particular, in image blur. Processing aimed at correcting this type of distortion is frequently referred to as *image deblurring*.

For this model, terms involved in the general Equation 8.16 for MMSE-optimal scalar Wiener filters are

$$AV_{\Omega_A} AV_{\Omega_N} \left(\alpha_{r,s} \beta_{r,s}^*\right) = AV_{\Omega_A} AV_{\Omega_N} \left[\alpha_{r,s} \left(\lambda_{r,s}^* \alpha_{r,s}^* + \nu_{r,s}^*\right)\right] = \lambda_{r,s}^* AV_{\Omega_A}\left(|\alpha_{r,s}|^2\right)$$

(8.32)

and

$$AV_{\Omega_A} AV_{\Omega_N} \left(|\beta_{r,s}|^2\right) = AV_{\Omega_A} AV_{\Omega_N} \left[\left(\lambda_{r,s} \alpha_{r,s} + \nu_{r,s}\right)\left(\lambda_{r,s}^* \alpha_{r,s}^* + \nu_{r,s}^*\right)\right]$$

$$= |\lambda_{r,s}|^2 AV_{\Omega_A}\left(|\alpha_{r,s}|^2\right) + AV_{\Omega_A}\left(|\nu_{r,s}|^2\right)$$

(8.33)

because for zero mean noise, $AV_{\Omega_N}(\nu_r^*) = AV_{\Omega_N}(\nu_r) = 0$. Therefore, coefficients $\left\{\eta_r^{(WF)}\right\}$ of scalar Wiener filter for restoration of blurred and noisy image are defined as

$$\eta_r^{(WF)} = \frac{\lambda_{r,s}^* AV_{\Omega_A}\left(\left|\alpha_r\right|^2\right)}{\left|\lambda_{r,s}\right|^2 AV_{\Omega_A}\left(\left|\alpha_r\right|^2\right) + AV_{\Omega_N}\left(\left|v_r\right|^2\right)} = \frac{1}{\lambda_{r,s}}\frac{SNR_{r,s}}{1 + SNR_{r,s}}, \quad (8.34)$$

where *SNR* is the signal-to-noise ratio at the output of the linear filter unit of the imaging system model:

$$SNR_{r,s} = \frac{\left|\lambda_{r,s}\right|^2 AV_{\Omega_A}\left(\left|\alpha_{r,s}\right|^2\right)}{AV_{\Omega_N}\left(\left|v_{r,s}\right|^2\right)}. \quad (8.35)$$

Correspondingly, the general deblurring empirical Wiener filter, deblurring empirical Wiener filter, and the rejecting filter for the case of white noise will be defined as follows:

$$\eta_r = \max\left[\frac{1}{\lambda_{r,s}}\frac{\overline{\left|\beta_{r,s}\right|^2} - AV_{\Omega_A}\left(\left|v_{r,s}\right|^2\right)}{\left|\beta_{r,s}\right|^2}; \quad 0\right], \quad (8.36)$$

$$\eta_r = \max\left[\frac{1}{\lambda_{r,s}}\frac{\overline{\left|\beta_{r,s}\right|^2} - \sigma_n^2}{\left|\beta_{r,s}\right|^2}; \quad 0\right], \quad (8.37)$$

and

$$\eta_r = \begin{cases} \dfrac{1}{\lambda_{r,s}}, & \text{if}\left|\beta_{r,s}\right|^2 \geq Thr \\ 0, & \text{otherwise} \end{cases}. \quad (8.38)$$

All these filters can be treated as two filters in a cascade: the filter with coefficients

$$\eta_r^{inv} = \frac{1}{\lambda_r} \quad (8.39)$$

called the *inverse filter* and signal denoising filters described by Equations 8.27 through 8.29. Inverse filter compensates for the distortions of signal frequency components in the imaging system while the denoising filter prevents from excessive amplification of noise and performs a sort of *regularization* of the inverse filter.

Figure 8.9a–h illustrates the image deblurring capability of the described scalar Wiener restoration filters.

The figure shows blurred test images with different noise levels (noise standard deviations 6.4, 25.6, and 85.3 in units of image grayscale range 0–255) and results of their deblurring using ideal Wiener filters, which characterize potentials of Wiener filtering, and empirical Wiener filters, which demonstrate the real deblurring capability of Wiener filtering. One can see from the figures that when the noise level in blurred images is relatively small, Wiener filters are capable of restoring text readability, and when the noise level increases, the deblurring capability of Wiener filtering deteriorates down to full incapability.

One of the most immediate applications of Wiener filtering is correcting distortions caused by the finite size of apertures of image sensors, image sampling devices, and image displays. We will refer to this processing as *aperture correction*.

Let an image sensor and sampling device be an array of light-sensitive elements with a square aperture of size $dx = dy = d^{(s)}$ (Figure 3.3) arranged over a square sampling grid with sampling interval Δx. Then, the frequency response of the individual sensor elements is

$$H(f_x, f_y) = \int_{-d^{(s)}/2}^{d^{(s)}/2} \exp(i2\pi f_x x)dx \int_{-d^{(s)}/2}^{d^{(s)}/2} \exp(i2\pi f_y y)dx$$

$$= \frac{\sin(\pi f_x d^{(s)}/2)}{\pi f_x d^{(s)}/2} \frac{\sin(\pi f_y d^{(s)})}{\pi f_y d^{(s)}} = \mathrm{sinc}(\pi f_x d^{(s)})\mathrm{sinc}(\pi f_x d^{(s)}) \quad (8.40)$$

or

$$H(f_x, f_y) = \mathrm{sinc}\left(\pi f_x \Delta x \overline{d}^{(s)}\right)\mathrm{sinc}\left(\pi f_x \Delta x \overline{d}^{(s)}\right), \quad (8.41)$$

where $\overline{d}^{(s)} = d^{(s)}/\Delta x$ is the sensor's fill factor. The discrete frequency response $\{\lambda_{r,s}\}$ of the sensor can be obtained as samples of $H(f_x, f_y)$ taken in frequency-domain sampling points $\{f_x = r\Delta f, f_y = s\Delta f\}$ with sampling interval Δf:

$$\lambda_{r,s} = \lambda_r \lambda_s = \mathrm{sinc}\left(\pi \overline{d}^{(s)} r/N\right) \cdot \mathrm{sinc}\left(\pi \overline{d}^{(s)} s/N\right), \quad (8.42)$$

where $N = 1/\Delta x \Delta f$ is the number of samples in both coordinates (assuming the cardinal sampling).

If the image display device also has a square aperture of size $d^{(r)} \times d^{(r)}$, the entire imaging system discrete frequency response is

$$\lambda_{r,s} = \lambda_r \lambda_s = \mathrm{sinc}\left(\pi \overline{d}^{(s)} r / N_x\right) \cdot \mathrm{sinc}\left(\pi \overline{d}^{(r)} r / N_x\right)\mathrm{sinc}\left(\pi \overline{d}^{(s)} s / N_y\right)\mathrm{sinc}\left(\pi \overline{d}^{(r)} s / N_y\right).$$

$$(8.43)$$

(a)
Bl and noisy image, BIF act = 0.2; PSNR = 40

(b)
Ideal Wiener filter

(c)
Empirical Wiener filter; Gamma = 2

(d)
Bl and noisy image, BIF act = 0.2; PSNR = 10

(e)
Ideal Wiener filter

(f)
Empirical Wiener filter; Gamma = 2

FIGURE 8.9

Image deblurring using scalar Wiener filtering for different noise levels: (a), (d), (g) blurred images with additive white noise of standard deviation 6.4, 25.6, and 85.3, respectively (in units of image gray level range 0–255); (b), (e), (h) images restored by ideal Wiener filters; and (c), (f), (i) corresponding images restored by empirical Wiener filters.

(g) (h)

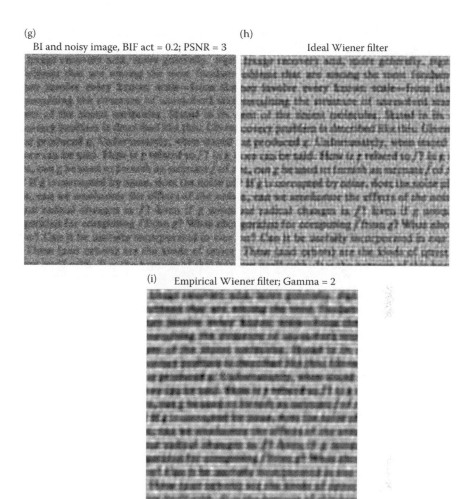

BI and noisy image, BIF act = 0.2; PSNR = 3

Ideal Wiener filter

(i)   Empirical Wiener filter; Gamma = 2

FIGURE 8.9
Continued.

Parameters $\overline{d}^{(s)}$, $d^{(r)}$, and Δx are imaging system design parameters that are usually specified in imaging system certificates. They can be used for aperture correction images in the computer before their display.

Figure 8.10 illustrates such an image aperture correction of a test air photograph. The middle image in this figure is obtained by applying to the left image an inverse filter for the system's frequency response defined by Equation 8.42 with $\overline{d}^{(s)} = \overline{d}^{(r)} = 1$ and the right image shows the correction term: the difference between initial and aperture-corrected images. The images were generated using the program invapert_CRC.m provided in Exercises.

FIGURE 8.10
Initial (left), aperture corrected (middle) images, and their difference (right).

Sliding Window Transform Domain Adaptive Image Restoration

Local Adaptive Filtering

For implementing image restoration using Wiener filtering, one should select a fashion in which images are to be processed. Given an image to be processed, filtering can be designed and carried out either over the entire available image frame or fragment-wise. We will refer to the former as *global filtering* and to the latter as *local filtering*.

The applicability of global and local filtering depends on whether images belong to classes of spatial homogeneous or spatial inhomogeneous images, which we mentioned in the section "Local Adaptive SCR-Optimal Correlators." As far as the design of empirical Wiener filters is concerned, spatial inhomogeneity of images in terms of their spectra is the issue. In the spatial homogeneous image, the local spectra do not vary substantially; otherwise, images are inhomogeneous. Figure 8.11 illustrates this, in addition to Figure 7.13.

Global filtering is suited for spatial homogeneous images. For spatial inhomogeneous filtering, local filtering is required.

The importance of local processing is also confirmed by the evolution of vision, which has resulted in an image analysis mechanism that works locally with relatively small fragments of the field of view. Human visual acuity is very uneven over the field of view. The field of view of the human vision is about 30°. The maximal resolving power of the vision is about one angular minute. However, such a relatively high resolving power is concentrated only within a small fraction of the field of view that has an angular size of about 2°. When viewing the image, the eye optical axis permanently jumps, apparently under the control of peripheral vision, over the field of view, which directs the high-resolution part of its light-sensitive retina called *fovea* to different places in the scene. Therefore, the brain analyzes visual scenes locally within the area of acute vision, which is about 1/15 the field of

(a)   Input image   (b)   Abs(Window DCT spectra).^0.5

FIGURE 8.11
An example of spatial inhomogeneous image of 256 × 256 pixels (a) separated in fragments of 16 × 16 pixels and DCT spectra of the corresponding fragments (b).

view (for images of 512 × 512 pixels, this means a window of roughly 33 × 33 pixels).

The theoretical framework for local filtering is provided by the *local criteria* of processing quality that evaluate the processing quality individually over image fragments centered at each pixel. In particular, for optimal local MMSE scalar filters acting in a rectangular window centered at pixel {k,l}, filter coefficients $\left\{\eta_r^{(k,l)}\right\}$ are, according to the local criteria, defined by the following modification of Equation 8.15:

$$\hat{\alpha}_r^{(k)} = \arg\min_{\{\eta_{r,s}\}}\left\{AV_{\Omega_A}AV_{\Omega_N}\left(\sum_{r=0}^{W_x-1}\sum_{s=0}^{W_y-1}\left|\alpha_{r,s}^{(k,l)} - \eta_{r,s}^{(k,l)}\beta_{r,s}^{(k,l)}\right|^2\right)\right\}, \tag{8.44}$$

where W_x and W_y are dimensions of the area, over which filtering performance is evaluated, and, respectively, of the filter window, $\left\{\beta_{r,s}^{(k,l)}\right\}$ are spectral coefficients of the input image samples within the window, and $\left\{\alpha_{r,s}^{(k,l)}\right\}$ are spectral coefficients of filter window samples of a hypothetical true image. Solutions obtained in the sections "Empirical Wiener Filters for Image Denoising" and "Empirical Wiener Filters for Image Deblurring" for empirical Wiener filters for image restoration can be straightforwardly extended to local filtering by replacing in Equations 8.37 through 8.30 and 8.34 through 8.38 signal spectra with corresponding spectra of the image window samples. Local empirical Wiener filtering is local adaptive because filter coefficients depend on local power spectra of image fragments.

The most straightforward way to implement local filtering is to perform it in a "hopping" window. *Hopping window processing* being very attractive

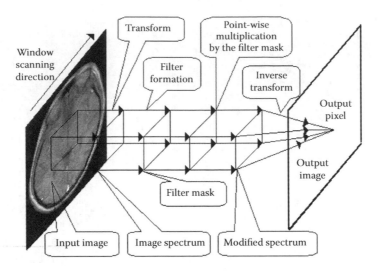

FIGURE 8.12
The principle of local adaptive filtering in a sliding window.

from the computational complexity point of view suffers, however, from *blocking effects*, artifacts in a form of discontinuities at the edges of the hopping window. Obviously, the ultimate solution of the "blocking effects" problem would be *sliding window processing*.

In the sliding window filtering, one should, for each position of the filter window, compute transform coefficients of the signal within the window, determine on this base the filter coefficients, modify the image transform coefficients accordingly, and then compute the inverse transform. With sliding window, inverse transform need not, in principle, be computed for all pixels within the window since only the central sample of the window has to be determined in order to form the output image in the process of image scanning with the filter window. Figure 8.12 shows the schematic flow diagram of this process.

Sliding Window DCT Transform Domain Filtering

The selection of orthogonal transforms for the implementation of the filters is governed by

- The required accuracy of approximation of general linear filtering with scalar filtering
- The availability of *a priori* knowledge regarding image spectra in the chosen base
- The accuracy of the empirical spectrum estimation from the observed data required for the adaptive filter design
- The computational complexity of the filter implementation

Among all known transforms, DCT proved to be one of the most appropriate ones for sliding window transform domain filtering. DCT exhibits good energy compaction capability, which is a key feature for the efficiency of filters. Being advantageous to DFT in terms of energy compaction capability, DCT can also be regarded as a good substitute for DFT in signal/image restoration tasks with imaging system specification in terms of their frequency responses. DCT is suitable for multicomponent signal/image processing as well. The use of DCT in sliding window has low computational complexity owing to the existence of recursive algorithms for computing DCT in sliding windows. In addition, note that, if the window size is an odd number $(2N_w + 1)$, the inverse DCT transform of local spectrum $\{\beta_r^{(k)}\}$ for computing window central pixel a_k involves only signal spectrum coefficients $\{\beta_{2s}^{(k)}\}$ with even indices (see Appendix)

$$a_k = \frac{1}{\sqrt{2N_w + 1}} \left[\beta_0^{(k)} + 2 \sum_{s=1}^{N_w} \beta_{2s}^{(k)} (-1)^s \right]. \qquad (8.45)$$

Therefore, only those spectral coefficients have to be computed, and the computational complexity of sliding window filtering in DCT domain is $O(N_w + 1)$ operations for 1D filtering in a window of $2N_w + 1$ samples and $O[(N_{w1} + 1)(N_{w2} + 1)]$ operations for 2D filtering in a rectangular window of $(2N_{w1} + 1) \times (2N_{w2} + 1)$ pixels.

Figure 8.13, generated using the program demo_lcdct2_CRC.m provided in Exercises, illustrates local adaptive rejecting filtering for denoising a piecewise constant test image. Image (a) is the test image contaminated by additive noise with standard deviation of 25 gray levels. Image (b) is the output image filtered with filter window of 7×7 pixels. Image (c) shows difference between input noisy and output images. It is the input image component that was filtered out. Its standard deviation is approximately equal to the noise standard deviation. Image (d) presents a map of the rejecting filters "transparence," for each window position; that is, the ratio of the number of ones in the array of the filter coefficients to the window size. Note that the latter is also the sparsity of output image local spectra. On average, the sparsity of output image local spectra for this particular test image and noise intensity turned out to be 0.157, that is, on an average, only 12–13 of 49 spectral coefficients of 7×7 pixel image fragments are left after the rejecting filtering.

One can see from this image that when the filter window seats inside of an image patch, where the image gray values are constant, the filter is almost completely opaque and preserves only local dc (local mean) image component. On the contrary, on the boundaries of the patches, the filter is almost completely transparent, thus preserving the edges of the patches at the expense of lower noise suppression. This feature of the local adaptive filters can be even better appreciated from magnified fragments of images (a),

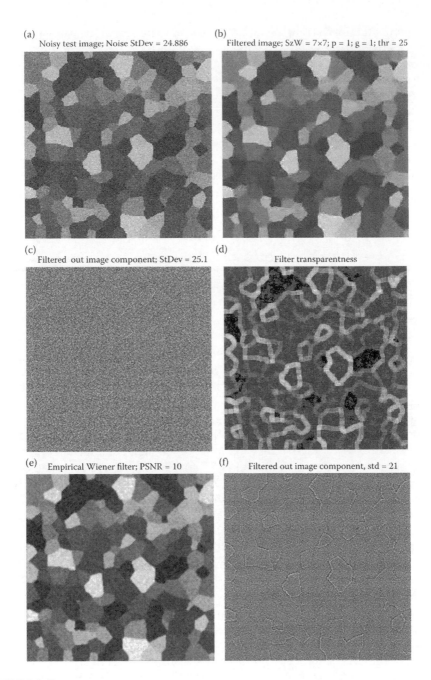

FIGURE 8.13
Local and global empirical Wiener filtering: (a) noisy test image; (b) result of local filtering in window 7×7 pixels; (c) difference between input noisy (a) and filtered (b) images; (d) map of local filter "transparentness"; (e) result of global filtering; (f) difference between input noisy (a) and filtered (e) images. Note image blur and residual noise correlation after global filtering.

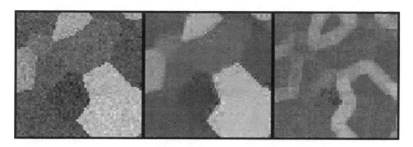

FIGURE 8.14
Magnified fragments of images of Figure 8.13a, b, and d respectively.

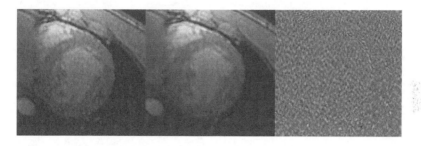

FIGURE 8.15
Denoising of a real MRI image by means of local adaptive filtering in DCT domain. Left to right: initial noisy image, filtered image, and difference between initial and filtered images.

(b), and (d) shown in Figure 8.14 and from results of global filtering shown in Figure 8.13e and f. The edge-preserving capability is an important advantage of local adaptive filtering compared to global filtering.

An example of the local adaptive DCT domain denoising of a real MRI image is shown in Figure 8.15. The left image in this figure is the initial image; the central image is the result of the filtering; and the right image is a difference image between initial and filtered ones. From the difference image, one can see that it looks almost chaotic and does not contain any visible details of the initial image.

Local adaptive filtering is also very well suited for denoising images with signal-dependent noise such as speckle noise. The characteristic feature of the speckle noise is that its standard deviation is proportional to the signal (see the section "Speckle Noise Model" in Chapter 2). For filtering speckle noise using local adaptive filters, one should set parameters σ_n and *Thr* of the filters of Equations 8.27 through 8.29 be proportional to the image local dc component $\beta_0^{(k,l)}$. Figure 8.16 shows an example of such a denoising of an ultrasound image.

Although the above discussion was limited to 2D filters, their extension to the 3D case of RGB images or video sequences is straightforward: 2D indices in all equations in the sections "Optimal Linear Filters for Image Restoration" and "Sliding Window Transform Domain Adaptive Image

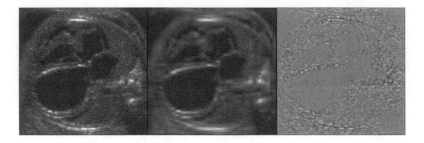

FIGURE 8.16
Local adaptive denoising of a speckled US image: left—initial image; middle—filtered image; right—difference between initial and filtered images. Image size is 256×256; filter window size is 25×25 pixels.

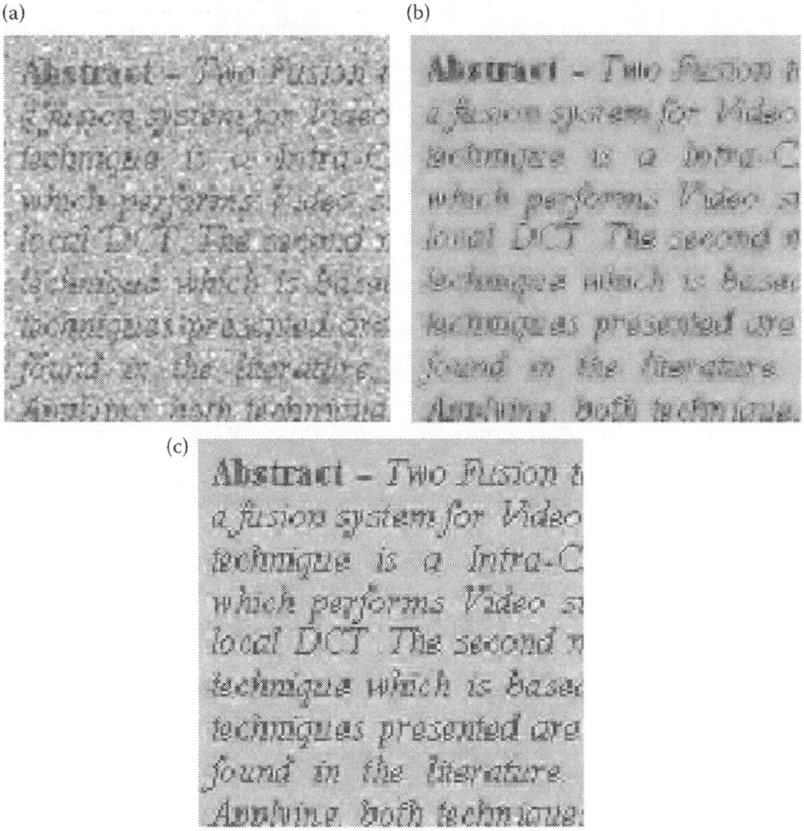

(a)

(b)

(c)

FIGURE 8.17
2D and 3D local adaptive filtering of a simulated video sequence: (a) one of the noisy frames of a test video sequence (image size 256×256 pixels); (b) a result of 2D local adaptive empirical Wiener filtering (filter window 5×5 pixels); (c) a result of 3D local adaptive empirical Wiener filtering (filter window 5×5 pixels $\times$ 5 frames).

FIGURE 8.18
3D local adaptive empirical Wiener filtering for denoising and deblurring of a real thermal video sequence: (a) a frame of initial video sequence; (b) a frame of restored video sequence. Filter window is 5×5 pixel $\times 5$ frames; image size 512×512 pixels.

Restoration" should be replaced by 3D ones (two for special coordinates and the third for channel index for RGB color images or frame index for video sequences) and 2D transforms should be replaced by the corresponding 3D transform. Figures 8.17 and 8.18 demonstrate the efficiency of 3D local adaptive empirical Wiener filtering in restoration (denoising and deblurring) of video sequences.

Hybrid DCT/Wavelet Filtering

An important practical issue in using local adaptive filters is selecting the filter window size. This selection is governed by the typical size of image details that should be preserved in the filtering. In general, optimal window size depends on the image and may vary even within the image. This requires filtering in multiple windows with an appropriate combination of the filtering results at each window position (for methods of combining, see the section "Multicomponent Image Restoration and Data Fusion").

One can avoid filtering in multiple windows by making use of the multi-resolution property of the wavelet image decomposition in the *hybrid filtering method*. In the hybrid filtering, hard or soft thresholding in each scale level of the wavelet filtering shown in a flow diagram of Figure 8.8 is replaced by the above-described local adaptive filtering with a fixed window size. Owing to wavelet transform multiscale image representation, this replacement imitates parallel local adaptive filtering of the input image with a set of windows of different sizes accordingly to the scales selected. The flow diagram of the hybrid filtering is shown in Figure 8.19.

As we mentioned, DCT transform is almost always the most appropriate for local adaptive filtering. For high-resolution images, DCT filtering in

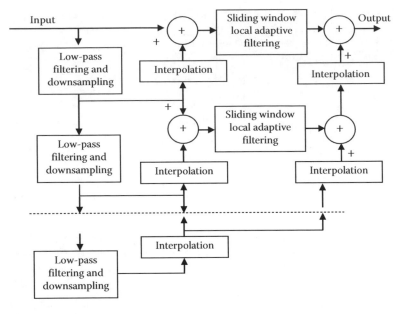

FIGURE 8.19
Flow diagram of the image hybrid denoising with wavelet decomposition and local adaptive of the decomposition components.

sliding window requires a window size of the smallest size of 3×3 pixels; otherwise, filtering may result in loss of tiny image details. This requirement limits the noise suppression capability of the local adaptive filtering that increases with window size. In the hybrid filtering, when the window 3×3 in the first scale is used, in scale 2, the effective window size is 6×6 pixels, in scale 3, it is 12×12, and so on.

In DCT filtering in the window of 3×3 pixels, only the following four basis functions of the DCT transform are involved:

$$\phi_{0,0}^{(DCT)} = \begin{bmatrix} 1 & 1 & 1 \\ 1 & 1 & 1 \\ 1 & 1 & 1 \end{bmatrix}; \quad \phi_{0,2}^{(DCT)} = \frac{\sqrt{2}}{2} \begin{bmatrix} -1 & -1 & -1 \\ +2 & +2 & +2 \\ -1 & -1 & -1 \end{bmatrix};$$

$$\phi_{2,0}^{(DCT)} = \frac{\sqrt{2}}{2} \begin{bmatrix} -1 & +2 & -1 \\ -1 & +2 & -1 \\ -1 & +2 & -1 \end{bmatrix}; \quad \phi_{2,2}^{(DCT)} = \frac{1}{2} \begin{bmatrix} -1 & -2 & -1 \\ -2 & +4 & -2 \\ -1 & -2 & -1 \end{bmatrix}. \quad (8.46)$$

The second and the third functions represent what is called vertical and horizontal *Laplacian operators*. The last function can be decomposed into a sum of diagonal Laplacians:

$$\phi_{2,2} = \frac{1}{2}\begin{bmatrix} -1 & -2 & -1 \\ -2 & +4 & -2 \\ -1 & -2 & -1 \end{bmatrix} = \frac{1}{2}\left\{ \begin{bmatrix} +2 & -1 & -1 \\ -1 & +2 & -1 \\ -1 & -1 & +2 \end{bmatrix} + \begin{bmatrix} -1 & -1 & +2 \\ -1 & +2 & -1 \\ +2 & -1 & -1 \end{bmatrix} \right\}. \quad (8.47)$$

Therefore, DCT filtering in the 3×3 window can be replaced by filtering in the domain of four directional Laplacians (not counting the basis function $\phi_{0,0}^{(DCT)}$, which is responsible for image local dc component (local mean)). Experimental experience proves that, for high-resolution images, such an implementation is advantageous to simple DCT since it produces less filtering artifacts. Its use in hybrid filtering promises additional advantages because it is equivalent to a corresponding increase in the number of effective basis functions.

In conclusion, note that, as we mentioned it in the section "Transforms in Sliding Window (Windowed Transforms) and Signal Subband Decomposition" in Chapter 2, local adaptive filters, wavelet shrinkage, and hybrid filtering methods can be treated as hard or soft thresholding of image subbands. The local adaptive filtering in the transform domain is filtering in subbands uniformly distributed in the base band with the number of the subbands equal to, in 1D case, (*Window Size* + 1)/2. Wavelet shrinkage is filtering in subbands arranged in the base band in a logarithmic scale with the number of subbands equal to the number of the resolution scales. Hybrid filtering combines logarithmic arrangement of subbands of wavelets with uniform arrangement of sub-subbands within wavelet subbands. It is interesting to note that this "logarithmic coarse-uniform fine" band arrangement resembles very much, from one hand, floating point representation of numbers in computers (logarithmic for order and uniform for mantissa) and, from the other hand, arrangement of tones and semitones in music (in Bach's equal-tempered scale, octaves are arranged in a logarithmic scale and 12 semitones are equally spaced within octaves).

Multicomponent Image Restoration and Data Fusion

The above theory of MSE-optimal scalar filtering can be extended to the restoration of multicomponent images such as color and multispectral images, video sequences, images of same objects generated by different imaging devices (*multimodality imaging*). Such a processing is called *data fusion*.

As an illustrative mathematical model, consider an M-channel imaging system, in which each of the M-component images can be described in the DFT domain by the model of Equation 8.31:

$$\left\{ \beta_r^{(m)} = \lambda_r^{(m)} \alpha_r^{(m)} + \nu_r^{(m)} \right\}, \quad (8.48)$$

where m is component index, $m = 1,2,\ldots,M$, $\left\{\beta_r^{(m)}\right\}$ and $\left\{\alpha_r^{(m)}\right\}$ are spectral coefficients of the observed and perfect image components, $\left\{\lambda_r^{(m)}\right\}$ are discrete frequency response coefficients of the system's m-th channel, and $\left\{v_r^{(m)}\right\}$ are spectral coefficients of additive signal-independent zero mean uncorrelated (white) noise with variance $\sigma_n^{2(m)}$ in the m-th channel, uncorrelated with noise components in other channels.

Suppose that image components are restored in the DFT domain by a linear combination of scalar filtered input image components:

$$\left\{\hat{\alpha}_r^{(m)} = \sum_{p=1}^{M} \eta_r^{(m,p)}\beta_r^{(p)}\right\}, \tag{8.49}$$

where $\left\{\beta_r^{(p)}\right\}$ and $\left\{\hat{\alpha}_r^{(m)}\right\}$ are DFT spectral coefficients of input and restored images in the p-th and m-th channel, correspondingly, and $\left\{\eta_r^{(m,l)}\right\}$ are coefficients of frequency response of a restoration scalar linear filter that generates a contribution to the m-th channel output image from the p-th channel.

From the MSE criterion

$$\hat{\alpha}_r^{(m)} = \underset{\left\{\eta_r^{(m,p)}\right\}}{\arg\min}\left\{AV_{\Omega_A}AV_{\Omega_N}\left(\sum_{r=0}^{N-1}\left|\alpha_r^{(m)} - \sum_{p=1}^{M}\eta_r^{(m,p)}\beta_r^{(p)}\right|^2\right)\right\}, \tag{8.50}$$

one can obtain the following system of M^2 equations for M^2 MMSE-optimal restoration multichannel filter coefficients $\left\{\eta_r^{(m,l)}\right\}$ (see Appendix):

$$\sum_{p=1}^{M} \eta_r^{(m,p)}AV_{\Omega_A}AV_{\Omega_N}\left(\beta_r^{(p)}\beta_r^{*(l)}\right) = AV_{\Omega_A}AV_{\Omega_N}\left(\hat{\alpha}_r^{(m)}\beta_r^{*(l)}\right). \tag{8.51}$$

Cross-correlations $AV_{\Omega_A}AV_{\Omega_N}\left(\beta_r^{(p)}\beta_r^{*(l)}\right)$ and $AV_{\Omega_A}AV_{\Omega_N}\left(\hat{\alpha}_r^{(m)}\beta_r^{*(l)}\right)$ are, according to Equation 8.48 (see Appendix),

$$AV_{\Omega_A}AV_{\Omega_N}\left(\beta_r^{(p)}\beta_r^{*(l)}\right) = \lambda_r^{(p)}\lambda_r^{*(l)}AV_{\Omega_A}\left(\alpha_r^{(p)}\alpha_r^{*(l)}\right) + \sigma_n^{2(p)}\delta(l - p) \tag{8.52}$$

$$AV_{\Omega_A}AV_{\Omega_N}\left(\hat{\alpha}_r^{(m)}\beta_r^{*(l)}\right) = \lambda_r^{*(l)}\,AV_{\Omega_A}\left(\hat{\alpha}_r^{(m)}\alpha_r^{*(l)}\right) \tag{8.53}$$

because channel noises are assumed to be zero mean and noncorrelated. Inserting Equations 8.52 and 8.53 into Equation 8.51, the following system of M^2 equations for multichannel filter coefficients $\left\{\eta_r^{(m,l)}\right\}$ is finally obtained:

$$\left\{ \sum_{p=1}^{M} \eta_r^{(m,p)} AV_{\Omega_A}\left(\tilde{\alpha}_r^{(p)}\tilde{\alpha}_r^{*(l)}\right) + \eta_r^{(m,l)}\sigma_n^{2(l)} = \frac{1}{\lambda_r^{(m)}} AV_{\Omega_A}\left(\tilde{\alpha}_r^{(m)}\tilde{\alpha}_r^{*(l)}\right) \right\}, \quad (8.54)$$

where

$$\left\{ \tilde{\alpha}_r^{(\cdot)} = \lambda_r^{(\cdot)}\alpha_r^{(\cdot)} \right\}. \quad (8.55)$$

In order to ease the analysis and interpretation of the solution and get an insight into how interchannel correlations and noise level determine channel contributions to the result of fusion, consider a special case of fusion of two channels ($l = 1,2$) into one ($m = 1,2$). In this case, we have a system of two pairs of equations for channel weight coefficients $\left\{ \eta_r^{(1,1)}, \eta_r^{(1,2)} \right\}$ and $\left\{ \eta_r^{(2,1)}, \eta_r^{(2,2)} \right\}$:

$$\begin{cases} AV_{\Omega_A}\left(\left|\tilde{\alpha}_r^{(1)}\right|^2\right)\eta_r^{(1,1)} + AV_{\Omega_A}\left(\tilde{\alpha}_r^{(2)}\tilde{\alpha}_r^{*(1)}\right)\eta_r^{(1,2)} + \sigma_n^{2(1)}\eta_r^{(1,1)} = \frac{1}{\lambda_r^{(1)}} AV_{\Omega_A}\left(\tilde{\alpha}_r^{(1)}\tilde{\alpha}_r^{*(1)}\right) \\[2mm] AV_{\Omega_A}\left(\tilde{\alpha}_r^{(1)}\tilde{\alpha}_r^{*(2)}\right)\eta_r^{(1,1)} + AV_{\Omega_A}\left(\left|\tilde{\alpha}_r^{(2)}\right|^2\right)\eta_r^{(1,2)} + \sigma_n^{2(2)}\eta_r^{(1,2)} = \frac{1}{\lambda_r^{(1)}} AV_{\Omega_A}\left(\tilde{\alpha}_r^{(1)}\tilde{\alpha}_r^{*(2)}\right) \end{cases}$$

$$\begin{cases} AV_{\Omega_A}\left(\left|\tilde{\alpha}_r^{(1)}\right|^2\right)\eta_r^{(2,1)} + AV_{\Omega_A}\left(\tilde{\alpha}_r^{(2)}\tilde{\alpha}_r^{*(1)}\right)\eta_r^{(2,2)} + \sigma_n^{2(1)}\eta_r^{(2,1)} = \frac{1}{\lambda_r^{(2)}} AV_{\Omega_A}\left(\tilde{\alpha}_r^{(2)}\tilde{\alpha}_r^{*(1)}\right) \\[2mm] AV_{\Omega_A}\left(\tilde{\alpha}_r^{(1)}\tilde{\alpha}_r^{*(2)}\right)\eta_r^{(2,1)} + AV_{\Omega_A}\left(\left|\tilde{\alpha}_r^{(2)}\right|^2\right)\eta_r^{(2,2)} + \sigma_n^{2(2)}\eta_r^{(2,2)} = \frac{1}{\lambda_r^{(2)}} AV_{\Omega_A}\left(\tilde{\alpha}_r^{(2)}\tilde{\alpha}_r^{*(2)}\right). \end{cases}$$

$$(8.56)$$

Consider in detail the first pair that can be rewritten as

$$\begin{aligned} \left(\overline{\left|\tilde{\alpha}_r^{(1)}\right|^2} + \sigma_n^{2(1)}\right)\eta_r^{(1,1)} + \overline{\tilde{\alpha}_r^{(2)}\tilde{\alpha}_r^{*(1)}}\eta_r^{(1,2)} &= \frac{1}{\lambda_r^{(1)}}\overline{\left|\tilde{\alpha}_r^{(1)}\right|^2} \\ \overline{\tilde{\alpha}_r^{(1)}\tilde{\alpha}_r^{*(2)}}\eta_r^{(1,1)} + \left(\overline{\left|\tilde{\alpha}_r^{(2)}\right|^2} + \sigma_n^{2(2)}\right)\eta_r^{(1,2)} &= \frac{1}{\lambda_r^{(1)}}\overline{\tilde{\alpha}_r^{(1)}\tilde{\alpha}_r^{*(2)}} \end{aligned}, \quad (8.57)$$

where $AV_{\Omega_A}(\cdot\cdot)$ is, for the sake of brevity, replaced by $\overline{(\cdot\cdot)}$.

Solutions of these equations are (see Appendix)

$$\eta_r^{(1,1)} = \frac{1}{\lambda_r^{(1)}} \frac{SNR_r^{(1)}\left[1 + \left(1 - \kappa_r^{(1,2)}\kappa_r^{(2,1)}\right)SNR_r^{(2)}\right]}{1 + SNR_r^{(1)} + SNR_r^{(2)} + \left(1 - \kappa_r^{(1,2)}\kappa_r^{(2,1)}\right)SNR_r^{(1)}SNR_r^{(2)}}$$

$$(8.58)$$

$$\eta_r^{(1,2)} = \frac{1}{\lambda_r^{(2)}} \frac{\kappa_r^{(2,1)}SNR_r^{(2)}}{1 + SNR_r^{(1)} + SNR_r^{(2)} + \left(1 - \kappa_r^{(1,2)}\kappa_r^{(2,1)}\right)SNR_r^{(1)}SNR_r^{(2)}},$$

where $\kappa_r^{(1,2)}$ and $\kappa_r^{(2,1)}$ are true signal interchannel cross-correlation coefficients

$$\kappa_r^{(1,2)} = \frac{\left|\overline{\left(\alpha_r^{(1)}\alpha_r^{*(2)}\right)}\right|}{\overline{\left|\alpha_r^{(1)}\right|^2}}, \quad \kappa_r^{(2,1)} = \frac{\left|\overline{\left(\alpha_r^{(1)}\hat{\alpha}_r^{*(2)}\right)}\right|}{\overline{\left|\alpha_r^{(2)}\right|^2}} \tag{8.59}$$

and

$$\left\{ SNR_r^{(m)} = \frac{AV_{\Omega_A}\left(\left|\tilde{\alpha}_r^{(m)}\right|^2\right)}{\sigma_r^{2(m)}} = \frac{\left|\lambda_r^{(m)}\right|^2 AV_{\Omega_A}\left(\left|\alpha_r^{(m)}\right|^2\right)}{\sigma_r^{2(m)}} \right\} \quad m = 1, 2 \tag{8.60}$$

are channel signal-to-noise ratios in image spectral components.

Equation 8.58 is an analog of Equation 8.34 for one-component image restoration and, similar to that, implies that each channel contributes to the fused output signal its inverse filtered signal weighted by a "regularization" factor that monotonically grows with the channel signal-to-noise ratio.

Two immediate important special cases of the described multicomponent image restoration are denoising of multiple images of the same object obtained with different sensors and superresolution from video.

If two input images are images of the same object that might be displaced with respect to each other, true signal interchannel cross-correlation coefficients are

$$\kappa_r^{(1,2)} = \kappa_r^{*(2,1)} = \exp(i2\pi k^{(1,2)}r), \quad \text{and} \quad \kappa_r^{(1,2)}\kappa_r^{*(2,1)} = 1, \tag{8.61}$$

where, according to the shift theorem for DFT, $k^{(1,2)}$ is spatial displacement between images in channels 1 and 2. Then the channel weight coefficients are

$$\eta_r^{(1,1)} = \frac{1}{\lambda_r^{(1)}} \frac{SNR_r^{(1)}}{1 + SNR_r^{(1)} + SNR_r^{(2)}}; \quad \eta_r^{(1,2)} = \frac{1}{\lambda_r^{(2)}} \frac{\kappa_r^{(2,1)}SNR_r^{(2)}}{1 + SNR_r^{(1)} + SNR_r^{(2)}}. \tag{8.62}$$

In general, Equation 8.54 is converted in this case to

$$\left\{ \left[\lambda_r^{*(l)} \sum_{p=1}^{M} \lambda_r^{(p)} \kappa_r^{(p,l)} \eta_r^{(m,p)} \right] \overline{\left|\alpha_r\right|^2} + \eta_r^{(m,l)}\sigma_n^{2(l)} = \lambda_r^{*(l)}\kappa_r^{(m,l)}\overline{\left|\alpha_r\right|^2} \right\} \tag{8.63}$$

or

$$\left\{ \sum_{p=1}^{M} \exp(i2\pi k^{(p,l)}r)\tilde{\eta}_r^{(m,p)} + \frac{\tilde{\eta}_r^{(m,l)}}{SNR_r^{(l)}} = \exp(i2\pi k^{(m,l)}r) \right\}, \tag{8.64}$$

where it is denoted

$$\tilde{\eta}_r^{(m,\cdot)} = \lambda_r^{(\cdot)}\eta_r^{(m,\cdot)}. \tag{8.65}$$

As one can verify, the solution of these equations is

$$\tilde{\eta}_r^{(m,l)} = \frac{\exp(i2\pi k^{(l,m)}r)SNR_r^{(l)}}{1 + \sum\limits_{s=1}^{M} SNR_r^{(s)}} \tag{8.66}$$

or

$$\eta_r^{(m,l)} = \frac{\exp(i2\pi k^{(l,m)}r)}{\lambda_r^{(l)}} \frac{SNR_r^{(l)}}{1 + \sum\limits_{s=1}^{M} SNR_r^{(s)}}. \tag{8.67}$$

Equation 8.67 implies that in this case multichannel image fusion consists of mutual alignment (factor $\exp(i2\pi k^{(l,m)}r)$) and inverse filtering (factor $1/\lambda_r^{(l)}$) of channel images and summing up the results with weights proportional to channel SNRs.

When mutual displacements of images are known, images can be aligned through their resampling with an appropriate interpolation using methods discussed in Chapter 6. In particular, if displacements $\{k_0^{(m,l)}\}$ are known with a subpixel accuracy, in the averaging, samples of images will be interlaced according to the mutual shifts of the images. Therefore, the resulting image will have accordingly higher sampling rate than that of the original image set and, hence, will have higher resolving power. This resolution increase is referred to as *superresolution* from multiple images. An example of superresolution that can be achieved in processing of turbulent videos is described in the section "Application Examples" in Chapter 5.

Mutual image lateral displacement is just a special case of general image affine geometric transformations, including scaling or rotation. Obviously, image alignment before fusing should allow for all those transformations.

When mutual image displacements or corresponding parameters are not known, they have to be determined. The estimation of image displacements or other geometrical parameters can be performed using methods of image parameter estimation discussed in Chapter 7. In particular, it is shown in Chapter 7 that, for the model of additive uncorrelated Gaussian noise in image components, mutual image lateral displacements can be determined by means of locating positions of image cross-correlation peaks. Image averaging with such an image correlational alignment is called *correlational averaging* or *correlational accumulation*. The correlation accumulation has found

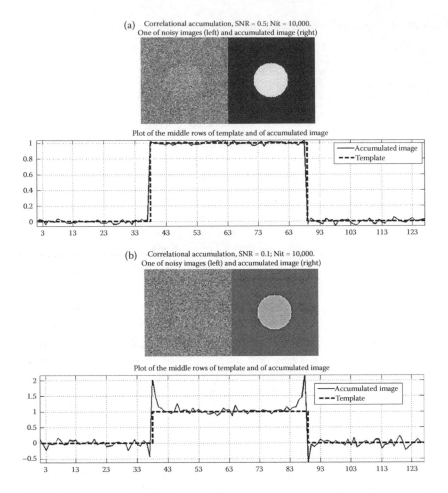

FIGURE 8.20
Correlational accumulation of 10,000 realizations of a circular template image mixed with additive noise for signal-to-noise ratios 0.5 (a) and 0.1 (b). Overshoots at the edges of the accumulated image seen on plots of image middle row for SNR = 0.1 are artifacts caused by anomalous errors in image alignments for the accumulation.

many practical applications, especially in high-resolution electrocardiography and similar biomedical signal processing and for processing electron micrographs of virus particles in which very many of the identical particles are simultaneously observed.

As it was shown in Chapter 7, in estimating lateral displacements of noisy images, two types of localization errors occur: normal errors and anomalous ones. Normal errors are small errors. In image correlational accumulation, they result in certain blur of the resulting fused image due to nonperfect alignment of component images. Anomalous localization errors appear, when the target image is erroneously identified with a realization of noise. In image fusing through correlational averaging, this means that some image components involved in the fusion will be not object images but just pieces of realizations of noise, which results in artifacts that are visually perceived as image edge emphasizing.

Figure 8.20 illustrates this phenomenon on simulation results of denoising, by means of correlational averaging, of noisy images containing a circular template in random position in the field of view for two values of signal-to-noise ratio 0.5 (a) and 0.1 (b).

One can see on the figure that for $SNR = 0.1$, template image is restored with substantial artifacts in the form of overshoots at the edges, while for $SNR = 0.5$, these artifacts are practically not noticeable, although certain blur of the template edges can be indicated.

Filtering Impulse Noise

Image distortions in the form of impulse noise or "pepper-and-salt" noise usually occur in image transmission via noisy digital communication channels and in the storage of digital imagery. Signal transmission errors result in losses, with certain probability, of individual image samples and in replacing their values by random values. Impulse noise may also appear as a residual noise after applying denoising filtering to noisy images with additive or speckle noise. In this case, impulse noise originates from large outbursts of additive or speckle noise that the filtering failed to remove. "Pepper-and-salt" noise usually occurs due to memory cell faults in digital image storage. The possibility of successful filtering of impulse and "pepper-and-salt" noise allows one to lower requirements to transmitter power or, given the transmitter power, to raise the transmission distance and to lower requirements to the reliability of digital storage.

As a factor that affects image perception, impulse noise is more destructive than additive normal noise of the same intensity. One can convince him(her) self in this by comparing, in Figure 8.21, the readability of a test image of printed text distorted by both types of noise of the same intensity, in terms of mean square deviation from the noise-free image.

Impulse noise is described by the ImpN-model introduced in the section "Models of Signal Random Interferences" in Chapter 2 (Equation 2.131):

$$b_{k,l} = (1 - e_{k,l}) \cdot a_{k,l} + e_{k,l} n_{k,l}, \tag{8.68}$$

where $\{b_{k,l}\}$ are samples of image contaminated by impulse noise, $\{a_{k,l}\}$ are the perfect image samples, $\{e_{k,l}\}$ are samples of a binary random sequence that determines the presence ($e_{k,l} = 1$) or the absence ($e_{k,l} = 0$) of impulse noise in the (k,l)-th sample, and $\{n_{k,l}\}$ are samples of yet another random sequence that replaces signal samples in cases when impulse interference takes place.

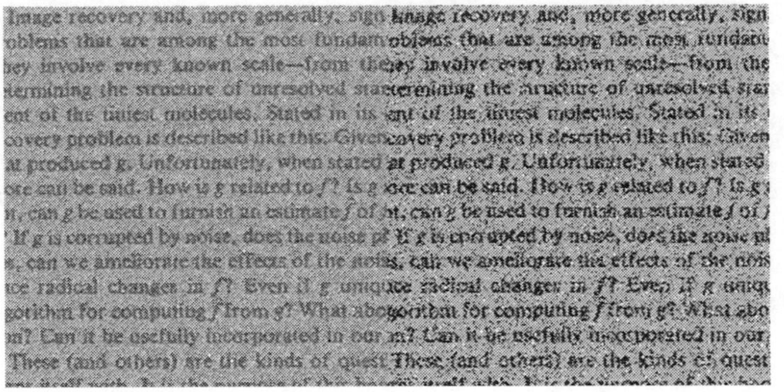

FIGURE 8.21
Test image with additive normal noise (left) and impulse noise (right) with mean squared deviation from noise free image equal to 50 (in the image range 0–255).

From this model, it follows that filtering impulse noise assumes two stages:

- Detection stage: detecting signal samples that have been lost due to the noise
- Correction stage: estimating values of those lost samples

In the case of image transmission over noisy digital channels using error detection codes, the positions of distorted pixels might be known, and no detection is required. Obviously, no special detection is required for "pepper-and-salt" noise either because distorted pixels are those that take extreme values in the image dynamic range.

If positions of error pixels are known, image restoration can, in principle, be achieved by means of, for instance, using the discrete sampling theorem-based iterative algorithm for band-limited approximation of images with omissions described in the section "Algorithms for Signal Recovery from Sparse Sampled Data." However, in practice, using much more simple algorithms is frequently sufficient, which replace error pixels by a certain averaging of those of 8 pixels in their 3×3 *pixel neighborhood* in the rectangular sampling grid (the closest three pixels in the row above from the given pixel, two pixels to the left and to the right from the given pixel in its row, and the closest three pixels in the row below it), which are known to be nondistorted. Examples of images restored using these two methods from the distorted image shown in Figure 8.22a are shown in Figure 8.22b and c, respectively.

When the positions of distorted pixels are not known, their detection represents the main problem. This task can be treated as a special case of the task of object localization in clutter images treated in Chapter 7. For localization of impulse noise samples, individual pixels replaced by noise are the localization targets. Hence, the target spectrum in Equation 7.45 that determines that the SCR-optimal adaptive correlator is uniform, and the correlator filter

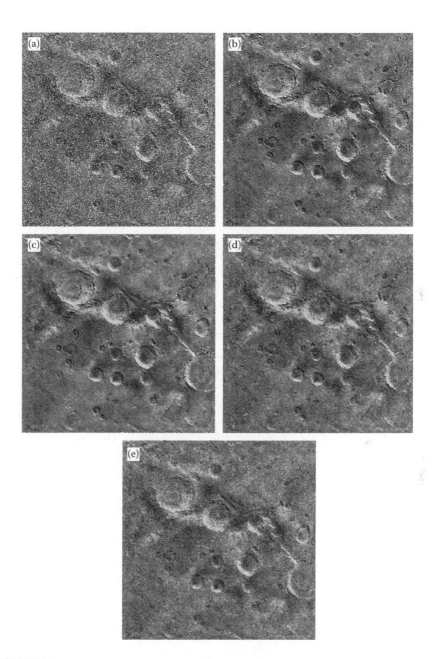

FIGURE 8.22

Filtering impulse noise: (a) noisy image with probability of error 0.3; (b) image denoised by the iterative band-limited reconstruction algorithm for known positions of error pixels; (c) image denoised by means of replacement of error pixels with mean over nondistorted pixels in the 3×3 neighborhood for known positions of error pixels; (d) image denoised by the iterative filtering algorithm for unknown positions of error pixels; (e) image denoised by the recursive filtering algorithm for unknown positions of error pixels.

frequency response will be in this case determined solely by an estimate $\overline{\left|\beta_{r,s}^{(bg,x_0,y_0)}\right|^2}$ of the power spectrum of the image background component:

$$\eta_{r,s}^{(opt)} = \frac{1}{\overline{\left|\beta_{r,s}^{(bg,x_0,y_0)}\right|^2}} = \frac{1}{\left(\overline{\left|\beta_{r,s}^{(bg,x_0,y_0)}\right|^2}\right)^{1/2}} \frac{1}{\left(\overline{\left|\beta_{r,s}^{(bg,x_0,y_0)}\right|^2}\right)^{1/2}}. \tag{8.69}$$

Equation 8.69 implies that image filtering using this filter is double whitening. As it is shown in the section "Object Localization and Edge Detection: Selection of Reference Objects for Target Tracking" (Chapter 7), double whitening can be approximated by two convolution of the image with *Laplacian* (the double Laplacian). Detection of impulse noise samples can then be implemented by comparing the output of such a filter in every pixel with a certain threshold. As the threshold, standard deviation of the filter output signal multiplied by a certain constant can be taken, the constant been selected so as to maximize the probability of correct detection, given the probability of false alarms that may occur due to small contrast details and object edges in the image. At the correction stage, image samples that have been detected as distorted can be replaced by "predicted" values found by one or another smoothing operation over those samples in their vicinity that are not marked as distorted. As the smoothing operation, arithmetic mean or its robust nonlinear versions discussed in the section "Nonlinear Filter Classification Principles" can be used. This algorithm can be mathematically formalized as

$$\hat{a}_{k,l} = \begin{cases} b_{k,l}, & \text{if } \left|dL_{k,l}\right| \le D\sigma_{dL_{k,l}} \\ \overline{b}_{k,l}, & \text{otherwise} \end{cases}, \tag{8.70}$$

where $\left\{\hat{a}_{k,l}\right\}$ are filtered image samples, $\{b_{k,l}\}$ are samples of the initial distorted image, $\{dL_{k,l}\}$ are samples of the result of applying to image $\{b_{k,l}\}$ the double Laplacian operator, D is a detection threshold constant multiplier to the standard deviation $\sigma_{dL_{k,l}}$ of $\{dL_{k,l}\}$, and $\left\{\overline{b}_{k,l}\right\}$ are the "predicted" values. D is a free parameter of the algorithm, which depends on the probability of pixel distortion: the larger the probability, the lower must be D.

In case when the probability of pixel distortion is sufficiently large, many clusters of two of more adjacent distorted pixels may occur, which hampers detection of individual distorted pixels by the filter aimed at the detection of individual distorted pixels. The reliability of the detection of distorted pixels can be increased if denoising is implemented in an iterative manner, using as input image, at each iteration, the image obtained on the previous iteration and with decreasing, from iteration to iteration, of the impulse noise detection threshold in order to remove and correct at earlier stages of iterations the most intensive impulse noise outbursts. This algorithm can be mathematically formulated by the equation

$$
\begin{cases}
\hat{a}_{k,l}^{(t-1)}, & \text{if } \left| d_{k,l}^{(t-1)} \right| \leq D^{(t-1)} \sigma_{d_{k,l}^{(t-1)}} \\
\\
\overline{\hat{a}_{k,l}^{(t-1)}} & \text{otherwise}
\end{cases}
, \tag{8.71}
$$

where $\left\{ \hat{a}_{k,l}^{(t)} \right\}$ are filtered image samples at the $(t{-}1)$-th iteration ($\left\{ \hat{a}_{k,l}^{(0)} = b_{k,l} \right\}$), $\left\{ dL_{k,l}^{(t-1)} \right\}$ are samples of the result of applying to image $\left\{ \hat{a}_{k,l}^{(t-1)} \right\}$ at the $(t{-}1)$-th iteration the double Laplacian operator, $D^{(t-1)}$ is the detection threshold constant multiplier to the standard deviation $\sigma_{d_{k,l}^{(t-1)}}$ of $\left\{ d_{k,l}^{(t-1)} \right\}$, and $\left\{ \overline{\hat{a}_{k,l}^{(t-1)}} \right\}$ are the pixel "predicted" values found at $(t-1)$-th iteration.

An even more simple and fast modification of this denoising filtering is recursive filtering in a sliding window, in which detection and estimation for each pixel are carried out in the process of image row-wise/column-wise scanning using, for detection, comparison of prediction error, found in the same way as in DPCM-coding (see the section "Outline of Image Compression Methods" in Chapter 3), with a certain detection threshold:

$$
\hat{a}_{k,l} =
\begin{cases}
b_{k,l}, & \text{if } \left| a_{k,l} - \overline{b}_{k,l} \right| \leq DetThr \\
\\
\overline{b}_{k,l}, & \text{otherwise}
\end{cases}
, \tag{8.72}
$$

where "predicted" values $\overline{b}_{k,l}$ of the (k,l)-th pixel for image scanning in the direction of index l are computed, in the same way as in DPCM coding, by weighted summation of pixels $\left\{ \hat{a}_{k-1,l-1}, \hat{a}_{k-1,l}, \hat{a}_{k-1,l+1}, \hat{a}_{k,l-1} \right\}$ on the preceding and the given row, which have already been processed, and *DetThr* is a detection threshold. In order to lower the probability of missing distorted pixels, *DetThr* should be lower, which increases the probability of false detection. This may result in blurring of image edges. In order to reduce the damage, it was found useful to modify the filter in the following way:

$$
\hat{a}_{k,l} =
\begin{cases}
b_{k,l}, & \text{if } \left| a_{k,l} - \overline{b}_{k,l} \right| \leq DetThr \\
\\
\overline{\hat{a}_{k,l}} + sign\left(a_{k,l} - \overline{b}_{k,l} \right)C, & \text{otherwise}
\end{cases}
, \tag{8.73}
$$

where C is a certain correcting constant, which is smaller than the visual detection threshold for small targets of the size of one pixel and larger than visual detection threshold for large targets. The practical values of C are roughly in the range of 10–20 gray levels (in the 0–255 scale). Images (d) and (e) in Figure 8.22 illustrate filtering impulse noise with the probability 0.3 by the described iterative and recursive filters. They were generated using the MATLAB program impulse_noise_filtering_demo_CRC.m (see Exercises).

The described impulse noise filtering algorithms belong to the family of nonlinear filters because they contain detection components, which are substantially nonlinear. In the section "Nonlinear Filters for Image Perfecting," we introduce a general framework for a family of nonlinear filters for image perfection, of which the described filters represent a special case.

Correcting Image Grayscale Nonlinear Distortions

Image grayscale reproduction is characterized by the imaging system transfer function. Normally it should be a linear function in the image intensity dynamic range from its minimum to its maximum. Deviations from linearity cause image grayscale distortions that need to be corrected. Imaging systems intended for image display are usually equipped for this purpose with special means for manual correction using buttons "brightness," "contrast," and *"gamma-correction,"* the "gamma" being a slope of the system transfer function plotted in a logarithmic scale. This is usually perfectly enough as far as visual image quality is concerned. When images are intended for metrological purposes, automatic and accurate correcting methods are needed.

Correcting grayscale distortions is obviously a special case of inverse problems. When system transfer function $TF(\cdot)$ is known, it can be corrected by means of applying to distorted images a correcting function $CF(\cdot) = TF^{-1}(\cdot)$ inverse to $TF(\cdot)$, provided noise and signal quantization effects are negligible.

Nonlinear correcting of quantized signals may result in conglutination of quantization levels as illustrated in Figure 8.23.

Therefore, quantization artifacts that were regarded negligible before nonlinear transformation may become intolerable and, generally, the

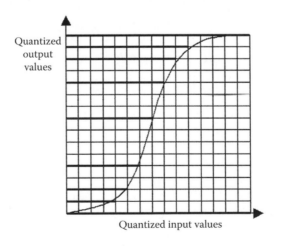

FIGURE 8.23
Nonlinear signal transformation and conglutination of quantization levels. Thin lines in the figure are bounds of 16 uniform quantization intervals. The curve shows transfer function of a nonlinear transformation. Bold lines indicate nine quantization levels left after the nonlinear transformation of 16 quantization levels of input values.

optimization of the correcting function over the image database or ensemble Ω_A is required:

$$CF(\cdot) = \arg\min_{CF(\cdot)} \left\{ AV_{\Omega_A} \left(\sum_{k=0}^{N_x-1} \sum_{l=0}^{N_y-1} LOSS \big| a_{k,l} - CF\{Q[TF(a_{k,l})]\} \big| \right) \right\}, \quad (8.74)$$

where $Q(\cdot)$ is the image scalar quantization transfer function and $LOSS|\cdot|$ is a loss function, for instance, quadratic one, that evaluates losses due to deviations of the image signal from its nondistorted original.

In applications, it very frequently happens that the imaging system transfer function that has to be corrected is not known. In some cases, it can be determined directly from an analysis of distorted images or calibrated using as a reference a certain standard image. This task is akin to the task of noise diagnostics discussed in Appendix.

Nonlinear point-wise distortion transfer function can be, for quantized signals, parameterized as a look-up table with the number of parameters (look-up table entries) equal to the number of signal quantization levels. It affects all pixels in the same way. Therefore, having enough pixels, one can, in principle, estimate those parameters from statistical characteristics of distorted images quite well. Specifically, distribution histograms of images, which are directly affected by the point-wise transformation, can be used for this purpose if normal histogram is known for the type of images, to which the image to be corrected belongs.

A typical example to illustrate such an opportunity is correcting nonlinear distortions in recording optical interferograms. Perfect interferogram is a phase-modulated sinusoidal signal. If the interferogram to be corrected contains many periods of a sinusoidal signal, the distribution function of its phase can be regarded to be uniform in the range $[-\pi,\pi]$. The probability distribution function of sinusoidal signals with a uniformly distributed phase is known. Therefore, the correcting nonlinear transformation can be found as a transformation that converts the histogram of the distorted interferogram into the histogram defined by the known probability distribution function of sinusoidal signals with uniformly distributed phase. Generally, correcting nonlinear transformation is the one that converts the histogram of the distorted image into a normal histogram for the given type of images. We will refer to such a transformation as *histogram matching*. Figure 8.24 illustrates the histogram matching for correcting nonlinear distortions of an interferogram using, as a reference, histogram of a simulated ideal interferogram. The MATLAB program Histo_standardization_CRC.m used for generating images in these figure is provided in Exercises.

Yet another example of an application, in which the histogram matching can be used for correcting unknown nonlinear image distortion, is creating image mosaics or panoramic images from sets of individually taken images.

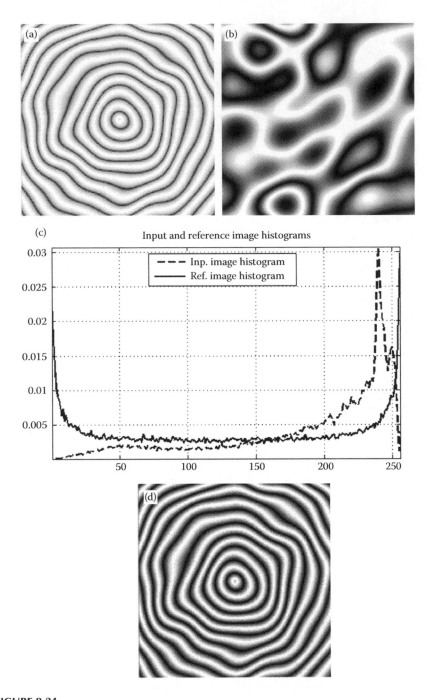

FIGURE 8.24

Histogram matching for correcting nonlinear distortions in imaging systems: (a) distorted interferogram; (b) reference interferogram; (c) histograms of the distorted and reference interferograms; (d) corrected interferogram.

Algorithmically, the histogram matching can be implemented through the *histogram equalization,* a nonlinear transformation that converts image histograms to uniform histograms. For the histogram matching, it is sufficient to find histogram equalization transformation for the image to be transformed and that for the image with the reference histogram. The required histogram matching transformation can then be obtained by combining histogram equalization transformation for the image to be transformed and the transformation inverse to the histogram equalization transformation for the reference histogram.

Histogram equalization is a special case of a large class of nonlinear filtering algorithms, which we discuss in the next section.

Nonlinear Filters for Image Perfecting

Digital linear filters for image perfecting introduced in the sections "Optimal Linear Filters for Image Restoration" and "Sliding Window Transform Domain Adaptive Image Restoration" are called linear because they are defined by Equation 4.7 of weighted summation of signal discrete representation coefficients (samples, when sampling is used for signal discretization), which is an equation of a straight line in the multidimensional signal space. Obviously, that, generally, filter input–output relationship might be nonlinear; hence, generally, nonlinear filters must considered. Since J.W. Tukey [5] introduced the *median filter,* a vast variety of nonlinear filters for image processing has been suggested. In order to ease navigating in this ocean of filters, we will provide in this chapter a classification of the nonlinear filters originated from the median filter and describe some most useful filters for image denoising and enhancement. Throughout almost the entire chapter, we will assume that images are single-component grayscale sampled images with quantized gray levels. Possible approaches to nonlinear filtering of multicomponent images will be briefly discussed in a separate section.

Nonlinear Filter Classification Principles

Equation 4.7 for linear filters can be considered as a special case of a general nonlinear relationship between filter output pixel $\hat{a}_{k,l}$ and its input pixels $\left\{ a_{m,n}^{(k,l)} \right\}$:

$$\hat{a}_{k,l} = \mathbf{ESTM}\{NBH(a_{k,l})\}, \tag{8.75}$$

where $\{k,l\}$ and $\{m,n\}$ are integer indices of image pixels, $NBH(a_{k,l})$ is a subset of image pixels $\left\{ a_{m,n}^{(k,l)} \right\}$, which we will call the *pixel neighborhood* of the pixel

$a_{k,l}$ that corresponds, in the input image, to the filter output sample $\hat{a}_{k,l}$, and **ESTM**{·} is an *estimation operation* that produces the filter output.

Pixel neighborhoods and estimation operation are key words in our classification of nonlinear filters. Such an approach enables unified treatment and comparison of nonlinear filters and their structuring that can be directly translated into algorithmic implementations of the filters.

The pixel neighborhood is formed on the base of *attributes* of pixels involved in the filtering. Obviously, primary pixel attributes that determine filtering operations are pixel gray values and their coordinates (indices). It turns out, however, that a number of attributes other than only these primary ones are essential for nonlinear filtering. Table 8.1 presents a sample list of pixel attributes that are involved in the design of nonlinear filters known from the literature.

Attributes *rank* and *cardinality* describe statistical properties of pixels in neighborhoods. They are interrelated and may actually be regarded as two faces of the same quality. While *rank* is associated, as it is described in the section "Principles of Statistical Treatment of Signals and Signal Transformations and Basic Definitions" in Chapter 2, with the variational row, that is, ordered in ascending order sequence of neighborhood pixel values, *cardinality* is associated with the histogram over the neighborhood. "Geometrical" attributes *gradient* and *Laplacian* describe properties of images as surfaces in 3D spaces, respectively.

TABLE 8.1

Sample List of Pixel Attributes

Attribute Gray Level	Denotation a	Definition Pixel Gray Level
Coordinates	$COORD_{Wnbh}(a_{k,l})$	Pixel coordinates (indices) in the filter window (window-neighborhood)
Cardinality	$HIST_{NBH}(a)$	The number of neighborhood pixels with the same gray level as that of element a: $$HIST_{NBH}(a) = \sum_{k,l \in NBH} \delta(a - a_{k,l})$$
Rank	$RANK_{NBH}(a)$	Number of neighborhood elements with values lower than a, or position of gray level a in the variational row $$RANK_{NBH}(a) = \sum_{v=a_{min}}^{a} HIST_{NBH}(v)$$
Gradient	$GRDT_{NBH}(k,l)$	Image gradient in position (k,l): $$GRDT_{NBH}(k,l) = \sqrt{\left(a_{k,l} - a_{k-1,l}\right)^2 + \left(a_{k,l} - a_{k,l-1}\right)^2}$$
Laplacian	$LPLC_{NBH}(k,l)$	Image Laplacian in position (k,l) (absolute value of convolution of the image with one of the kernels defined by Equation 7.65): $$LPLC_{NBH}(k,l) = \left| L \circ \{a\}(k,l) \right|$$

Pixel neighborhoods are formed as subsets of filter window pixels selected according to their certain attributes. The formation of pixel neighborhoods can be interpreted as segmentation of image fragment within the filter window for separating pixels that will be involved in generating filter output from those that are irrelevant for this. The primary neighborhood is the set of all pixels of the filter window. We will call this neighborhood the *window-neighborhood Wnbh(k,l)*, where (k,l) are indices of the window central pixel that specify the window position. Other neighborhoods are subsets of *Wnbh*. In their denotations, we will indicate the attribute, upon which the neighborhood is formed, and its numerical parameters: *NBH (Attribute, Parameters)*.

The process of neighborhood formation can be multistage using as an attribute, at each stage, a certain estimate over the neighborhood formed at the previous stage. One can distinguish the following groups of neighborhoods in terms of pixel attributes, upon which they are formed:

- *C*-neighborhoods built according to pixel coordinates within the filter window.
- *V*-neighborhoods built according to pixel gray levels.
- *R*-neighborhoods built according to pixel ranks, that is, pixel positions in variational row.
- *H*-neighborhoods built according to pixel cardinalities.
- *G*-neighborhoods built according to pixel "geometrical" attributes, such as gradient or Laplacian, that are associated with properties of images regarded as 2D surfaces.

Table 8.2 presents a list of types of pixel neighborhoods most relevant for the design of nonlinear filters.

The selection of the neighborhood for the design of a particular filter is governed by *a priori* assumption regarding properties of objects in the image: their size and shape (*C-neighborhoods*), the spread of their gray levels (*EV-neighborhoods*), the size and gray-level spread (*KNV-neighborhood*), and its counterparts (*R-neighborhoods*). Figure 8.25 illustrates forming *EV*-neighborhood for a fragment of an image of angiogram. It was generated using the MATLAB program neighborhoods_CRC.m provided in Exercises.

CL-neighborhood is an option that can be used when no *a priori* knowledge concerning spread of objects gray level is available and, from the other side, it is known that objects gray level may form "clusters," that is, more or less well-distinguished hills, in image histograms as illustrated in Figure 8.26. Note that for better distinguishability of histogram clusters, it is frequently advisable to smooth window histograms using one or another data smoothing filters, such as those described below in the section "Filter Classification Tables and Particular Examples." This figure was also generated using the MATLAB program neighborhoods_CRC.m.

TABLE 8.2

Types of Neighborhoods Used in Nonlinear Filtering Algorithms

C-Neighborhoods: Pixel Coordinates as Attributes	
Window-neighborhood: *Wnbh*	The primary neighborhood composed of all pixels of the filter window
KSN-neighborhood: $KSNnbh(a_{k,l};K)$	Neighborhood composed of a given number K of pixels spatially closest to the given one
Shape-neighborhoods: $SHnbh(a_{k,l},S)$	Spatial neighborhood of a certain spatial shape and size of S pixels
Weighted C-neighborhoods $WKSNnbh(a_{k,l};\{w_{m,n}\})$ $WSHnbh(a_{k,l};\{w_{m,n}\})$ $RpKSNnbh(a_{k,l};\{w_{m,n}\})$ $RpKSNnbh(a_{k,l};\{w_{m,n}\})$	KSN- and shape-neighborhoods with weight coefficients $\{w_{m,n}\}$ assigned to their pixels according to certain pixel attributes. If weight coefficients $\{w_{m,n}\}$ are noninteger numbers, pixels gray levels are multiplied by those coefficients ($WKSNnbh$). If weight coefficients $\{w_{m,n}\}$ are integer numbers, pixel gray levels are replicated accordingly thus producing correspondingly enlarged set of pixels ($RpKSNnbh$)
V-Neighborhoods: Pixel Gray Level as Attributes	
Epsilon-*V*-Neighborhood of pixel a_k: $EVnbh(a_{k,l};\varepsilon_v^+;\varepsilon_v^-)$	A subset of filter window pixels with values $\{a\}$ that satisfy the inequality: $$a_{k,l}-\varepsilon_v^- \le a \le a_{k,l}+\varepsilon_v^+.$$
Weighted *V*- (*WV*-) neighborhood: $WVnbh(a_{k,l};w(v-a_{k,l}))$	Filter window pixels with weights $\{w(v-a_{k,l})\}$ assigned according to their gray levels
Range-neighborhood: $RNGnbh(k,l;V_{min},V_{max})$	A subset of filter window pixels with gray levels $a_{m,n}$ within a specified range $[V_{min} \le a_{m,n} \le V_{max}]$
"*K nearest-by-value*"-neighborhood of pixel $a_{k,l}$: $KNVnbh(a_{k,l},K)$	A subset of K filter window pixels with gray levels closest to that of pixel $a_{k,l}$
R-Neighborhoods: Pixel Ranks as Attributes	
"*Epsilon-R*"-neighborhood: $ERnbh(a_{k,l};\varepsilon_R^+;\varepsilon_R^-)$	A subset of filter window pixels with ranks $\{R(a_{m,n})\}$ that satisfy inequality: $$R(a_{k,l})-\varepsilon_R^- \le R(a_{m,n}) \le R(a_{k,l})+\varepsilon_R^+$$
Quantil-neighborhood: $Qnbh(NBH,R_{left},R_{right})$	Filter window order statistics $\{a^{(R_n)}\}$, whose ranks $\{R_n\}$ satisfy inequality $R_{left} < R_n < R_{right}$
H-Neighborhoods: Pixel Cardinalities as Attributes	
"Cluster" neighborhood of pixel $a_{k,l}$: $CLnbh(NBH;a_k)$	Neighborhood elements that belong to the same cluster of the histogram over the neighborhood as that of element $a_{k,l}$.
G-Neighborhoods: Geometrical Attributes	
"Flat"-neighborhoods: $FLnbh_L(a_{k,l},Thr)\ FLnbh_G(a_{k,l},Thr)$	Filter window pixels with values of absolute value of Laplacian (or module of gradient) lower than a certain threshold

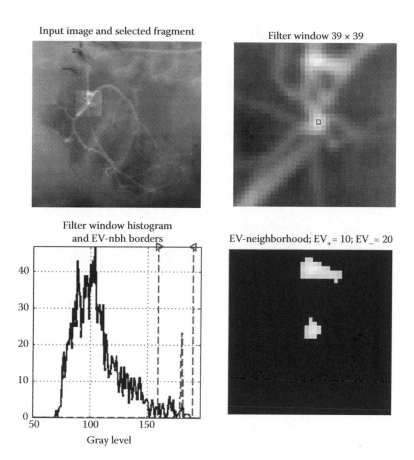

FIGURE 8.25

An illustrative example of EV-neighborhood for a fragment (upper right image) of an image of angiogram (upper left image with the fragment highlighted). The bottom left graph shows image fragment histogram with EV-neighborhood borders (dotted line with triangle markers) and window central pixel gray level indicated by dotted lines without marker. The bottom right image represents the selected EV-neighborhood; pixels that do not belong to it are shown dark black.

G-neighborhoods such as *"Flat"-neighborhood* can be used to unify pixels that belong to more or less "flat" portions of images regarded as 2D surfaces.

Typical estimation operations used in known nonlinear filters are listed in Table 8.3. They are split into two groups: data smoothing operations and operations that evaluate data spread in the neighborhood.

In the filter design, the selection of estimation operation is, in general, governed by requirements of statistical or other optimality of the estimate at the filter output. For instance, if data are observations of a single value distorted by an additive uncorrelated Gaussian random noise, arithmetic *MEAN* over the data is an optimal MAP of this value. It is also an estimate that minimizes mean squared deviation of the estimate from the data. *PROD* is an

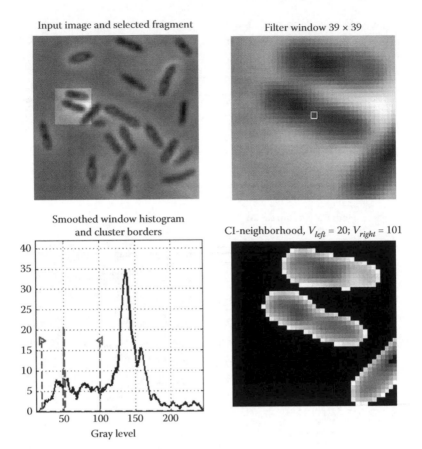

FIGURE 8.26

An illustrative example of CL-neighborhood for a fragment (upper right image) of a microscopic image (upper left image with the fragment highlighted). The bottom left plot presents image fragment histogram smoothed for better distinguishability of histograms clutters. Dotted lines with triangle markers show left and right borders of the histogram clutter that corresponds to dark particles in the image. Dotted lines without marker indicate gray level of the fragment central pixel outlined in the upper right image by the white square. The bottom right image represents the selected CL-neighborhood; pixels that do not belong to it are shown dark black.

operation homomorphic to the addition involved in *MEAN* operation: the sum of logarithms of a set of values is the logarithm of their product.

ROS operations might be optimal MAP-estimations for models other than additive Gaussian noise models. For instance, if neighborhood elements are observations of a single value distorted by the addition of independent random values with exponential distribution density, *MEDN* over the data is known to be optimal MAP-estimation of this value. It is also an estimate that provides the minimum to average of absolute values of its deviation

TABLE 8.3

Estimation Operations

Operation	Denotation	Definition
SMTH (NBH): Data Smoothing Operations		
Arithmetic and geometric mean	MEAN(*NBH*)	Arithmetic mean of samples of the neighborhood
	GMEAN(*NBH*)	Product of samples of the neighborhood
Order statistics	K_ROS(*NBH*)	Value that occupies the *K*-th place (has *rank K*) in the variational row over the neighborhood. Special cases:
	MIN(*NBH*)	Minimum over the neighborhood (the first term of the variational row)
	MEDN(*NBH*)	Central element (median) of the variational row
	MAX(*NBH*)	Maximum over the neighborhood (the last term of the variational row)
Histogram mode	MODE(*NBH*)	Value of the neighborhood element with the highest cardinality: $$\text{MODE}(NBH) = \arg_a \max(\text{HIST}_{NBH}(a))$$
Random number over neighborhood	RAND(*NBH*)	A pseudorandom number taken from an ensemble with the same gray level distribution histogram as that of elements of the neighborhood
SPRD (NBH): Operations that Evaluate the Spread of Data within the Neighborhood		
Standard deviation	STDEV(*NBH*)	Pixel gray level standard deviation over the neighborhood
Interquantil distance (quasispread)	IQDIST(*NBH*)	R_ROS(*NBH*) – L_ROS(*NBH*), where $1 \leq L < R \leq \text{SIZE}(NBH)$
Range	RNG(*NBH*)	MAX(*NBH*) – MIN(*NBH*)
Neighborhood size	SIZE(*NBH*)	Number of elements of the neighborhood

from the data. If additive noise distribution is one sided, *MIN* or *MAX* might be optimal estimations. *MODE* can be regarded an operation of obtaining MAP-estimation if the distribution histogram is considered to be *a posteriori* distribution of signal gray levels.

RAND is a "stochastic" estimation operation. It generates an estimate that, statistically, is equivalent to all the above "deterministic" estimates: it has the same mean, standard deviation, and all other distribution moments. Images presented in Figure 8.27 clearly show that the replacement of image samples with pseudorandom numbers by *RAND*-operation may provide a reasonably good estimate of pixel gray levels from pixel neighborhoods.

It is especially true for higher-order neighborhoods. For instance, one can hardly distinguish image visually (a) and image (c) obtained from pseudorandom

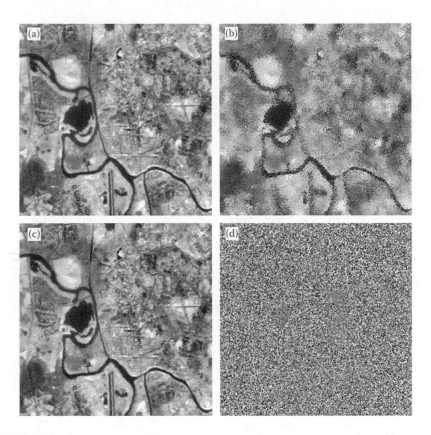

FIGURE 8.27
Illustration of RAND(*NBH*) operation: (a) initial test image with 256 quantization levels; (b) result of applying operation to the initial image RAND(*KSN*) in the filter window of 7×7 pixels; (c) result of applying to the initial image operation RAND(*EV-nbh*) in the filter window of 7×7 pixels and $\varepsilon_v^+ = \varepsilon_v^- = 10$; (d) difference between images (a) and (c); standard deviation of the difference is 5.6 (in units of the image range 0–255).

numbers that have the same local histograms over EV-neighborhood for $\varepsilon_v^+ = \varepsilon_v^- = 10$ gray levels as the image (a). The difference (d) between these two images looks almost chaotic, which means that no substantially important visual information is lost in the image (c).

All the above-mentioned operations belong to a class of data smoothing operations. We will use a common denotation SMTH estimation operation for them. The rest of the operations generate numerical measures of neighborhood data spread. We will use a common denotation SPRD operation for them. Their two modifications, "interquantil distance" IQDIST and "range" RNG, are recommended as a replacement for standard deviation for the evaluation of spread of data. The SIZE operation computes the number of samples in the neighborhood, when it does not directly follow from the neighborhood definition.

Filter Classification Tables and Particular Examples

Tables 8.4 and 8.5 list the most known nonlinear filters. Filters presented in Table 8.4 are intended for cleaning images from noise and other foreign contaminations. The simplest *local mean filter*, MSE-optimal and local adaptive linear filters described in the sections "Optimal Linear Filters for Image Restoration" and "Sliding Window Transform Domain Adaptive Image Restoration" can be regarded in such a classification as special cases of general nonlinear filters. Other *C-neighborhood data smoothing filters*: *"percentile" filters* and *adaptive mode quantization filter* are robust to outliers alternatives to the local mean filter. The most popular from them are *median filter* and *morphological filters*. The most representative from *V-neighborhood data smoothing filters* are *KNV-mean* and *EV-mean*, or the *sigma filter*. The so-called *bilateral filter* is a kind of a "softened" modification of the "sigma" filter, in which uniform weights in the range $\left[a_{k,l} - \varepsilon_v^- \leq a \leq a_{k,l} + \varepsilon_v^+ \right]$ are replaced by "soft" weights $w(v - a_{k,l}) = \exp\left[(v - a_{k,l})^2 / 2\sigma_v^2 \right] / \sqrt{2\pi}\sigma_v$.

Examples of robust to outliers data smoothing filters, which reject outliers on the basis of pixel ranks, are *R-neighborhood filters*: *alpha-trimmed mean* and *alpha-trimmed median filters*.

In principle, the specification of filter parameters requires knowledge of true pixels gray levels, which is, of course, unavailable as far as images are contaminated with noise that should be cleaned out by the filtering. Escape out of this vicious circle lies in iterative filtering:

$$\hat{a}_k^{(t)} = \mathbf{ESTM}(NBH^{(t-1)}), \qquad (8.76)$$

where subscript t is an iteration index. With infinitely large number of iterations, filter outputs may arrive at the so-called *filter "fixed" points*. Filter-fixed point is an image, which the filter does not change. It is a kind of an "eigen" function of the filter. Trivial fixed points of all described smoothing filters are, obviously, constants. Nontrivial fixed points of many of the smoothing filters are piece-wise constant images, as it is illustrated for filter *MODE(Wnbh)* in Figure 8.28.

Filter *MODE(Wnbh)* is a very hard smoother. Its natural application is image segmentation. Other smoothing filters listed in Table 8.4 are softer and can be used for edge-preserving noise cleaning. The simplest of them and very popular median filter, as well as other "percentile" filters, apparently do not have a nontrivial fixed point. They can be used for noise cleaning in a non-iterative manner. In particular, median filter is especially efficient as robust substitutes for local mean filters in cleaning images from additive noise with outliers and for the detection of outliers in images by means of comparing with a threshold of difference between images and their smoothed copies.

Figure 8.29 illustrates the application of the median filter for removing banding noise in an image that contains contrast fiducial marks that must be

TABLE 8.4

Image Smoothing Filters

C-Neighborhood Filters

Moving average (local mean) filter	$\hat{a}_k = \text{MEAN}(Wnbh)$

MSE-optimal and local adaptive linear filters

$\hat{a}_k = \text{MEAN}(WKSNnbh)$

weight coefficients of WKSN-neighborhood are determined by correlation functions (or power spectra) of images and noise as it is described in the sections "Optimal Linear Filters for Image Restoration" and "Sliding Window Transform Domain Adaptive Image Restoration"

"Percentile" filters

$\hat{a}_k = \text{K\_ROS}(KSNnbh)$

Median filter	$\hat{a}_k = \text{MEDN}(Wnbh)$
MAX filter	$\hat{a}_k = \text{MAX}(Wnbh)$
MIN filter	$\hat{a}_k = \text{MIN}(Wnbh)$
Weighted ROS- (median, min, max, …) filters	$\hat{a}_k = \text{K\_ROS}(RpKSNnbh)$
"Morphological" Erosion filters	$\hat{a}_k = \text{MAX}(SHnbh)$
Dilation	$\hat{a}_k = \text{MIN}(SHnbh)$

Adaptive mode quantization filter

$\hat{a}_k = \text{MODE}(Wnbh)$

V-Neighborhood Filters

KNV-mean filter

$\hat{a}_k = \text{MEAN}(KNV(a_k; K)); \quad K$ is a filter parameter

EV-mean filter (the "Sigma" filter)

$\hat{a}_k = \text{MEAN}(EVnbh(Wnbh; a_k; \varepsilon_V^+; \varepsilon_V^-)); \quad \varepsilon_V^+; \varepsilon_V^-$ are filter parameters

C-Neighborhood Filters

"Bilateral" filter

$\hat{a}_k = \text{MEAN}(WVnbh(Wnbh; a_{k,l}; w(v - a_{k,l})));$

$w(v - a_{k,l}) = \exp\left[(v - a_{k,l})^2 / 2\sigma_v^2\right] / \sqrt{2\pi}\sigma_v; \quad \sigma_v$ is a filter parameter

"Modified" trimmed mean filter

$\hat{a}_k = \text{MEAN}(EV(\text{MEDN}(Wnbh); \varepsilon_V^+; \varepsilon_V^-))); \quad \varepsilon_V^+; \varepsilon_V^-$ are filter parameters

R-Neighborhood Filters

"Alpha"-trimmed mean, median

$\hat{a}_k = \text{MEAN}(Qnbh(Wnbh, R_{left}, R_{right}));$

$\hat{a}_k = \text{MEDN}(Qnbh(Wnbh, R_{left}, R_{right})), \quad R_{left}, R_{right}$ are filter parameters

---

TABLE 8.5

Image Enhancement Filters

Local Contrast Enhancement Filter	
"Unsharp masking"	$\hat{a}_{k,l} = (1 - g)a_{k,l} + g\left[a_{k,l} - \text{SMTH}(NBH, a_{k,l})\right]$; g is user defined local contrast amplification parameter
Local histogram equalization	$\hat{a}_{k,l} = \dfrac{a_{\max} - a_{\min}}{\text{SIZE}(NBH)} \text{RANK}(NBH) + a_{\min}$
Weighted local histogram equalization	$\hat{a}_{k,l} = \dfrac{a_{\max} - a_{\min}}{\text{SIZE}(RpKSNnbh)} \text{RANK}(RpKSNnbh) + a_{\min}$
Local *p*-histogram equalization	$\hat{a}_{k,l} = \dfrac{a_{\max} - a_{\min}}{pK} \text{pRANK}(NBH) + a_{\min}$, where $$\text{pRANK}(NBH) = \sum_{v=a_{\min}}^{a} \text{HIST}_{NBH}^{p}(a); \quad pK = \sum_{v=a_{\min}}^{a_{\max}} \text{HIST}^{p}(a);$$ p is a user-defined enhancement parameter ($0 \le p \le 1$)
Edge Extraction Filters	
Local variance filter	$\hat{a}_k = \text{STDEV}(Wnbh)$
Quasirange filter; Max-min filter	$\hat{a}_{k,l} = \text{QSRNG}(NBH) = R\_ROS(Wnbh) - L\_ROS(Wnbh)$ $\hat{a}_k = \text{MAX}(NBH) - \text{MIN}(NBH)$
Size-EV filter	$\hat{a}_{k,l} = \text{SIZE}(EVnbh(Wnbh; a_{k,l}; \varepsilon_V^+; \varepsilon_V^-))$
Local cardinality filter	$\hat{a}_k = \text{HIST}(Wnbh, a_k)$

preserved. The filtering algorithm is similar to that illustrated in Figure 8.4, except the row-wise mean is replaced by the local median in row-wise- and column-wise-oriented rectangular windows.

More efficient image noise cleaners are *V*-neighborhood filters, especially the *"sigma" filter*. One can compare the noise-cleaning capability of iterative median and sigma filters in Figure 8.30.

The fixed points of the "sigma" filter are piece-wise images, provided appropriately selected filter parameters (filter window and borders of EV-interval). Therefore, "sigma" filters, when applied as noise cleaners, tend to generate piece-wise constant images. Another limitation of the applicability of "sigma" filters as noise cleaners is that they are incapable of cleaning impulse noise, because EV-neighborhood for pixels distorted by impulse noise may not contain any pixel except the given one if noise outburst contrast exceeds the EV interval of the filter.

The described algorithms are most suitable for smoothing image contaminations that can be regarded as being additive. In cases when a model of impulse noise is more appropriate for noise contaminations, noise cleaning filters should be built as two stages: (i) contamination detection and (ii) smoothing.

(a)   Mode(Wnbh): initial (left) and iterated (right) images; 50-th iteration

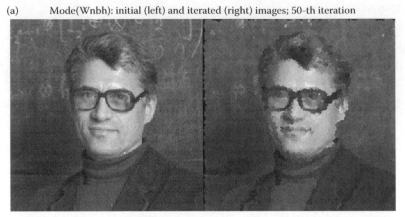

(b)   Model(Wnbh): StDev(INPIMG-OUTIMG)

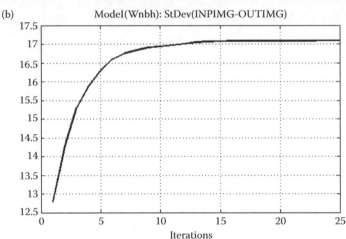

Iterations

FIGURE 8.28
Piece-wise constant image (a) (right) obtained after 25 iteration of filter *MODE(Wnbh)* for filter window of 3×3 pixels applied to an initial image (a) (left) and plot of standard deviation of the difference between initial and iterated images (b) as a function of the number of iterations, which proves the filter convergence to a fixed point.

This principle is described by the equations

$$\hat{a}_{k,l} = sw \cdot a_{k,l} + (1 - sw) \cdot \text{SMTH}_{\text{est}}(NBH(a_{k,l})), \qquad (8.77)$$

where *sw* is a binary switch variable

$$sw = \begin{cases} 1, & a_{k,l} \in NBH(\text{SMTH}_{\text{det}}(a_{k,l})) \\ 0, & \text{otherwise} \end{cases}, \qquad (8.78)$$

Initial (left), vertical medn 170×3-filtered (middle) and horizontal medn 3×257-filtered (right) images 512×512 pxls

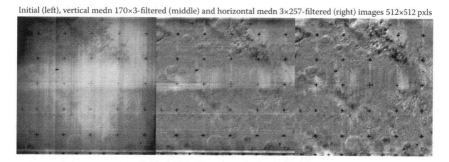

FIGURE 8.29

Fiducial marks (dark crosses) robust filtering banding noise in an image of Mars surface returned from the Soviet space probe Mars-4: left—initial distorted image; middle—image after removal vertical bands by subtracting from the initial image its local median in the window of 170 × 3 pixels; right—image after subsequent removal of horizontal bands by subtracting from the previous image its local median in the window of 3 × 257 pixels (image size is 512 × 512 pixels).

where $\text{SMTH}_{est}(\cdot)$ and $\text{SMTH}_{det}(\cdot)$ are, respectively, estimation and detection smoothing operations, for which one of the smoothing filters from Table 8.4 can be used, and $NBH_{det}(\text{SMTH}_{det}(a_{k,l}))$ is a "detection" neighborhood formed around the (k,l)-th pixel smoothed by the detection smoothing operation.

Typical image enhancement nonlinear filters are listed in Table 8.5. The filters are classified into two groups: *local contrast enhancement filters* and *edge extraction filters*. Local contrast enhancement filters are intended to make visible objects that might not be visible in the initial images because their contrast with respect to their background is lower than the contrast sensitivity of vision. There are at least two ways to achieve this goal. The first one is *"unsharp" masking*. With unsharp masking, the difference is amplified between initial images and their smoothed copies, which are used as estimates of objects' background. For the smoothing, one of the smoothing filters described in Table 8.4 can be used. Images in Figure 8.31 illustrate this method.

The second group of methods for local contrast enhancement form methods in which image gray levels are subjected to a point-wise transformation with a transfer function, whose slope, for a given gray level, is determined by the image histogram: the greater the gray level rate in the image the greater is the amplification of local gray level contrasts for these gray levels. The most important representative of this group is *p-histogram equalization*, in which the transfer function slope is proportional to p-th certain power of the image histogram. It can be applied both globally to the entire image frame and locally in sliding window. In the latter case, we call them *local p-histogram equalization*. The term "equalization" stems from the fact that when $p = 1$ such transformation converts images with an arbitrary histogram to images with the uniform histogram (in fact, almost but not exactly uniform due to effects of quantization). This particular transformation of image gray levels

(a)

FIGURE 8.30

(a) Comparison of noise-cleaning capabilities of iterative application of median and sigma filters. First (from top to bottom) row: a test image (left) and its copy with additive uniformly distributed in the range (0–20) noise (right). Second row: iterated median-filtered image after 75 iterations (left) with filter window 3×3 pixels and its difference from the initial noisy image (right); note the difference in image bright points from corners of patches, which were removed from the image. Third row: iterated median-filtered image after 75 iterations with filter window 5×5 pixels (left) and its difference from the initial noisy image (right); note further losses in patch corners and corresponding image blur. Fourth and fifths rows: iterated sigma-filtered images after 75 iterations with $\varepsilon_V^+ = \varepsilon_V^- = 12$ gray levels and filter window 3×3 and 5×5 pixels, respectively (left) and their corresponding differences from the initial noisy image (right); note the preservation of edges of patches that have sufficient contrast. (b) Plot of the difference between initial noisy and iterated sigma-filtered image for 3×3 filter window that demonstrates that after 75 iteration filtering has reached the filter-fixed point.

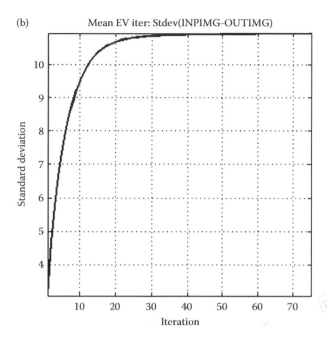

FIGURE 8.30
Continued.

is called histogram equalization. One of the immediate applications of histogram equalization is image dynamic range blind calibration, which we mentioned in the section "Correcting Image Grayscale Nonlinear Distortions."

In the case $p = 0$, the transformation stretches linearly the dynamic range of the image to the entire dynamic range available for image display:

$$\hat{a}_{k,l} = \frac{a_{max} - a_{min}}{MAX - MIN}(a_{k,l} - MIN) + a_{min}, \tag{8.79}$$

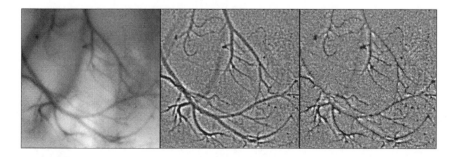

FIGURE 8.31
Local contrast enhancement of a test image of 256×256 pixels (left) by means of unsharp masking using local mean (middle) and local median (right) in the window 7×7 pixels with parameter g set to 1.

where (a_{max}, a_{min}) are maximal and minimal gray levels for image display (usually 0 and 255) and (MAX, MIN) are maximal and minimal gray levels of the input image, or, respectively, image fragment in the filter window if the filter is applied locally in a sliding window. Intermediate values of p enable flexible, using only one parameter, control of image local contrast enhancement.

The histogram equalization can be treated as converting image gray-level contrasts into contrast in terms of pixel ranks. As we indicated in section "Design of Optimal Quantizers" (Chapter 3), histogram p-equalization and, in particular, histogram equalization can also be interpreted in terms of signal optimal quantization.

We illustrate image local contrast enhancement by means of global and local p-histogram equalization in windows of different sizes in Figure 8.32. Figure 8.32 shows how image local histogram equalization can make visible invisible low-contrast texture in a test image and how processing with differently oriented windows reproduces differently oriented contrast patches.

Figure 8.33 demonstrates the effects of applying global and local histogram p-equalization for improving the visibility of blood vessels and other low-contrast details in a real x-ray image of a woman's breast (mammogram).

Edge extraction nonlinear filters are exemplified by filters that measure the local spread of data: the *local variance filter* and its robust to outliers substitute local *quasispread filter*, the *max-min filter*, which is a special case of the quasispread filter, the *size-EV filter* and the local *cardinality filter*, which, in its turn, can be regarded as a special case of the size-EV filter, when its EV-interval is set to zero. Figure 8.34 illustrates the performance of these filters on an example of detection of edges in a microscopic image of a crystalline structure.

Images for Figures 8.32 through 8.34 were generated using the MATLAB program enhancement_demo_CRC.m provided in Exercises.

Nonlinear Filters for Multicomponent Images

The above-described classes of nonlinear filters are intended for working with single-component grayscale images. An immediate and trivial option for their usage for multicomponent images is separable, that is, component-wise, filtering. For nonseparable applications, notions of pixel attributes, neighborhood, and estimation operation must be revisited. Some of them are equally relevant to both single-component and multicomponent images. These are attributes "coordinates" and "cardinality," C-neighborhoods ("window-", KSN-, and "cluster"-neighborhoods), arithmetic mean MEAN(*NBH*), histogram mode MODE(*NBH*), and RAND(*NBH*). The design of corresponding unseparable nonlinear filters based on these pixel attributes, neighborhoods, and estimation operations is straightforward. Most of the others pixel attributes, neighborhoods, and estimation operations need to be defined, an introduction of certain scalar measures over image components that would

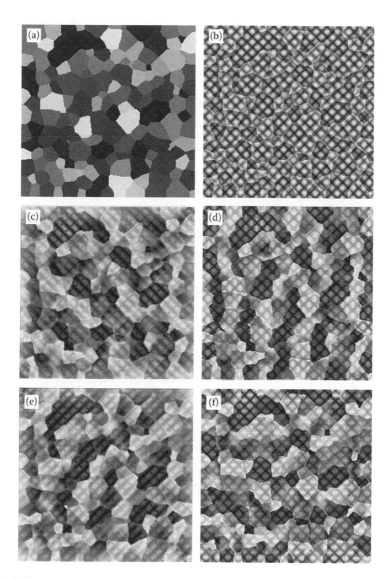

FIGURE 8.32
Amplification of local contrasts by means of local histogram equalization in windows of approximately same size and different shape: (a) test image that contains an invisible low-contrast periodical texture with maximal amplitude 5 gray levels (out of 256); (b)–(f) results of local histogram equalization in different windows: square of 15×15, a rectangle of 71×3 pixels rotated by 45°, a horizontal rectangle of 3×71 pixels, a rectangle of 71×3 pixels rotated by –45° and vertical rectangle of 71×3 pixels, respectively.

numerically evaluate variations of vector of components. These measures are essentially determined by the specificity of image components. No general approach seems feasible to their definitions and each particular case must be treated ad hoc.

(a)   (b)

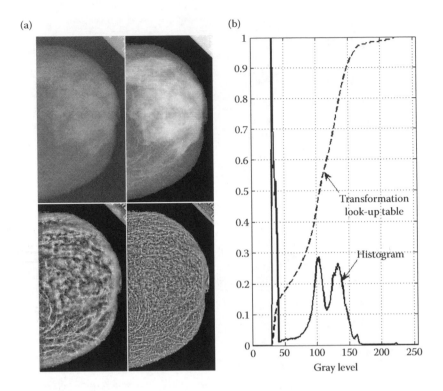

Gray level

FIGURE 8.33
Enhancement of local contrasts of a mammogram (a, upper left image). (a) Global (upper right image) and local p-histogram equalization in windows 91×91 and 31×31 pixels (accordingly, bottom left and right images); (b) image histogram and transformation transfer function (look-up table), both normalized to one. $P = 0.5$, image size 892×470 pixels.

Display Options for Image Enhancement

The way, in which images are displayed, is a very important aspect of image enhancement. As it was mentioned, in image enhancement, all capabilities of human visual system must be employed: the capability to distinguish colors, the capability of 3D perception through stereo vision, and the capability to sense temporal changes. These capabilities open up to five channels to the human brain: three RGB (red–green–blue) channels, two stereoscopic channels through two eyes, and one temporal channel. When displaying grayscale images for visual analysis, one can employ these channels to convey information to the visual cortex of the brain. This can be achieved using the following methods of such a multichannel displaying grayscale images:

- Pseudocolor display
- Image colorization
- Image stereo-visualization
- Image animation

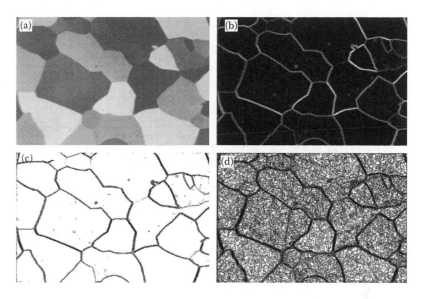

FIGURE 8.34
Edge extraction on an example of a microscopic image of a crystalline structure: (a) initial image; (b) max-min filter; (c) size-EV filter; (d) local cardinality filter. In all cases, filter window size 3×3 pixels; borders of EV-neighborhood for the size-EV are (+7, − 7) gray levels.

In pseudocolor display method, grayscale images are reproduced as artificial RGB color images by means of encoding image gray levels within their dynamic range, say 0–255, by a certain color, that is, by one of 256 appropriately selected combinations of gray levels of the artificial RGB components. Thanks to the high sensitivity of human visual system to color variations, this method helps to visually distinguish grayscale image variations otherwise not visible by virtue of limited grayscale contrast sensitivity of vision. Figure 8.35 illustrates how the pseudocolor representation of the test image from Figure 8.32a makes visible an invisible low-contrast periodic texture of the test image.

Image colorization is a method of simultaneous displaying in one synthetic color image up to three versions of image enhancement results by means of assigning each version to one of three RGB components of the synthetic color image. This way of displaying enables parallel analysis of different versions of image enhancement that may represent different features of images useful for visual image analysis. Figure 8.36 illustrates this method on an example of displaying different versions of enhancement of the mammogram shown in Figure 8.33.

Stereo-visualization is a method of simultaneous representation of two versions of enhancement or feature extraction of an image. For this, from initial two image versions, an artificial stereoscopic pair is generated, in which the intensity of images is defined by the first input image version and parallax, or mutual displacement between of corresponding points of images of the pair, is defined by the second input image version treated as an image depth map.

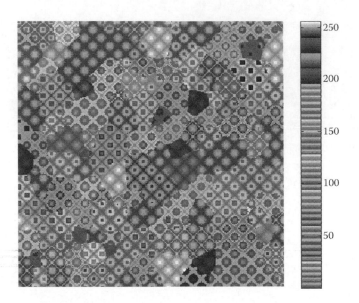

FIGURE 8.35
(See color insert.) Pseudocolor representation of the test image of Figure 8.32a, which contains an invisible low-contrast periodical texture. Color bar to the right shows color coding table for image gray levels.

Image animation enables visual analysis of many versions of image enhancement, for instance, local histogram p-equalization in windows of different sizes, by means of representing different image versions as frames of an artificial video sequence.

Obviously, these visualization methods can be used in combination. Methodologically, these methods can be treated as representatives of a class of nonlinear filters, which change the number of image components.

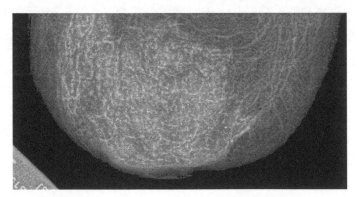

FIGURE 8.36
(See color insert.) Colorization of enhanced versions of mammogram shown in Figure 8.33: the lower right version is shown in red, the upper right version is shown in green, and the initial image (upper left) is shown in blue.

Appendix

Derivation of Equation 8.16

$$AV_{\Omega_A}AV_{\Omega_N}\left(\sum_{r=0}^{N-1}|\alpha_r - \eta_r\beta_r|^2\right)$$

$$= AV_{\Omega_A}AV_{\Omega_N}\left[\sum_{r=0}^{N-1}(\alpha_r - \eta_r\beta_r)(\alpha_r^* - \eta_r^*\beta_r^*)\right]$$

$$= AV_{\Omega_A}AV_{\Omega_N}\left\{\sum_{r=0}^{N-1}\left(|\alpha_r|^2 - \eta_r\alpha_r^*\beta_r - \eta_r^*\alpha_r\beta_r^* + |\eta_r|^2|\beta_r|^2\right)\right\}$$

$$= AV_{\Omega_A}AV_{\Omega_N}\left\{\left[\sum_{r=0}^{N-1}\left(|\alpha_r|^2 - \left(\eta_r^{(re)} + i\eta_r^{(im)}\right)\alpha_r^*\beta_r - \left(\eta_r^{(re)} - i\eta_r^{(im)}\right)\alpha_r\beta_r^*\right.\right.$$

$$\left.\left. + \left(\eta_r^{(re)2} + \eta_r^{(im)2}\right)|\beta_r|^2\right)\right]\right\}. \tag{A8.1}$$

Compute partial derivatives of this expression over real $\eta_r^{(re)}$ and imaginary $\eta_r^{(im)}$ parts of η_r and equal those to zero to obtain $\left\{\eta_r^{(opt)}\right\}$ that minimize it:

$$\frac{\partial}{\partial\eta_r^{(re)}}AV_{\Omega_A}AV_{\Omega_N}\left(\sum_{r=0}^{N-1}|\alpha_r - \eta_r\beta_r|^2\right)$$

$$= AV_{\Omega_A}AV_{\Omega_N}\left(-\alpha_r^*\beta_r - \alpha_r\beta_r^* + 2\eta_r^{(re)}|\beta_r|^2\right)$$

$$= AV_{\Omega_A}AV_{\Omega_N}\left[-2\,\mathrm{Re}(\alpha_r\beta_r^*) + 2\eta_r^{(re)}|\beta_r|^2\right]$$

$$= 2\eta_r^{(re)}AV_{\Omega_A}AV_{\Omega_N}\left(|\beta_r|^2\right) - 2\,\mathrm{Re}\left[AV_{\Omega_A}AV_{\Omega_N}(\alpha_r\beta_r^*)\right] = 0 \tag{A8.2}$$

from which it follows that

$$\eta_r^{(re,opt)} = \frac{\mathrm{Re}\left[AV_{\Omega_A}AV_{\Omega_N}(\alpha_r\beta_r^*)\right]}{AV_{\Omega_A}AV_{\Omega_N}\left(|\beta_r|^2\right)}. \tag{A8.3}$$

Similarly, one can obtain

$$\eta_r^{(im,opt)} = \frac{\text{Im}\left[AV_{\Omega_A}AV_{\Omega_N}(\alpha_r\beta_r^*)\right]}{AV_{\Omega_A}AV_{\Omega_N}\left(\left|\beta_r\right|^2\right)}. \tag{A8.4}$$

Therefore

$$\eta_r^{(opt)} = \frac{AV_{\Omega_A}AV_{\Omega_N}(\alpha_r\beta_r^*)}{AV_{\Omega_A}AV_{\Omega_N}\left(\left|\beta_r\right|^2\right)}. \tag{A8.5}$$

Empirical Estimation of Variance of Additive Signal-Independent Broad Band Noise in Images

Additive signal-independent zero mean noise with normal distribution is fully specified by its standard deviation and autocorrelation function. If the noise is uncorrelated or weakly correlated (broad band), as is often the case, then its variance and correlation function can be found by means of the following simple algorithm based on detecting and measuring abnormalities in the covariance function of the observed image.

By the definition (Equation 2.110), the correlation function of signal $\{b_k\}$ obtained as an additive mixture

$$b_k = a_k + n_k \tag{A8.6}$$

of an "ideal" signal $\{a_k\}$ and signal-independent noise zero mean noise $\{n_k\}$ is

$$R_b(\kappa) = AV_{\Omega_A}\left[AV_{\Omega_N}\left(\frac{1}{N}\sum_{k=0}^{N-1}b_k a_{k+\kappa}^*\right)\right]$$

$$= AV_{\Omega_A}\left(\frac{1}{N}\sum_{k=0}^{N-1}a_k a_{k+\kappa}^*\right) + AV_{\Omega_N}\left(\frac{1}{N}\sum_{k=0}^{N-1}n_k n_{k+\kappa}^*\right)$$

$$+ \frac{1}{N}\sum_{k=0}^{N-1}AV_{\Omega_A}(a_k)AV_{\Omega_N}(n_{k+\kappa}^*) + \frac{1}{N}\sum_{k=0}^{N-1}AV_{\Omega_A}(a_k^*)AV_{\Omega_N}(n_{k+r})$$

$$= R_a(\kappa) + R_n(\kappa), \tag{A8.7}$$

where $AV_{\Omega_N}(n_{k+\kappa}) = AV_{\Omega_N}(n_{k+\kappa}^*) = 0$ as mean values of zero mean noise, $R_a(\kappa)$ is the correlation function of the noiseless signal, and $R_n(\kappa)$ is the correlation function of the noise.

Consider the uncorrelated (white) noise case when

$$R_n(\lambda) = \sigma_n^2 \delta(\kappa) = \begin{cases} \sigma_n^2, & \kappa = 0 \\ 0, & \kappa > 0 \end{cases}, \tag{A8.8}$$

where σ_n^2 is the noise variance. In this case, the correlation function $R_b(\kappa)$ of the observed image deviates from the correlation function $R_a(\kappa)$ of the noiseless image in the coordinate origin only:

$$R_b(\kappa) = \begin{cases} R_a(\kappa) + \sigma_n^2, & \kappa = 0 \\ R_a(\kappa), & \kappa > 0 \end{cases}. \tag{A8.9}$$

Therefore, one can estimate the value $R_a(0)$ for $k = 0$ by means of extrapolating values of $R_b(\kappa) = R_a(\kappa)$ for $\kappa > 0$ and then use the extrapolated value $\hat{R}_a(0)$ to compute an estimate $\hat{\sigma}_n^2$ of noise variance as

$$\hat{\sigma}_n^2 = R_b(0) - \hat{R}_a(0). \tag{A8.10}$$

This is well illustrated in Figure A8.1 generated using the program NoiseVar_CRC.m (Exercises). Correlation functions of images are normally quite smooth functions in the vicinity of the coordinate origin ($\kappa = 0$). This property enables obtaining sufficiently accurate estimates of the noise variance using, for extrapolation, a few samples of $R_b(\kappa)$. For instance, the data presented in Figure 8.2 for noisy images were obtained by means of evaluating 1D correlation functions of images through inverse DFT of the mean of power spectra of image rows, which implements image ensemble averaging $AV_{\Omega_A}\left((1/N)\Sigma_{k=0}^{N-1} a_k a_{k+\kappa}^*\right)$, and subsequent extrapolation of $R_b(0)$ from $R_b(1)$ and $R_b(2)$ using a parabolic approximation of $R_b(\kappa)$ in the vicinity of point $\kappa = 0$:

$$\hat{R}_a(0) = \frac{4}{3} R_b(1) - \frac{1}{3} R_b(2). \tag{A8.11}$$

A similar approach can be used for the estimation of the correlation function of a weakly correlated (broad band) noise, that is, a noise whose correlation function $R_n(\kappa)$ is known to be concentrated in only a few first samples of $R_n(\kappa)$. Estimates of these samples can be obtained as differences between image empirical correlation function $R_b(\kappa)$ in those samples and their values extrapolated from the rest of the samples of $R_b(\kappa)$, in which noise correlation function values are known to be negligible.

Derivation of Equation 8.45

The DCT spectra coefficients of signal samples $\{a_k\}$ in the sliding window of $(2N_w + 1)$ samples centered at the k-th sample are defined as

$$\beta_r^{(k)} = \frac{1}{\sqrt{2N_w + 1}} \sum_{n=0}^{2N_w} a_{k-N_w+n} \cos\left(\pi \frac{n + 1/2}{2N_w + 1} r \right).$$

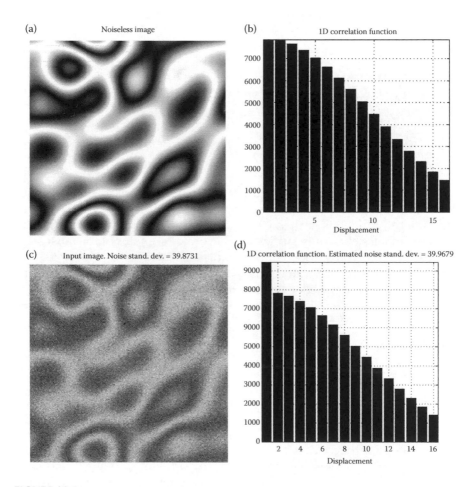

FIGURE A8.1
Noiseless and noisy images (left column) and their row-wise 1D correlation functions (right column); for the noisy image, empirical estimate of standard deviation of noise is indicated in the image title.

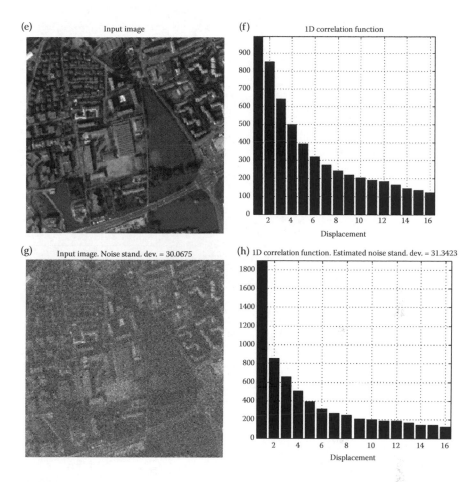

FIGURE A8.1
Continued.

By the definition of inverse DCT:

$$a_{k-N_w+n} = \frac{1}{\sqrt{2N_w+1}}\left[\beta_0^{(k)} + 2\sum_{r=1}^{2N_w}\beta_r^{(k)}\cos\left(\pi\frac{n+1/2}{2N_w+1}r\right)\right].$$

Therefore, the window central pixel a_k can be found as

$$a_k = \frac{1}{\sqrt{2N_w+1}}\left[\beta_0^{(k)} + 2\sum_{r=1}^{2N_w}\beta_r^{(k)}\cos\left(\pi\frac{N_w+1/2}{2N_w+1}r\right)\right]$$

$$= \frac{1}{\sqrt{2N_w+1}}\left[\beta_0^{(k)} + 2\sum_{r=1}^{2N_w}\beta_r^{(k)}\cos\left(\frac{\pi r}{2}\right)\right] = \frac{1}{\sqrt{2N_w+1}}\left[\beta_0^{(k)} + 2\sum_{s=1}^{N_w}\beta_{2s}^{(k)}(-1)^s\right].$$

Derivation of Equation 8.51

Find the partial derivatives of functional

$$\left\{ AV_{\Omega_A} AV_{\Omega_N} \left(\left| \sum_{r=0}^{N-1} \hat{\alpha}_r^{(m)} - \sum_{n=1}^{M} \eta_r^{(m,n)} \beta_r^{(n)} \right|^2 \right) \right\}$$

over real and imaginary parts of thought weight coefficients $\left\{ \eta_r^{(m,n)} \right\}$ and equal them to zero:

$$\frac{\partial}{\partial \eta_r^{re(m,l)}} \left\{ AV_{\Omega_A} AV_{\Omega_N} \left(\sum_{r=0}^{N-1} \left(\alpha_r^{(m)} - \sum_{p=1}^{M} \eta_r^{(m,p)} \beta_r^{(p)} \right) \left(\alpha_r^{*(m)} - \sum_{p=1}^{M} \eta_r^{*(m,p)} \beta_r^{*(p)} \right) \right) \right\}$$

$$= \frac{\partial}{\partial \eta_r^{re(m,l)}} \left\{ AV_{\Omega_A} AV_{\Omega_N} \left[\sum_{r=0}^{N-1} \left(\alpha_r^{(m)} - \sum_{p=1}^{M} \eta_r^{re(m,p)} \beta_r^{(p)} - i \sum_{p=1}^{M} \eta_r^{im(m,p)} \beta_r^{(p)} \right) \right. \right.$$

$$\times \left. \left. \left(\alpha_r^{*(m)} - \sum_{p=1}^{M} \eta_r^{re(m,p)} \beta_r^{*(p)} + i \sum_{p=1}^{M} \eta_r^{im(m,p)} \beta_r^{*(p)} \right) \right] \right\}$$

$$= AV_{\Omega_A} AV_{\Omega_N} \left[-\beta_r^{(l)} \left(\alpha_r^{*(m)} - \sum_{p=1}^{M} \eta_r^{re(m,p)} \beta_r^{*(p)} + i \sum_{p=1}^{M} \eta_r^{im(m,p)} \beta_r^{*(p)} \right) \right.$$

$$\left. - \beta_r^{*(l)} \left(\alpha_r^{(m)} - \sum_{p=1}^{M} \eta_r^{re(m,p)} \beta_r^{(p)} - i \sum_{p=1}^{M} \eta_r^{im(m,p)} \beta_r^{(p)} \right) \right]$$

$$= -AV_{\Omega_A} AV_{\Omega_N} \left(\alpha_r^{*(m)} \beta_r^{(l)} + \alpha_r^{(m)} \beta_r^{*(l)} \right)$$

$$+ \sum_{p=1}^{M} \eta_r^{re(m,p)} AV_{\Omega_A} AV_{\Omega_N} \left(\beta_r^{(l)} \beta_r^{*(p)} + \beta_r^{*(l)} \beta_r^{(p)} \right)$$

$$- i \sum_{p=1}^{M} \eta_r^{im(m,p)} AV_{\Omega_A} AV_{\Omega_N} \left(\beta_r^{(l)} \beta_r^{*(p)} - \beta_r^{*(l)} \beta_r^{(p)} \right). \tag{A8.12}$$

From this equation, it follows that

$$\sum_{p=1}^{M} \eta_r^{re(m,p)} AV_{\Omega_A} AV_{\Omega_N} \left(\text{Re} \left(\beta_r^{*(p)} \beta_r^{(l)} \right) \right) + \sum_{p=1}^{M} \eta_r^{im(m,p)} AV_{\Omega_A} AV_{\Omega_N} \left(\text{Im} \left(\beta_r^{*(p)} \beta_r^{(l)} \right) \right)$$

$$= AV_{\Omega_A} AV_{\Omega_N} \left(\text{Re} \left(\hat{\alpha}_r^{*(m)} \beta_r^{(l)} \right) \right) \tag{A8.13}$$

$$\frac{\partial}{\partial \eta_r^{im(m,l)}} \left\{ AV_{\Omega_A} AV_{\Omega_N} \left(\sum_{r=0}^{N-1} \left(\alpha_r^{(m)} - \sum_{p=1}^{M} \eta_r^{(m,p)} \beta_r^{(p)} \right) \left(\alpha_r^{*(m)} - \sum_{p=1}^{M} \eta_r^{*(m,p)} \beta_r^{*(p)} \right) \right) \right\}$$

$$= \frac{\partial}{\partial \eta_r^{im(m,l)}} \left\{ AV_{\Omega_A} AV_{\Omega_N} \left[\sum_{r=0}^{N-1} \left(\alpha_r^{(m)} - \sum_{p=1}^{M} \eta_r^{re(m,p)} \beta_r^{(p)} - i \sum_{p=1}^{M} \eta_r^{im(m,p)} \beta_r^{(p)} \right) \right. \right.$$

$$\left. \left. \times \left(\alpha_r^{*(m)} - \sum_{p=1}^{M} \eta_r^{re(m,p)} \beta_r^{*(p)} + i \sum_{p=1}^{M} \eta_r^{im(m,p)} \beta_r^{*(p)} \right) \right] \right\}$$

$$= AV_{\Omega_A} AV_{\Omega_N} \left[-i\beta_r^{(l)} \left(\alpha_r^{*(m)} - \sum_{p=1}^{M} \eta_r^{re(m,p)} \beta_r^{*(p)} + i \sum_{p=1}^{M} \eta_r^{im(m,p)} \beta_r^{*(p)} \right) \right.$$

$$\left. + i\beta_r^{*(l)} \left(\alpha_r^{(m)} - \sum_{p=1}^{M} \eta_r^{re(m,p)} \beta_r^{(p)} - i \sum_{p=1}^{M} \eta_r^{im(m,p)} \beta_r^{(p)} \right) \right]$$

$$= -iAV_{\Omega_A} AV_{\Omega_N} \left(\alpha_r^{*(m)} \beta_r^{(l)} - \alpha_r^{(m)} \beta_r^{*(l)} \right)$$

$$+ i \sum_{p=1}^{M} \eta_r^{re(m,p)} AV_{\Omega_A} AV_{\Omega_N} \left(\beta_r^{(l)} \beta_r^{*(p)} - \beta_r^{*(l)} \beta_r^{(p)} \right)$$

$$+ \sum_{p=1}^{M} \eta_r^{im(m,p)} AV_{\Omega_A} AV_{\Omega_N} \left(\beta_r^{(l)} \beta_r^{*(p)} + \beta_r^{*(l)} \beta_r^{(p)} \right)$$

$$= -iAV_{\Omega_A} AV_{\Omega_N} \left(2i \operatorname{Im} \left(\alpha_r^{*(m)} \beta_r^{(l)} \right) \right) + i \sum_{p=1}^{M} \eta_r^{re(m,p)} AV_{\Omega_A} AV_{\Omega_N} \left(2i \operatorname{Im} \left(\beta_r^{(l)} \beta_r^{*(p)} \right) \right)$$

$$+ \sum_{p=1}^{M} \eta_r^{im(m,p)} AV_{\Omega_A} AV_{\Omega_N} \left(2 \operatorname{Re} \left(\beta_r^{(l)} \beta_r^{*(p)} \right) \right)$$

$$= 2AV_{\Omega_A} AV_{\Omega_N} \left(\operatorname{Im} \left(\alpha_r^{*(m)} \beta_r^{(l)} \right) \right) - 2 \sum_{p=1}^{M} \eta_r^{re(m,p)} AV_{\Omega_A} AV_{\Omega_N} \left(\operatorname{Im} \left(\beta_r^{(l)} \beta_r^{*(p)} \right) \right)$$

$$+ 2 \sum_{p=1}^{M} \eta_r^{im(m,p)} AV_{\Omega_A} AV_{\Omega_N} \left(\operatorname{Re} \left(\beta_r^{(l)} \beta_r^{*(p)} \right) \right) = 0 \tag{A8.14}$$

from which it follows that

$$
\sum_{p=1}^{M} \eta_r^{re(m,p)} AV_{\Omega_A} AV_{\Omega_N} \left(\mathrm{Im}\left(\beta_r^{(l)} \beta_r^{*(p)} \right) \right) - \sum_{p=1}^{M} \eta_r^{im(m,p)} AV_{\Omega_A} AV_{\Omega_N} \left(\mathrm{Re}\left(\beta_r^{(l)} \beta_r^{*(p)} \right) \right)
$$

$$
= AV_{\Omega_A} AV_{\Omega_N} \left(\mathrm{Im}\left(\alpha_r^{*(m)} \beta_r^{(l)} \right) \right). \tag{A8.15}
$$

Multiplying this equation by $(-i)$ and combining this with Equation A8.13, obtain

$$
\sum_{p=1}^{M} \eta_r^{re(m,p)} AV_{\Omega_A} AV_{\Omega_N} \left(\mathrm{Re}\left(\beta_r^{*(p)} \beta_r^{(l)} \right) - i\left(\beta_r^{*(p)} \beta_r^{(l)} \right) \right)
$$

$$
+ \sum_{p=1}^{M} \eta_r^{im(m,p)} AV_{\Omega_A} AV_{\Omega_N} \left(i\,\mathrm{Re}\left(\beta_r^{(l)} \beta_r^{*(p)} \right) + \mathrm{Im}\left(\beta_r^{*(p)} \beta_r^{(l)} \right) \right)
$$

$$
= AV_{\Omega_A} AV_{\Omega_N} \left(\mathrm{Re}\left(\alpha_r^{*(m)} \beta_r^{(l)} \right) - i\,\mathrm{Im}\left(\alpha_r^{*(m)} \beta_r^{(l)} \right) \right)
$$

or

$$
\sum_{p=1}^{M} \eta_r^{re(m,p)} AV_{\Omega_A} AV_{\Omega_N} \left(\beta_r^{(p)} \beta_r^{*(l)} \right) + \sum_{p=1}^{M} i\eta_r^{im(m,p)} AV_{\Omega_A} AV_{\Omega_N} \left(\beta_r^{(p)} \beta_r^{*(l)} \right)
$$

$$
= AV_{\Omega_A} AV_{\Omega_N} \left(\alpha_r^{(m)} \beta_r^{*(l)} \right)
$$

and finally

$$
\sum_{p=1}^{M} \eta_r^{(m,p)} AV_{\Omega_A} AV_{\Omega_N} \left(\beta_r^{(p)} \beta_r^{*(l)} \right) = AV_{\Omega_A} AV_{\Omega_N} \left(\alpha_r^{(m)} \beta_r^{*(l)} \right). \tag{A8.16}
$$

Cross-correlations $AV_{\Omega_A} AV_{\Omega_N} \left(\beta_r^{(p)} \beta_r^{*(l)} \right)$ and $AV_{\Omega_A} AV_{\Omega_N} \left(\alpha_r^{(m)} \beta_r^{*(l)} \right)$ are, according to Equation 8.57

$$
AV_{\Omega_A} AV_{\Omega_N} \left(\beta_r^{(p)} \beta_r^{*(l)} \right)
$$

$$
= AV_{\Omega_A} AV_{\Omega_N} \left(\left(\lambda_r^{(p)} \alpha_r^{(p)} + v_r^{(p)} \right) \left(\lambda_r^{*(l)} \alpha_r^{*(l)} + v_r^{*(l)} \right) \right)
$$

$$
= AV_{\Omega_A} AV_{\Omega_N} \left(\lambda_r^{(p)} \lambda_r^{*(l)} \alpha_r^{(p)} \alpha_r^{*(l)} + \lambda_r^{*(l)} v_r^{(p)} \alpha_r^{*(l)} + \lambda_r^{(p)} v_r^{*(l)} \alpha_r^{(p)} + v_r^{(p)} v_r^{*(l)} \right)
$$

$$
= \lambda_r^{(p)} \lambda_r^{*(l)} AV_{\Omega_A} \left(\alpha_r^{(p)} \alpha_r^{*(l)} \right) + \lambda_r^{*(l)} AV_{\Omega_A} \left(\alpha_r^{*(l)} \right) AV_{\Omega_N} \left(v_r^{(p)} \right)
$$

$$
+ \lambda_r^{(p)} AV_{\Omega_A} \left(\alpha_r^{(p)} \right) AV_{\Omega_N} \left(v_r^{*(l)} \right) + AV_{\Omega_N} \left(v_r^{(p)} v_r^{*(l)} \right)
$$

$$
= \lambda_r^{(p)} \lambda_r^{*(l)} AV_{\Omega_A} \left(\alpha_r^{(p)} \alpha_r^{*(l)} \right) + \sigma_n^{2(p)} \delta(l - p) \tag{A8.17}
$$

$$AV_{\Omega_A} AV_{\Omega_N}(\alpha_r^{(m)}\beta_r^{*(l)}) = AV_{\Omega_A} AV_{\Omega_N}\left[\alpha_r^{(m)}\left(\lambda_r^{*(l)}\alpha_r^{*(l)} + v_r^{*(l)}\right)\right]$$

$$= \lambda_r^{*(l)} AV_{\Omega_A}\left(\alpha_r^{(m)}\alpha_r^{*(l)}\right) \tag{A8.18}$$

because noise is assumed to be zero mean and noncorrelated. Inserting Equations A8.17 and A8.18 into Equation A8.16, obtain

$$\left\{\sum_{p=1}^{M}\eta_r^{(m,p)}\left[\lambda_r^{(p)}\lambda_r^{*(l)} AV_{\Omega_A}\left(\alpha_r^{(p)}\alpha_r^{*(l)}\right) + \sigma_n^{2(p)}\delta(l-p)\right] = \lambda_r^{*(l)} AV_{\Omega_A}\left(\alpha_r^{(m)}\alpha_r^{*(l)}\right)\right\} \tag{A8.19}$$

or

$$\left\{\sum_{p=1}^{M}\eta_r^{(m,p)} AV_{\Omega_A}\left(\tilde{\alpha}_r^{(p)}\tilde{\alpha}_r^{*(l)}\right) + \eta_r^{(m,l)}\sigma_n^{2(l)} = \frac{1}{\lambda_r^{(m)}} AV_{\Omega_A}\left(\tilde{\alpha}_r^{(m)}\tilde{\alpha}_r^{*(l)}\right)\right\}, \tag{A8.20}$$

where $\tilde{\alpha}_r^{(p)} = \lambda_r^{(p)}\alpha_r^{(p)}$ and $\tilde{\alpha}_r^{(l)} = \lambda_r^{(l)}\alpha_r^{(l)}$ denote spectral coefficients of images in corresponding channels after their linear distortions. For every l, this is a system of M equation for finding MMSE-optimal weight coefficients $\{\eta_r^{(m,l)}\}$.

In order to ease the analysis and interpretation of the solution and get an insight into how interchannel correlations and noise level determine channel contributions to the result of fusion, consider a special case of fusion. For the case of two channels ($l = 1,2$) into one ($m = 1,2$), we have

$$\begin{cases} AV_{\Omega_A}\left(\left|\tilde{\alpha}_r^{(1)}\right|^2\right)\eta_r^{(1,1)} + AV_{\Omega_A}\left(\tilde{\alpha}_r^{(2)}\tilde{\alpha}_r^{*(1)}\right)\eta_r^{(1,2)} + \sigma_n^{2(1)}\eta_r^{(1,1)} = \dfrac{1}{\lambda_r^{(1)}} AV_{\Omega_A}\left(\tilde{\alpha}_r^{(1)}\tilde{\alpha}_r^{*(1)}\right) \\[2ex] AV_{\Omega_A}\left(\tilde{\alpha}_r^{(1)}\tilde{\alpha}_r^{*(2)}\right)\eta_r^{(1,1)} + AV_{\Omega_A}\left(\left|\tilde{\alpha}_r^{(2)}\right|^2\right)\eta_r^{(1,2)} + \sigma_n^{2(2)}\eta_r^{(1,2)} = \dfrac{1}{\lambda_r^{(1)}} AV_{\Omega_A}\left(\tilde{\alpha}_r^{(1)}\tilde{\alpha}_r^{*(2)}\right) \\[2ex] AV_{\Omega_A}\left(\left|\tilde{\alpha}_r^{(1)}\right|^2\right)\eta_r^{(2,1)} + AV_{\Omega_A}(\tilde{\alpha}_r^{(2)}\tilde{\alpha}_r^{*(1)})\eta_r^{(2,2)} + \sigma_n^{2(1)}\eta_r^{(2,1)} = \dfrac{1}{\lambda_r^{(2)}} AV_{\Omega_A}\left(\tilde{\alpha}_r^{(2)}\tilde{\alpha}_r^{*(1)}\right) \\[2ex] \eta_r^{(2,1)} AV_{\Omega_A}\left(\tilde{\alpha}_r^{(1)}\tilde{\alpha}_r^{*(2)}\right) + \eta_r^{(2,2)} + AV_{\Omega_A}\left(\left|\tilde{\alpha}_r^{(2)}\right|^2\right) + \sigma_n^{2(2)}\eta_r^{(2,2)} = \dfrac{1}{\lambda_r^{(2)}} AV_{\Omega_A}\left(\left|\tilde{\alpha}_r^{(2)}\right|^2\right). \end{cases} \tag{A8.21}$$

Consider in detail the first pair of equations that can be rewritten as

$$\begin{aligned} \left(\overline{\left|\tilde{\alpha}_r^{(1)}\right|^2} + \sigma_n^{2(1)}\right)\eta_r^{(1,1)} + \overline{\tilde{\alpha}_r^{(2)}\tilde{\alpha}_r^{*(1)}}\eta_r^{(1,2)} &= \frac{1}{\lambda_r^{(1)}}\overline{\left|\tilde{\alpha}_r^{(1)}\right|^2} \\[2ex] \overline{\tilde{\alpha}_r^{(1)}\tilde{\alpha}_r^{*(2)}}\eta_r^{(1,1)} + \left(\overline{\left|\tilde{\alpha}_r^{(2)}\right|^2} + \sigma_n^{2(2)}\right)\eta_r^{(1,2)} &= \frac{1}{\lambda_r^{(1)}}\overline{\tilde{\alpha}_r^{(1)}\tilde{\alpha}_r^{*(2)}}, \end{aligned} \tag{A8.22}$$

where $AV_{\Omega_A}(\cdot\cdot)$ is, for the sake of brevity, replaced by $(\overline{\cdot\cdot})$.

The solutions of these equations are

$$
\eta_r^{(1,1)} = \frac{1}{\lambda_r^{(1)}} \frac{\overline{\left|\tilde{\alpha}_r^{(1)}\right|^2}\left(\overline{\left|\tilde{\alpha}_r^{(2)}\right|^2} + \sigma_n^{2(2)}\right) - \overline{\left|\left(\tilde{\alpha}_r^{(1)}\tilde{\alpha}_r^{*(2)}\right)\right|^2}}{\left(\overline{\left|\tilde{\alpha}_r^{(1)}\right|^2} + \sigma_n^{2(1)}\right)\left(\overline{\left|\tilde{\alpha}_r^{(2)}\right|^2} + \sigma_n^{2(2)}\right) - \overline{\left|\left(\tilde{\alpha}_r^{(1)}\tilde{\alpha}_r^{*(2)}\right)\right|^2}}
$$

$$
= \frac{1}{\lambda_r^{(1)}} \frac{\left(\overline{\left|\tilde{\alpha}_r^{(1)}\right|^2\left|\tilde{\alpha}_r^{(2)}\right|^2} - \overline{\left|\left(\tilde{\alpha}_r^{(1)}\tilde{\alpha}_r^{*(2)}\right)\right|^2}\right) + \sigma_n^{2(2)}\overline{\left|\tilde{\alpha}_r^{(1)}\right|^2}}{\left(\overline{\left|\tilde{\alpha}_r^{(1)}\right|^2\left|\tilde{\alpha}_r^{(2)}\right|^2} - \overline{\left|\left(\tilde{\alpha}_r^{(1)}\tilde{\alpha}_r^{*(2)}\right)\right|^2}\right) + \sigma_n^{2(1)}\overline{\left|\tilde{\alpha}_r^{(2)}\right|^2} + \sigma_n^{2(2)}\overline{\left|\tilde{\alpha}_r^{(1)}\right|^2} + \sigma_n^{2(1)}\sigma_n^{2(1)}}
$$

$$
= \frac{1}{\lambda_r^{(1)}} \frac{\left(1 - \frac{\overline{\left|\left(\tilde{\alpha}_r^{(1)}\tilde{\alpha}_r^{*(2)}\right)\right|^2}}{\overline{\left|\tilde{\alpha}_r^{(1)}\right|^2}\overline{\left|\tilde{\alpha}_r^{(2)}\right|^2}}\right) + \frac{\sigma_n^{2(2)}}{\overline{\left|\tilde{\alpha}_r^{(2)}\right|^2}}}{\left(1 - \frac{\overline{\left|\left(\tilde{\alpha}_r^{(1)}\tilde{\alpha}_r^{*(2)}\right)\right|^2}}{\overline{\left|\tilde{\alpha}_r^{(1)}\right|^2}\overline{\left|\tilde{\alpha}_r^{(2)}\right|^2}}\right) + \frac{\sigma_n^{2(1)}}{\overline{\left|\tilde{\alpha}_r^{(1)}\right|^2}} + \frac{\sigma_n^{2(2)}}{\overline{\left|\tilde{\alpha}_r^{(2)}\right|^2}} + \frac{\sigma_n^{2(1)}}{\overline{\left|\tilde{\alpha}_r^{(1)}\right|^2}}\frac{\sigma_n^{2(2)}}{\overline{\left|\tilde{\alpha}_r^{(2)}\right|^2}}}
$$

$$
= \frac{1}{\lambda_r^{(1)}} \frac{\left(1 - \kappa_r^{(1,2)}\kappa_r^{(2,1)}\right) + 1/SNR_r^{(2)}}{\left(1 - \kappa_r^{(1,2)}\kappa_r^{(2,1)}\right) + 1/SNR_r^{(1)} + 1/SNR_r^{(2)} + 1/SNR_r^{(1)}SNR_r^{(2)}}
$$

$$
= \frac{1}{\lambda_r^{(1)}} \frac{SNR_r^{(1)}\left[1 + \left(1 - \kappa_r^{(1,2)}\kappa_r^{(2,1)}\right)SNR_r^{(2)}\right]}{1 + \left(1 - \kappa_r^{(1,2)}\kappa_r^{(2,1)}\right)SNR_r^{(1)}SNR_r^{(2)} + SNR_r^{(2)} + SNR_r^{(1)}} \qquad (A8.23)
$$

and

$$
\eta_r^{(1,2)} = \frac{1}{\lambda_r^{(1)}} \frac{\overline{\left(\tilde{\alpha}_r^{(1)}\tilde{\alpha}_r^{*(2)}\right)}\left(\overline{\left|\tilde{\alpha}_r^{(1)}\right|^2} + \sigma_n^{2(1)}\right) - \overline{\left|\tilde{\alpha}_r^{(1)}\right|^2}\overline{\left(\tilde{\alpha}_r^{(1)}\tilde{\alpha}_r^{*(2)}\right)}}{\left(\overline{\left|\tilde{\alpha}_r^{(1)}\right|^2} + \sigma_n^{2(1)}\right)\left(\overline{\left|\tilde{\alpha}_r^{(2)}\right|^2} + \sigma_n^{2(2)}\right) - \overline{\left|\left(\tilde{\alpha}_r^{(1)}\tilde{\alpha}_r^{*(2)}\right)\right|^2}}
$$

$$
= \frac{1}{\lambda_r^{(1)}} \frac{\overline{\left(\tilde{\alpha}_r^{(1)}\tilde{\alpha}_r^{*(2)}\right)}\sigma_n^{2(1)}}{\left(\overline{\left|\tilde{\alpha}_r^{(1)}\right|^2} + \sigma_n^{2(1)}\right)\left(\overline{\left|\tilde{\alpha}_r^{(2)}\right|^2} + \sigma_n^{2(2)}\right) - \overline{\left|\left(\tilde{\alpha}_r^{(1)}\tilde{\alpha}_r^{*(2)}\right)\right|^2}}
$$

$$
= \frac{1}{\lambda_r^{(1)}} \frac{\overline{\left(\tilde{\alpha}_r^{(1)}\tilde{\alpha}_r^{*(2)}\right)}\sigma_n^{2(1)}}{\sigma_n^{2(1)}\sigma_n^{2(2)}\left(1 + SNR_r^{(1)}\right)\left(1 + SNR_r^{(2)}\right) - \overline{\left|\left(\tilde{\alpha}_r^{(1)}\tilde{\alpha}_r^{*(2)}\right)\right|^2}}
$$

$$= \frac{1}{\lambda_r^{(1)}} \frac{\left(\overline{\tilde{\alpha}_r^{(1)}\tilde{\alpha}_r^{*(2)}}\right)\sigma_n^{2(1)}}{\sigma_n^{2(1)}\sigma_n^{2(2)}\left(1 + SNR_r^{(1)}\right)\left(1 + SNR_r^{(2)}\right) - \left|\tilde{\alpha}_r^{(1)}\right|^2\left|\tilde{\alpha}_r^{(2)}\right|^2\left|\overline{\left(\tilde{\alpha}_r^{(1)}\tilde{\alpha}_r^{*(2)}\right)}\right|^2 \Big/ \left|\tilde{\alpha}_r^{(1)}\right|^2\left|\tilde{\alpha}_r^{(2)}\right|^2}$$

$$= \frac{1}{\lambda_r^{(1)}} \frac{SNR_r^{(2)}\left(\overline{\tilde{\alpha}_r^{(1)}\tilde{\alpha}_r^{*(2)}}\right)\Big/\left|\tilde{\alpha}_r^{(2)}\right|^2}{\left(1 + SNR_r^{(1)}\right)\left(1 + SNR_r^{(2)}\right) - SNR_r^{(1)}SNR_r^{(2)}\left|\overline{\left(\tilde{\alpha}_r^{(1)}\tilde{\alpha}_r^{*(2)}\right)}\right|^2 \Big/ \left|\tilde{\alpha}_r^{(1)}\right|^2\left|\tilde{\alpha}_r^{(2)}\right|^2}$$

$$= \frac{1}{\lambda_r^{(2)}} \frac{SNR_r^{(2)}\left(\overline{\alpha_r^{(1)}\alpha_r^{*(2)}}\right)\Big/\left|\alpha_r^{(2)}\right|^2}{\left(1 + SNR_r^{(1)}\right)\left(1 + SNR_r^{(2)}\right) - SNR_r^{(1)}SNR_r^{(2)}\left|\overline{\left(\tilde{\alpha}_r^{(1)}\tilde{\alpha}_r^{*(2)}\right)}\right|^2 \Big/ \left|\tilde{\alpha}_r^{(1)}\right|^2\left|\tilde{\alpha}_r^{(2)}\right|^2}$$

$$= \frac{1}{\lambda_r^{(2)}} \frac{\kappa_r^{(2,1)}SNR_r^{(2)}}{\left(1 + SNR_r^{(1)}\right)\left(1 + SNR_r^{(2)}\right) - \kappa_r^{(1,2)}\kappa_r^{(2,1)}SNR_r^{(1)}SNR_r^{(2)}}$$

$$= \frac{1}{\lambda_r^{(2)}} \frac{\kappa_r^{(2,1)}SNR_r^{(2)}}{1 + SNR_r^{(1)} + SNR_r^{(2)} + \left(1 - \kappa_r^{(1,2)}\kappa_r^{(2,1)}\right)SNR_r^{(1)}SNR_r^{(2)}}, \qquad \text{(A8.24)}$$

where

$$\kappa_r^{(1,2)} = \frac{\left|\overline{\left(\alpha_r^{(1)}\alpha_r^{*(2)}\right)}\right|}{\left|\alpha_r^{(1)}\right|^2}; \qquad \kappa_r^{(2,1)} = \frac{\left|\overline{\left(\alpha_r^{(1)}\alpha_r^{*(2)}\right)}\right|}{\left|\alpha_r^{(2)}\right|^2} \qquad \text{(A8.25)}$$

and

$$\left\{SNR_r^{(m)} = \frac{\left|\lambda_r^{(m)}\right|^2 AV_{\Omega_A}\left(\left|\alpha_r^{(m)}\right|^2\right)}{AV_{\Omega_n}\left(\left|\nu_r^{(m)}\right|^2\right)}\right\}, \qquad m = 1, 2. \qquad \text{(A8.26)}$$

Verification of Equation 8.66

Prove that the solution defined by Equation 8.66 satisfies Equation 8.64. First, find $\sum_{p=1}^{M} \exp(i2\pi k^{(p,1)}r)\tilde{\eta}_r^{(m,p)}$ for $\tilde{\eta}_r^{(m,p)}$ defined by Equation 8.66:

$$\sum_{p=1}^{M} \exp(i2\pi k^{(p,1)}r)\tilde{\eta}_r^{(m,p)} = \sum_{p=1}^{M} \exp(i2\pi k^{(p,1)}r)\frac{\exp(i2\pi k^{(m,p)}r)SNR_r^{(p)}}{1 + \sum_{s=1}^{M} SNR_r^{(s)}}$$

$$= \sum_{p=1}^{M} \exp(i2\pi[k^{(p,l)} - k^{(p,m)}]r) \frac{SNR_r^{(p)}}{1 + \sum_{s=1}^{M} SNR_r^{(s)}}$$

$$= \exp(i2\pi k^{(m,l)}r) \sum_{p=1}^{M} \frac{SNR_r^{(p)}}{1 + \sum_{p=1}^{M} SNR_r^{(s)}}$$

$$= \exp\left(i2\pi k^{(l,m)}r\right) \frac{\sum_{p=1}^{M} SNR_r^{(p)}}{1 + \sum_{s=1}^{M} SNR_r^{(s)}}$$

because $k^{(p,l)} - k^{(p,m)} = k^{(l,m)}$. Inserting this in Equation 8.64, we arrive at the identity

$$\exp(i2\pi k^{(l,m)}r) \frac{\sum_{p=1}^{M} SNR_r^{(p)}}{1 + \sum_{s=1}^{M} SNR_r^{(s)}} + \frac{\frac{\exp(i2\pi k^{(l,m)}r)SNR_r^{(l)}}{1 + \sum_{s=1}^{M} SNR_r^{(s)}}}{SNR_r^{(l)}}$$

$$= \exp(i2\pi k^{(l,m)}r) \frac{\sum_{p=1}^{M} SNR_r^{(p)}}{1 + \sum_{s=1}^{M} SNR_r^{(s)}} + \exp(i2\pi k^{(l,m)}r) \frac{1}{1 + \sum_{s=1}^{M} SNR_r^{(s)}}$$

$$= \exp(i2\pi k^{(l,m)}r) \frac{1 + \sum_{p=1}^{M} SNR_r^{(p)}}{1 + \sum_{s=1}^{M} SNR_r^{(s)}} = \exp(i2\pi k^{(l,m)}r),$$

which completes the proof.

Exercises

demoire1_CRC.m
WienerFilter_CRC.m
filtering_stripes_CRC.m
demo_lcdct2_CRC.m
impulse_noise_filtering_demo_CRC.m
histo_standardization_CRC.m
neighborhoods_CRC.m
enhancement_demo_CRC.m
noiseVar_CRC.m

References

1. N. Wiener, *The Interpolation, Extrapolation and Smoothing of Stationary Times Series*, Wiley, New York, 1949.
2. A. N. Kolmogorov, Sur l'interpolation de suits stationaires, *C. R. Acad. Sci.*, (Paris) 208, 2043–2045, 1939.
3. C. E. Shannon, Communication in the presence of noise, *Proc. IRE*, 1, 10–21, 1949.
4. V. A. Kotelnikov, *The Theory of Optimum Noise Immunity* (Translated from Russian), McGraw Hill, New York, 1959.
5. J. W. Tukey, *Exploratory Data Analysis*, Addison Wesley, Boston, Massachusetts, 1971.

Index